Proceedings

TINA '97 — Global Convergence of Telecommunications and Distributed Object Computing

ALCATEL 97

NTT

CONICYT

CTC

Compañía de Telecomunicaciones de Chile

Sun microsystems

TANDEM CHILE S.A.

Integradores de Sistemas

EDIBANK

Proceedings

TINA '97 — Global Convergence of Telecommunications and Distributed Object Computing

November 17 - 20, 1997

Santiago, Chile

Sponsored by

IEEE Communications Society
The Presidency of the Republic of Chile
IEICE Japan
Object Management Group (OMG)
Chilean Computer Society

IEEE COMPUTER SOCIETY

Los Alamitos, California

Washington • Brussels • Tokyo

IEEE Computer Society Order Number PR08335
Library of Congress Number 97-77716
ISBN 0-8186-8335-X
ISBN 0-8186-8337-6 (microfiche)
IEEE Order Plan Catalog Number 97TB100218

Additional copies may be ordered from:

IEEE Computer Society
Customer Service Center
10662 Los Vaqueros Circle
P.O. Box 3014
Los Alamitos, CA 90720-1314
Tel: + 1-714-821-8380
Fax: + 1-714-821-4641
E-mail: cs.books@computer.org

IEEE Service Center
445 Hoes Lane
P.O. Box 1331
Piscataway, NJ 08855-1331
Tel: + 1-908-981-1393
Fax: + 1-908-981-9667
mis.custserv@computer.org

IEEE Computer Society
13, Avenue de l'Aquilon
B-1200 Brussels
BELGIUM
Tel: + 32-2-770-2198
Fax: + 32-2-770-8505
euro.ofc@computer.org

IEEE Computer Society
Ooshima Building
2-19-1 Minami-Aoyama
Minato-ku, Tokyo 107
JAPAN
Tel: + 81-3-3408-3118
Fax: + 81-3-3408-3553
tokyo.ofc@computer.org

Editorial production by Regina Spencer Sipple

Cover art production by Joseph Daigle/Studio Productions

Printed in the United States of America by The Printing House

IEEE
COMPUTER
SOCIETY

IEEE

Table of Contents

Session 4: Internet II

Chair: Emmanuel Darmois, ALCATEL

Session 5: Network Management

Chair: Kazumi Endo, Fujitsu

Session 6: Security

Chair: Alvaro Campos, PUC-Chile

Session 7: Experience Reports II

Chair: Kenichi Kitami, NTT

Introduction

The TINA Consortium is a remarkable global alliance for the research and development of advanced communications, led by the principal telecom operators in the world in a joint collaboration with the major manufacturers of telecommunications and computer equipment. The consortium was established at the end of 1992 as a five-year project to develop a universal software layer (middleware), which would allow the utilization of distributed computing over telecommunication network platforms on a planetary scale. As the TINA Consortium reaches the end of this first phase (1993-1997), 1997's conference offered a unique opportunity to analyze and evaluate the results of this effort, and to debate the main challenges ahead in the second phase of the consortium (1998-2000).

The University of Chile was particularly proud to organize this conference, TINA'97, held for the first time in a developing country. Indeed, it is both a great honor and a big challenge for our country to help to introduce TINA-based solutions into the regional telecommunication market. This is a role Chile is well prepared to undertake, given the privatization of the telecommunications industry one decade ago as well as being the first country in the region to have achieved full deregulation of its market. Indeed, it is a well-known fact that, over the last few years, Chile has become a world telecommunications laboratory, where advanced services can be deployed and tested for later transfer to the South American region. As a small country, we are fully aware that our only competitive advantage lies in the speed with which we can innovate and develop advanced services first.

Nevertheless, despite this vision, as an emerging country, we still lack the solid R&D infrastructure necessary to foster the development of future telecommunications services. In this aspect, we are firmly committed to play an active role in international collaboration efforts such as TINA. The AccessNova Project, begun in November 1994 by the University of Chile and Nippon Telegraph and Telephone Corporation (NTT) as a creative attempt to trigger the early introduction and development of broadband services into the region, is a good example of this approach. This program built Chile's first experimental high-speed network, allowing the broadcasting of uncompressed digital video using TCP/IP protocols over an ATM platform in early 1996 and linking Santiago and Tokyo through a high-speed digital link in August 1997, among other experimental milestones. At the completion of its third year, AccessNova also includes the Federico Santa Maria Technical University in Valparaíso and has the corporate support of Compañía de Telecomunicaciones de Chile (CTC), Alcatel, Sun Microsystems, Tandem Chile, Edibank and CONICYT, all proud sponsors of this conference.

The TINA architecture is based on advanced information technology concepts in the areas of object-oriented computing and distributed processing. As a result, an ever increasing synergy is established between the telecommunications and the computing industries. Thus, the global convergence of telecommunications and distributed object computing is rapidly making us enter the global information age, where the boundaries between many disciplines and different cultures are constantly blurring as digital technology dominates and multimedia information services grow. However, telecommunication operators need to update their business concepts and to rethink their commercial strategies as completely new paradigms surface due to the radical redefinition that the industry is currently undergoing. TINA'97 provides a unique chance to examine and discuss these issues with the people who are actually producing them.

The conference program included a mixture of network and applications oriented research results both in the TINA (**T**elecommunications **I**nformation **N**etworking **A**rchitecture) area itself and in other aspects of distributed computing. The papers in the technical sessions range from the description of real world applications to basic studies on topics such as network management, connection management, software architectures, protocols, mobility and other aspects of advanced distributed middleware environments. TINA'97 reflects the vitality of the infocommunication community, in which the best research in object orientation and distributed computing are coming together within the framework of the telecommunications information architecture. The technical program, which contains 38 papers selected from over 90 high-quality submissions, is an outstanding one and it included a panel that addressed some of the toughest challenges ahead that the telecommunications industry faces in the next stage of the TINA Consortium. The program also offered four tutorials on TINA, CORBA, and Java, as well as five prototype demonstrations of TINA-based results. Participants experienced four days of stimulating presentations, vigorous debate and new ideas, together with a unique opportunity to visit one of South America's most beautiful countries, well known for its superb wines that were certainly enjoyed throughout the participants' stay.

Eduardo Vera
Director, AccessNova Project
ICT, University of Chile

Committees

Conference Chair

Eduardo Vera, *University of Chile, Chile*

Program Committee Chairs

Kenichi Kitami, *NTT, Japan*
Richard Soley, *OMG, USA*

Program Committee Members

Ricardo Baeza-Yates, *University of Chile, Chile*
Pier Giorgio Bosco, *CSELT, Italy*
Alvaro Campos, *Catholic University of Chile, Chile*
Eunho Choi, *Korea Telecom, Korea*
Emmanuel Darmois, *Alcatel, France*
Fabrice Dupuy, *CNET, France*
Kazumi Endo, *Fujitsu, Japan*
Andrew Herbert, *AMP, UK*
Choong Seon Hong, *Korea Telecom, Korea*
Hiroshi Ishii, *NTT, Japan*
Martine Lapierre, *Alcatel, France*
Aurel A. Lazar, *Columbia University, USA*
Armando Limongiello, *Telecom Italia, Italy*
Laurent Marchand, *Ericsson Research Canada, Canada*
Raul Monge, *Federico Santa Maria University, Chile*
Hiroaki Nisikawa, *University of Tsukuba, Japan*
Toshikane Oda, *KDD, Japan*
Declan O'Sullivan, *IONA Technologies, Ireland*
Manoel Camillo Penna, *CEFECT-PR, Brazil*
Kerry Raymond, *DSTC, Australia*
Peter Richardson, *Object Oriented Pty., Australia*
Tom Rutt, *Lucent Technologies, USA*
Gerd Schuermann, *GMD Fokus, Germany*
Max Sevcik, *Siemens AG, Germany*
Steve Sim, *BT, UK*
Dave Stringer, *Nortel, UK*
Jean Bernard Stefani, *CNET, France*
Joe Sventek, *Hewlett Packard Laboratories, USA*
Mario Tokoro, *Sony, Japan*
Paul Vickers, *Hewlett Packard Laboratories, UK*

Organizing Committee Board

Eduardo Aninat, *Minister of Finance, Chile*
Claudio Hohmann, *Minister of Transport and Telecommunications, Chile*
Mauricio Sarrazin, *President, University of Chile*
Jaime Lavados, *President, University of Chile*
Víctor Pérez, *Dean, School of Engineering and Science, University of Chile*
Francisco Brieva, *Director, ICT, School of Engineering and Science, University of Chile*
Marcel Mafille, *Executive Manager, Alcatel de Chile S.A.*
Jacinto Díaz, *Executive Manager, Compañía de Telecomunicaciones de Chile (CTC)*
Juan Carlos Macuada, *Executive Manager, Sun Microsystems de Chile S.A.*
Carlos Busso, *Executive Manager, Tandem Chile, S.A.*
Hernán Herrera, *Executive Manager, Edibank S.A.*
Yuji Inoue, *Vice President, NTT Global Business Headquarters*

Organizing Committee Members

Eduardo Vera, *ICT-AccessNova, University of Chile*
Iván Tabkha, *ICT-AccessNova, University of Chile*
Juan Nazal, *UNTEC, University of Chile*
Gimena Gonzalez, *UNTEC, University of Chile*

Keynote Address

Directions of Research and Development for Multimedia Networks

Toshiharu Aoki, Vice President, NTT

Keynote Address

Directions of Research and Development for Multimedia Networks

Toshiharu Aoki
Senior Executive Vice President, Nippon Telegraph and Telephone Corp.
and Executive manager of NTT R&D Headquarters, Japan

Improvements in multimedia networks enable customers to select from a wider variety of network services. In order to provide multimedia network services effectively, high-speed infrastructures, i.e., access lines and transit systems need to be installed. Additionally, advanced service applications need to be provided over a high-speed infrastructure through network middleware.

Telecommunication operators need to update their newer business concepts and also need to have new research and development strategy to provide such advanced services. "The information distribution business" allows the smooth spread of information throughout networks. An example of this is distributing information from a customer to a supplier.

To achieve innovations in the information distribution business, research and development should have three balanced objectives; multimedia applications, network services, and new infrastructures. Multimedia applications should provide a cyberspace where 'encounter and discovery' and even 'atmosphere' between customers is done in virtual space.

Network services should provide multimedia applications easier, faster, and inexpensively. New network infrastructures should meet the requirements for the next generation's network services.

The multimedia era will mature in the early 21st century, and the effect of changes brought about by this will reach and spread beyond the workplace to all areas of society. This maturity will not occur without innovative research and development. To achieve maturity at the earliest possible date, it is important for R&D to be advanced by utilizing contributions received from around the world.

0-8186-8335-X/98 $10.00 © 1998 IEEE

Session 1

Internet I

Jose Miguel Piquer, University of Chile

Applying TINA-C Service Architecture to the Internet and Intranets

Chris Smith
British Telecommunications (BT)

Abstract

This paper describes some of the more urgent problems faced by developers of multiparty, interactive services for the Intranet and Internet. It then goes on to show how applying the TINA architecture, concepts, principles and technology could solve these problems and bring the rapid development of interactive, multiparty services to the Internet.

Research within BT Labs has implemented the Experimental Service Platform (ESP) to allow rapid creation of multiparty services for use within its intranet. The platform comprises both a service runtime, as well as a service developers kit containing tools and APIs.

1. Introduction

In recent years, the Internet, and more recently, Intranets, have experienced explosive growth and the World Wide Web (WWW) has brought information and communication services to the masses. These services are based on textual and pictorial content, and even after the advent of technologies such as Java, they are not very interactive. In addition to this the services are difficult to develop, and service support systems (billing, authentication, subscription etc.) are custom built for each service. In order to reap the full rewards of the 'Information Revolution', interactive, multi-party services must become available, and in such a world, the time-to-market for services will be critical. In order to gain a competitive advantage in such a market, the service development process must be made significantly simpler and quicker.

The Telecommunications Information Networking Architecture Consortium (TINA-C) is defining an architecture that exploits the convergence of the telecommunications, computing and entertainment industries. TINA-C has so far assumed that its work is to be applied to traditional telecommunication systems.

This paper proposes that the TINA-C service architecture and concepts can be applied to the Internet and Intranets, and that they solve some of the most urgent problems currently faced by Internet/Intranet Service Developers. This is based on experiences gained in a BT project which attempted to tackle these problems through building an Intranet based TINA trial platform.

The paper assumes some familiarity with the TINA-C Service Architecture[1], although Section 2 gives a brief overview of the concepts and technologies involved.
Section 3 then describes some of the most important problems faced by both developers and users of interactive internet/intranet services, and shows how they can be solved using TINA concepts and technologies.

2. The TINA-C Architecture

The TINA-C Architecture contains many parts, but can be broadly broken down into two layers - the service architecture and the network architecture. The service architecture[1] is concerned with the problems of service control - authentication, identification, subscription to services, use of services, and charging for service usage. The network architecture is concerned with the creation of network connections at the request of services, and the management of the network (e.g. Faults and QoS).
This paper concentrates on the concepts in the TINA Service Architecture, although some of the connection request interfaces introduced in the Network Architecture do have an impact, and will be discussed.

2.1 TINA-C Business Model

TINA-C has a business role model [2] which defines 4 roles in the information provision market - Retailer,

0-8186-8335-X/98 $10.00 © 1998 IEEE

Service Provider, Consumer and Connectivity Provider (There is a fifth role - broker - but it has been left out of this discussion for simplicity).

Consumer
A stakeholder which consumes services - this could be a private individual, or a small or large company.

Retailer
A stakeholder who sells services to the consumers, providing a single access mechanism, and bills for the services which are supplied. The services supplied are sourced from service providers

Service Provider
A stakeholder who provides services to retailers for onward supply to consumers.

Connectivity Provider
A stakeholder who provides network transport and connectivity services to other stakeholders.

The roles and their relationships are shown in Fig 1.

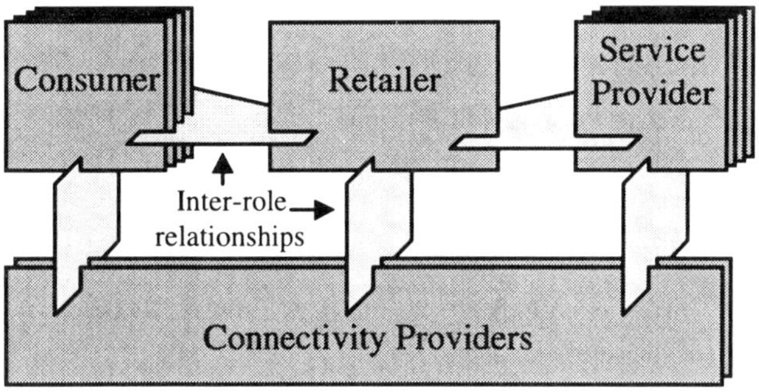

Figure 1 - The Business roles and Relationships

Note that these are Business <u>**roles**</u>, and any particular business enterprise could take on any (or all) of these business roles. The links between the roles signify the relationships between them. These include both runtime relationships between objects, as well as contractual relationships dealing with how the roles interact.

2.2 Service Architecture Concepts

The TINA service architecture itself is split into two layers - the access layer and the service usage layer (Figure 2). The access layer is concerned with the negotiation of an authenticated context between two roles, which is called an 'access session' in the TINA architecture. From this access session, instances of services can be instantiated and used. The usage layer is concerned with how the various parts of this service instance interact to provide service to the consumer.

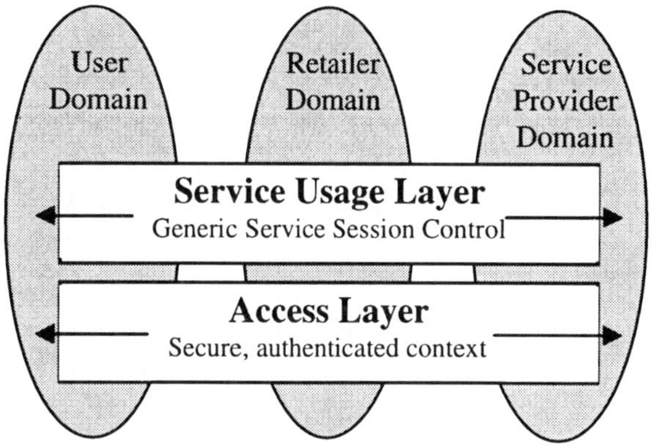

Figure 2 - The Access and Usage Layers

2.3 Underlying Technology

The TINA-C Architecture assumes that software components comprising a TINA implementation will interact using distributing computing technologies and more specifically OMG CORBA [3]. Specification of objects (and indeed Inter-domain reference points) are written in OMG Interface Definition Language (IDL). The existence of an underlying distributed computing platform is one of the key assumptions made in the TINA architecture.

3. How TINA-C Concepts solve the Internet Service Problems

This section describes the problems faced by both service developers and by service users, and in each case shows how TINA concepts, designs and technologies could solve the problems.

3.1 Basic Problems for Developers

Most users make use of the Internet/Intranets through the WWW, on which Information and services mainly take one of the following forms:

- Hypertext-based pages embedded with images, sounds and video clips. These are HTML based, and are easy to create, but can be uninteresting, even with animated pictures.
- Forms based pages which can be completed and then submitted. These are implemented using HTML Forms and CGI (Common Gateway Interface) scripts which do not allow true interactivity, as they work by generating new HTML forms dynamically.
- Small embedded games or applications such as a Calculators etc. These are generally implemented using Java which on its own does not provide any

communications support other than 'sockets', although Java-RMI and CORBA are now beginning to become more prevalent.

- 'Push' technology which is used to provide dynamic updating of information on clients from servers. Unfortunately this technology, although still developing, is currently proprietary and quite immature.

Some interactive services are now becoming available, but they are few and far between. One of the reasons for this is that they are very time-consuming and difficult to produce. In addition, support systems to manage user records, and billing systems to record service usage must be created.

It can be seen that service development can be a painful process.

Summary - Basic Problems for Developers

The problems faced by Intranet/Internet service developers are summarised below, along with descriptions of how TINA can help to solve then.

1. Support systems have to be hand-built for each service and each system must often be managed separately.

   TINA separates the business roles of service provision and service retailing. Service provision is only concerned with those parts of a service which are specific to the particular service in question, whereas service retailing is concerned with the generic functions and systems which can be shared between many services. The TINA architecture also shows through the definition of reference points [4] and component specifications, how separate services can take advantage of these shared support systems.

2. Service development is done at a level of abstraction that is not natural to service developers. The service developer would like to deal in terms of instances of a service, and in participation in that service. Instead, the developer must deal with the generation of Web pages or socket-based servers.
   The TINA Generic session model defined by the service graph and by the Ret Reference points [4] provides a paradigm for service developers which deals with the types of concepts and objects which are meaningful (e.g. session, participations etc.). Developers are therefore removed from the level of sockets and pages of content and can make use of generic session control through higher level APIs.

3. Communication of information between client programs (e.g. Browsers) and servers is also done using non-ideal paradigms (e.g. HTML and sockets). In order to reduce development time and increase ease of maintenance, communication between the Browser and server (at least for interactive applications) should be done in an intuitive object oriented manner. Instead it must currently be done by generating HTML pages or designing application specific socket protocols (although with the advent of RMI and Java-CORBA this is changing).

   By taking advantage of CORBA technology, different parts of distributed services can communicate in ways which are intuitive for the programmer. Location and distribution transparency makes service development much easier and removes the need for the developer to deal with network programming.

3.2 Basic Problems for Users

Use of services is often based on registering with the provider of the service (e.g. Electronic Newspapers etc.), and so a user of several services has to have a multitude of login names and passwords. Payment for these services is made directly to each provider, normally via Credit Card. This has several problems, firstly due to the security issues of sending credit card numbers unencrypted over the web, and secondly and probably more significantly, because the user may not have any knowledge of how trustworthy the provider is.

Summary - Basic Problem for Users

The main problem that users face when using internet services is summarised below, along with an explanation of how TINA could solve it.

1. Services are obtained through many points of contact, each with their own authentication system, and payment mechanism. This causes inconvenience to users, and is a substantial disincentive to the take up of such services.

   It is again the separation of the business roles of Retailer and Service Provider that could help users with this factor. If users obtain their services from a small number of retailers, they can use these retailers as their point of access to

their portfolios of services. This means that in addition, they would be able to obtain bills from the retailers and not from a multitude of service providers. Indeed, if the Retailers were fully federated, then all the services a user would want could be obtained indirectly through a single retailer.

4. Applying TINA-C Service Architecture in Practice

In a BT research project called ESP (Experimental Service Platform), some of the TINA-C ideas, concepts and principles have been used to create an Intranet/Internet-based TINA platform. This platform is under trial and is not simply a prototype demonstration like other previous TINA validation experiments[6]. The platform comes with a service development kit, tutorials, and API documentation for use by developers.

A simplified diagram of the system architecture is shown in Figure 3.

The following sections will describe the APIs and tools developed and the impact upon the TINA-C architecture. Furthermore it will discuss what issues remain which might prevent the deployment of this type of technology into Internets and Intranets, including issues concerning service naming, and data/video/audio streaming.

5. APIs and Tools

5.1 Service Designer APIs

When a service designer builds a service, they should be able to use well designed, self-contained APIs which are expressed in terms of concepts which are meaningful to them. Currently TINA only specifies inter-domain reference points (which are written in IDL), and descriptive guidelines and specifications for components that might lie behind those reference points.

The reference points and guidelines specify the interfaces and behaviours for generic service session control which any TINA service will use, but it will probably not be the job of a service designer to

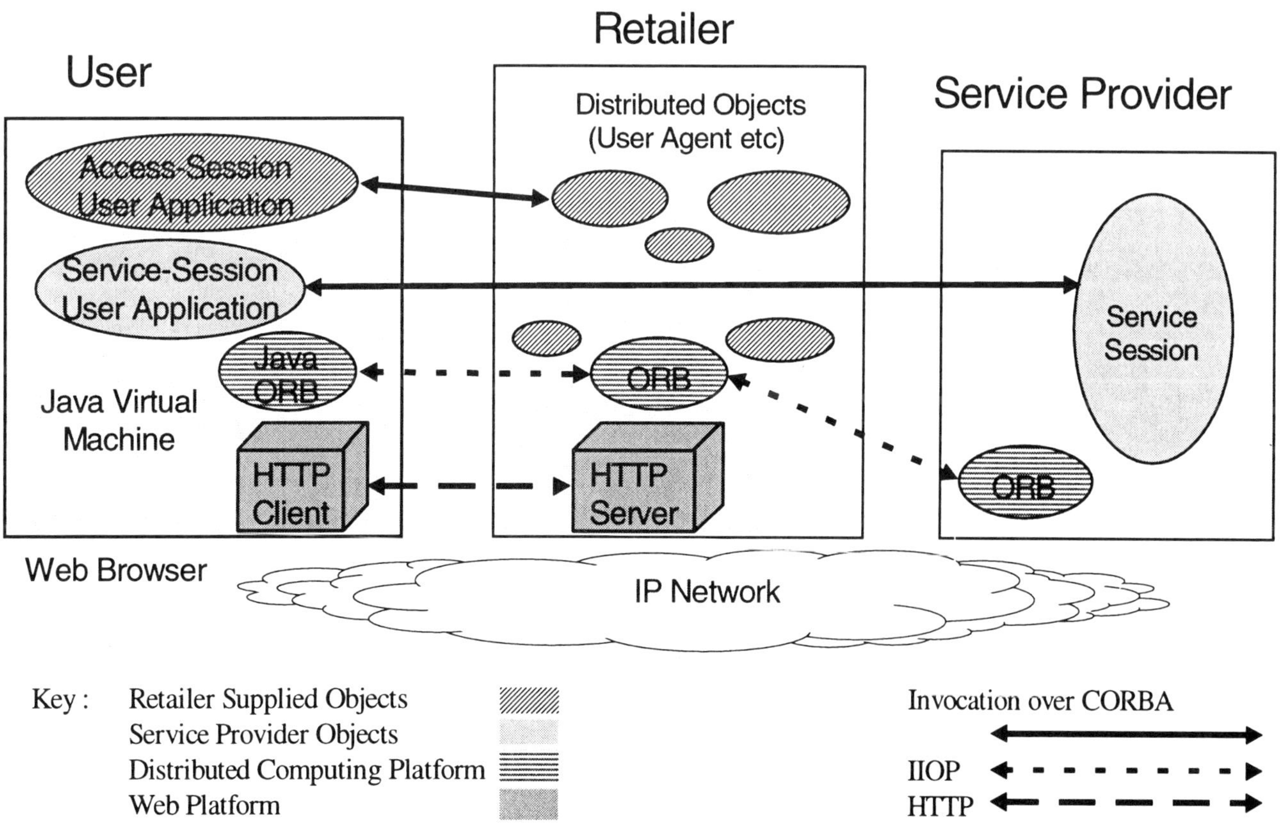

Figure 3 - A simplified view of the ESP platform

implement this generic service control. The generic control should be implemented only a few times, and this should be reused by many service developers.
ESP implements a generic service session control toolkit which is encapsulated behind well-defined and easy to use APIs.
The features currently supported by these APIs include:

- Multiparty behaviour (e.g. start, invite, join, leave etc.)
- Navigation of participants
- Suspend and Resume participation
- Storage of service specific user preferences
- IDL binding (see below)
- Generation of billable events

In order to explain these APIs, they are shown both from a TINA viewpoint, and from a service developer viewpoint.

The ESP project coined two new terms - the 'front' and 'back' of a service. The 'back' contains the parts of the service session manager (SSM) and user service session manager (USM) which are designed and built by the service developer, and the 'front' is the part of the user application (UAP) which is designed and built by the service developer. Figure 4 shows the Front and Back of a service, the Generic Session Control toolkits, and the APIs separating them.

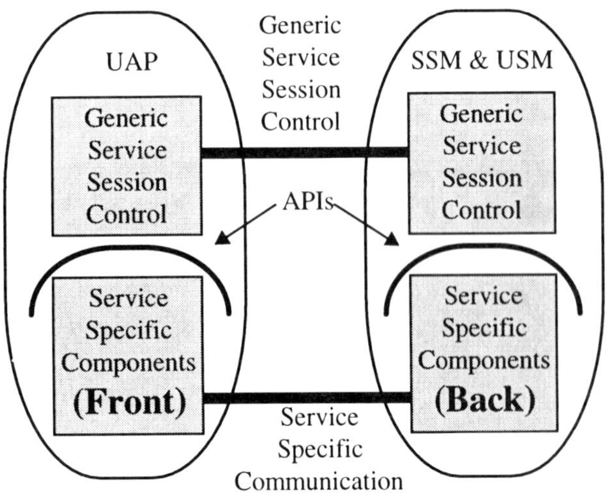

Figure 4 - Separation of Generic and Specific functionality via APIs

The service developers view of the platform is as shown in Figure 5. All that is perceived is the ESP platform as a black box system, and the 'back' and 'front' of the service.

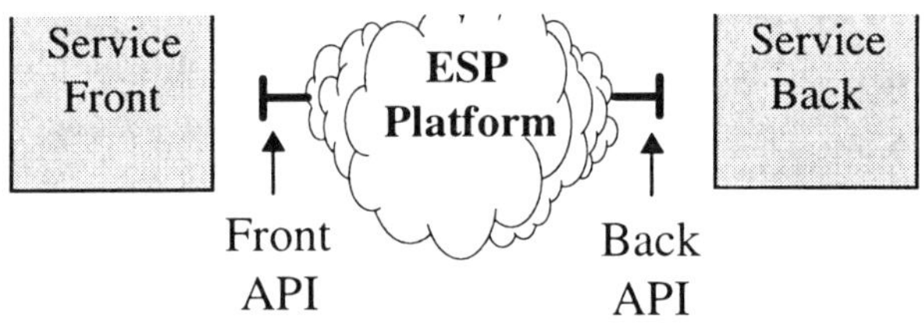

Figure 5 - Service Developer view of ESP

These 'fronts' and 'backs' are fitted into the ESP service framework to form a full TINA service.

The most important part of any toolkit such as those in ESP, is the API between the toolkit and the service specific parts of the service. Previous TINA related work has speculated about two models for this API - the inheritance and delegation models - both based on OO languages and principles.
The ESP APIs use a combination of these two approaches. 'Hooking-in' of service specific behaviour is done by inheriting from an ESP class which has certain 'callback' methods. The default 'callbacks' have no behaviour but by inheriting and overriding, service specific behaviour can be added. Note that the generic session control cannot be overridden, only added to. Concrete methods to perform actions such as ending participation in the session are also available on this class, but the implementions of these methods 'redirect' the calls onto the real session control. In this way a very clear separation between generic session control and service specific code is ensured. (See Figure 6)

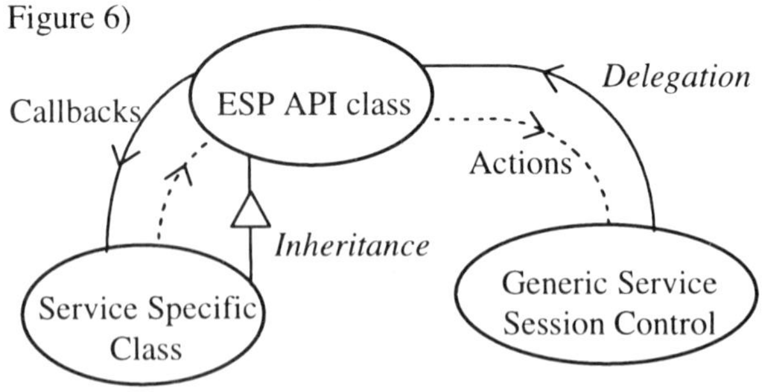

Figure 6 - The ESP API Framework

The APIs, which continue to be iterated based on developer feedback, are now stable and are being used by several developers to build trial services. The Front API is available just in Java, and the Back API in both Java and C++.

5.2 Development Tools

Within the ESP project, one of the goals was to provide a toolset to developers such that they could design, build, test and deploy a service to the platform without the manual intervention of the individuals running the platform. Several of the previous TINA implementations constructed in BT, and within auxiliary projects have not had this capability - services have had to be manually added to the retailing platform, often with a high degree of manual configuration.

The platform is complemented by a service developers kit (SDK) which contains the following things :

- API documentation (in HTML) for the 'front' and 'back' APIs
- Online tutorials on how to use the APIs, the tools, and guidelines on how to develop TINA/ESP services
- 6 simple example services
- Development tools including
  - generation of sample/stub code
  - Automated generation of IDL bindings (see below)
  - Uploading of service onto platform
  - Creation of test accounts
  - Obtaining debugging information

5.3 IDL Bindings

Within the TINA architecture, it is expected that service specific communication between UAPs, Service Session Managers and User Service Session Managers could use distributed computing in the form of a CORBA (TINA) DPE.

However, when designing the ESP platform, there were two conflicting requirements on the use of CORBA. Firstly, the use of CORBA would reduce the amount of work the developers would have to do when providing service specific communication between their UAPs, and SSM/USMs. However, the developers being targeted by the platform were **not** expected to be skilled or knowledgeable in OMG CORBA as this would significantly (even drastically) reduce the number of service developers able to use the platform. Because of these two contradicting requirements, some extra functionality was added to the platform and to the APIs - a feature we refer to as an 'IDL binding'. (See figure 7)

The IDL binding is a feature which simplifies greatly the use of CORBA, in return for some constraints on its use. It provides the ability to automatically create and bootstrap service specific IDL interfaces on the 'front' and 'back' of a service. All the service developer has to do is fill in the operations on these two interfaces in an auto-generated IDL file, and the object references, IDL compilers, libraries, stubs etc. are all hidden by the platform. Experience indicates that this is a very useful mechanism which provides the power of distributed computing semantics in a constrained environment without a large learning curve, and that developers grasp and use the IDL concepts in hours.

Figure 7 - IDL bindings in the ESP Framework

6. Platform Trial

Two aspects of the platform have so far been trialled - a service developer trial and a user trial.

6.1 Service Developer Trial

The service developer kit has been made available to intranet/internet service developers within BT. In order to make the trial as real as possible, only minimal support via email was provided to developers.

The results have so far been very encouraging. The developers have not needed much more support, and they have provided excellent feedback on the level of value-add provided by the platform and the toolsets.

To date the following set of interactive multiparty applications have been developed against the platform API.

- Multi-user chatrooms
- Multiparty Crossword championships
- Multiuser shared-whiteboards
- Text->phone messaging
- Two player, n spectator Chess
- Games (e.g. TicTacToe etc.)
- Bat and Ball Games
- Online subscription services
- Whitepages (e.g. Finding users)

The tests suggest that services can be *prototyped* within 10 days, and full services can be created with around 20-30 days of development effort.

6.2 User Trial

Initially, a small scale validation test of concentrated usage was conducted. 22 users were given uncontrolled usage to the platform and asked to exercise it for 90 minutes. The users seemed to enjoy the experience of using the platform, and Figure 8 shows a graph of users logged in, service sessions, and user service sessions.

Efforts were made to reduce the time required to download the 'ESP environment' into Web Browsers from which ESP is accessed. The ESP download environment now measures between 200 and 400Kbytes depending on the Browser being used, and this is downloaded as a single file. Access from LANs takes between 5 and 10 seconds (including the Browser starting up the Java Virtual Machine), and over a Modem dial-up link, takes about 1 minute. This kind of performance seems to be acceptable to users (e.g. comparable to the time required to load a highly graphical Web page).

7. Issues for TINA

7.1 APIs

The APIs between the service specific code written by service developers and the generic session control written either by retailers, or even by separate toolkit vendors are not defined by TINA-C. The service component specs of components such as User Service Session Manager could represent these APIs, but such components are not currently stable, nor in our opinion are they suitable for us as high level APIs separating generic session control from service specifics.

Constructing proprietary APIs in order to lock in service developer customers may be desirable for platform providers, but it does nothing to help with the creation of an open service market and is not beneficial to service developers. Once a service has been written to a particular proprietary service toolkit API, the developer is faced with either being locked in to that platform provider for the foreseeable future, or moving to a different platform and rewriting a large amount of the service to use different APIs.

Figure 8 - Usage of ESP during validation test

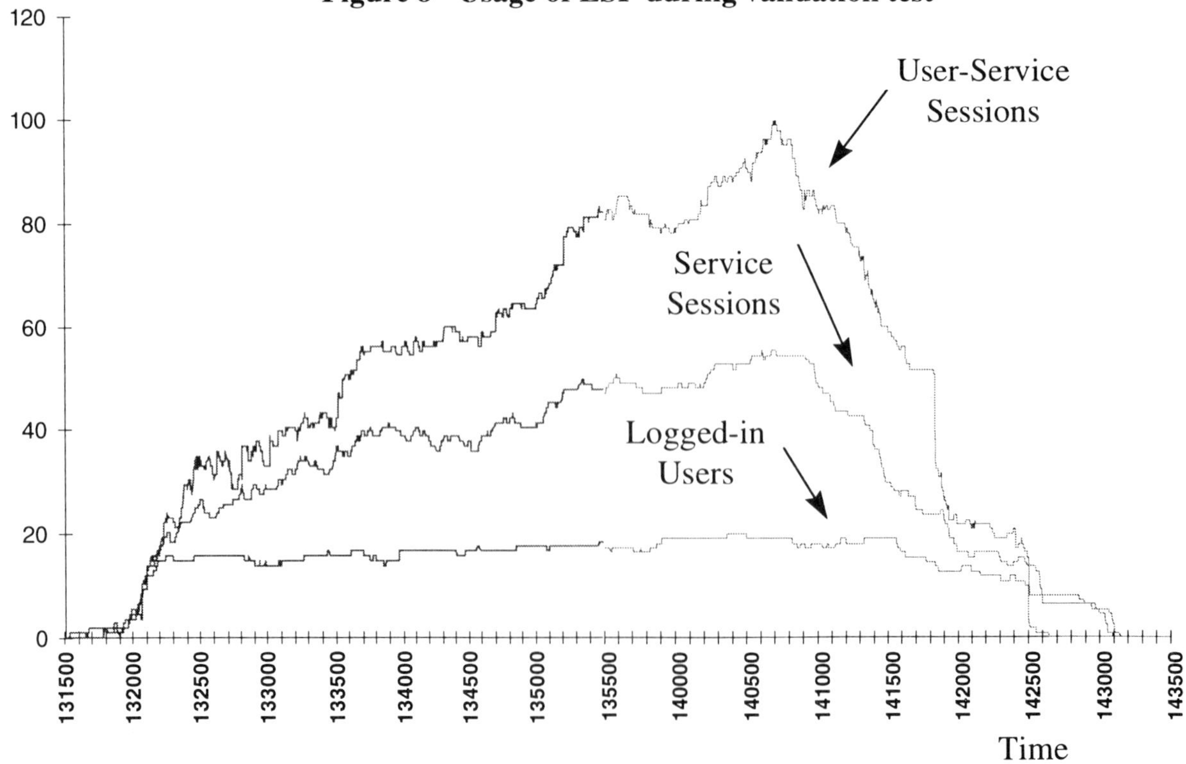

It is felt that one of the factors preventing the success of TINA-C is the lack of such APIs. Without them, take up of TINA-C amongst service developers will not occur. It seems fairly certain that at least in the Internet arena, service and application developer support is critical to the success of a particular paradigm.

7.2 Naming

As in any distributed platform or service environment, naming is of course a big issue. However, when building ESP, certain issues concerning naming in a TINA service platform were raised that we had not perceived before.

We have experimented with using Java as a language for implementing TINA components, especially for those on the end-user terminal, because of its 'write-once, run-everywhere' characteristics. However, if you run user applications for services in this environment, you must be very careful that these UAPs cannot affect each other. Due to the shared runtime nature of the Java Virtual Machine, if two services implement classes with the same name (including package name etc.), they might interfere with each other, and so the Java namespace must be segmented into partitioned spaces for each service.

Similarly, the CORBA namespace must also be partitioned so that one service implementing a certain IDL interface cannot overwrite an IDL interface in the Interface Repository specified by another service.

These issues have implications for both interworking of TINA environments and for 3rd party and Retailer to Retailer federation.

7.3 Stream Connections

Because one of the objectives of ESP was to be available on as many desktops as possible, use of a particular network technology (particularly those not in widespread use such as ATM) was not an option. Services in the ESP environment arc constrained to just make use of TCP/IP streams. Currently the native TCP/IP APIs are used for these streams, but we are currently working on simplified, network technology independent APIs which could then be implemented using TCP/IP, ATM or other network technologies.

It would perhaps be helpful to have a 'standard' TINA way of doing proxy/UNI based Connection Control underneath a Connection Graph [5] interface. The principle has been talked about and considered for many years now, but there is still no standard specification for how it would be done.

8. Business Context

Utilizing this platform in a real environment will probably depend on the business assumptions and rules being applied in order to generate revenue. There are 4 basic situations in which the platform could be applied, although it is important to note that due to the speed of development in the Internet arena, these views can become obsolete very quickly.
The four scenarios are shown below - each of them exhibits a particular mapping of TINA business roles to particular real-world companies.

In the following scenarios, the IP access is ignored - it is assumed that consumers obtain access from a supplier and pay directly.

Direct Service Provider Use

The first way in which a platform such as ESP could be used is the very situation described in this paper. Any company which supplies more than 1 or 2 services to users could run the platform themselves, and therefore take advantage of the reuse of the common retail facilities and fast service provision. (Fig 9)

Figure 9 - Use of ESP by Service Providers

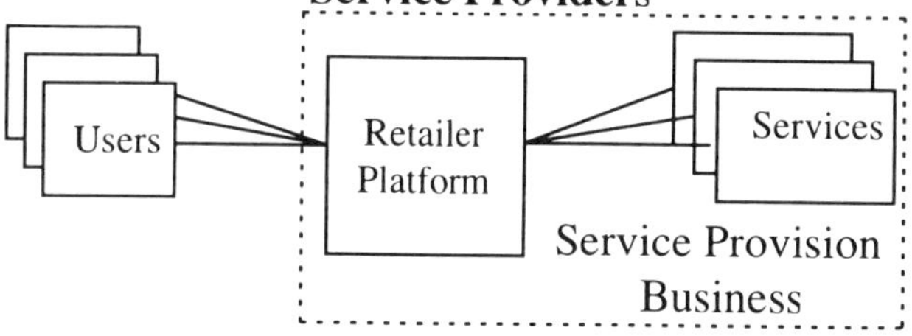

Separate Service Retailing

The second way in which it could be used is as a service platform run by a separate business - an actual retailing company. Many service providers would supply their services to users through this retailer. (Fig 10)

Figure 10 - ESP as a Separate Retailing Business

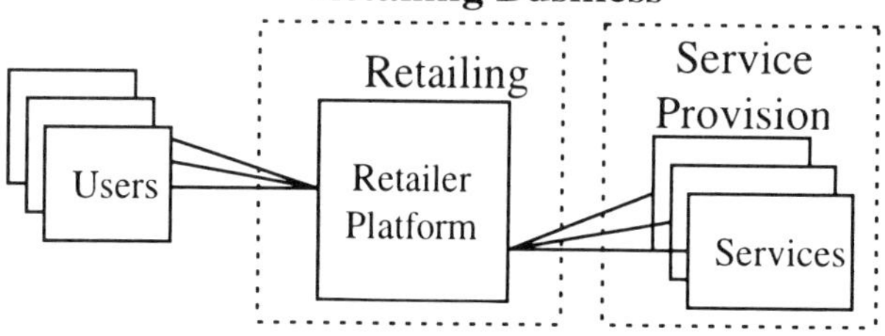

Intranet Service Provision

The company owning the Intranet could run the platform itself, providing value-add to its in-house business-application developers, or bought in applications. (Fig 11)

Figure 11 - ESP in a corporate Intranet

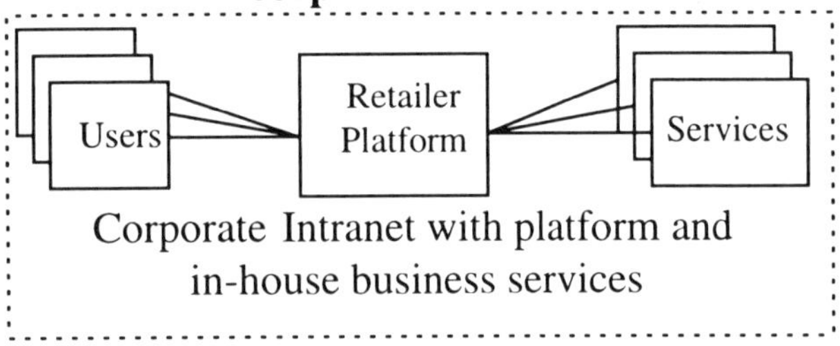

Outsourced Intranet Service Platform

Alternatively, a company could outsource the running of its service platform, just as many companies outsource their data networks today, thus creating a market for companies who will run and manage Intranet Service platforms (Fig 12). It is interesting to note, that running a TINA platform in an Intranet, actually makes it is easier to extend it to an extranet - i.e. giving customers and suppliers limited access via a subscription mechanism.

Figure 12 - ESP as an outsourced Service platform

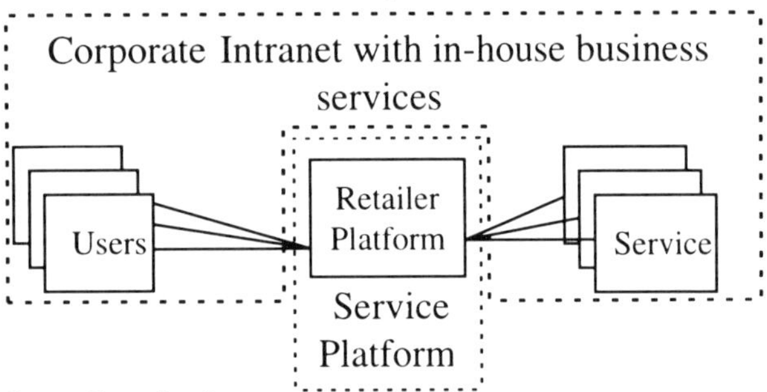

9. Conclusions

An operational trial service platform has been constructed from distributed object technologies, and developer tools for the platform API have been used successfully to implement a dozen services. Reusing the features embedded with the platform enabled complex multiparty service development times measured in days. Uncontrolled usage during a small scale validation test has proved the platform to be robust, with satisfactory performance for users. Further, larger scale trials are planned.

Significant advantage can be gained by translating TINA concepts into platforms specifically targeted at multiparty service creation in Intranet/Internet. The rapid change in the Internet arena means that this window of opportunity will not be long lived.

However, in order for TINA to impact this arena, it must do more work to allow services built for one TINA platform to work on another. The TINA reference points are not defined in the right place or at the right level of abstraction to allow this kind of portability of services between retailing platforms, and so high level service designers APIs must be defined.

10. Acknowledgments

The author would like to thank the current and past members of the ESP team without whose efforts the ESP platform would never have come about. These include Martin Ellis, Duncan James-Bell, Dan Creswell, Jeff Farr, Nigel Warren, Mike Hosking, Patrick Farley, Martin Yates, Pete Barnsley and Ian Marshall.

11. About the Author

Chris Smith is a Senior Software Designer/Engineer at BT Laboratories, Martlesham Heath, England. He has worked on TINA-C validation and design for four years, initially in the area of Connection Control, but for the last three years on the Service Architecture. He has been responsible for much input into the TINA consortium as a result of validation work, and has been one of the foremost contributors in the Retailer Reference Point RFR. He was a major contributor to the TIMMAP TINA Auxiliary project, and is a CORBA and Java expert. He currently represents BT at the OMG.

12. References

[1] TINA-C Service Architecture
Version 5.0 16th July 1997 C. Abarca et al.

[2] TINA Business Model and Reference Points
Version 4.0 22nd May 1997 Martin Yates et al.

[3] Object Management Architecture Guide
Revision 3.0 13th June 1995 Richard Soley et al.

[4] TINA Ret Reference Point Specifications
Draft 1 16th July 1997 Patrick Farley et al.

[5] TINA Network Resource Architecture
Version 3.0 10th February 1997 C. Abarca et al.

[6] Information Services Supermarket
International Conference on Distributed Processing. Dresden '96. M.J.Ellis et al..

Value-Added Internet: a Pragmatic TINA-Based Path to the Internet and PSTN Integration

G. De Zen, M. A. Marsiglia, G. Ricagni, L. Vezzoli: ITALTEL SPA

H. Hussmann, H. Schoenbauer, M. Sevcik, A. Zoernack: SIEMENS AG

Abstract

This paper deals with the issue of integrating the Internet with the PSTN in order to devise an architecture encompassing the advantages of both networks. Key TINA concepts like the Session Concept and the Service Architecture computational model are used to provide the glue for such an integration. The result is the enhancement of the Internet with real-time packet delivery capability as well as advanced billing features that successfully exploit existing billing systems. A comparison with other approaches to quality of service over the Internet is provided. The prototype that is currently being developed within the SISTINA (Solutions for Integrated Services à la TINA) project to demonstrate the concepts outlined in the paper is described.

1. Introduction

In the current telecommunication scenario the provision of new services and the enrichment of existing services with value added features is becoming a key issue for the Public Network Operator (PNO) success. This is a very important objective for PNOs as the wide diffusion of Internet is endangering their position as network service providers potentially reducing their role to pure connectivity providers. The Internet phenomenon is causing a migration of network intelligence from the core network to the peripheral systems. Lost revenue for PNOs is the result.

Within the SISTINA (Solutions for Integrated Services a la TINA) project, Italtel and Siemens have jointly devised a TINA based solution capable of exploiting the advantages of two apparently contrasting worlds such as the Internet and the PSTN in order to provide value added services and to keep a share of network intelligence in a PNO's network.

Starting from the analysis of the state of the art as far as the Internet and telecommunication worlds are concerned, a new attractive service deployment architecture has been identified. The proposed architecture derives from the combination of the strengths of each world, which allows to overcome the weaknesses of the other.

The services offered by telecommunication service providers are usually characterized by a guaranteed quality of service and by the utilization of payment mechanisms based on a monthly bill. Capabilities like these, typical of the PNO world, are very attractive also to WWW application developers. However, these cannot be provided on the Internet without the cooperation of a telecommunication service provider who has a direct control on its network infrastructure.

On the other hand, the wide Internet acceptance and the large diffusion of WWW applications, does not allow to realistically envisage a network scenario in which advanced services are provided using a different paradigm. The simple and easy to use graphical user interface of Web applications has become mandatory for every type of service provided through the network. As a consequence, only an architecture capable of enhancing existing WWW applications with value added features can be envisaged. WWW applications have to be provided with new features in such a way that they do not need to be redesigned but they can simply take advantage of new APIs that enrich the

0-8186-8335-X/98 $10.00 © 1998 IEEE

set of supported capabilities. In this scenario TINA provides the enabling technologies and concepts that allow to successfully merge these two different worlds.

The result of the SISTINA project is a TINA-based distributed architecture. The TINA Service Architecture and the TINA Computing Architecture have provided the basis for the design of such a solution.

TINA Service Architecture concepts such as the session concept (in terms of access, service and communication session) are integrated in a service platform which can be used by all kinds of standard Internet applications. SISTINA enhances them through four key value added features:

1. Session control: to overcome the limitations of the stateless pure WWW approach
2. Guaranteed Quality of Service (QoS): when high bandwidth and/or real time communication is required within a service session, dedicated switched end to end connections are established between users and service providers, bypassing the Internet or at least its bottlenecks
3. Integrated billing, in terms of
    - flexible/reverse charging of the dedicated real-time connections
    - charging of online service usage and goods purchased through electronic commerce applications on the monthly telephone bill
4. Confidence

This outstanding bundle of value added features can give an answer to some of the major Internet open issues and provide a new central role for PNOs in the Internet era.

2. SISTINA Architecture

The architecture devised for the SISTINA project is depicted in Figure 1. This architecture involves some of the Business Roles identified by TINA, such as the Consumer (End User and Internet Application Provider), the Retailer and Connectivity Provider (SISTINA Retailer), and the 3rd Party Service Provider (Confidence Provider). The task of the SISTINA Retailer in this scenario is to add value to normal existing Internet applications, through features that only Public Network Operators can provide.

Network Operators can play an important role in providing these features as they have the required know how and the capability to control the telecommunication network. Existing TMN/IN Billing Systems can be used to support the billing mechanisms foreseen by SISTINA .

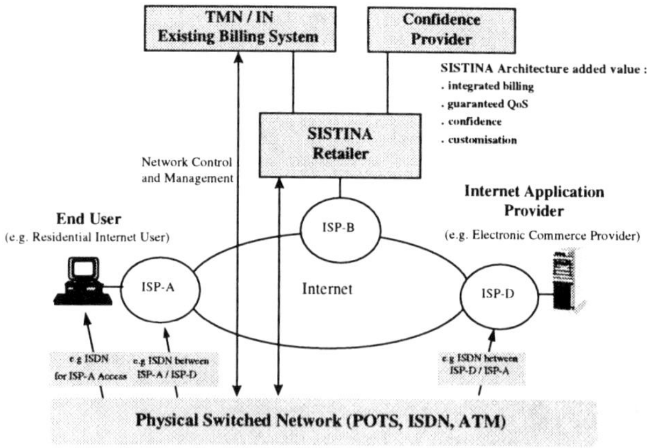

Figure 1: SISTINA architecture

Figure 1 shows the normal Internet business roles: the End User, the Internet Application Providers (IAPs) and the Internet Service Providers (ISPs) that provide the Internet Access. The End User uses a standard CPE to access the Internet (e.g. a PC). The ISPs may or may not operate an own subnetwork and are interconnected via the global Internet.

With respect to this scenario a new additional role is played by the SISTINA Retailer. The SISTINA Retailer provides service independent Session Management capabilities in support of the interactions between an End User and one or more Internet Application Providers. In the context of such a session the value added features of the SISTINA Retailer can be used by the participants. By establishing a session, which includes authentication and authorization of all participants, and by providing a subscription mechanism, the SISTINA Retailer acts as a "Retailer" in the TINA sense, as it mediates between End Users and Providers of Internet Applications. All the functionality of the SISTINA Retailer is independent of the services offered by the various Internet Application Providers which can range from Video Shopping, to Electronic Malls or simple HTML pages.

In order to provide a guaranteed Quality of Service it is required that the owner of the SISTINA Retailer has control over a telecommunication network and that the possibility exists to provide a "direct" (i.e. switched) communication path between an End User and an Internet Application Provider. In this role the owner of the SISTINA Retailer also acts as "Connectivity Provider", in the TINA sense.

The overall architecture makes extensive use of TINA Computing Architecture and Service Architecture principles and concepts (e.g. DPE, Session Concept, ...).

Two major phases during a service session can be distinguished: the Access Phase where the session is established and the Usage Phase where the End User transparently interacts with the Internet Application Provider. For both phases, TINA-DPE communication via the Internet is used. For this purpose the normal Internet End User Application and Internet Applications must be made TINA-DPE capable. During the usage phase, if required, dedicated end-to-end connections between End Users and Internet Application Providers can be established.

In the SISTINA scenario five major stakeholders are identified:

- End User
- SISTINA Retailer
- Internet Application Provider
- Confidence Provider
- Billing Provider

2.1 The End User

The End User is a classical Internet user (in a private or business environment). To enable the End User to enjoy the added value provided by the SISTINA Retailer it is necessary to download additional Java software in order to make his browser TINA capable. This software incorporates a Java CORBA client, for functional interaction. In brief, the SISTINA software executed in the user's browser environment supports:

- the IIOP communication with the SISTINA server;
- the invocation of remote methods on the SISTINA server objects interfaces as a consequence of the users actions on the SISTINA applet GUI in order to realize session management, integrated billing and connectivity control related operations;
- the support for the remote invocation, addressed by the SISTINA server, of operations related to session management, integrated billing and connectivity control that influence the user application.

The goal is to minimize specific requirements for the hardware and software of the End User's CPE in order to widen the base of SISTINA suitable terminals.

2.2 The SISTINA Retailer

The SISTINA Retailer acts both as a TINA Retailer as well as a TINA Connectivity Provider. As a Retailer it provides generic service independent features like access session capabilities (authentication, authorization, billing), service session capabilities (service session control) and communication session features (connection control). As a Connectivity Provider it provides access to the resources of the physical switched network.

The SISTINA Retailer offers to the End User and to the Internet Application Providers the following service features:

1. Connectivity Control for a guaranteed QoS
2. Integrated Billing
3. Confidence
4. Service Customization

The Connectivity Control service feature allows for the establishment of dedicated switched connections with guaranteed QoS between two points in the path between CPEs and Internet Application providers, bypassing Internet bottlenecks. This feature is described in more detail in the section 5.

The Integrated Billing service feature offers the possibility of integrating the SISTINA payment system with the public network operator charging and billing system. This integration can be exploited both for:

2a. the payment of goods purchased on an Electronic Commerce platform or of on-line services offered by a generic IAP;

2b. the payment of connectivity resources (e.g. connections) used to have a guaranteed quality of service.

The Confidence service feature plays an important role in the business relationships between users and IAPs. It allows the SISTINA Retailer to provide a second level of authorization that is applied to both users and IAPs service requests. For example, a IAP can know he has a business relationship with a paying user and vice versa the user knows he can trust the IAP. The SISTINA Retailer makes use of the Confidence 3rd pty service provider to provide this service feature.

Finally, the Customization service feature allows the IAP to customize the services on the basis of the knowledge of the user's identity.

2.3 The Internet Application Provider

The Internet Application Provider can be any kind of provider of existing WWW-based applications (e.g. an Electronic Mall) or simply a set of HTML pages. After having signed a contract with the SISTINA Retailer, the Internet Application Provider becomes a SISTINA affiliated IAP.

Interactions between server applications and the SISTINA Retailer are CORBA based. Internet server applications require only minor enhancements in order to allow for their exploitation of the added value features provided by the SISTINA Retailer.

Basically, this can be achieved in two ways:

1. In a scenario where the traditional Web technology is used, the best solution is to install on the Web server a SISTINA software package which is in charge both of supporting the IIOP based communication with the SISTINA server and of providing a very simple high level API, that can be used by Web application designers to enrich their services with new functionality. Methods can be called inside CGI BINs executed upon HTTP requests. This package completely hides the details of the SISTINA environment from developers and allows designers to continue utilizing the Web technology they are familiar with.
2. In a more advanced Internet scenario in which the utilization of IIOP is already foreseen, Web application designers can directly implement a SISTINA enhanced Web service as a distributed service using the CORBA and Java technology. In any case, the effort for the support of the interaction between the Web server and the SISTINA server could be minimized by delegating these operations to a SISTINA software package installed in the Web server.

From an high level point of view, the SISTINA software package installed on the IAP Web server mainly supports:

- the IIOP communication with the SISTINA server;
- the invocation of remote methods on the SISTINA server objects interfaces in relation with the operation defined by the API;
- the retrieval of information related to the Web server involvement in the SISTINA services provision;
- the discrimination between Internet users having an active service session with a SISTINA IAP and normal Internet user.

2.4 The Confidence Provider

The Confidence Provider is an independent organization in charge of evaluating the reliability of both end users and Internet Application Providers. For example End Users can be considered to be reliable if they have always paid their bills, while IAPs can be evaluated on the basis of the quality of their service (in terms of both the quality of their goods and the quality of their shipping service, in cases where this applies). Unreliable end users and IAPs are inhibited from logging in.

2.5 The Billing Provider

As far as billing and charging is concerned, existing TMN/Billing systems can be integrated in the overall architecture. In this scenario the Billing Provider role can be played by the PNO, who takes care of both the charging of the dedicated real-time connections and the payments of goods/on-line services. For those PNOs who are not willing to take care of payments of goods and on-line services, a scenario in which the SISTINA Retailer also interfaces with different stakeholders such as banks/credit card companies has been devised.

3. SISTINA service scenario: user's point of view

The user can start his navigation as a "classical" Internet user on the Web Pages of a SISTINA affiliated Internet Application Provider (a Service Provider who has signed a contract with a SISTINA Retailer).

In any case, in order to exploit the SISTINA value added service features, the user must enter the SISTINA world, that is he must give his username and password. On-line subscription to SISTINA is foreseen. After the authentication procedure the user knows he will enjoy all the value added features he has subscribed to.

The Internet Application Provider can decide to propose the user to enter the SISTINA world:

- at the beginning of his navigation, if the IAP is providing a customized service (and therefore needs to know about the identity of the user in order to provide him with a customized navigation), or
- just before the utilization of a particular value added service feature, if the IAP is providing a service which makes only use of the connectivity control and/or integrated billing SISTINA Service Features.

The decision about when proposing the user to enter the SISTINA world depends upon the type of service the Web application developer has designed. At that point the user is presented with a choice between continuing the navigation as a normal user or accepting to enter SISTINA world to enjoy some enhanced features. It means that the user has to explicitly accept to identify himself in order to be recognized as a SISTINA subscriber and consequently to have access to something more. Once the user has accepted to access the SISTINA world and he has been authenticated and authorized, he can take advantage of a single SISTINA value-added service feature or of any their combination. For example, the user can be asked to provide his username and password:

- before entering a payment phase that allows the charging of the purchased items on the telephone bill or
- before starting a high quality video conference on a dedicated connection with a sales person, paid by an Electronic Commerce Application Provider or
- before initiating the navigation on a Web server that is able to personalize the content of the pages for SISTINA users. The cost of any purchase is added to the user telephone bill without requesting him a new identification but only the acceptance of the operation.

In this approach, the logic of the service being provided resides within the Web application, while the SISTINA value added service features are service independent capabilities implemented in the SISTINA Retailer.

After the user has enjoyed the SISTINA value added service features foreseen by the service offered by a first IAP, he can be invited in following links that point to new SISTINA affiliated IAPs. The interaction between the user and the new SP can occur in the context of the same access session (i.e. the access session between the user and the SISTINA Retailer is maintained): the user does not need to log in again.

It is also possible that the SISTINA Retailer has a Web site and acts as an IAP. Such a site can be used to advertise all SISTINA services and IAPs. The SISTINA Retailer Web pages allow a SISTINA subscriber to enter the SISTINA world as a named user or to subscribe SISTINA services.

4. SISTINA computational model

The SISTINA service features, described in section 2.2, can be provided by one or more stakeholders. The scenario devised by the SISTINA project foresees the involvement of two different stakeholders:

- the SISTINA Retailer: it offers service features 1, 2 and 4;
- the Confidence Provider: it offers service feature 3;

Service features 2a and 2b can be split, and the scenario can be easily extended with the involvement of a third stakeholder (e.g. a bank or a credit card company) taking care for the payment of goods.

In the SISTINA target architecture, depicted in figure 1, both the End User and the Internet Application Provider play the Consumer role in the TINA sense, since they access and use services provided by the SISTINA Retailer. The Consumer and the SISTINA Retailer are in a "User to Provider" relationship.

The SISTINA Retailer plays both the Retailer role, since it provides the Consumer with access to services and uses other providers to support the provision of services to Consumer, and the Connectivity Provider role, as it has to control network resources to provide an end-to-end quality of service. In this sense, the other provider, the Confidence Provider, plays the Third Party Service Provider role, in the TINA sense. The SISTINA Retailer and the other Service Providers are in a "User to Provider" relationship.

The SISTINA Retailer is the only "point of contact" of both the End User and the IAP. It interacts with the Confidence Provider to check authorization rights of the End User and of the IAP (parties), while it interacts with the legacy billing systems when the cost of a purchased good or the cost of a dedicated connection has to be added to the consumer's telephone bill .

The TINA-based SISTINA computational model is depicted in figure 2.

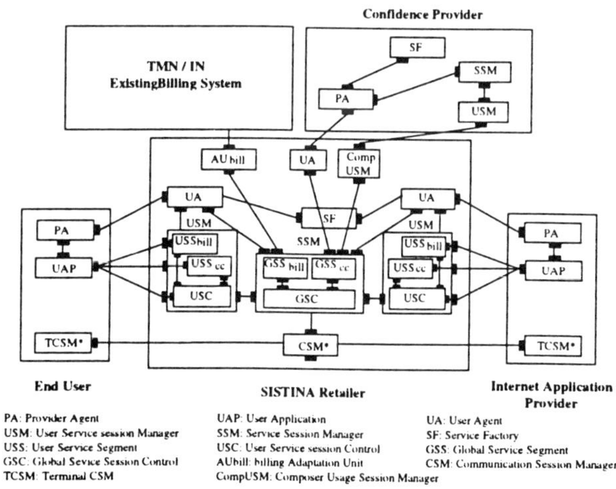

Figure 2 - SISTINA computational model

The SISTINA Retailer USM comprises the service-specific and generic session control segments of a provider domain user service session. It is composed by one common service component, the USC, and by one or more service components specific for the supported service features, the USSbill and the USScc.

The SISTINA Retailer SSM comprises the service-specific and generic session control segments of a provider service session. It is composed by one common service component, the GSC, and by one or more service components specific for the supported service features, the GSSbill and the GSScc. The last two service components are in charge of interacting respectively with the AUbill and the CSM* to request the actuation of the billing and connectivity control functionality. Both the AUbill and the CSM* are in charge of translating SISTINA requests respectively of integrated billing and connectivity control into operations supported by existing billing systems and existing communication systems.

5. Provision of additional QoS by means of the SISTINA architecture

The SISTINA service architecture aims at providing added value to pure INTERNET-based services, such as the capability to offer the users a "on demand" better Quality of Service. Typical examples of services which can take advantage of increased QoS provisioning are:

- download of big files such as pictures and videos
- real-time audio or audio/video interaction (depending on the CPE's capabilities) between customer and sales people/customer support people
- real time provision of multimedia contents
- upload of a file to a printing or CD burning shop
- deferred access to the service at reduced cost using network load information (i.e. downloading copies during night time or updating proxies during night time)

Thanks to the existence of the service session maintained by the SISTINA Retailer dedicated switched connections can be established between users and service providers, taking advantage of different transport technologies.

On the basis of the knowledge of CPEs capabilities, in terms of type of access to the Public Network (POTS / ISDN and ATM/ADSL), of service provider capabilities and of the characteristics of the network deployed between users and service providers, the SISTINA service platform is capable of selecting the most appropriate transport technology and of establishing the dedicated connection required in order to achieve the desired QoS. Service providers can be expected to have multiple different connections to several kinds of Public Networks, to cater for the different types of user terminals.

The SISTINA Retailer is the stakeholder in charge of instructing the network elements involved in the service session on the way to establish the additional connection, to cut off Internet bottlenecks.

Different technical realization can be envisaged. One promising possibility is the utilization of third party call control mechanisms. Other possible alternatives are currently under investigation (i.e. connections established by terminals or by other network elements).

In figure 3 a general scenario is illustrated. Two transmission paths are indicated: C1 is a transmission path from the Client towards the Server through the Internet , C2 is the dedicated connection established on the basis of a request coming from the SISTINA Retailer in order to offer increased QoS to the Client.

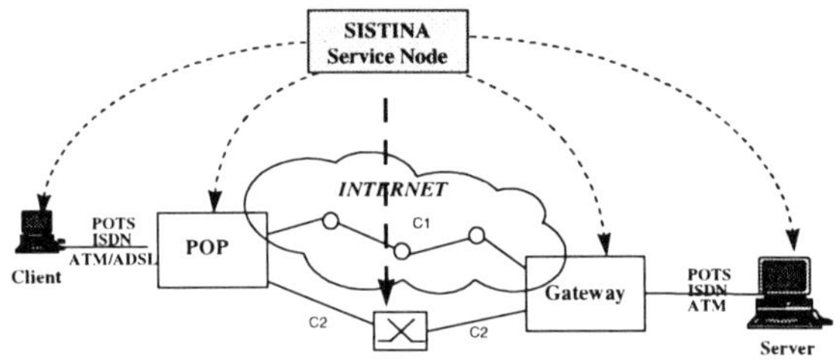

Figure 3: A general scenario for guaranteed QoS provision

This feature not only provides IAPs with the possibility to reach Internet users through Quality of Service-guaranteed connections, but it also gives the possibility to integrate at the service level Internet and traditional connection-oriented transport mechanism.

This feature can be combined with flexible charging policies in order to offer the End User an homogeneous view of the services he is accessing.

For example an additional connection established to achieve better Quality of Service could be paid by the IAP in case of electronic commerce applications, implementing a sort of Internet-based multimedia freephone service to advertise products. This is allowed by the existence of a single complex telecommunication session, which involves one or more service providers and at least one network provider, which detains all the information related to the service provisioning, including that related to network connectivity provision and to billing.

6. Comparison with other approaches

Mulimedia applications typically require two different kinds of communication mechanisms:

- Control messages, that are best conveyed on a datagram-based network, in that they do not require a continuous data flow between a source and a destination, but only the rapid delivery of small chunks of information;
- Video/audio streams, that are best conveyed by connection oriented networks as the concept of flow is inherently tied to the concept of resource allocation over a (statically or dynamically) defined path.

In any of these cases a strict delay requirement exists: delays in multimedia services like videoconferencing, tele-education, networked games, and in general in all interactive services, is a major factor of perceived quality.

The SISTINA approach is that of integrating, at the service level, PSTN with the Internet, trying to use each network to convey the type of traffic it is best suited for. Routed datagram networks are very cost-effective in transporting small chunks of information, and that is the purpose they are used for within SISTINA. Traditional switched networks have proven to be very efficient in conveying real-time, fixed bandwidth, traffic, therefore whenever the provision of a stream-oriented data flow is required, a real-time switched connection is transparently established end-to-end, cutting off bottlenecks. QOS is provided on-demand, whenever really needed and whenever there's someone willing to pay for it.

The recent overwhelming success of datagram-based networks, such as the Internet, has proven their ability to provide multimedia-like services at a fraction of the cost required to provide the same services on a traditional connection oriented telecommunication network. The huge success of the Internet has pushed researchers to investigate the possibility to effectively convey stream-oriented traffic with strict bandwidth and delay requirements on it.

A number of different approaches have been devised and new protocols, among which RTP and RSVP have been defined.

This section will highlight the advantages of the SISTINA approach with respect to such protocols.

6.1 RTP

RTP is a protocol suite made up of two components:

- a protocol for the actual transmission of real-time data (called itself RTP)
- a control component (RTP Control Protocol, RTCP) to transmit QoS information and session control information to all participants to a RTP session

Despite its name, RTP doesn't really provide any guarantee of delay in the packet delivery: only because of its being UDP and not TCP-based, packets with high delay can be simply discarded and no retransmission of lost or corrupted packets is supported (it would simply be pointless in most of real time applications).

RTCP provides information, in the form of control packets, to all participants to the audio/video session. From the QOS point of view, the most important data it conveys is a feedback on the quality of data distribution. This feedback reports help in controlling congestion which is one of the biggest causes of delay [1] In other words, RTP simply warns the applications when something went wrong, but cannot (and doesn't want to) do anything to remedy the problems [2].

In any case, even if technology is quickly progressing, IP routing still tends to introduce higher delays than traditional switching within connection-oriented networks. Even if and when routers with pass-trough delays comparable or inferior to those of traditional switches will be available, it will take years before they will be ubiquitously deployed. SISTINA, on the other hand, by exploiting the huge installed base of telecommunication switches, is realistically capable of meeting strict delay requirements without any revolution of the installed base of equipment.

6.2 RSVP

RSVP is an attempt to provide a signalling-like protocol to reserve resources on the links connecting sources and destinations of a data flow.

RSVP has a number of major flaws:

- There is no guarantee the data packets for which resources have been allocated will actually follow the route where the allocation has been performed: in case of route oscillations, data packets will be routed in a different way PATH messages have been routed [2]. Some work on protocol behaviors [3] show that route oscillations are not a seldom event in the Internet.

- Soft states put inherently at risk the allocation of resources: in the case where several successive RESV messages are lost, some useful resources could be freed leading to disruption in the communications
- It will take years before all routers will be RSVP compliant. In the meanwhile, if a data flow encounters a non-RSVP compliant island, there's no guarantee of QOS. As RSVP does not influence routing, no mechanisms are defined to try to at least circumnavigate such islands.

The SISTINA approach has an answer to all these RSVP open points. QOS in SISTINA is guaranteed in that fixed bandwidth switched connections are used to convey stream oriented traffic. SISTINA makes use of hard states, as resources are allocated by means of traditional signalling mechanisms. Last but not least SISTINA relies on those switched telecommunication networks that have been proving QOS for years. Again, no revolution is needed: the installed base of an equipment that has been proving for years to provide those features we are looking for, can be successfully exploited.

7. SISTINA integrated billing

Billing systems are something very valuable in the context of an on-line service business scenario. The combination of electronic commerce servers with advanced billing systems and capillary billing organizations such as those run by PNOs, can represent a very powerful business machine.

Within SISTINA it is possible for the IAP:

1. to pay for the cost of the additional connection(s) established in order to increase the QOS, or at least for part of it
2. to take over the cost of the telephone call from the customer to his ISP, for the duration of a SISTINA session
3. to offer discounts on the End User's telephone bill (e.g. in case the customer purchases goods for a certain amount of money).
4. to ask the PNO to charge the cost of the goods purchased through the Electronic Commerce application, or the cost of the services enjoyed by the consumer, on his monthly telephone bill

In order to achieve these goals, the SISTINA Retailer has the capability to:

- flexibly split the charge of the additional connection(s), established in order to increase the QOS, between the End User and the IAP. A particular case within this framework is that of the Application Provider paying for the entire cost of the connection.
- charge the cost of a connection the End User is using at that particular moment, such as the connection to the Internet POP, on a third party.
- decrease the End user's telephone bill, independently of the cost of the connections he is using at that particular moment
- charge on the End User's telephone bill, or on a separate document which can be shipped together with the telephone bill, the cost of the goods purchased through the Internet Application, or the on-line services enjoyed by the customer

In order to safeguard SISTINA Retailers against insolvent consumers an expense limit is supported (Retailers could always decide not to use this feature).

Within this context it is possible to perform an on-line agreement between End User and SISTINA Retailer on the monthly expenditure ceiling (off-line procedures can follow: for example the SISTINA Retailer could ask for a deposit to be paid by consumers). The amount of the monthly expenditure ceiling can be later re-negotiated (again, off-line procedures can follow)

It is possible for End users to check their monthly expenses. The difference between their current expense and the monthly expenditure ceiling is explicitly stated. The on-line visualization of personal data related to the consumer (e.g. the monthly telephone bill, the log of outgoing PSTN calls etc.) could represent an interesting enhancement of this feature

The SISTINA Retailer checks the expenditure ceiling before authorizing 3pty service provider to sell goods or services.

8. Conclusions

The SISTINA network architecture adds value to the existing WWW Internet applications by interworking with the standard connection-oriented PSTN services. Key TINA concepts like the session concept or parts of the business model have been exploited.

For a PNO a SISTINA-based solution results in the ability to generate extra revenue and to provide Internet users with advanced features like quality of service and accounting.

The possibility of charging the cost of goods electronically purchased on the monthly telephone bill can be regarded as a major value added feature, in that not only it easily solves the problem of payments over the Internet, but it also decreases the cost of transactions, giving way to a whole new range of goods and services to be purchased on the Internet (e.g. low cost goods and services).

The possibility to apply flexible billing to the additional connections established to increase the QoS, to reverse the charge, or to apply discounts on connections in case they lead to successful business transactions, can overcome the well-known Internet user unwillingness to pay for whatsoever services or goods.

The same extremely successful business paradigm exploited by PSTN freephone numbers applies here, where the stakeholder asking for the service (the additional connection) is different from the one paying for it, and the latter is willing to pay as the service itself can increase his revenues.

We believe that this is one of the very pragmatic approaches which makes use of the TINA principles in the Internet context.

The Siemens/Italtel team is developing a prototype for the SISTINA architecture.

9. References

[1] H. Schulzrinne et al., "RTP: A Tansport Protocol fo Real-time Applications" RFC 1889, January 1996

[2] D. Hutchinson, R. El-Marakby, L. Mathy, A Critique of Modern Internet Protocols: The issue of Support for Multimedia

[3] K. Varadhan, R. Govindan, D. Estrin, "Persistant Route Oscillations in Internet Domain Routing", Technical report, USC/ISI, USA, February 1996

TINA and the Internet: an Evaluation of Some Scenarios

C.A. Licciardi, R. Minerva, C. Moiso, G. Spinelli, M. Spinolo
CSELT S.p.A.
Via G. Reiss Romoli 274
I-10148 Torino, Italy
roberto.minerva@cselt.it

Abstract

This paper is an attempt to analyze the applicability of TINA in an Internet context considering the ***usefulness and the need*** *of TINA solutions from a business point of view and not in its applicability "per se" For this reason some applicability scenarios have been chosen. They describe the applicability in a general framework more than describing the technicalities. The results are that TINA should pay more attention to the developments of the Internet in terms of protocols, APIs and products so that it could be a useful "component based architecture" for building servers and services over the Internet.*

Prologue

It was a sunny September afternoon in Heidelberg. The TINA 96 Conference was smoothly coming to an end. A few of the participants were discussing the *hot topics* of that time: the integration of TINA [1] in existing networks and the future *relationships between TINA and the Internet*. The issue posed by someone (laughing at the others) was a striking one for the TINA community "the real question -he said - is: *How to integrate the Legacy of TINA into existing Internet applications*". Some of the authors of the paper were astonished by the rough, but straight question. The paper is all about this topic. The people involved in this paper have dealt with TINA stuff since a while, so a stunning exercise was to be open-minded in thinking of some fundamental issues on the applicability of TINA to the Internet.

1. Introduction

As many may have noticed, the world is divided into two distinct categories: the TINA *addicts* and the real world. Usually the former ones try to apply the TINA concepts *per se*[1] without considering the merits or benefits of it, the latter ones get annoyed as soon as the word TINA is pronounced and usually make questions related to the performance issues of TINA and its long term applicability. We believe that there is something in between these opposing trends. This paper is an attempt to analyze the applicability of TINA without being *addicted* of it and without the pre-opposition for any TINA solution. The starting point for such an analysis is in the **usefulness and the need** from a business point of view of the application of TINA in an Internet context and not in its applicability "per se" For this reason we have chosen to describe some applicability scenarios more than describing the technicalities in the application of TINA to the Internet. We do not claim that the presented scenarios cover all the cases in which TINA can be applied to the Internet, however a good number of scenarios have been considered and described.

The paper discusses also some misconcepts related to TINA and the Internet considering the following issues:

- is there a real need for TINA in order to provide QoS within Internet transport ?
- what services are to be realized according to the TINA architecture in an Internet context ?
- how to progress the TINA specifications in order to converge with the Internet ?

The first question is dealt with in section 2, considering some of the current results in the realm of Internet QoS. The second one is discussed in section 3 considering classes of services that could benefit from TINA. The third one is answered in section 4 and in the Conclusions, where a more business centered approach in the application of TINA is promoted.

[1] An example of this was the discussion within the core-team about the conformance of the number three (exactly the integer of cardinalyty three) to the USCM model.

0-8186-8335-X/98 $10.00 © 1998 IEEE

2. TINA Network Aspects and the Internet

Nowadays the Internet is seen as a way to promote the multimedia market. Public Network Operators (PNOs) are facing two alternatives: to use the Internet paradigm for their multimedia offerings or to adopt a "classical" telecom paradigm (i.e., DAVIC, B-ISDN approach). An example that shows how difficult it is to promote the multimedia market is represented by the *case* of Video on Demand. Currently many Public Network Operators (PNOs) deem that there is no real value in providing Interactive Multimedia services using traditional call based networks. As a matter of fact, PNOs are trying to understand whether there is a business case behind the Internet. Several network and services scenarios are currently analyzed aiming at clarifying how TINA can be applied in an Internet based environment. A first result is that there is an on-going effort for improving the QoS in the Internet connectivity in order to guarantee some levels of efficiency to multi-media services (to be paid for by the customers). On the other hand, TINA claims to be an architecture made for telecommunications as well as information networks. It becomes obvious to investigate the question "What TINA can offer in order to improve the *Internet Connectivity* ?". Two different scenarios are here considered: the first is related to the use of TINA technology in a network environment made of routers (it is essentially a management scenario), the second one focuses on how TINA can be applied in real-time control of connectivity in an IP based network.

2.1 Management of Routers

Many PNOs are strengthening their IP infrastructure by adding to the public network infrastructure an overlay network made of routers interconnected by Frame Relay or even ATM networks. The goal is to build a powerful backbone supporting the increasing (actually the overwhelming) IP traffic. This effort is not problem-free, the IP backbone is becoming complex and expensive, e.g., moving from a few mono-vendor nodes to an heterogeneous infrastructure with stringent requirements on the interoperability. There is the clear need to manage such an infrastructure and to take advantage of the investments by offering new services to large customers (e.g., exploiting the emerging market of the Intranet).

With respect to the applicability of TINA to the Internet, major differences between the two approaches for the control of connectivity are to be considered:

- mechanisms supporting the connectivity. The Internet connectivity is based on a "best-effort" approach and on a connection-less[2] network service (in OSI terms) provided by the IP protocol. This is quite different from a TINA approach that better fits a traditional telecom view on connectivity. This view is based on the control of complex configurations of end-to-end circuits. In this respect the problem of applying TINA for supporting IP connectivity is not much different from the ongoing debate related to IP over ATM. With respect to signalling, it should be noted that TINA is much more Third Party Control prone. This emphasize the difference with the Internet.
- network element characterization. On the Internet side, any network system can be defined as an *autonomous system*. A router, for instance, is implemented so that it can choose the best route to forward IP packets in the path from a source to a destination. The decision is taken on the basis of the view of the router on the network topology. This requires a lot of intelligence within the system and a way to share the network knowledge between systems. On the TINA side, instead, each network element is controlled by a higher level intelligence that determines what the connectivity needs are. In this sense the network element at the lowest level can be defined simple (or dumb). In addition, the current implementations of routers do not support APIs for controlling the system and there is not much interest in opening up the routers interfaces for the purpose of a more granular control of them.

Considering these characteristics, the TINA approach seems to be hierarchical and, in some sense, centralized, while the Internet one can be defined as highly distributed and based on autonomous systems. The above features need to be taken into account in order to try to apply TINA to the Internet. In the following sections some scenarios related to the applicability of TINA to the IP connectivity are briefly discussed.

2.1.1 Objectives

A possible application case for TINA [2] is in the realm of the management of systems supporting the IP traffic. The network is made of routers and dedicated (ATM) switches. The objective is to define a set of management functionalities enabling the PNO:

- to manage in an integrated way a heterogeneous network of routers and switches,
- to support the ability to configure the infrastructure so that some customer driven requirements for QoS can be fulfilled from a management point of view,
- eventually to support customer control functionalities in order to have the users to configure their IP connectivity over the public infrastructure.

[2] Obviously on top of a connection-less service, a connection oriented service can be provided. This is actually the case for TCP/IP stack.

2.1.2 Technical Description

The TINA management system, depicted in Figure 1, is based on a DPE that supports the intercommunication between components.

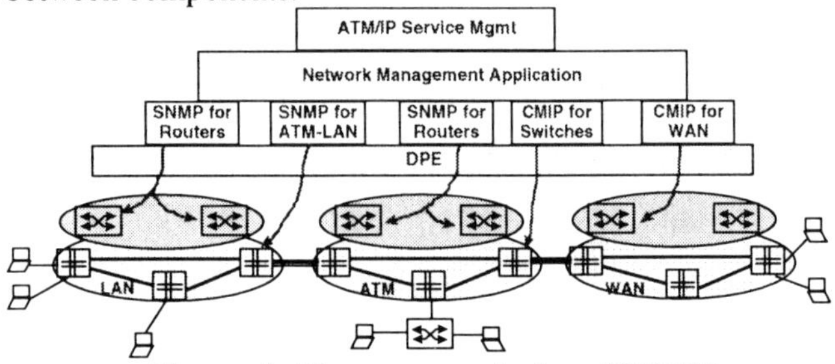

Figure 1: Management of an IP/ATM network

Several Adaptation Units allow the TINA infrastructure to manage a heterogeneous network of routers and network elements. There is a sort of Interoperability and Legacy problem, the routers typically offer a management interface based on SNMP (the MIB-II definition), while the switches can offer both SNMP or CMIP based interfaces. In addition, the IP backbone could become really multi-vendor (each vendor providing its own management interface). These adaptation units can be based on the definition given by P.508 [3]. On top of the adaptation level, TINA based applications for network and service management can be created.

The network management of ATM network elements can be TINA compliant, whilst the management functionalities for routers are still to be defined in the TINA architecture. The foreseen functionalities are related to all the FCAPS functions. In particular the connection management could be used for setting up and tearing down connections between routers according to the TINA paradigm.

What seems to be of the paramount importance is to build an infrastructure supporting the IP/ATM Service Management. IP/ATM service management sits on top of network related functionalities. It offers the ability to build in a TINA based manner services that allow the PNO to supervise different services or components of its IP/ATM network. An example could be: the dynamic configuration of sites and related connections supporting Intranet connectivity, the related ability to account and charge the customer on a per-use base, or the ability to monitor whether the performance of the Intranet are consistent with those subscribed by the customer. A possible extension is the ability to offer to customers Web based interfaces so that they are able to manage their own (part of the public) IP network.

2.1.2.1 An Intranet Example

The scenario described before can be extended to the case of a large company that wants to manage its connectivity needs. CORBA based communication mechanisms can be used in order to allow a customer to dynamically request to vary some configuration parameters of the IP over ATM network in order to satisfy particular communication requirements (e.g., based on time of the day, day of the week policy). TINA based mechanisms, i.e., the Access Session, can be used for recognizing the customer and a specialized session control can be instantiated in order to accommodate the user requests in terms of control over network resources, i.e., a management service compliant to the TINA service session concept. Figure 2 depicts such a scenario.

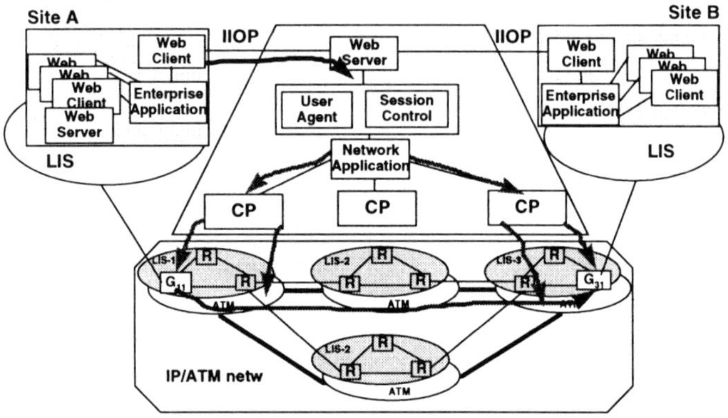

Figure 2: An Intranet Scenario

2.1.3 Evaluation

The advantages of the application of TINA in this scenario are:

- the ability to integrate into a single (network) management layer the functionalities needed to manage an IP network and the related switching infrastructure. It should be noted that, even if the target is the management of an IP over ATM network, such a solution can be applied to a heterogeneous network infrastructure (e.g., Frame Relay and ATM infrastructure);
- the reuse of the legacy in terms of management interfaces offered by different equipment (i.e., routers);
- the ability to build service management applications that allow the dynamic configuration and the supervision of Intranets on a per customer base;
- the ability to offer to customers advanced service management capabilities by means of Web based mechanisms in order to supervise the IP based connectivity.

These features could be realized in a TINA like fashion[3]. For instance there is a chance for the TINA connection management to find out a practical application in this kind of management applications. However there are still missing parts in the TINA architecture that are to be completed. Once more the need for a TINA framework for Service Management is stressed out. A major advantage of the applicability of TINA to this scenario is the good mix between the business value, the timeliness and the lack of existing or competing solutions. TINA solutions in this scenario allow the enhancement of the intelligence within the IP over ATM infrastructure. We assume the ATM connectivity is more likely for large corporate customers than for residential customers, for this reason the solution is for a restricted number of users that do require a high level of programmability of their services. This offers TINA-C a window of opportunity in order to promote the consortium technology within the Internet community.

2.2 Connectivity with QoS in the Internet

Almost everybody is experiencing the long response time of the Internet. It seems appropriate to investigate how the IP connectivity can be improved in order to offer at run-time an acceptable QoS. Different trends are emerging in the debate. Two of them are investigated in the following sections. They are: the Next Hop Resolution Protocol, NHRP (a more telecom oriented solution) and the Reservation Protocol, RSVP (more aligned to the Internet philosophy). Other approaches, e.g., the use of P-NNI, are not investigated in this document.

2.2.1 Objectives

The goal of this analysis is to figure out whether and how TINA can support a higher degree of QoS for the Internet. Finding out solutions in this area can enforce the interest around TINA solutions in a wider audience (i.e., the Internet market). Under this respect two themes are considered: the use of TINA connection management for setting up connections with quality of service and the ability of TINA to deal and enrich the capability of address resolution. The first capability can be considered as a sort of pre-requisite. In fact it is supposed that the TINA connection management (or similar approaches such as X-Bind [4]) could be adopted in order to support real-time connectivity in ATM based networks. For an evaluation of the applicability of TINA connection management, the reader can refer to [3]. Some generic ways for improving the QoS of connectivity within the Internet are:

- establishing end-to-end connections between Edge routers (i.e., the routers nearest to the end-users),
- cutting-through the network of routers. The cut-through approach tries to minimise the time spent on routing packets in favour of the switching of packets. An example of such a technique is implemented by IP Switches [5]. They determine that packets to be forwarded to a destination belong to a flow so instead of routing packets to the next router they set up connections to a destinations cutting through a set of routes.

The application of distributed processing within routers and Addressing servers and the advantages of having a DPE and related services (such as the trader) could put TINA technology in a good position to improve the addressing resolution mechanisms of the Internet providing enhanced services. As a matter of fact, the address resolution in the Internet is mainly related to a single domain or an individual *Logical Internet Subnetwork* (LIS). This means that addresses (e.g., the translation from IP to ATM addresses) are solved within a domain without considering the possibility of solving them on a global scale or applying a dynamic resolution policy. An example could be the ability to implement routing policies depending on the time of the day. Within an IP over ATM network there could be the case to dynamically route IP packets towards the less loaded routers or to solve IP addresses to ATM ones not limiting the resolution within a single LIS. The application of TINA to address resolution and the control of connections between routers (Figure 3) is possible. Two scenarios are briefly discussed in the following sections.

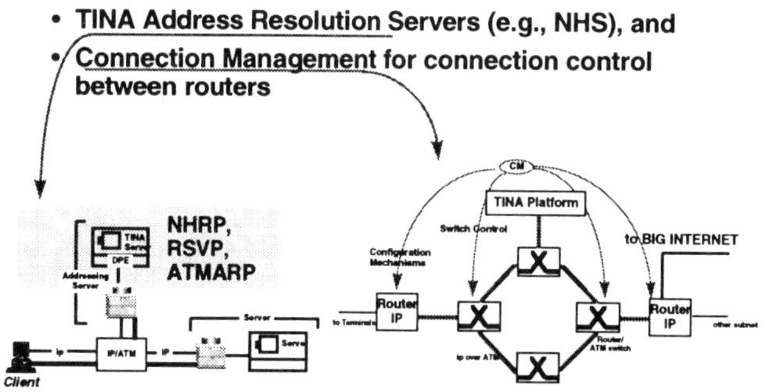

Figure 3: TINA and Internet Connectivity

2.2.2 The Next Hop Resolution Protocol (NHRP) Scenario

This scenario considers the application of TINA for allowing the translation between IP addresses and ATM ones as defined for the Next Hop Protocol (NHRP). This protocol assumes that within each Logical Internet Subnetwork (LIS) there is a server, called Next Hop Server (NHS), which is able to determine the ATM address of the next host or IP router along the path from a source to a destination. Once the IP address has been mapped onto an ATM address a VC can be set up in order to support the connectivity.

[3] However other architectures could be equally considered.

Such a scenario can be realized using some TINA solutions. Figure 4 shows a number of TINA NHS. They communicate by means of the DPE and do implement a sort of distributed "routing logic". In other words, NHSs could be implemented as CORBA servers. In order to resolve addresses, they access distributed data bases, or as an alternative, they could be implemented as an extension to existing DNS servers. This scenario allows an extension to the usual definition of the NHRP, in fact the address resolution can be considered not only with respect to a single LIS, but the addresses resolution could be extended across several LISs, i.e., those comprised in network domain controlled by the single network operator.

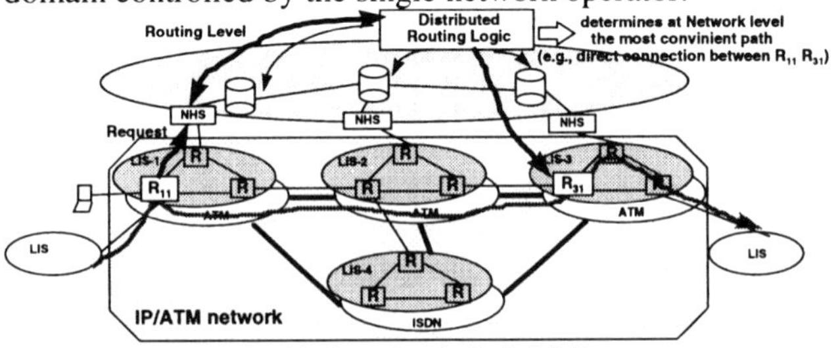

• NHS servers query a distributed data base to determine the physical address
• The Routing Logic considers the whole ATM network
• Different Policies can be implemented in order to determine the physical address of next hop
• The Logic is used to orchestrate the creation of connections between R_{11} e R_{31} (e.g., by means of TINA Connection Management)

Figure 4: A TINA implementation of the NHS

In addition TINA connection management can be used in order to create connections between the routers themselves or the routers and the end-systems. By means of the application of the TINA technology, the policies for solving addresses are directly "programmable" by the PNOs that can optimize the use of the ATM backbone. Such a scenario can be further extended considering different types of networks, e.g., if the destination IP number corresponds to an ISDN number, than the TINA structured control can interwork with the ISDN (through the signalling adapters discussed in [3]) while it is creating connections through an ATM network by means of the connection management or using Third Party Call Control mechanisms (as described in [3]).

2.2.3 The Resource Reservation Protocol (RSVP) Scenario

This scenario considers the use of the Resource Reservation Protocol [6]. It is designed so that end-user systems can request the reservation of network resources in order to satisfy communication needs. Resources are reserved along the path to the user in a non exclusive fashion.

This scenario assumes that special gateways (between IP networks and a TINA based network) are built using TINA solutions. They gather the RSVP requests and are able to map them into requests to set up connection within a Subnetwork (e.g., an ATM Subnetwork) according to the requested QoS.

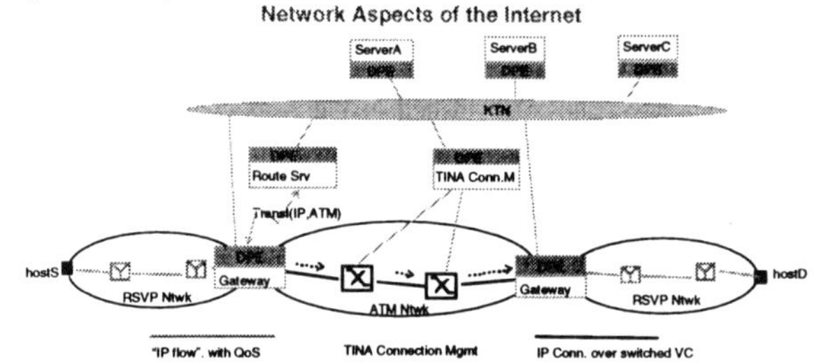

Gateways can request the creation of switched/permanent links (e.g., TINA Connection Management) in order to satisfy QoS requests received by means of RSVP
Gateways route datagrams also in a "best-effort" fashion
Gateways can be programmed for creating VCs/VPs to particular routers or Servers in the same subnet at a particular time of the day

Figure 5: Application of TINA to RSVP

Figure 5 depicts the case in which the gateways are able to communicate with TINA components (e.g., a TINA Service Session Manager) in order to map the IP destination to an ATM address. In this case the TINA connection management can be used in order to create ATM connections within the Subnetwork. Other solutions such as Third Party Call Control could be applied.

The RSVP scenario could be generalized considering the RSVP gateway as a more general device that guarantees the interoperability between different physical and logical networks. It would be possible to consider the (TINA) network comprised between the RSVP gateways as a sort of Virtual Router able to treat IP packets providing a certain level of QoS. There is a certain interest within the standardization for solutions based on Multiple Protocol over ATM (MPOA). These solutions consider a Subnetwork as a sort of Virtual router that supports interfaces for different physical networks. The Virtual Router can control the connectivity within a Subnetwork. A TINA structured Virtual Router (e.g., an implementation of MPOA) is an interesting scenario from a technical point of view, in fact it could demonstrate the use of TINA connection management for supporting connectivity within the Virtual Router. This could show benefits of open interfaces for controlling network elements. On the other side, such a scenario could force the TINA specifications to consider or design some connections related APIs that are based on the current standardization of the P-NNI (that is going to be used for supporting the MPOA).

2.2.4 Evaluation

In principle the TINA Connection Management can be considered as a generalized "cut through mechanism" that allows taking advantage of switching capabilities instead of routing mechanisms. TINA connection management based solutions can be implemented on currently available ATM open switches. However the TINA connection management has not yet reached the level of performance required for being applied in real-time control of connections. An evaluation of the scenarios yields to the following considerations:

- it is not clear whether the NHRP will be really adopted; RSVP seems to have more chances being more aligned with the Internet philosophy;
- the use of the DPE for supporting the communication between the NHSs or RSVP gateways and TINA based servers enhances the level of programmability in the resolution of addresses and the ability to control connectivity;
- a flexible routing logic can be easily programmed using TINA based solutions; TINA servers for address resolution are points of flexibility with an IP over ATM network;
- an MPOA solution based on TINA architecture could prove feasible;
- the flexibility and programmability of the TINA connection management could be exploited;
- it is likely that routers with NHRP or RSVP capabilities built-in will emerge much before than TINA based routers will do. This means that PNOs will not introduce TINA based solutions in the medium term;
- alternative solutions to the TINA connection management are emerging now. IP switching and ATM Forum signalling already offering real time control.

As a general remark it should be stated that non-TINA solutions in this area are on their way to the market. It is likely that in a two year time frame they will emerge as robust products. Currently there is no effort within TINA to cope with these issues, and there are only few products promoting the adoption of TINA connection management at an industrial level. So it is unlikely that TINA based products will emerge in this area in a useful time-frame. For these reasons it could be stated that there is not a TINA business case in the area of QoS and real-time connectivity for the Internet.

3. TINA Service Aspects and the Internet

TINA solutions are often coupled with the use of Internet based mechanisms. A typical example is the use of Web Browser for accessing to TINA service architecture objects and functionalities.

Analyzing the use of IP from a telecom point of view, it can be used for conveying information flows **and** control messages. Therefore IP communication will be more and more used for requesting services and for low-cost and delay independent transport of information.

In the TINA community, it has been stated that IP solutions can be used for providing services in a "best-effort" fashion, while the TINA architecture can improve the service provision offering connectivity with guaranteed QoS on demand. Figure 6 depicts the use of IP and TINA for transport and control.

- **IP can be used for transporting Control and Information Flows with a Best Effort Policy towards/from TINA based systems**
- **TINA can provide end-to-end connectivity with guaranteed QoS on demand**

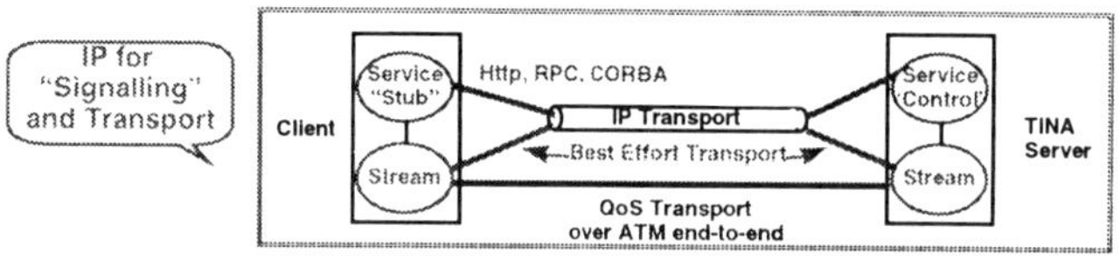

Figure 6: IP "best effort" transport and TINA "guaranteed" transport

This section is an investigation on whether and how TINA can provide services ensuring "QoS on demand" for Internet connectivity.

It should be considered that TINA scenarios usually consider end-to-end connectivity provided by a Broadband Network (see for example the session model definition). In the real world this means that an ATM network has been deployed. In addition TINA does not pay attention to issues related to the access network (e.g., hybrid configurations in which the fiber is not deployed to the "home" and not even to the "curb"). Such a perspective implies a pervasive deployment of ATM that coincides with an "ATM to the desktop" scenario. On the other hand, the IP protocol does not pose these constraints, in fact it is designed to be used on several networks with different maximum bandwidth. The conjunction of IP and TINA should be framed in a precise network scenario. Such a scenario seems to be interesting for many European PNOs.

3.1 Service Access Using Browser Capability

One of the major claim of TINA related to the integration with the Internet has been the easiness of accessing the TINA service architecture by means of a Web Browser. Actually the Web Browsers do support a sort of Universal Graphical User Interface. So, for sure, TINA has to make the best out of it. The concept of Access Session seems to be a good concept to be promoted within the Internet and to be required in the PoP/Web Server products in order to provide better services to the Internet Service Providers. The enabling factor for this enhanced communication between the client (the terminal) and the server (the network server) is the terminal ability to support complex interactions with "servers". In the future this capability can yield to richer and more complex telecom services than current ones. This is represented in Figure 7.

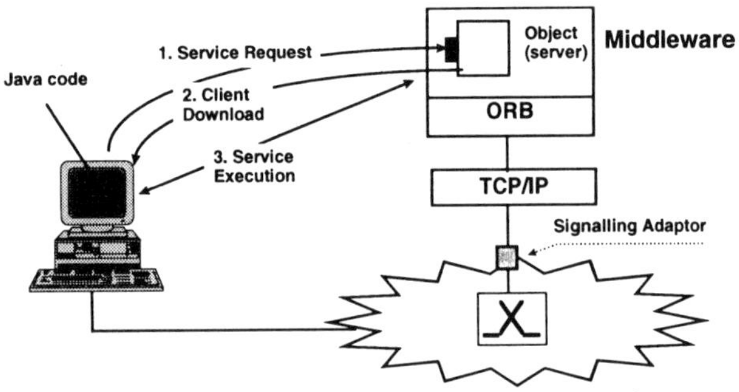

Figure 7: Terminal-Network Interactions

TINA promotes this paradigm and proposes the User Agent / Provider Agent paradigm for service access also in an Internet context.

3.1.1 Objectives

The scenario tries to evaluate the benefits of the introduction of the User Agent / Provider Agent paradigm as proposed by TINA and the ability to launch TINA applications in a Web Server context. The goal of this exercise is to assess the need for requiring/implementing TINA solutions for providing service access on a large scale and not only for TINA demonstration purposes. A particularly open Web Server, Netscape ONE, is considered. It uses IIOP protocol (interoperability with TINA DPEs is possible), and it offers the ability to introduce TINA objects within its own application architecture. Netscape ONE is just one example of how Web Servers could support the introduction of the TINA Service Architecture.

3.1.2 Technical Description

With reference to the Netscape ONE architecture (see Figure 8), it is possible to build applications that are completely TINA compliant using services and interface supported by the Netscape Server's architecture. In addition it is possible to introduce (at the level of Foundations classes) some TINA service architecture objects. This approach could be extended even further by providing a much more service rich environment. In fact, some client objects could be implemented as JAVA applets and be compliant to TINA User Agent / Provider Agent specifications. In this sense, a Provider Agent object could be downloaded to the client realizing one of the major concepts of the TINA architecture. Another feasible option is related to the ability to launch TINA applications. Also other Web Servers are able (through several mechanisms: CGI, Java RMI or CORBA based solutions) to activate applications as a response to client requests. This enables the possibility to "launch" TINA services on user request.

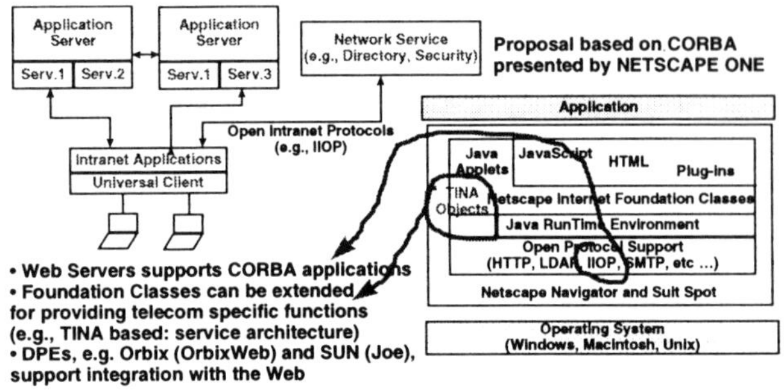

Figure 8: Netscape ONE Architecture

In this perspective a user could access to services through TINA Access Session mechanisms. It would be identified and authorized according to the TINA service architecture, i.e., by means of the Initial Agent, then a Provider Agent could be downloaded to the user terminal and the User Agent be activated. On user request, a list of executable services could be shown (by means of HTML pages). The user could equally request to launch a TINA application (e.g., a Multimedia Conference Service) or an Internet application (e.g., a whiteboard application). However it should be noted that this product (and its competitors) already do provide many service access functionalities. For instance security and authentication mechanisms are built-in according to standards or proprietary (and well established) solutions. In addition they do support some levels of customization and programmability that can be compared with those (hypothetically) offered by a TINA product. So it is debatable the need for introducing TINA solutions.

3.1.3 Evaluation

Web Servers have not been designed for TINA and many products have been developed for efficiently supporting the Service Access and many more functions (think to the PoP/Web Server products already used from many Internet Service Providers worldwide). Applying TINA to these products, on one side, means to modify/update established products, and, on the other side, does not guarantees better or improved functionalities of the PoP/Web Servers. Currently Web Server can support much more advanced identification and security policies for the authorization of users than those offered by TINA architecture, namely by the Access Session. So it is not too much appropriate to advocate the flexibility of TINA versus the rigidity of these systems. Viceversa, this scenario demonstrates the usefulness of state of the art Web technology. **The flexibility of CGI or Java RMI offers an easy mechanism for implementing mainly the access session of TINA**. The Internet Service Provider is free to store service and user related information by means of different mechanism, so existing databases can be used for this case according to the desired template. Some major conclusions can be drawn by the presented scenario:

- Web Technology is now mature enough to be of value to TINA
- The Web Browsers are to be used as universal GUIs
- There is no clear advantage for Service Providers in using TINA technology instead of Internet technology in Web Servers today
- "TINA Applications" could be provided using Web mechanisms for interacting with the user
- Having a TINA structured Web server has value if it is related to the provision of a service based on the TINA architecture (see next sections)
- TINA or Internet applications can be indifferently supported and executed by a Web Server.

The main conclusion from this scenario is that TINA has more to gain from adopting today's existing Internet service software suites than the Internet has from using TINA. This due to the fact that the Internet technology has now reached a much more mature state.

3.2 TINA Services with QoS for Users

There is an increasing user demand for "simple-to-use" but customized and personalized services that integrate "Internet" capabilities with the wide availability of the telephone services. Competing approaches are possible for fulfilling the demand for multi-media services: on one side, the increased terminal capabilities coupled with the deployment of specialized servers (at the *edge* of the public network) will allow the offering of "multi-media" services according to the Internet Paradigm. In this case the Public Network is seen as a "pipe". On the other side, the public network can offer connectivity with guaranteed QoS and specialized functions (and related network systems) that enhance the value of the transport service providing for flexible routing, flexible charging and supporting the *mobility* of users and services. So far these two approaches have been perceived as opposing each other. However there is a new trend that fosters the possibility of co-operation of two different paradigms and network infrastructures (i.e., packet and circuit switching networks) in order to integrate Telecom and Internet services.

In the short-medium term, an Internet based communication paradigm can be used for providing **new** Telecom services. The synergy between the Internet and the telephone network enables a rich set of ***mixed services*** supporting multi-media capabilities. Services are provided by network servers deployed within the public network infrastructure (as shown in Figure 9). These Internet/TINA based servers are open and programmable by definition. This yields to important consequences from a software architecture point of view:

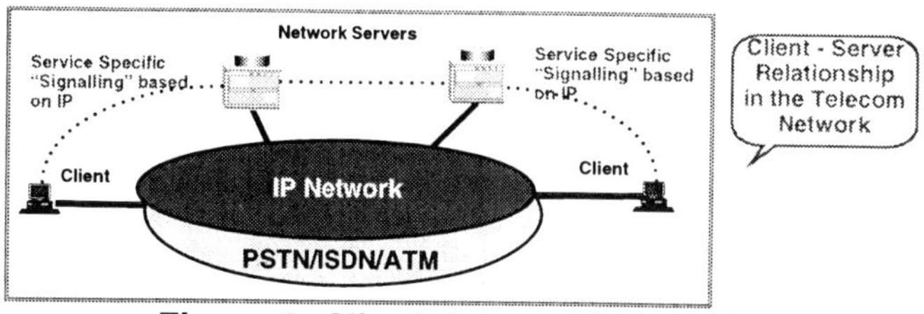

Figure 9: Client-Server relationship between terminals and Network Servers

- a client-server relationship between terminals and the network, a shift from the paradigm that all the intelligence is within the network (and terminals are dumb)
- the opening up of interfaces, more exactly Application Programming Interfaces, within network servers.

This scenario considers the possibility of navigating through the service offering using Internet mechanisms (e.g., the Web Browsers discussed in the previous sections). The TINA architecture is hidden behind the "universal" graphical interfaces offered by Web Browsers.

3.2.1 Objectives

The goal of the scenario is to investigate whether TINA solutions can be applied within an Internet framework for providing services with QoS. This scenario can be termed "QoS on Demand". End-users navigate in the Internet with the usual mechanisms, but when they want to retrieve information (e.g., a MPEG file) with guaranteed QoS they have just to "click" on the QoS button associated with the information. This scenario is deemed a valuable one because it associates the well-known Internet capabilities with the TINA flexibility to provide complex interactive multimedia services.

3.2.2 Technical Description

The DPE technology offers the ability to integrate CORBA applications with Internet Browsers. This means that it is relatively simple to build a Web Server[4] that acts as a front-end to the users for the TINA architecture. This scenario is depicted in Figure 10.

[4] This does not necessarily have to be in the server. Next generation clients will have native IIOP support or even OrbixWeb in the meantime.

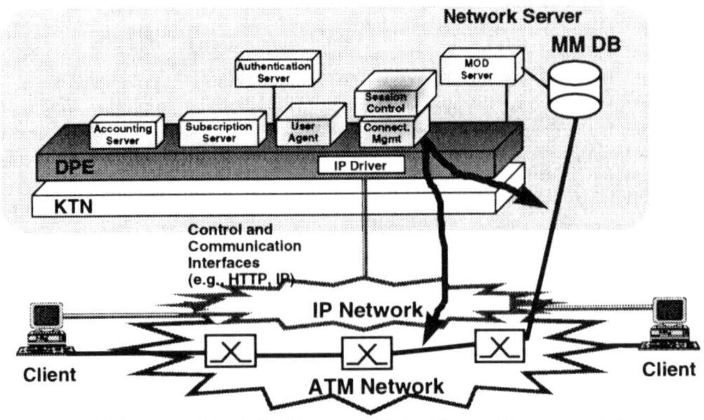

Figure 10: TINA and "QoS on Demand"

The communication between the user and the service is supported by the HTTP protocol. In this case the user can be presented "HTML pages" that are associated to TINA (service) architecture components. For instance, when a user logs on the TINA Web Server, an HTML page is presented. It requests the user to type a user name and a password. This information is passed to a User Agent that recognizes the user. The list of subscribed services can be created and displayed to the user by means of another HTML page. The user can so interact with the TINA Web Server. When the user requests a particular service (e.g. a Video on Demand Service), new HTML pages are presented in order to support the user in choosing a specific movie. Finally a choice between the ability to watch the movie with low quality (i.e., sending the information flow in the "best effort" fashion) or with high quality is presented to the user. If the choice is for high quality, the TINA session model and connection management can be used in order to orchestrate the needed network resources. The information flow is now carried over ATM links with negotiable and guaranteed QoS. Accounting and billing can be implemented according to TINA solutions.

This example can be extended considering that the needed "service stubs" could be downloaded to the user by means of TINA mechanisms (i.e., the Provider Agent) or simply downloading Java applets[5]. It is also possible to exploit the TINA session control mechanism. In fact the user can suspend a service session, move to another terminal, log on (using the procedure described before) and resume the session. This scenario shows how TINA services can be activated by means of Web Browsers. The main point is not the technical feasibility of this scenario (almost any TINA company has already developed this within its Labs), but indeed the need/benefit of having TINA structured services in this context. The scenario "QoS on demand", in fact, can be implemented by current Web Servers without using TINA. The end-to-end QoS can be controlled by means of existing mechanisms such as Winsock (Winsock 2 promotes the integration of sockets mechanisms with RSVP and/or ATM signalling). So the business case for TINA in this area has to be found in the ability of the Architecture in providing better services (which is not always the case) or in new and different classes of services. The following section is an attempt to describe some emerging services that could be provided or integrate within a TINA framework.

3.2.3 Sample Services

It should be noted that for many services the currently available bandwidth (e.g., 64 Kbits/s) is not necessarily a limitation. The bid/claim for the need for more bandwidth in order to support multimedia services is, in this case, only a face of the coin. From the public network perspective, advanced services are to be supported by a higher level of flexibility in the control of network systems. Services integrating the Web capabilities with existing telecom services are acquiring an increasing interest. Several proposals (e.g., see [7]) are emerging in this field. These services seem to be based on the general scenario described in the previous section, but they don not necessarily require a broadband infrastructure.

3.2.3.1 Web Telephony

Under the term Web Telephony, different services are identified. Examples are: users accessing fax servers by means of Internet Web pages or users placing telephone calls to Network Computers or users that are visiting a Web Site of a Company placing a call to a customer representative without dropping the Internet session and many more. The Web telephony scenario could be implemented using some TINA based solutions. The service is based on the ability of a user to browse the web and, on demand, to make a telephone call through the browser to a user. Different options exist, they are briefly sketched in the sequel.

A first option (Figure 11) considers a Web Server (hiding a TINA infrastructure) able to provide to users a service similar to White Pages. The assumption here is that the originating terminal is able to treat, in a integrated way, IP based and telephone based communication (e.g., using TAPI based mechanisms to place a call from a computer). The service is provided according to the following phases: user A searches for a telephone number of user B, the Web server gives back the telephone number of B, the user A makes a telephone call to B.

[5] Java is a promising technology and for sure it will solve in the near term problems related to the control of local resources by means of applets downloaded by a server and executed in the Web Browser context.

1. Browsing
2. Address Resolution
3. Return address
4. Make Call

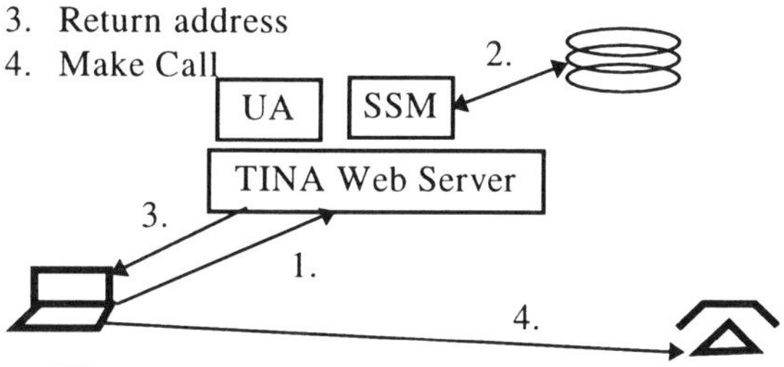

Figure 11: A TINA Web Server resolving telephone addresses

The second option (Figure 12) considers a Web Server able to control by means of Computer Telephony Integration (CTI) or other third party call control mechanisms (e.g. programmable switches) the connectivity between the different users. The originating terminal is able to deal with IP communication and Telephony communication. In a first phase the user A browses in the White pages, then s/he selects a user to call. The TINA Web server uses CTI mechanisms in order to place a call towards user A and B. The users are finally connected.

1. Browsing
2. Address Resolution
3. Make distinct Calls
4. Cross connect

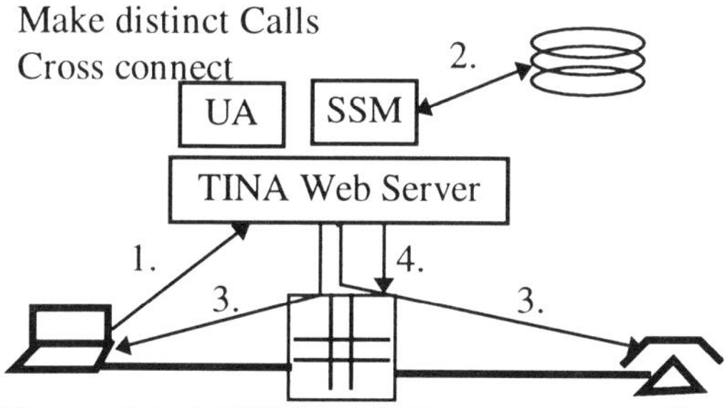

Figure 12: A TINA Web Server controlling the connectivity

The third option (Figure 13) considers a TINA Web server able to use CTI mechanisms to connect a user and to special resources, and to control conversion systems that transform voice into IP packets and vice-versa (IP telephony).

1. Browsing
2. Address Resolution
3. Allocate Converter
4. Make Call to B
5. Cross connect

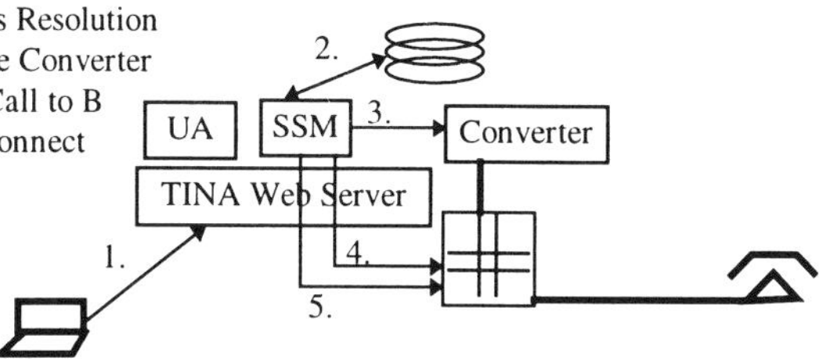

Figure 13: A TINA Web Server converting Internet to Telephony

In this case the originating terminal is only supporting IP based communication. In this case, when user B is selected, the TINA Web server allocates a conversion unit, makes a call to user B and forwards the packets from/to the converter.

All these options are feasible with TINA. There could be some business value in pursuing the Web Telephony by means of TINA. Another example of this kind of services is depicted in Figure 14.

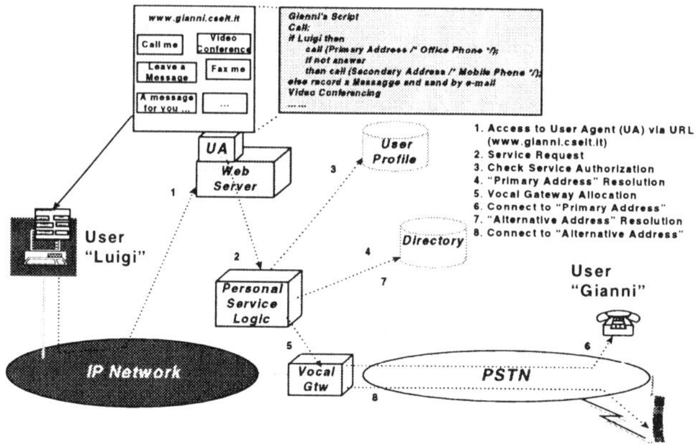

Figure 14: Telecom Services provided by means of User Agents

A User (called *Luigi*) is seeking information about his friend *Gianni*. *Luigi* connects the Web page of *Gianni* and select the "Call me" option. *Gianni* has defined a personalized logic that is activated at the press of button "call me". The "Personal Service Logic (PSL) " goes through several steps:

- it looks up a Directory for determining the network address associated to the Primary Address, i.e., the first address to be called (it is *Gianni's* Office Phone);
- then it requests a special resource to place a call to the Office Phone in order to enable a voice communication between *Gianni* (on the PSTN) and *Luigi* (on the Internet);
- on "no answer" the Personal Service Logic decides to try an alternative address (i.e., the Mobile Phone);
- the PSL requests the Special Resource to connect to a second network address.

3.2.3.2 Mobile IP and the TINA User Agent

It is quite surprising that not too much attention has been paid from the TINA community to emerging solutions related to Mobile IP [8]. As a matter of fact, the current proposal recalls, at the logical level, some functionalities that TINA Service Architecture associates (or at least should) with the User Agent. Figure 15 depicts a simplified view of the Logical Functions proposed for Mobile IP. In it a set of specialized Agent cooperates in order to trace the user location.

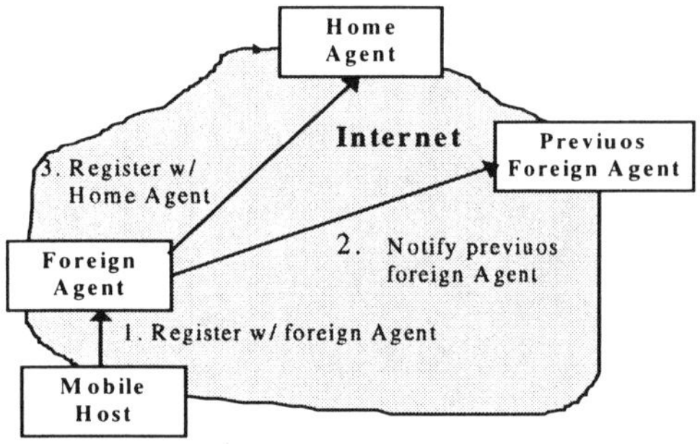

Figure 15: Logical Functions of Mobile IP

A similar organization has been considered within the VoIP Forum in order to promote services based on mobility of users. The service proposed allows to retrieve the user in any place and independently from the terminal being used in that moment. Figure 16 shows in a pictorial way the service functionalities [9].

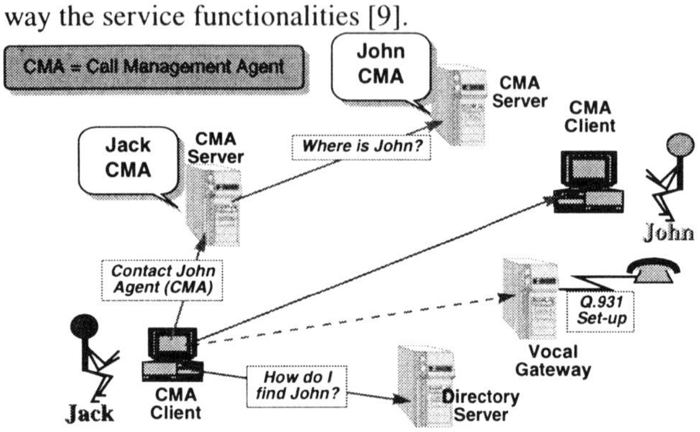

Figure 16: Call Management Agent service

A mobile user is traced wherever s/he is and a "call" can be placed independently from the terminals being used. A service like this could be provided by a TINA infrastructure if appropriated interfaces and APIs (derived from the protocol under specification from IETF and VoIP Forum) are adopted or promoted within the TINA architecture.

3.2.3.3 Multimedia Conferencing

Another valuable scenario is related to the use of Internet based protocols for controlling Audio and Video Conferences. IETF has specified a suite of protocol under the name MMUSIC [10] for dealing with conferencing. At the same time products compliant to ITU standards (e.g., H323) for multimedia communication are emerging. Notably H.323 products are acquiring interest within the Internet community, and major vendors are providing APIs for building applications on top of their products.

There are two options for TINA with respect to this: to integrate already existing products into the TINA Service Architecture, and/or to map the primitives of protocols developed so far (e.g., the simple session control protocol SSCP) onto Service Architecture interfaces (e.g., at the level of SSM). These options seem to be profitable ones for TINA, because it should be relatively easy to integrate these solutions, APIs and products. In fact, TINA should be a software architecture able to accommodate for products without requiring modifications to components.

An example of integration services, an Internet Conference is depicted in Figure 17.

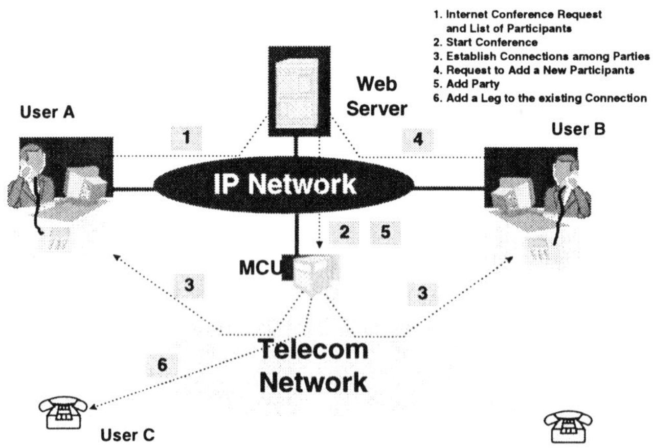

Figure 17: Internet Conferencing

User A can book an Internet Conference interacting with a Web Server: User A provides a list of participants and the time of the day for starting the session. At the scheduled time, the Web Server instructs a Multimedia Conference Unit to add the parties involved in the session. Users having "two access means" (e.g., a PC and a mobile telephone or two ISDN lines) can use one "line" for quality voice and the other for conveying data information. User B requests to add new participants to the conference (the Web Server is so used for controlling the conference session). The application used are based on standards, the role of the TINA architecture should be the identification of users, the retrieval of the best server for providing services, and the allocation of multicasting resources. It should be noted that Users could also access the session without using switched resources (only by means of the Internet.

3.2.4 Evaluation

These scenarios are characterized by the fact that the interactions between the user and the service provider are based on the use of HTML and HTTP or other Internet protocols instead of signalling or CORBA based ones.

This seems to be a very promising approach for the application of TINA with respect to Internet. However it should be pointed out that similar scenarios can be built with "poorer" technology. In fact it is possible to built dedicated servers using existing protocols that could make available the same type of services (maybe in an *ad hoc* and simpler manner). Another issue to consider is that the more powerful scenario (QoS on demand) is inherently tied to a Broadband environment in which the ability to establish end-to-end connections with guaranteed QoS is provided. It seems more likely that in the future only a part of the Internet backbone will be based on ATM technology (or Frame Relay). Few users will have the possibility of having a direct access to such a backbone (mainly business users). That means that such a TINA scenario (in the most favorable condition) is for a few users. There could be some value in considering services related to the integration of Telecom and the Internet provided by a TINA structured Web Server. In addition it is worth to show how the TINA architecture can encapsulate, or accommodate for existing products.

3.3 A step further: TINA Services and the use of Packet vs. Circuit Switching

The scenario presented in this section is built on the assumption that in the long run, users could **seamlessly access to services independently from the transport network**. The IP protocol will be used for conveying from/to terminals control flows and **also** information flows with guaranteed QoS. In principle this is possible due to the twofold capability of the IP based communication (see Figure 6). For delay sensitive information (e.g., voice) Circuit Switching is still more appropriate. It is possible to conceive a combined use of packet and circuit switching for providing better services to users. This concept is expressed by V. Cerf's Inversion [11]: *"Today you go through a circuit switching network to get to packet switching network, tomorrow you'll use a packet switching network to get to a circuit switching network"*. In the future, there will be the need for hybrid access server supporting IP communication with users terminals and able to forward protocol data units towards a packet and a switched network. Under this perspective there is a chance for using TINA technology to govern both interfaces and being able to select the most appropriate transport mechanisms. This scenario, even if it is for the long term, aims at verifying whether TINA is able to address the challenge posed by the Cerf's Inversion.

3.3.1 Objectives

The goals of the scenario are:

- to provide a long term perspective on the evolution of (access) networks and the use of TINA Service Architecture;
- to demonstrate that TINA service components are able to support a scenario of "hybrid transport";
- to provoke within the TINA community a discussion on related topics.

3.3.2 Technical Description

Figure 18 depicts the case in which terminals use IP for conveying different information flows (the hypothesis is made that different flows are tagged, like in IPv6, or are recognizable at the IP level). An Access Server is able to "deflect" the different flows towards a Packet Switching Network (for Control and Best-Effort flows) and a Circuit Switching Network for flows that have QoS requirements.

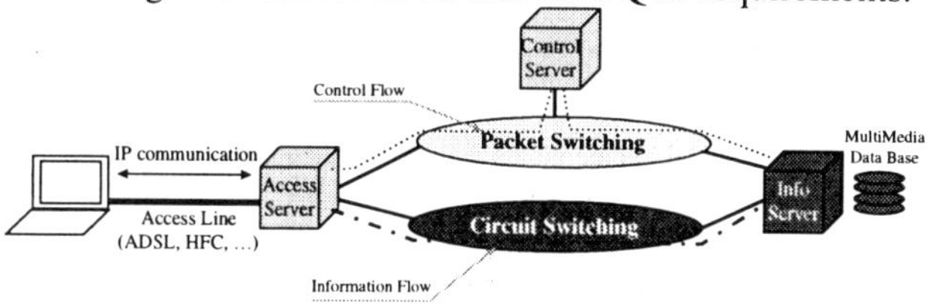

Figure 18: Combined use of Packet and Circuit Switching

The Access Server could be seen as a set of Terminal Agents. They are able to represent and deal with the direct communication towards the terminal. They can decide whether to use a cheap transport (based on best-effort mechanisms) or an expensive one (but with guaranteed QoS). These terminal agents should offer some levels of programmability for the control infrastructure and even for the end-user. The Control Server is a nickname for an undercover TINA Architecture.

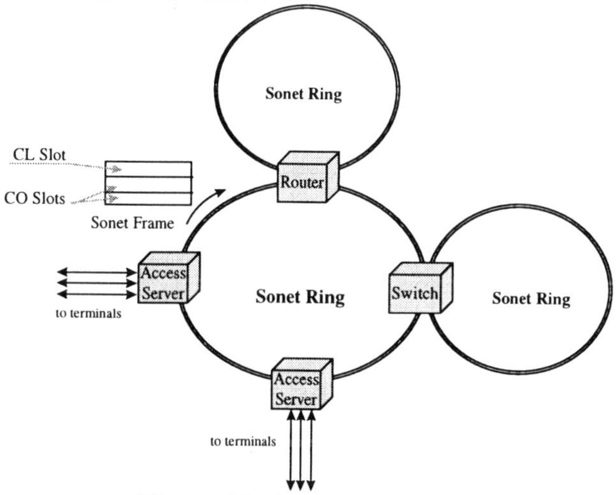

Figure 19: Access Servers as programmable Add-Drop Multiplexers

The Service Architecture, as well as the Connection Management Architecture, could be used in order to provide services and end-to-end connectivity. Users can be identified and supported by User Agents that are to be dynamically associated to Terminal Agents. The Access Server is based on concepts similar to those proposed by the ROADS architecture for Add-Drop Multiplexers. A high level representation of the network infrastructure is represented in Figure 19.

The network infrastructure is based on a simple model, different servers are located over interconnected Sonet Rings. Some nodes have add-drop functions (in order to fill the Sonet Frame[6] with Connection-less packets for Routers, or with Connection Oriented packets for switches). The Router extracts the packets (from ATM VPs/VCs) and tries to route them in an appropriate way. The switch ensure that CO packets are delivered according to the negotiated QoS.

3.3.3 Evaluation

In spite of the future applicability (if any!), the scenario sketched in Figures 18 and 19 is a possible application of the TINA architecture. It is characterized by the joint (and optimized) use of different transport techniques. It is more than likely that in the future solid and industry supported "hybrid transport" scenarios will emerge, but by then we hope that the TINA community will have considered how to cope with the Cerf's Inversions.

4. Where should TINA go from here

As said, many of the scenarios and technical solutions presented according to TINA, could also be realized with other architectures. This points directly to the question of what TINA is and what it is for.

In our view, it should be much more a Software Architecture than a Network Architecture. One of its goals should be the definition and specification of mechanisms and solutions for supporting the communication and the event passing between objects comprised in a session. A session has to be regarded as a means to send and receive events from/to a set of communicating objects having the cooperative goal of supporting a service or an application. In this definition, there is nothing in common between a Session and a Call Model. However, some objects associated to a session could themselves implement Call Control functions according to some models (e.g., the current definition of a multi-party and multi-connection TINA session model, or a IN CS.1 call model). A good representation of an event based session model is the ECTF framework [12]. Objects are associated to the session (i.e., instantiated/created or allocated). Each new created object registers to the events that it is able to receive and deal with and, conversely, notifies to other objects the events that occur during the execution of its methods (and behavior). A generic Session Manager is an object that ensure the smooth functioning of the session, guaranteeing that any object associated to the session is up and running. Some session objects could represent network resources, other could represent User related Info, others could be implementing, as said, a call control. Their interfaces and methods should be defined according to **existing** APIs, or as an implementation of an API related to some communication protocol.

[6] ATM layers are still used.

If we look back to the effort of TINA in imposing a set of Reference Points (e.g., the Ret reference point), we have learned the lesson that Reference Points (and related protocols or APIs) are defined because there is a business case behind them and an industry push to define the **minimum** set of operations that guarantee the Interoperability between systems.

TINA specifications have usually disregarded existing APIs or have not tried to define APIs related to relevant control protocols. In our view, instead of the definition of new reference points, TINA should put effort in specifying APIs strongly related to existing or already specified Reference Points and related protocols. Cases for the Internet could be the Resource Reservation Protocol, the Session Control for H.323 or MMUSIC applications and the like. In a general way it would be quite useful having a specification of an API for SS.7 applications compliant to a TINA session model. This would enable the implementation of services that integrate Telecom and Internet capabilities. In other terms there is a strong requirement for having TINA accommodating a large set of APIs (related to existing protocols and reference points elsewhere defined) in order to allow interoperability of different solutions in a similar fashion as Java is trying to do.

5. Overall Evaluation & Conclusions

All the presented scenarios can be realized with the TINA technology. There are no doubts that the application of TINA can enhance the flexibility and the programmability of systems. But there is skepticism with respect to the availability of TINA products supporting those scenarios in the short term. Moreover there are other technologies that are supported by products and that offer capabilities similar to those of TINA. This means that the adoption of TINA is not so likely in the short to medium term in an Internet context. As seen, we feel that there is only a niche market for TINA solutions within the Internet. This is due to:

- TINA tends to be designed for "classical" Telecom applications for broadband networks;
- the different pace between the specification and development of solutions in Internet and TINA;
- Internet is more "industry and market" oriented than TINA.

Roughly speaking, it could be stated that TINA has not played an important role with respect to the Internet and will persist in this if important (from a business perspective) applications/services are disregarded. In spite of this perspective, there are a few things that are worthwhile to consider. The most meaningful scenario is related to the management of routers, it offers the ability for the PNO to enable an integrated management for an IP based network. An enabling technology that could offer some chances for the TINA applicability is Java in conjunction with the network computing (e.g., the downloading capabilities). There is a sort of alignment between this approach and the TINA PA/UA paradigm. This convergence should be promoted building TINA components that offer Java APIs and that constitute a TINA based infrastructure used to specify/create services to end-users. This requires a much stronger liaison with Java activities (in the service area as well as in the DPE area).

All the scenarios based on the TINA Connection Management are deemed not likely. Other approaches (even if, in principle less flexible) have the advantage of being backward compatible with the signalling approach.

Scenario		Time frame
• **Management of a IP/ATM Network**	➡	• Useful by '99
• **Provision of end-to-end connectivity with QoS**	➡	• Unlikely till 2000 and beyond
• **Access to TINA services by means of Browser technology**	➡	• Feasible now, useful in conjunction with TINA deployed services
• **Provision of TINA Services with guaranteed QoS**	➡	• Unlikely till beyond 2000
• **Network Computing**	➡	• Possible by '99. The PA /UA paradigm should be promoted within Internet

Figure 20: Time frame for TINA solutions

Figure 20 shows an evaluation of the time frame in which TINA could be adopted for some scenarios sketched in this document. They are based on a guess on when products (TINA or Internet) will be available. Under this respect there are very small opportunities for TINA to be faster than the Internet, however some possibilities still exist.

As a final remark, the reader should consider that the telecom world is evolving so fast that to predict the TINA applicability in a five year time frame casts a shadow on its usefulness. To show that TINA is useful and will be applied sooner .....is a big challenge for the TINA community.

Epilogue and Acknowledgments

One year and a half has passed from that sunny and warm afternoon in Heidelberg: TINA people are getting older and (hopefully) wiser; the number of TINA addicts has increased as well as the applicability scenarios to existing networks; a few TINA products have been announced; and the TINA Core-team will be soon dismissed. What do we have learned from this ? Well, as President Mao Tse Tung once said: "Revolution men have to be like fish in the sea" so TINA should be a component architecture based on the existing (sea of) Internet (but not only) APIs and protocols.

Our grateful acknowledgments go to Telma Mota and Harald Korte whose suggestions and ideas triggered the need and some themes of this paper. To our P.508 colleagues and Eurescom Staff (especially David Muldowney) to whom we are grateful for the open-minded exchange of opinions and the continuos support. We also acknowledge CSELT and Telecom Italia people that (even if they do not necessarily share our view) have allowed to collect some provoking ideas in a paper. And finally we would like to thank the readers that were so kind and patient to read the whole paper.

References

[1] *Overall Concepts and Principles.* TINA C, February 1996.

[2] K. Koyanagi, et alii. "*Technology Initiatives for the near future in Internetworking*", IEEE Communication Magazine, May 1997, vol. 35, No. 5, page 108.

[3] EURESCOM P.508 Document suite on "*Evolution, Migration Paths and Interworking to TINA*", January 1997

[4] X-Bind Documentation available on line at http://comet.ctr.columbia.edu/xbind/

[5] P. Newman, et alii. "*IP Switching and Giga Routers*", IEEE Communication Magazine, January 1997, vol. 35, No. 1, page 64.

[6] P.P. White, "*RSVP and Integrated Services in the Internet*", IEEE Communication Magazine, May 1997, vol. 35, No. 5, page 100.

[7] C. Low, "*Integrating Communication Services*", IEEE Communication Magazine, June 1997, vol. 35, No. 6, page 164.

[8] C. Perkins, "*Mobile IP*", IEEE Communication Magazine, May 1997, vol. 35, No. 5, page 84.

[9] S. Petrack, et alii "*Call Management Agent system requirements functions architecture and protocol*", Vocaltec contribution to the Seattle meeting of the IMTC Voice over IP Forum, January 15th 1997

[10] V. Cerf, "*Cerf's UP*", MCI internet site: http://www.mci.com

[11] Mmusic Documentation available on line at http://www.ietf.cnri.reston.va.us/ids.by.wg/mmusic.html

[12] ECTF Documentation available on line at http://www.ectf.com

Session 2

Mobility

D.O. O'Sullivan, IONA

Terminal mobility support in TINA

Do van Thanh[a] and **Jan Audestad**[b]

[a]*Ericsson AS, Olav Brunborgsvei 6, P.O Box 40, N-1361 Billingstad, Norway. Phone: +47 66 84 16 68, e-mail: etodvt@eto.ericsson.se*

[b]*Telenor AS, P.O. Box 6701 St.Olavs plass, N-0130 Oslo, Norway. Phone: 47 22 77 99 52, e-mail: Jan-Arild.Audestad@s.hk.telenor.no*

Abstract - As shown in [1], Access and Location transparencies defined for Open Distributed Processing (ODP) and TINA are unsufficient to support terminal mobility since interoperability between DPE platforms is not guaranteed at all times. In this paper, we study all the functions and computational objects necessary to support terminal mobility. Both forms of interactions, operational and stream-oriented, are considered. The solution proposed is generic and can be customised for both continuous and discrete mobility in either wireline or wireless networks.

BACKGROUND

The researched work described in this paper results from the mobility part of the TELEcom REsearch Programme TELEREP, carried out at the Center for Technology at Kjeller, UNIK [2]. The programme has been established to obtain practical experience with ODP/TINA methods and the implementation of TMN and IN functionality, and to study how security could be provided in the DPE environment. The programme started in 1995 and will run over several years. Initially the programme decided to base its work on a public domain version of CORBA [3], and will be re-evaluated at a later stage. One is currently studying both theoretically and practically, as part of a Ph.D. programme, how mobility and security can be provided transparently by the platform environment.

I. INTRODUCTION

There are defined two forms of interactions between computational objects: operational and stream-oriented [4]. Supporting terminal mobility consists therefore of enabling both forms of interactions between a Computational Object CO1 belonging to the terminal domain and another one CO2 belonging to the telecom system domain (See Figure 1)

II. ENABLING OPERATIONS BETWEEN THE TERMINAL DOMAIN AND THE TELECOM SYSTEM DOMAIN

Prior to any operational interaction (operation invocation), a binding must exist between the two objects. This is called implicit binding, i.e objects do not explicitly request establishment of the binding but the establishment is done by the DPE on the kernel transport network (kTN). This kTN is logically separated from the transport network supporting stream flows. Such a separation allows evolution of kernel communication independently of the evolution of the technologies used for stream flows transport [5] (see Figure 2).

It is observed that operational interactions between computational objects are only possible if the kernel transport network is operative and there is always connectivity between two arbitrary DPE nodes, i.e every DPE node is reachable from every other DPE node.

For a wired network the connectivity is ensured at configuration time, i.e necessary links between nodes are defined such that connectivity remains even if some nodes and links fails. The topology of the network is defined statically and will only be altered in case of a failure or reconfiguration. A DPE node can always determine how to reach another DPE node.

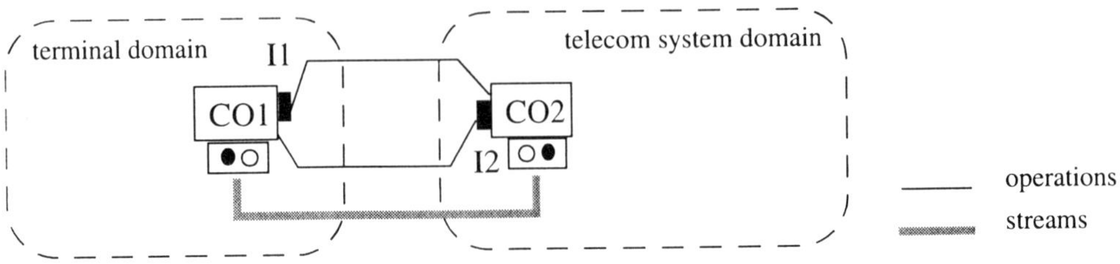

Fig.1 Interactions between COs residing the terminal and telecom system domains

0-8186-8335-X/98 $10.00 © 1998 IEEE

This is not always true with terminal mobility. The mobile terminal is a particular DPE node with special behaviour:

- It changes frequently the node with which it has direct link.
- It may just disappear from a node and reappear later at any other node

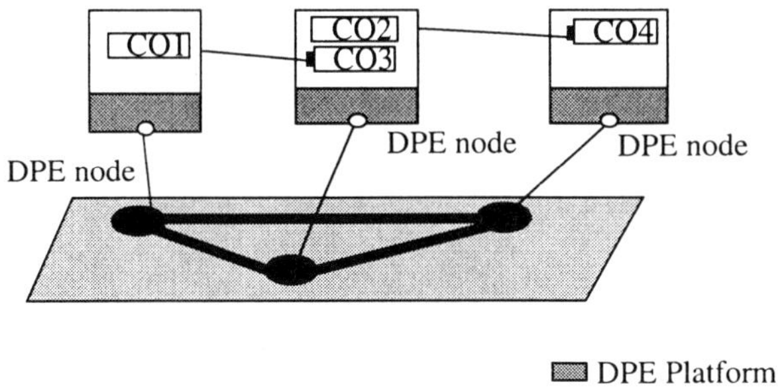

Fig.2 The kernel Transport Network

The topology of the kernel transport network for systems containing mobile DPE nodes is changing dynamically and more seriously, it can also be in an undetermined state in the sense that the connectivity with such a mobile DPE node is not always ensured unless additional functionally is inserted in the DPE.

We propose to consider the kTN as consisting of two parts:

- The fixed part comprising all fixed DPE nodes.
- The mobile part comprising all mobile DPE nodes.

At the boundary of the fixed part of the kTN there are several Network Access Points (NAP), i.e points where mobile DPE nodes (terminals) can link themselves to the fixed kTN (see Figure 3). It is worth noting that the NAP notion is different from the Network Termination Point (NTP) which designates the access point to the Transport network used for stream flows. As specified before the kTN and the Transport network are logically separate networks.

An **NAP** object is introduced to represent an NAP on the kTN. An **NAP** object is an interceptor which stands at the boundary between the terminal domain and the telecom system domain and is responsible for checking, transforming and forwarding of interactions that cross the boundary. An **NAP** object has two communication interfaces, one with the mobile DPE and one with the fixed kTN. Several **NAP**s can be located at the same DPE node.

Each mobile DPE node may have one or several Terminal Access Points (TAP), i.e. the points where the mobile DPE node can exchange operations with other DPE nodes. Here again, it is important to differentiate between a TAP and a Terminal Termination Point (TTP) which is used for the connection of stream flows. A TAP will be represented by a **TAP** object in the terminal domain.

Prior to any operational interaction the mobile DPE, i.e one of its TAPs, must be attached to an NAP. Operational interactions between a computational object residing on a mobile DPE and a computational object residing on a fixed DPE always go through a TAP and an NAP.

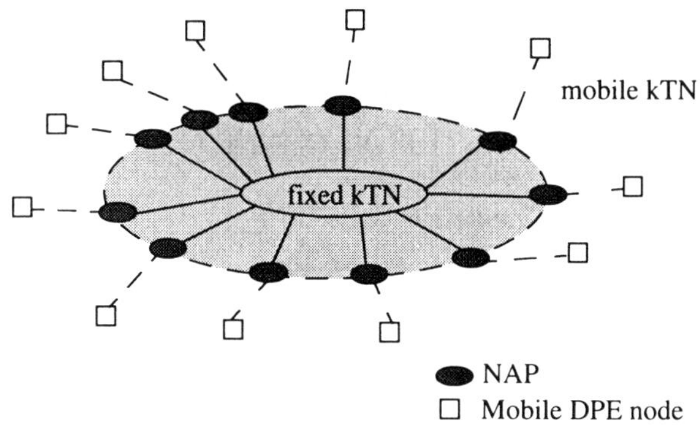

Fig.3 The kTN consisting of a fixed and a mobile part

A. Interactions initiated by the fixed part

When **CO2** invokes an operation opY on **CO1** (see Figure 4), the operation must be sent via the correct **NAP** object, i.e the one that is currently connected with the **TAP** of the mobile DPE. Since the terminal is moving, the **NAP** to which the **TAP** is connected may change from time to time.

To unburden **CO2** with mobility-related functions, a **Terminal_Agent** Object (**TA**) is introduced to undertake a kind of relocation function as shown in Figure 4. It keeps track of which is the current **NAP**. This relocation function is somewhat different from a relocation function defined in ODP. The latter records the change in location of an object when its is moved from one computing node to another. But in fact, recording the current **NAP** is semantically the same as recording the location of the terminal and hence the location of object **CO1** because the location of a terminal can be deduced from the location of the **NAP**.

For each mobile DPE node (or terminal) one instance of the **Terminal_Agent** object will be instantiated.

Instead of issuing an operation request directly to **CO1**, **CO2** issues a request to the **Terminal_Agent**. The **Terminal_Agent** will then forward it to the appropriate **NAP**. The **NAP** transfers it to the **TAP** which finally delivers it to **CO1**. To support interactions initiated by the fixed part, the **Terminal_Agent** object is therefore required.

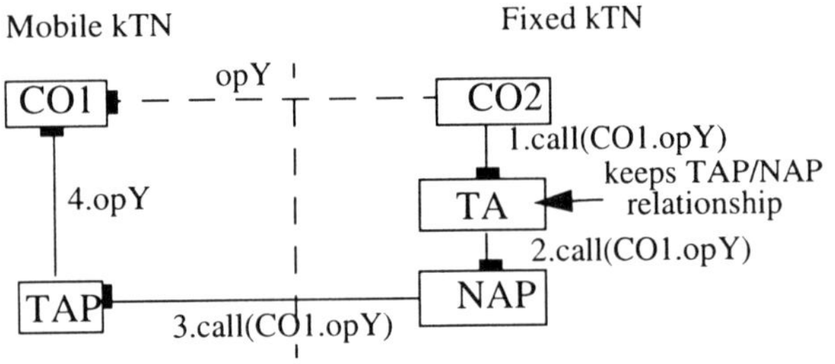

Fig.4 Interactions initiated by the fixed part

B. Interactions initiated by the mobile part

Suppose now that **CO1** wants to invoke an operation opX on **CO2** (Figure 5). First, the invocation must be conveyed to a **TAP**. This can be easily done assuming location transparency locally on the mobile DPE node. The **TAP** must then know to which **NAP** the operation should be forwarded. From the **NAP** the invocation can be conveyed to the **TA** (**Terminal_Agent**) and then to **CO2**, using access and location transparencies in the fixed kTN.

It is crucial that the **TAP** knows which **NAP** instance to communicate with in order to send the invocation to it. Since the **TAP** resides in a mobile DPE, the current **NAP** may be changing from time to time. The terminal mobility management consists in keeping track of the correct **NAP** instance. This responsibility may be allocated to the **SPA** object (Service Provider Agent).

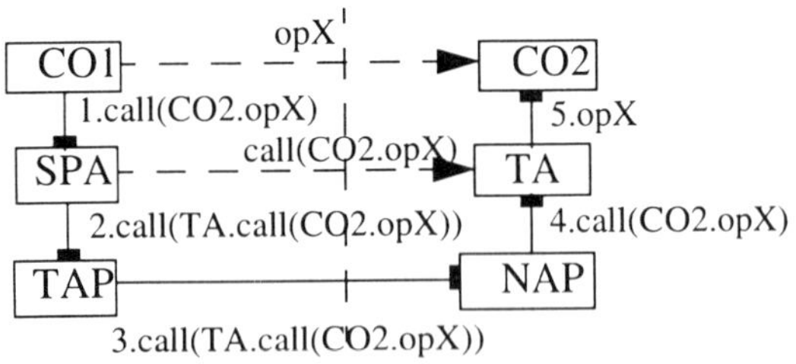

Fig.5 Interactions initiated by the mobile part

The **SPA** is thus entrusted with two responsibilities: supporting security functions and keeping location information. In this way, only one interceptor object is required in the terminal for managing both security and location updating. The introduction of the **SPA** is also convenient to keep the **TAP** hidden from the application objects. Instead of issuing an operation invocation to **CO2**, **CO1** issues a request to the **SPA**. The reason is that a mobile DPE, for example a PABX may in fact have several **TAP**s which should be transparent to the application objects. The **SPA** will ensure this transparency. Interactions initiated from the mobile DPE are realised as shown in Figure 5:

We may let the **SPA** involve in interactions initiated by the fixed part in order to obtain a symmetric interceptor configuration. Combining both types of interactions, the conveyance of operations across the boundary between the mobile DPE and the fixed DPE will be as shown in Figure 6.

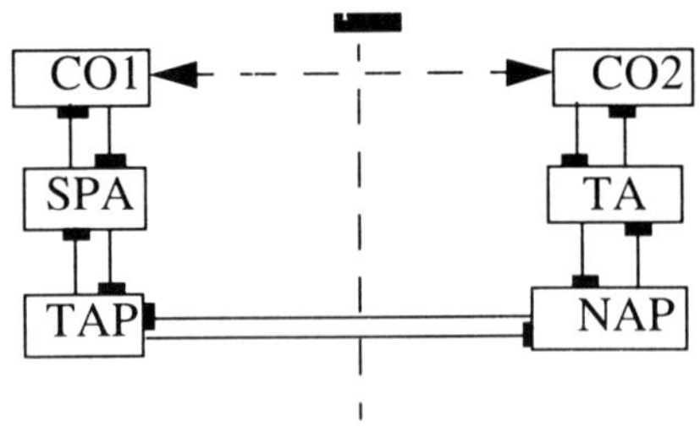

Fig.6 The conveyance of operations between the mobile DPE and the fixed DPE

C. Location registration and deregistration

It is observed that the conditions for success of both types of interactions is the definition of the association between the **TAP** and the **NAP** and the definition of the association between the **TA** and the **NAP**. If one or both of the two associations are undefined, the interaction will be unsuccessful. Second, the two associations must be consistent with each other. If the **TAP** is associated with an instance of **NAP**, then the **TA** must be associated with the same instance of **NAP**. An inconsistency will lead to failure.

When the mobile DPE is moving, the **TAP** - **NAP** association and the **NAP** - **TA** association must change simultaneously and consistently, and may sometimes be undefined. The operations necessary to determine these two associations are commonly referred to as **location registration** and **location deregistration** [6]. The various methods used to determine these two associations constitute therefore the different strategies for location registration and location deregistration. Below we shall look at several such methods and identify how they can be supported by the objects proposed above. Some of the methods will require additional objects. We will, however, show that the basic structure proposed above will be retained also in all cases we have identified. However, the different methods will require that the **TAP**, **NAP** and **TA** objects are equipped with the appropriate algorithms depending on the method adopted for a given system.

0.1 The wireline case

In the wireline case, the association between the **TAP** and the **NAP** is directly reflected by and is equivalent to the physical link between the mobile DPE and the telecom system domain. The association between the **TAP** and the **NAP** is also equivalent to the association between the **TA** and the **NAP**. We have:

Physical link <=> Association **TAP - NAP** <=> Association **TA - NAP**

A change in state of the physical link results in the change of the association between the **TAP** and the **NAP** and consequently a change of the association between the **TA** and the **NAP**. The wireline terminal can be in one of two states: *registered* when the physical link is up and the association between the **TAP** and the **NAP** and the association between the **TA** and the **NAP** are defined; *deregistered* when the physical link is down and both associations are undefined. There is no intermediate state.

Terminal state	**Physical link**	**Ass.TAP-NAP Λ Ass. TA-NAP**
registered	Up	Defined
deregistered	Down	Undefined

Fig.7 States of the wireline terminal

The "physical link surveillance" method

A change of the physical link can be used to trigger both a transition in terminal state and a location registration or deregistration. The transition of the terminal state is shown in figure below:

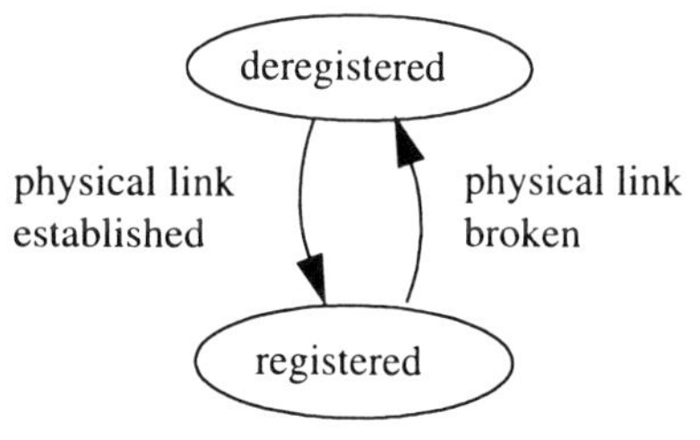

Fig.8 The state transition diagram of a wireline terminal

When the physical link is broken, i.e the terminal is unplugged from the socket or switched off or the line is broken, the association between the **TAP** and the **NAP** is removed. The **NAP** will immediately discover the situation and notify the **TA**. The **TA** set its **NAP** pointer to Nil. From now on, any attempts to invoke operations on the mobile terminal will be denied by the **TA**. The terminal is in the *deregistered* state.

When the physical link is established, i.e the terminal is plugged into the socket or switched on, the association between the **TAP** and the **NAP** is defined. The **NAP** will immediately discover the situation and notify the **TA**. The **TA** set its **NAP** pointer to the corresponding **NAP**. Interactions are then enabled. The terminal is in the *registered* state.

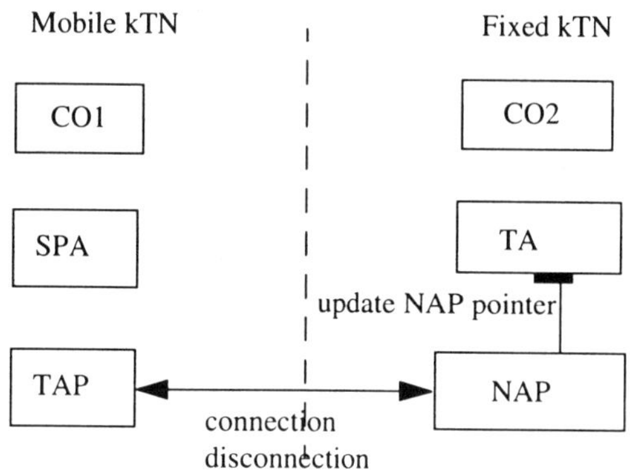

Fig.9 Location updating for wireline terminal

D. The wireless case

The wireless terminal stays most of the time in the disconnected state. When the physical link is down, it does not necessarily mean that the association between the **TAP** and the **NAP** is removed. The mobile DPE may still be there and, upon request, the physical link can be re-established. The mobile DPE may also have moved out the area. Hence, the state change of the physical link does no longer reflect the state change of the association between the **TAP** and the **NAP**.

Other methods to detect and update the change of the association between the **TAP** and the **NAP** are required. There are two alternatives: the initiative is taken either by the **NAP** or by the mobile DPE.

1) Initiative by the NAP

The NAP's initiative method

The detection of the changes of the association between the **TAP** and the **NAP** means that the **NAP** is able to find out which mobile DPEs are located in its area. The **NAP** must therefore have the capability to store the identifier of the **TAP**s which have been previously connected to it, i.e before the physical link is "broken".

Periodically, the **NAP** broadcasts a "hello" in its area as shown in Figure 10. Any mobile DPE present in the **NAP**'s area will respond to the broadcast with its **TAP** identifier. The **NAP** will then compare the received **TAP** identifiers with the stored ones. For mobile DPEs previously registered, there is no change in the state of the association between the **TAP** and the **NAP** and no action is necessary. For mobile DPEs which have just arrived, the **NAP** issues an update call to the corresponding **TA**s. The **TA**s will set their **NAP** pointer to the **NAP**. For mobile DPEs which have left, the **NAP** also initiates an update of the corresponding **TA**s indicating that the mobile DPE did not respond. The **TA**s will reset their **NAP** pointer to Nil.

If the intervals between consecutive detection processes are short enough, the representation of the association between the **TAP** and the **NAP** in the telecom system domain can be considered to be in agreement with the real association. The association is also determined for all the terminals, i.e the telecom system domain has the status of all the terminals. A terminal is registered when the **TAP** and the **NAP** association is defined and deregistered when the association is undefined. In registered state all interactions from and to the mobile DPE are possible. In deregistered state no interaction is possible.

This solution can be used for systems where the status of the terminals are important. It can be used in systems with larger number of terminals but less geographically distributed, i.e small number of **NAP**s. For systems with large number of **NAP**s, this solution has many disadvantages. First, the **NAP**s must have storage capacity and consequently must keep the stored data, i.e **TAP** identifiers of mobile DPEs present in its area, up-to-date and consistent. Second, there are periodically much processing activity in every **NAP**. The third disadvantage is the inefficient use of the radio frequency or infrared access channel to the **NAP** which is a scarce resource shared by all mobile DPEs for both data transmission and signalling. When all the present mobile DPEs answer, much capacity is used just for detection of the DPEs.

2) Initiative by the mobile DPE

a. The Periodic method

The detection and updating of the change of the association between the **TAP** and the **NAP** can also be initiated by the mobile DPE. Each mobile DPE can periodically report itself to the nearest **NAP**. The **NAP** will then send a register request to the corresponding TAs. In order to detect the silent terminals, i.e those that have disappeared without a deregistration, each **TA** can be equipped with a timeout. If a mobile DPE does not register itself within a period of time t_0, its **TA** will set it as deregistered. By this method, the status of all the terminals are always known by the telecom system domain. This method is suitable for systems with small number of terminals but which are more geographically distributed (larger number of **NAP**s).

It generates, however, activity both on the access channel and in the telecom system domain.

b. The method based on location changes

It is observed that the changes in the association between the **TAP** and the **NAP** are caused by the mobility of the terminal. If the mobile DPE knows that it has moved from one **NAP** to another, it also knows that the association between the **TAP** and the **NAP** has to be updated. For this to work the mobile DPE need to have the capacity to store the identifier of the previous **NAP** and the capability to obtain the identifier of the current **NAP**.

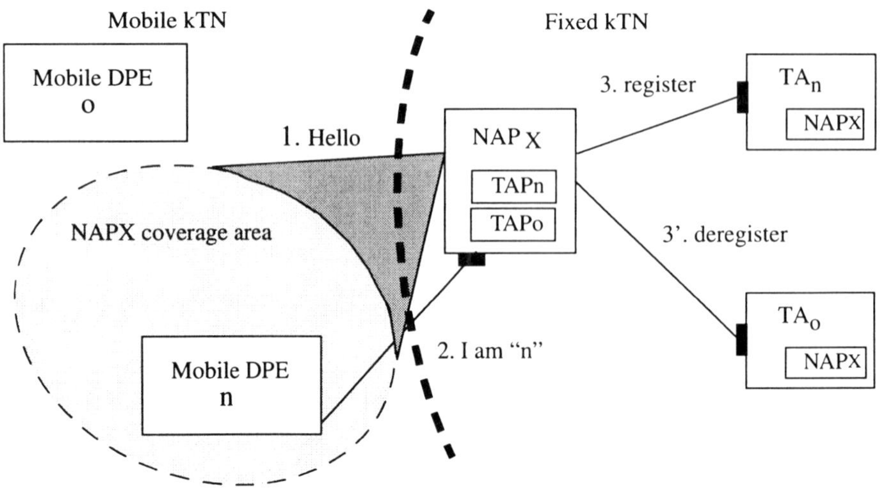

Fig.10 The case of detection function assumed by the NAP

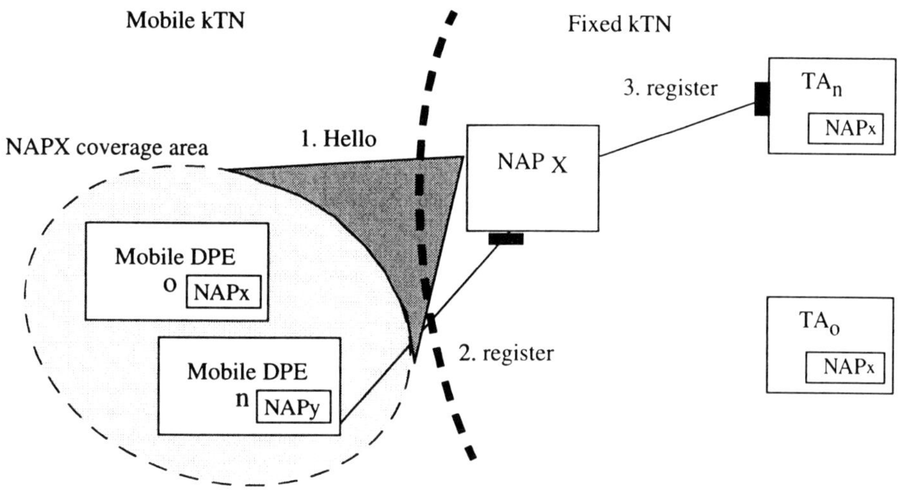

Fig.11 The case of detection function assumed by the mobile DPE

As shown in Figure 11, an **NAP** is associated with a geographic area called NAP coverage area. The **NAP** broadcasts periodically its identifier in this coverage area. Every mobile station in the coverage area reads periodically this **NAP** identifier and compares it with the one stored previously. If they are similar, there is no change in the association between the **TAP** and the **NAP** and no updating is necessary. If they are different, there is a change in the association. The mobile DPE updates its local **NAP** pointer and requests a `register` operation on its **TA**. The request is sent to the **NAP** which forwards it to the corresponding **TA**. The **TA** sets its **NAP** pointer to the current **NAP**.

This alternative requires that the mobile DPE has some storage capacity and some intelligence to decide whether a registration is necessary or not. On the other hand, the **NAP** does not have to store and process the information about the mobile DPEs present in its area. The use of the access channel to detect and update a change in the association between the **TAP** and the **NAP** is more optimal, since the probability that all the mobile DPEs will change **NAP** is very small and hence a total registration of all DPEs will be very infrequent.

The solution used in GSM [7] is slightly different from the one presented above. A base station can be mapped to an **NAP** in our model. There is however an intermediary object between the **NAP** and the **TA**. As shown in Figure 12 this object can be modelled as a group of **NAP**s, **GNAP** or as a "mirror" **TA**, **MTA**, holding some data for the **TA**. The location updating is executed only when the mobile station has moved from one **GNAP** or **MTA** to another. When required, e.g for initiating a call to the mobile DPE, the determination of the **NAP** is done "on the fly" by the **GNAP** or **MTA**..

There is, however, a problem. The updating of the associations between the **TAP** and the **NAP** and the association between the **TA** and the **NAP** are initiated by the mobile DPE when it knows that a change has occurred. It is not always true that the mobile DPE is aware of such a change and capable of notifying the telecom system domain. It can be switched off and brought to another area, a fault may occur in the mobile DPE, or it may have run out of battery power. It can also move out of the coverage area, loose all contact with the telecom system domain and be unable to initiate the updating.

In such a situation there may or may not be a mismatch between the real association between the **TAP** and the **NAP** and the perception of the telecom system domain, i.e the state of the association stored in the telecom system domain. When a terminal is registered, i.e the associations between the **TA** and the **NAP** and the association between the **TAP** and the **NAP** are defined, it is not guaranteed that interactions from and to the mobile DPE are possible. Only when the physical link is in operation, i.e there is activities to or from the terminal, is the state well-defined. When the physical link is down, nothing more is known about the location of the mobile DPE than its previous registration.

In the telecom system domain it is useful to differentiate between the two registered states of the terminal, namely **registered and confirmed** state and **registered and unconfirmed** state. In addition to the **NAP** pointer the **TA** needs also to save the terminal state. On the mobile DPE side there is no need for additional state but the registered state and the deregistered state. The **NAP** pointer is sufficient to represent these two states.

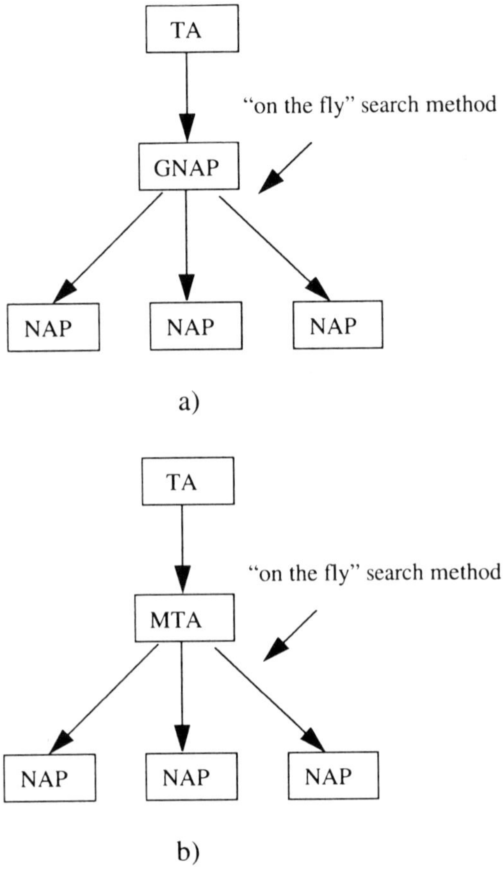

Fig.12 Alternative models of GSM

Seen from the telecom system domain, a wireless terminal may therefore be in one of the following three states:

- **deregistered**: the mobile station has actively notified that it has been taken temporarily out of service.
- **registered and unconfirmed**: the mobile station is registered at an NAP but there is no activity on it, i.e the physical link is down.
- **registered and confirmed**: the mobile station is registered at an NAP and there is activity going on, i.e the physical link is up.

The states of the wireless terminal is thus defined by the state of the physical link, the association between the **TAP** and the **NAP**, and the association between the **TA** and the **NAP** as follows:

Terminal state	**Ass.TAP - NAP ∧ Ass. TA - NAP**	**Physical link**
deregistered	Undefined	Down
registered & unconfirmed	Defined	Down
registered & confirmed	Defined	Up

Fig.13 States of the wireless terminal

From the registered and confirmed state the mobile DPE changes its state to the registered and unconfirmed state when all the activities have ceased. It will remain in that state until some successful activity takes place between the mobile DPE and the fixed DPE. The terminal state is then reset to the registered and confirmed state. If the request is unsuccessful, the terminal state will be changed to the deregistered state and will remain there until a registration takes place. From the registered and confirmed state, the terminal state can be changed to the deregistered state when the mobile DPE has explicitly made a deregistration request or has disappeared, i.e does not respond to an invocation initiated by the fixed DPE.

The transitions between the terminal states are shown in Figure 14. It is worth noting that in addition to normal transitions such as registration, activation, etc. there are two special transitions shown in Figure 14: handover and location updating. In fact, the handover and location updating can be regarded as a transition because the associations **TAP - NAP** and **TA - NAP** do change, but remain defined. Location updating may be regarded as a transition from the registered and unconfirmed state back to the registered and unconfirmed state. Similarly, handover may be regarded as a transition from the registered and confirmed state to the registered and confirmed state. A handover may follow a location updating or may be executed alone. From one registered and confirmed state, the terminal may also migrate to another registered and confirmed state after a location updating alone without a handover.

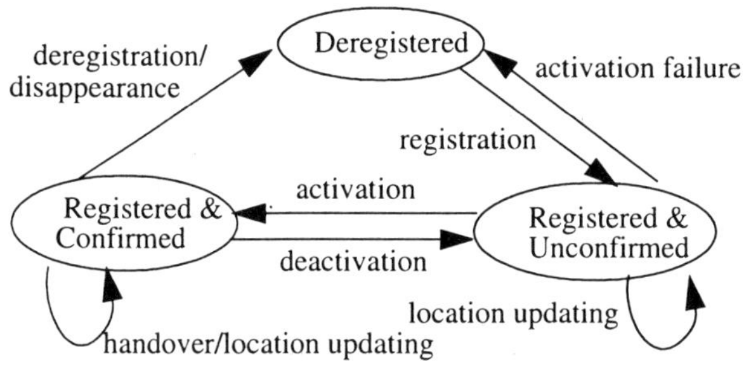

Fig.14 The state transition diagram of the wireless terminal

III. ENABLING STREAM BINDING WITH THE MOBILE TERMINAL

We assume now that operational interactions can be established between the mobile terminal and the telecom system domain. From now on, in order to reduce the complexity and make the explanations easier, operations across the boundary between the mobile DPE and the fixed DPE will be denoted as if they are directly addressed between peer objects, although they still have to go through the objects **TA**, **NAP**, **TAP** and **SPA**.

The transport network used for stream flow can be modelled as network delimited by two or more NTPs (Network Termination Points) and capable of transporting bit stream between two NTPs (see Figure 15). An NTP is the point at which a bit stream is accepted and/or delivered. Associated with an NTP is the characteristic of the stream accepted/delivered at the NTP, such as frame structure, QoS, etc. Different NTPs may have different characteristics.

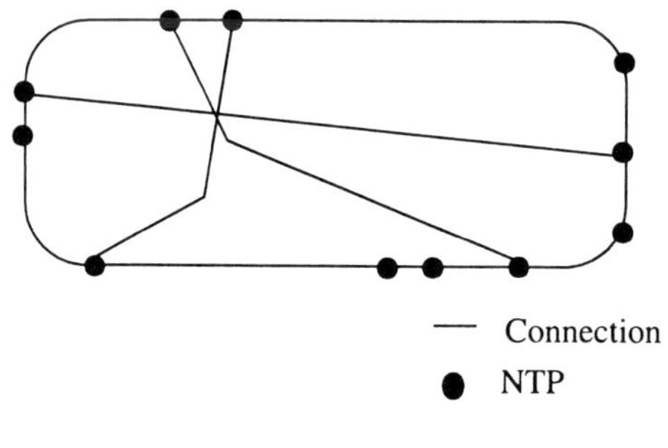

Fig.15 The transport network for stream flow

TINA has defined a similar model called the connectivity layer network [8]. However, the term Access Point is used instead of NTP. There is actually no difference between these two terms. The reason that the term NTP is used in this paper is just to differentiate from the NAP (Network Access Point) which is the access point of the kTN (kernel Transport Network). It is very important to make a distinction between the kTN and the transport network for stream. As mentioned in [5], these networks may be physically separate or share the same physical resources and represent only different concepts of the network. By separating the two network concepts, more freedom of choice for the kernel transport network is obtained and the evolution of the kernel transport network is made independent of the evolution of the technologies used for stream flow transport. The NTPs can be grouped into an NTPPool for example according to their geographical location. The area covered by the telecom system domain can hence be divided into smaller areas, each one served by one NTPPool.

As shown in Figure 16, when the mobile terminal is in area A, a stream flow between an object residing in the terminal and another object residing in the telecom system domain goes through a given NTPa belonging to $NTPPool_a$. When the terminal leaves area A and enters, for example area B, then its stream flows must go through NTPb belonging to $NTPPool_b$. $\mathbf{TCSM_t}$ must somehow be aware that the terminal has moved and that another NTP belonging to another NTPPool should be used.

For each NTPPool let us define an object called **NTP_Manager**. The **NTP_Manager** is in charge of an area and has information about all NTPs in that area, i.e which NTPs are available and their characteristics (QoS parameters). The **NTP_Manager** is periodically broadcasting its identifier in its area. The **NTP_Manager** identifier will be used by the $\mathbf{TCSM_t}$ when it wants to communicate with the **NTP_Manager**.

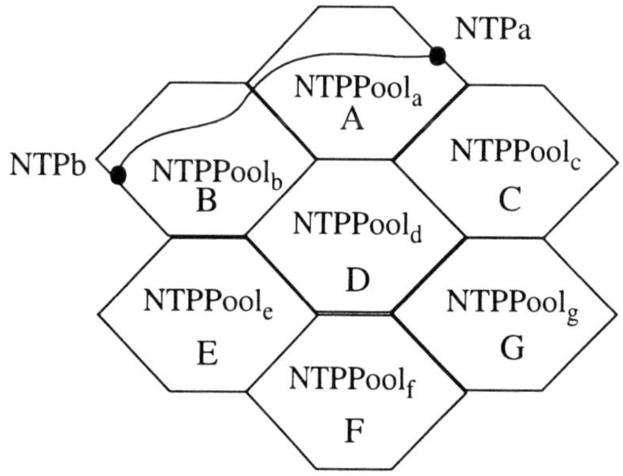

Fig.16 The coverage area divided into area served by NTPPools

When the terminal t is entering area A, the **$TCSM_t$** will get the identifier of the **$NTP\_Manager_A$**, which is received through the **TAP**. Note that the kTN is used for the broadcast of NTP information. Upon a connection request from **CSM**, **$TCSM_t$** will ask the **NTP_Manager** to allocate an NTP for its TTP. The **$TCSM_t$** will return the identity of this NTP back to **CSM**. **CSM** can then request the **CC** to establish a connection between this NTP to be used for **CO_t** object and the NTP to be used for **CO_n**.

Actually, the **TCSM** of the fixed node closest to the area A may be in charge of the allocation of NTPs. However, it is better to place all the connectivity functions which are specially related to terminal mobility in a dedicate object such as the **NTP_Manager** which can be modified without any consequence for the **TCSM**. Furthermore, according to TINA [9], a **TCSM** is defined for a node in the transport network and the notion of coverage area presented here has nothing to do with a node. A coverage area may or may not correspond to a node. A node may in fact contain several areas when it has numerous NTPs covering a large geographical area which need to be grouped into smaller areas.

The area covered by the **NTP_Manager** may or may not correspond to the area covered by the **NAP** since the two objects may be responsible for the connectivity with two separate physical networks, the kernel transport network and the transport network supporting streams.

There are three correspondence alternatives:

1. an NAP's area corresponds to a **NTP_Manager**'s area.
2. an NAP's area corresponds to several **NTP_Manager**'s areas.
3. an **NTP_Manager**'s area corresponds to several **NAP**'s areas.

The choice of the appropriate correspondence alternative for a particular system is a matter of dimensioning and configuration and will not be studied further here.

A. Handover

When the mobile DPE moves from one NTP coverage area to another, there are two ways to support the availability of service:

- **Continuous terminal mobility** ensures the continuity of service by using the continuous handover mechanism. GSM is an example of this type of terminal mobility [7].
- **Discrete terminal mobility** ensures only the continuity of the service sessions but not the continuity of service, i.e the service sessions are suspended before and during the transition and resumed when the mobile DPE has reached the new NTP coverage area. This is referred as discrete handover or session mobility.

We shall now consider both cases.

B. Continuous handover

To ensure that stream flows are not disrupted when the mobile terminal is moving from one NTP coverage area to another, there must be an overlap between coverage areas as shown in Figure 17. In the intersection area $A \cap B$ of two NTP coverage areas A and B, the mobile terminal has contact with both **NTP_Managers** and stream flows can be established with NTPs belonging to both NTPPools.

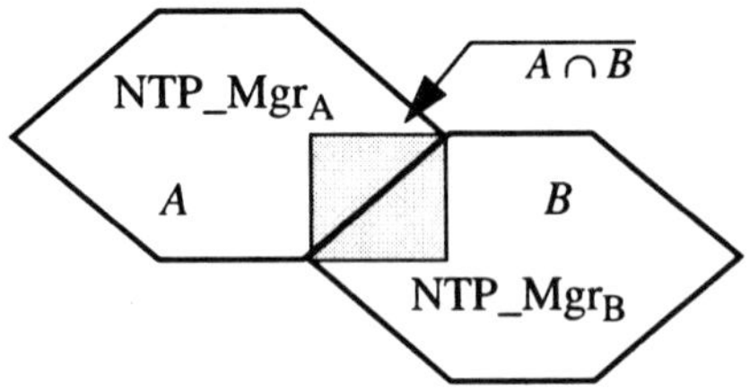

Fig.17 Overlap between NTP coverage areas

1) Handover procedure

A handover procedure consists of the following steps:

1. To detect that handover may be necessary. This can be done via the detection that the mobile DPE is in an intersection area such as $A \cap B$ or $A \cap B \cap C$, etc.
2. To decide that handover must be executed. The decision can be taken based on different criteria such as transmission quality, distance between the mobile DPE and the NTP, etc. The criteria are system dependent and cannot be implemented in a generic way. Furthermore, the decision for handover can be made by the mobile DPE or by the telecom system domain or also by both.
3. To perform the handover in the fastest and most secure way.
4. To move to a new stable state.

2) Necessary computational objects

In order to realise the procedure described above, several new objects are required in the system.

First, an object called **Handover_Initiator** is introduced to decide and supervise the handover procedure. The **Handover_Initiator** can be created by the **TCSM** on the mobile DPE if the handover decision should be made by the mobile terminal. It can also be created by the **NTP_Manager** on the fixed DPE if the handover decision should be made by the telecom system domain. The **Handover_Initiator** object encapsulates the functions necessary to realise a specific decision criterion. Hence, for a specific system using a specific criterion, a specific version of the **Handover_Initator** is needed. All the **Handover_Initiator** objects have, however, the same operational interface toward other objects in the system.

Second, two objects called **Handover_Source** and **Handover_Sink** are introduced. Each object has at least three stream interfaces. The **Handover_Source** object has the ability to connect an input stream interface with two output interfaces and to duplicate the information flow form the input stream interface. The **Handover_Sink** object has the ability to connect two input stream interfaces into one output stream interface and to discard the duplicated information from two identical input streams. In the case where there is a bidirectional stream flow between two objects, it is more convenient to use an object called **Handover** having the functionality of both the **Handover_Source** and **Handover_Sink** as shown in Figure 18. The binding of stream interfaces belonging to the same object is described in [10].

The **Handover_Source** should be inserted after the source object using the stream direction as reference and the **Handover_Sink** object should be inserted before the sink object.

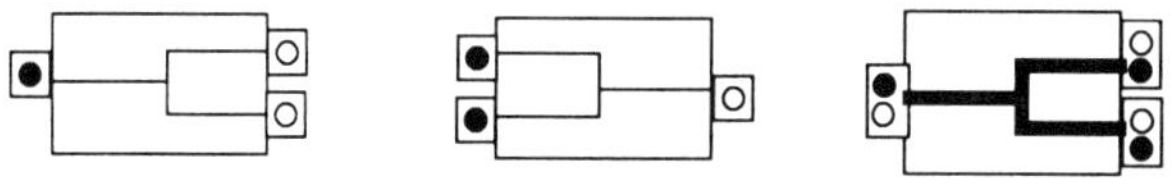

a. Handover_Source b. Handover_sink c. Handover

Fig.18 The Handover objects

It is worth noting that the **Handover** objects in fact represent a combination of both hardware and software modules which are specific to a system. Although these objects may have different internal functions and mechanisms they will have the same external interfaces.

3) Example of handover

Suppose that the mobile terminal is originally in area A and there is a stream flow between its object $\mathbf{CO_t}$ and another object $\mathbf{CO_n}$ in the telecom system domain. The stream goes through TTP_a, NTP_{AX} and NTP_u.We introduce now two objects $\mathbf{Handover_t}$ having three stream interfaces St1, St2, St3 and $\mathbf{Handover_n}$ having three stream interfaces Sn1, Sn2 and Sn3. We let the mobile DPE make the handover decision through a **Handover_Initiator** object residing in the mobile DPE.

a. Before handover

When the **CSM** receives the request to bind $\mathbf{CO_t}$ and $\mathbf{CO_n}$, it knows that $\mathbf{CO_t}$ is residing on a mobile DPE by reading the name of $\mathbf{CO_t}$. It may then deduce that a handover object is necessary. It creates or requests a Service Factory object to create an object $\mathbf{Handover_n}$ and proceeds with the procedure to establish a stream between $\mathbf{CO_n}$ and the stream interface Sn1 of $\mathbf{Handover_n}$. This is done without problems since both $\mathbf{CO_t}$ and $\mathbf{Handover_n}$ are residing in the fixed kTN. **CSM** requests $\mathbf{Handover_n}$ to bind internally the stream interfaces Sn1 and Sn2 as shown in Figure 19

Towards the mobile side, **CSM** requests $\mathbf{TCSM_t}$ to establish a local connection for $\mathbf{CO_t}$. $\mathbf{TCSM_t}$ knows that it is residing on a mobile DPE and that a handover object is necessary. It creates or requests a Service Factory object to create an object $\mathbf{Handover_t}$. It will then establish a stream between $\mathbf{CO_t}$ and stream interface St1 of $\mathbf{Handover_t}$. $\mathbf{TCSM_t}$ will then ask $\mathbf{Handover_t}$ to bind internally the stream interfaces St1 and St2. $\mathbf{TCSM_t}$ requests also $\mathbf{Handover_t}$ to bind St2 to TTP_a. $\mathbf{TCSM_t}$ returns the corresponding NTP identifier to which TTP_a is connected, namely NTP_{AX}, to the **CSM**.

The **CSM** then connects the $\mathbf{Handover_n}$ object to NTP_{AX}. When this is accomplished, a stream is established between $\mathbf{CO_t}$ and $\mathbf{CO_n}$.

When the terminal moves into the intersection area $A \cap B$, $\mathbf{TCSM_t}$ will receive the identifier of the $\mathbf{NTP\_Mgr_B}$ and know that a handover may be necessary. It creates or requests a Service Factory to create a **Handover_Initiator** object in the mobile DPE.

The **Handover_Initiator** encapsulates the mechanism for handover decision and starts to operate immediately after its creation.

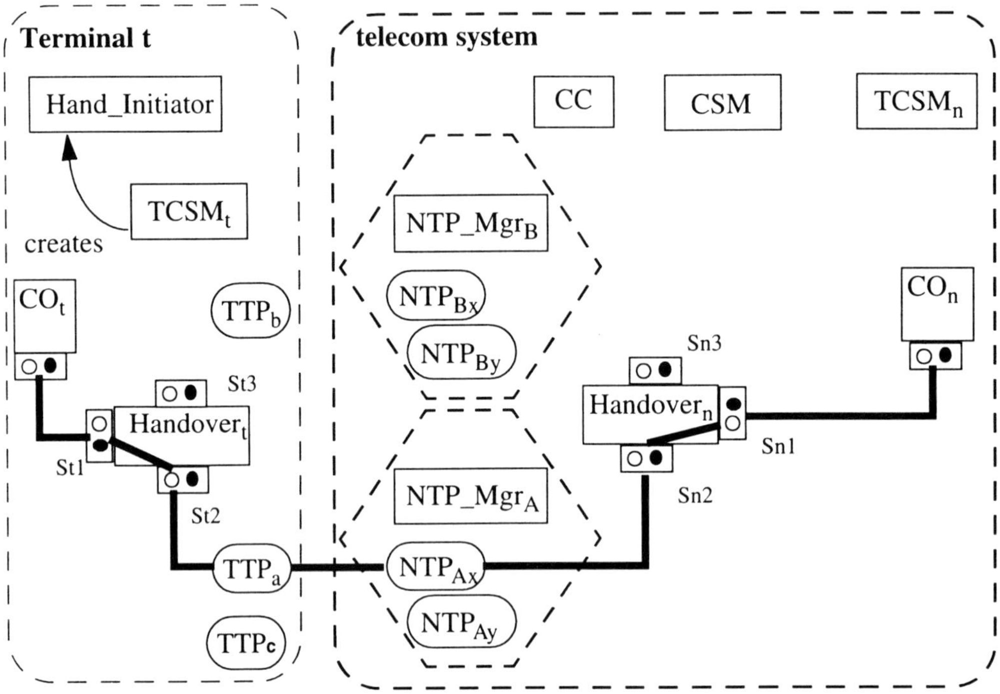

Fig.19 The stream flow before handover

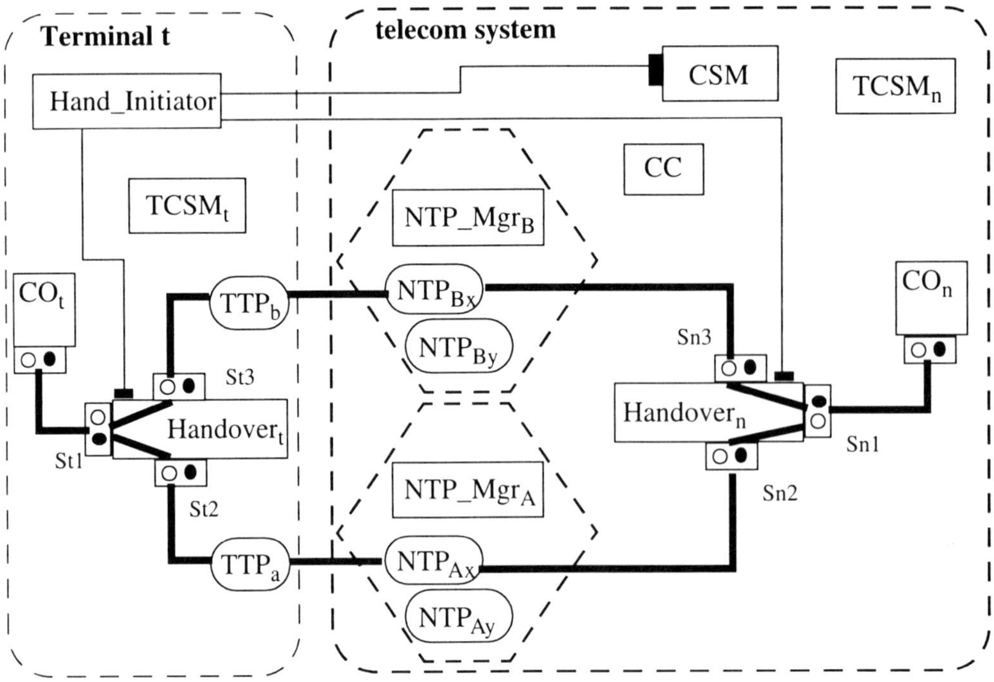

Fig.20 The stream flow during handover

b. During handover

When the criterion for handover is fulfilled, the **Handover_Initiator** will start the handover procedure. It requests **CSM** to establish a stream between the interface St3 of $\textbf{Handover}_t$ and the interface Sn3 of $\textbf{Handover}_n$. **CSM** requests the $\textbf{TCSM}_t$ to bind St3 to a TTP. The $\textbf{TCSM}_t$ binds St3 to TTP_b. $\textbf{TCSM}_t$ returns the corresponding NTP identifier to which TTP_b is connected, namely NTP_{BX}, to the **CSM**. **CSM** then connects the $\textbf{Handover}_n$ object to NTP-BX.

Everything has now been prepared for the switchover. The **Handover_Initiator** requests $\textbf{Handover}_t$ to bind internally the interfaces St1 and St3, and $\textbf{Handover}_n$ to bind internally the interfaces Sn1 and Sn3. The **Handover_Initiator** then requests **CSM** to release the stream between the interfaces St2 of $\textbf{Handover}_t$ object and Sn2 of $\textbf{Handover}_n$ object. It also requests $\textbf{Handover}_t$ object to unbind internally St1 and St2 and $\textbf{Handover}_n$ to unbind Sn1 and Sn2.

In Figure 20 we just show that there are interactions between the **Handover_Initiator** and the **CSM**, and $\textbf{Handover}_t$ and $\textbf{Handover}_n$ without any further specification in order to preserve clarity.

c. After Handover

As shown in Figure 21, after the accomplishment of the handover procedure the **Handover_Initiator** object will terminate itself. The stream between the objects $\textbf{CO}_t$ and $\textbf{CO}_n$ follows now the new path without disruption.

D. Discrete handover

Without overlap between the NTP coverage areas, the mobile DPE will loose all contact with the telecom system domain when it moves out of an NTP coverage area. The continuity of service cannot be ensured. However, the service sessions in use can be suspended in order to be resumed later when the mobile DPE arrives at the new NTP coverage area. In a way, the service sessions can be considered as non-disrupted. The continuity of sessions requires the intervention of the user, i.e the user must explicitly order the suspension of the sessions in use before moving to another NTP area. He must also explicitly resume these sessions later on.

V. CONCLUSION

We have shown how terminal mobility can be supported in TINA architecture. To support mobility, the following mobility computational objects are required:

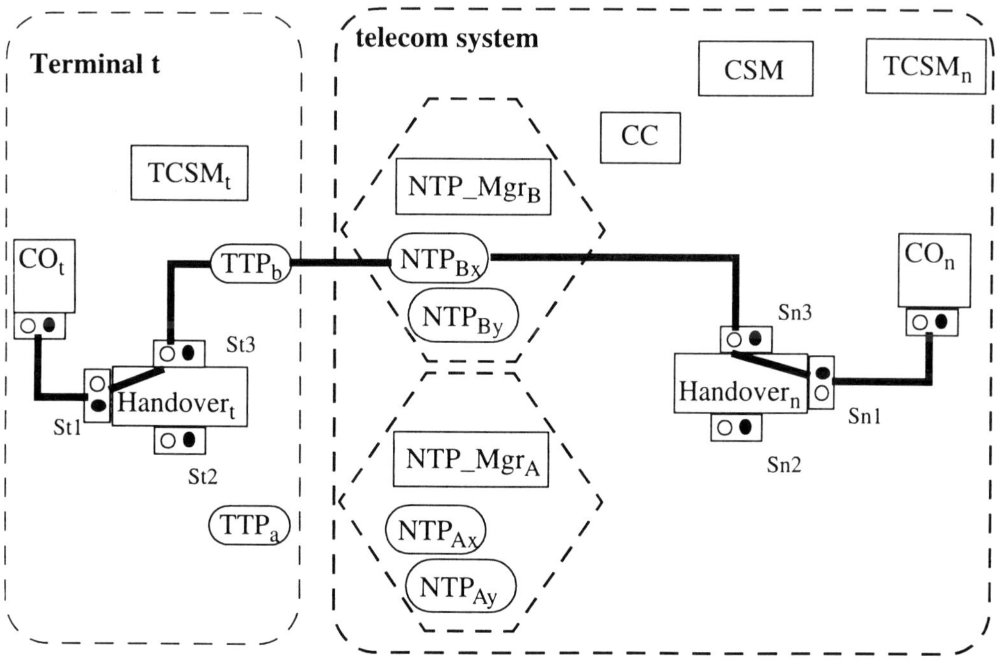

Fig.21 The stream flow after handover

a. To support operational interactions:

- In the mobile DPE:
  - `SPA`
  - `TAP`
- In the telecom system:
  - `TA`
  - `NAP`
  - `Terminal_Data`

b. To support stream flows:

- In the mobile DPE:
  - `Handover` objects
  - `Handover_Initiator`
- In the telecom system
  - `NTP_Manager`
  - `Handover` objects
  - `Handover_Initiator`

In addition, the following objects defined by the TINA Connection Management need to have enhanced functionality in order to cooperate with the mobility-related objects:

- `CSM`
- `TCSM`

These two objects must interact with the mobility objects such as the `TA`, the `SPA`, etc. in order to support streams. They must have the ability to distinguish between objects residing on a mobile DPE and objects residing on the fixed DPE, in order to decide about the creation of the `Handover` objects for streams spanning the terminal domain and the telecom system domain. The `TCSM` must also have the ability to detect that a handover may be necessary in order to create the `Handover_Initiator` object which controls the handover procedure.

REFERENCES

[1] v. T. Do, Jan A. Audestad. Mobility and TINA. In *Proceedings of the TINA 96 Conference*, volume ISBN 0. TINA-C, September 1996. Heidelberg, Germany.

[2] P. Spilling. The telecom research program telerep at unik. *UNIK Aarsrapport 1996*, 1996. http:www.unik.no/ paal/ telerep.html.

[3] Object Management Group, Inc. *The Common Object Request Broker: Architecture and Specification*. Document No. 91.12.1, Revision 1.1, ISBN 0-471-58792-3.

[4] TINA-C. *Computational Modelling Concepts*, January 1995.

[5] J. A. Audestad. Introduction to distributed processing in telecommunications systems. Norwegian University of Science and Technology, Departement of Computer Systems and Telematics, February 1996. Lecture notes - Course 45356 Communication in distributed systems.

[6] ETSI TC-SMG. *European digital cellular telecommunications system phase 2; Location registration procedures*, september 1994.

[7] ETSI TC-SMG. *European digital cellular telecommunications system phase 2; Mobile Application Part (MAP)*, july 1994.

[8] TINA-C. *Unification of Connection/session Graphs, Stream Interfaces and Channel Model*, April 1996. TR_PL.001_1.3_95.

[9] TINA-C. *Connection Management Architecture*, March 1995.

[10] v. T. Do, Jan A. Audestad. Experiences using corba to build telecommunications applications. In *Proceedings of the IFIP/ICCC International Conference on Information Network and Data Communication*, number ISBN 0 412 75750 8. Chapman & Hall, June 1996. Trondheim, Norway.

BIOGRAPHY

Dr. Do van Thanh received a M.Sc. degree in Electronical Engineering and Computer Science from the University of Trondheim, The Norwegian Institute of Technology, Norway in 1984 and Ph.D degree in Informatics from the University of Oslo, Norway in 1997. He joined Ericsson in 1991 after having worked seven years for Norsk Data, a minicomputer manufacturer in Oslo. He is currently a senior research scientist at Ericsson Applied Research Center in Norway and is in charge for Mobility and Distributed Computing research activity.

Jan A. Audestad is a senior advisor to Corporate Management of Telenor and part-time Professor in Distributed Systems at the Norwegian University of Science and Technology. He has thirty year research experiences in telecommunications, covering Satellite Systems, land Mobile Systems (GSM), Intelligent Networks, Security and Telecommunications Policy. He has chaired a number of standardisation and research groups in ITU, ETSI and Eurescom.

Service Architecture Support of Personal Mobility in a Multi-Domain Environment

A. Lombardo, P. Nicosia, S. Palazzo, M. Samarotto

Istituto di Informatica e Telecomunicazioni, Università di Catania
V.le A. Doria, 6 - 95125 Catania (Italy)
{lombardo, nicosia, palazzo, msamaro} @iit.unict.it

Abstract

This paper introduces some proposals for the support of personal mobility services in the TINA environment. In particular, the reasons for maintaining terminal information inside the network and the need to support services across different Provider Domains by handling them through multiple User Agents are presented. The effectiveness of User Agent replication is analysed by comparing the signalling load generated on the network in the hypotheses of: i) accessing remote user data associated with a single User Agent and, ii) transferring these data to the visited domain where the user is actually accessing the service.

1. Introduction

Service Architectures are evolving in order to support the next generation of mobile, wireless services. The target services, known as Personal Communication Services (PCS), have the aim of supporting both terminal and personal mobility. Whilst a big effort has already been devoted to supporting terminal mobility in Open Service Architectures, the impact of personal mobility in this area is still under study and some key issues remain open.

In this paper two of these issues are analysed.

The first concerns the association between user and terminal which, in a personal mobility environment, is not a fixed one-to-one association, but provides a user with the capability to register with different terminals for different services, thus allowing a terminal to be shared by different users. In this case, the need arises to distinguish between user and terminal information as two separate entities which can dynamically be associated with each other.

The second issue analysed here concerns the support of personal mobility in a wide multi-domain perspective. In such an environment it is not realistic to hypothesise the presence of a single universal service provider or network operator, since personal mobility aims at providing the user with universal service access and supporting him with the capability to be known through his personal identifier, irrespective of his location in the network or the terminal he is actually using. In order for these requirements to be met, and with such a wide coverage, it is necessary to provide distributed agents to act on behalf of the users and to identify effective mechanisms for provider co-operation irrespective of the multi-domain nature of the network.

In the paper these two issues are studied according to the TINA model [1], which is emerging as a de-facto standard in the area of Open Service Architectures for telecommunication systems, and their main impacts on the TINA architecture are identified in the following two aspects:

- where to maintain the terminal-related data, namely the Terminal Profile,
- how to handle service accesses by users roaming in domains different from the one where the subscription has been made and associated with different terminals for different services.

Some solutions to these questions are identified and proposed, and a comparison between them is made in terms of estimation of the signalling load generated on the network by significant personal mobility procedures.

2. Terminal Data Location

In a personal mobility environment where the association between user and terminal can be dynamically changed through the user registration procedure, it is necessary to consider terminal and user information as two distinct entities. Currently in TINA, information about terminal location is maintained inside subordinated objects associated with the User Agents of all the users

0-8186-8335-X/98 $10.00

registered on that terminal (the so called Usage Contexts UCxt's), and terminal characteristics are maintained inside the terminal itself [2]. This choice, which may be appropriate for terminal mobility provision, is not effective when personal mobility has to be supported because, due to the dynamic association between user and terminal, the following problems arise.

- **Load on the air interface.** Every time a user wants to associate with a terminal (service activation, registration, etc.), it is necessary to check the user subscribed conditions against the terminal capabilities and restrictions: this information would have to be retrieved from the terminal itself, thus loading the air interface.
- **Complex domain updating procedure.** As more than one user can be registered on the same terminal and the only entities where the terminal location information is maintained are the UCtx's of these users, every time the terminal moves from one domain to another, all these UCtx's would have to be updated, thus resulting in a complex procedure.
- **Impossibility of remote registration**. If a remote registration (i.e. a registration on a terminal different from the one currently in use for accessing the service) has to be carried out, the terminal location has to be determined in order to check the terminal capabilities against the user subscribed conditions: this is not possible if the terminal location information can only be found in the UCtx's of all the users registered on that terminal because, in this case, the user performing the remote registration would not know which users are already registered on the terminal he wants to register on.

To solve the aforementioned issues, we propose to introduce into the network an agent associated with the owner of the terminal, referred to as the Terminal Owner Agent (TOA). This agent should be located in the domain where the owner of the terminal has made subscription, namely the Terminal Owner Domain, so that it can always be retrieved through the terminal identity. The presence of this entity, regarded as separate from the agents of the users registered on the terminal, definitely solves the problem of achieving remote registration and minimises the complexity of the domain updating procedure. Moreover, in order to guarantee the availability of updated terminal information inside the network and to minimise traffic on the air interface, it is more effective to associate the following information with the TOA:

- Terminal location information (in terms of the domain where the terminal is located);
- Registration policies;
- List of users registered on the terminal;
- Terminal capability information.

The first two information elements need to be maintained inside the network in order to support remote registration. Relating this information to the terminal owner makes it possible to locate the terminal without any knowledge of the users registered on it. The last two types of information could, alternatively, be stored in the terminal itself. However, it would be more efficient to maintain this information inside the network, thus reducing the traffic on the terminal access link. The dynamic associations between a user and the terminals he has registered on for the various services are maintained within the UCtx associated with the User Agent, whilst, as shown in Figure 1, the information about each terminal location, together with the specific owner preferences and restrictions, are maintained within suitable subordinated objects of the different Terminal Owner Agents, referred to as Terminal Equipment Agent (TE-A), Subscription Agent (SubAgt) and Personal Profile (PPrf), respectively.

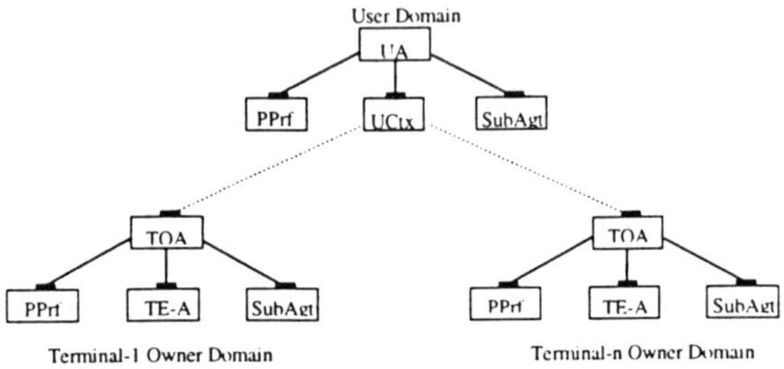

Figure 1: User Agent / Terminal Owner Agents Association

3. Multiple User Agents

In the multi-provider domain perspective, different types of domain can be identified as relevant when supporting users roaming from one domain to another, who associate themselves with different terminals for different services. These types of domain can be identified as follows:

- User Home Domain: this is the domain where the user originally subscribed and where his policies and restrictions, referred to as the User Profile, are maintained;
- User Visited Domain: this is the domain where the user is currently roaming and activating or receiving services;
- Terminal Owner Domain: this is the domain where the owner of the terminal has subscribed and where the terminal information is maintained.

The existence of the User Home and Visited Domains is strongly related to the policies for maintaining user-related data in current cellular systems like GSM. It is, in fact, always necessary to maintain, in the User Home Domain, information related to the user subscribed conditions and to the association of the user with the different terminals he is registered on: these data need to be located and accessed using the user's personal identifier. The existence of a Visited Domain is related to the user's capability to move from one domain to another and to receive and activate services from these domains. A peculiar issue in this multi-domain perspective is the need to identify a Terminal Owner Domain as distinct from the other domains. The reasons for this choice were highlighted in the previous section and are strongly related to the distinction between user and terminal which characterises personal mobility services.

According to the TINA Service Architecture [1], when a user subscribes to a service with a provider, the latter creates a User Agent inside the User Home Domain where the user and the service subscription characteristics, and the association between user and terminal are maintained inside subordinated objects (SubAgt, PPrf and UCtx). This agent must be located and accessed, by means of the user identity, whenever the user has to be located and invited to join a service session.

In view of supporting terminal and personal mobility in a multiple domain environment, it must be considered that the existence of a single User Agent may not be an effective solution. In fact it might be burdensome to interact with such an agent from a remote domain where the user is roaming, whenever, for example, it is necessary to check the user capabilities for activating or receiving a service. For this reason the hypothesis of making the User Agents migrate to the domain where the user is currently roaming while accessing a service should be investigated as an effective alternative to the solution of maintaining a fixed User Agent in the User Home Domain.

User Agent replication can be performed according to one of three different strategies:

1. The User Agent, including the User Profile, migrates to the Visited Domain.
2. Only the User Profile migrates and a local User Agent is created in the Visited Domain.
3. The User Agent, including the User Profile, is maintained in the Home Domain.

Each of these possible alternatives has its pros and cons.

The first provides control of user-related data in a remote domain as if it were performed by the provider with which the user has subscribed. The disadvantage of this approach is to require a suitable platform for agent migration. In this sense we can regard the problem as being strictly connected with the emerging Mobile Agent technology [3]: in this view this approach can be regarded as a matter of service customisation where the proper User Agent migrates close to the user's current location [4], in order to allow the user to access the required service according to the subscribed characteristics. The Mobile Agent application can therefore support the user with the capability to associate himself with different terminals for different services. Through User Agent migration, the user is provided with a customised interface to the specific service he is registered for on the specific terminal. In this way the user can access the service from any terminal according to the characteristics subscribed with his provider (if compatible with the terminal and network capabilities).

In the second alternative control of the User Profile is exercised by User Agents created in the domains where the user is active. In this case the User Agent in the Home Domain can be considered as simply resulting in a database maintaining pointers to User Agents created in the Visited Domains.

The third alternative is the opposite of the first one. In fact, by maintaining and controlling data in the original domain, no transfer is necessary, but a continuous interaction with the provider where subscription has been made is required.

An effective comparison between these alternatives can be made by considering proper user traffic models, as the signalling traffic actually depends on the service access rate. A method for the evaluation of this signalling traffic is described in the next section.

4. Signalling load analysis

In this section we compare the User Agent handling alternatives proposed above by calculating the signalling load generated by personal mobility procedures.

The differences between the alternatives compared mainly depend on the amount of transferred data relating to User Profile (and User Agent) migration. The range of variation for this data has been fixed according to the hypotheses of either maintaining both the User Agent and the User Profile in the User Home Domain, or transferring the User Profile (and the User Agent) to the Visited Domain. The resulting signalling load is associated with those personal mobility procedures [5,6] which most interact with Terminal and User Profiles. The amount of signalling load that crosses different domains is of particular importance as it implies agreement between different providers and, more frequently, makes use of long-distance signalling links, which are expensive and tend to be of lower bandwidth than intra-network signalling links.

The methodology used to carry out the signalling load analysis is the following.

1. Defining the event traces for the procedures under study, to identify the messages that flow between the various Computational Objects which act in these procedures and to calculate the length in bytes of these

messages.

2. Fixing a certain Service Network Configuration (i.e. allocation of Computational Objects to Service Nodes), in order to identify which messages contribute to the signalling load across different domains (T_D), which contribute to the signalling load between different nodes within the same domain (T_N) and which can be neglected because exchanged between elements within the same Service Node and therefore do not influence the signalling load evaluation.
3. Determining the rates at which certain procedures are invoked.
4. Deriving expressions for the two signalling loads, T_D and T_N, as the sum of the signalling load generated by the individual procedures under study.

The analysis is carried out by comparing the results of these expressions, evaluated in the different system implementations.

Only the traffic generated by roaming users is taken into account in the analysis, because only this type of traffic requires migration or remote access to profiles and is therefore significant for our comparison.

As regards the location of Terminal Profiles, we assume all terminal-related data to be associated with the Terminal Owner Agent (TOA) in the Terminal Owner Domain, because the alternative solution of maintaining the terminal-related data on the terminal would generate a heavy load on the air interface, which is considered an expensive, bandwidth-limited and non-secure resource.

In the following sections, the results obtained will be presented through proper diagrams, where the signalling load is shown as a function of the number of service invocations and the amount of data constituting the User Profile, respectively. From a comparison of these results the conditions for which migration is an effective solution will be identified.

4.1 Step 1: Procedure descriptions

The description of the personal mobility procedures is based on previous studies on mobility developed within European ACTS projects [5] and, in particular, on the results obtained within the DOLMEN project [6] which is evaluating the impact of mobility on Open Service Architectures compliant with the TINA model. The descriptions of the individual procedures were developed according to the Message Sequence Chart technique [7] and show interaction between the Computational Objects supporting the functions under study. The personal mobility procedures selected in this analysis are those which more frequently access data stored in the User Profile. They are:

- User Access Session and Start Service Session, which access the User Profile in order to authenticate the user and to check his subscribed conditions.
- User Registration and User Remote Registration, which access the User Profile to verify whether the required association between user and terminal for a specific service is compliant with the user subscribed conditions.
- Invite to a Service Session and End Service Session, which access the User Profile to check whether the user is enabled to receive an incoming service or to release a service session.

A specific terminal mobility procedure, the Domain Updating procedure, has also been taken into account, as it requires profile updating.

For each single message, significant parameters have been identified and the corresponding message length estimated.

4.2 Step 2: Definition of a Reference Service Network Configuration

Starting from the descriptions of the personal mobility procedures made in terms of Message Sequence Charts, it is necessary to fix a certain Service Network Configuration. Such a configuration is obtained by mapping Computational Objects onto Service Nodes and then defining which Service Nodes are present in each domain considered. In this way it is possible to estimate the signalling load across different domains and the signalling load between different nodes within the same domain and to neglect messages exchanged between objects located within the same Service Node which do not influence the signalling load evaluation.

With this aim a comparison has been made between possible Service Network Configurations. The configurations analysed have been defined on the main principle of specialising Service Nodes for dedicated functions as in [8], thus providing a distributed architecture where the processing logic is independent of the switching elements and the specific nodes can easily be configured inside the network as needed. The main factors which have been taken into account in order to identify the reference Service Network Configuration are:

- User Demand: Computational Objects most often accessed by users, such as User Agents, have been located at Service Nodes in the vicinity of the users.
- Information Ownership: Computational Objects managing information have been located in Service Nodes having direct access to the information sources they exploit. A significant example is given by the User Agent and the User Profile data.
- Commonality: Computational Objects that are closely related to each other and thus interact heavily should

be placed in the same Service Node since communication cost has been considered as the most important distribution criteria.

In this way specific Service Nodes have been identified [9]: the Access Node, which represents the first node accessed inside the network where the bridging functions are located; the Session Node where the functions which handle the service sessions are stored; the Location Node, where specific functions related to Location Updating procedure are stored. Another Service Node of particular interest for the characterisation of the selected reference configuration is represented by the User Node, which contains the User Agent and its subordinated objects. This node, by co-locating both the User Agent and the associated objects constituting the User Profile, minimises the signalling load due to the continuous interaction between this agent and its related data. In the hypothesis of transferring user-related data to the Visited Domain, both a User Node in the User Home Domain and a User Node in the Visited Domain will be provided for in the reference configuration.

It must be stressed that the reference configuration envisaged only considers Computational Objects involved in the personal mobility procedures under study. This means that the reference configuration does not represent a realistic Service Network Configuration, as, in principle, in a complete Service Network Configuration, other functions should be included inside the Service Nodes identified. Nevertheless, the reference configuration identified is meaningful for our evaluation, which is not devoted to finding an optimised Service Network Configuration, but just to comparing different strategies for User Agent handling.

4.3 Step 3: Procedure invocation rates

In order to evaluate the signalling load associated with personal mobility procedures it is necessary to determine the rates at which these procedures are activated, i.e. to assume proper user and traffic models.

Again only the traffic generated by roaming users is taken into account, because only this type of traffic requires migration or remote access to profiles and is therefore significant for our comparison.

The analysis is carried out by considering the input parameters summarised in Table 1.

The values have been derived from a previous study on PCS systems carried out in [10]. Among them, the one representing the probability of being in a boundary location area has been evaluated by assuming a uniform probability distribution for the user position over all the location areas. Such a parameter is important for the evaluation of the domain updating procedure invocation rate. Two parameters have been considered as varying: the rate of service invocations within the same session, and the amount of data constituting the User Profile. The first parameter is strongly related to the user model assumed, and the second reflects the different alternatives of agent migration described above. Table 2 specifies how the procedure invocation rates are calculated for the entire population of users.

For all of them, with the exclusion of domain updating, the procedure invocation rates are given by the product of the corresponding rates per single user assumed in Table 1 and the number of users. As regards domain updating, the procedure invocation rate is given by the product of the domain boundary crossing rate and the probability of users being in a boundary location area.

Input parameter	Value
Number of roaming users per domain, S	1.5 million
Density of users, ρ	150/sq.mile
Location area perimeter, L_{la}	20 miles
Number of location areas per domain, N_{la}	20000
Probability of being in a boundary location area, P	0.2
Speed of mobile users, v	50 m.p.h.
Service access rate per user, r_{acc}	2 per day per user
Service session set-up rate per user, r_{sup}	variable
Registration procedure invocation rate per user, r_{re}	2 per day per user
Incoming service rate per user, r_{is}	variable
Service session release rate per user, r_{sr}	variable
User Profile size, S_{up}	variable

Table 1 Input parameters

Description	Equation
Service access rate	$r_{acc} * S$
Registration rate	$r_{re} * S$
Service session set-up rate	$r_{sup} * S$
Service session release rate	$r_{sr} * S$
Incoming service rate	$r_{is} * S$
Domain boundary crossing rate	$(\rho * v * L_{la} / \pi) * P$

Table 2. Procedure invocation rates

4.4 Step 4: Signalling load evaluation

By assuming the reference configuration defined in Section 4.2, and determining both the procedure invocation rates and the number of bytes generated in each procedure, we can calculate:

- the signalling load across different domains (T_D),
- the signalling load between different nodes within the same domain (T_N),
- the total signalling load (T).

The general expression for the signalling load due to all mobility management procedures is given by:

$$T = T_D + T_N = T_{as} + T_{is} + T_{ur} + T_{rr} + T_{du} + T_{ss} + T_{es}$$

where

T_{as} is the total signalling load due to a user-initiated User Access Session (outgoing service);

T_{is} is the total signalling load due to an Invite to a Service Session (incoming service);

T_{ur} is the total signalling load due to User Registration;

T_{rr} is the total signalling load due to User Remote Registration;

T_{du} is the total signalling load due to Domain Update;

T_{ss} is the total signalling load due to Start Service Session;

T_{es} is the total signalling load due to End Service Session.

Each of these terms is evaluated through the following formula:

$$T_x = \sum_i N * a_i + \sum_j I * b_j + \sum_k E * c_k$$

where

i varies in the range corresponding to the number of messages exchanged between objects within the same node;

j varies in the range corresponding to the number of messages exchanged between nodes within the same domain;

k varies in the range corresponding to the number of messages exchanged between nodes located in different domains;

N is the weight associated with a message exchanged between two objects within the same node (in the following, N will be considered equal to 0);

I is the weight associated with a message exchanged between two nodes within the same domain;

E is the weight associated with a message exchanged between two nodes located in different domains;

a_i is the number of bytes of a message exchanged between two objects within the same node;

b_j is the number of bytes of a message exchanged between two nodes within the same domain;

c_k is the number of bytes of a message exchanged between two nodes located in different domains.

The rough values assumed by a_i , b_j , c_k , were fixed without making any assumption on the control fields added by the protocol in the underlying transport network.

A further assumption is that, as different domains are involved in our study, we evaluate this expression by assuming users to be equally distributed over all domains; this allows us to address a single domain D. In D we consider the traffic generated for incoming and outgoing services, registration and domain updating by users visiting D, as well as by users who subscribed in that domain and are roaming in another domain. A complete list of all these cases would be tedious for the reader. Therefore, as an illustrative example, we only present the terms of the expression needed for the evaluation of Tas:

$$T_{as} = K * [T_{asrv} + T_{asrh}];$$

where:

K is the percentage of outgoing services with respect to the total amount of services generated,

T_{asrv} is the signalling load generated by users who subscribed in the domain under study and are roaming and requiring services in another domain,

T_{asrh} is the signalling load generated by users roaming in the domain under study.

5. Results

The calculations performed according to the method described above result in the graphs in Figures 2 and 3 where the signalling loads generated by the personal mobility procedures under study are evaluated according to the two alternatives of either transferring the User Profiles or maintaining them fixed in the Home Domain. They show a comparison between the signalling load generated in the case of fixed profiles and the load when the User Profile is transferred, the size of profiles being 500 and 5000 bytes respectively.

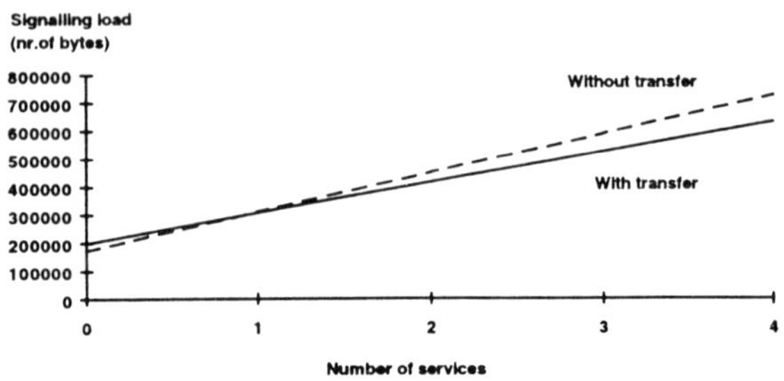

Figure 2: Total Signalling load (User Profile size = 500 bytes)

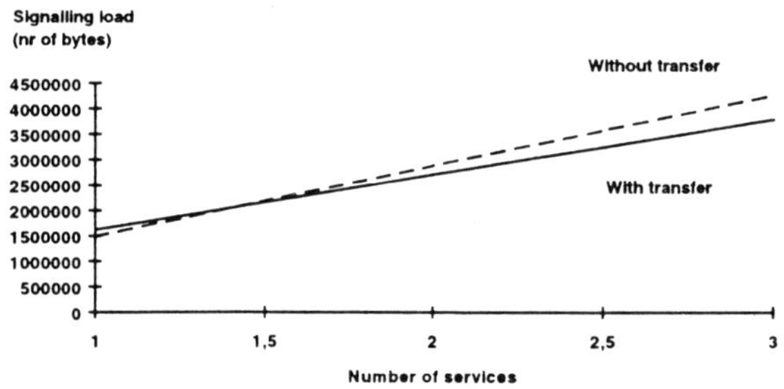

Figure 3: Total signalling load (User Profile size = 5000 bytes)

It can be observed that, even with a User Profile size of 5000 bytes, when the number of services required is around 1.5 services per session, it is still convenient to transfer User Profiles to the Visited Domain. Similar results were also obtained for larger sizes of User Profiles. For example, with 10000 bytes, User Profile transfer turns out to be convenient for a number of services greater than 3.

Extended to a wider range of profile size transfers, the above analysis produces the graph shown in Figure 4, where the convenience region for profile transferring is identified by the shadowed area.

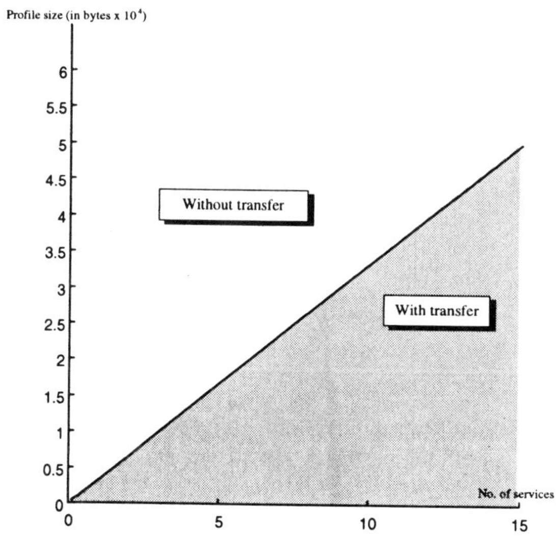

Figure 4: Convenience regions for profile transfer

We can therefore confidently conclude that the results obtained, adopting the signalling load as a cost measurement, demonstrate User Profile transfer to be an effective approach. As a consequence it could be reasonable to investigate this solution further by considering security aspects related to profile transfer across different domains, which have till now been neglected in our study. In addition, as the effectiveness of profile transfer depends on the number of services activated within a session, the hypothesis of maintaining the User Agent in the Visited Domain for the whole period during which the registration on the specific terminal is active (rather than for the time the access session is active) should be investigated. This hypothesis should be evaluated by taking into account the domain updating procedure which, in this case, would increase its complexity by requiring the transfer of all User Agents associated with the user when he is registered on different terminals for different services.

6. Conclusions

In this paper some issues which are relevant to the support of personal mobility in Open Service Architectures, and specifically in TINA-compliant systems, have been analysed. In particular they concern terminal data location and handling of User Agents in a multi-provider domain environment. The first issue results in the introduction of a Terminal Owner Agent, located in the domain where the owner of the terminal has made the subscription. The second causes the introduction of User Agents in the Visited Domains, distinct from the User Agent in the Home Domain. The User Agent in the Visited Domain is in charge of maintaining relevant user subscription information in the domain where the user is roaming. With respect to this, an analysis of the effectiveness of transferring the User Profile to the User Visited Domain has been carried out through evaluation of the signalling load generated on the network by significant personal mobility procedures. The results have shown that, even when the amount of transferred data is large, it is still convenient to transfer User Profiles close to the user location instead of accessing them in the remote domain. Security aspects related to this transfer have not been taken into account till now in this study but should be investigated in view of a "retailer federation" which should be envisaged in the perspective of wide, efficient PCS support. As regards terminal data location, further studies should also be carried out to enhance the TINA Service Architecture in order to include the terminal provider role, thus distinguishing between user and terminal concepts in the business model as well as in the computational model.

Acknowledgements

This work has been carried out by a team from the University of Catania within the framework of the ACTS DOLMEN project whose target is to develop, demonstrate, access and promote an Open Service

Architecture for a fixed and mobile environment in which PCS services are deployed. The authors wish to thank all colleagues involved in the DOLMEN project for the helpful discussions.

References

[1] TINA-C Deliverable: "Service Architecture", Version 5.0, June 16, 1997.

[2] TINA-C: "Terminal Mobility", Version 1.0 Draft, 16 July, 1996

[3] T. Magedanz, K. Rothermel, S. Krause "Intelligent Agents: An Emerging Technology for Next Generation Telecommunications?" , Proc. IEEE INFOCOM '96, San Francisco, March 1996.

[4] R. Ramjee, T.F. La Porta: "The Use of Network-Based Migrating User Agents for Personal Communication Services", IEEE Personal Communications, Vol.2, No. 6, December 1995.

[5] RACE MONET - R2066, Deliverable R2066/BT/PM2/DS/P/113/a4, UMTS System Structure Document (Revised), 1995

[6] ACTS- DOLMEN - AC036, Intermediate Deliverable AC036/FUB/ASD2/vol. 2, Open Service Architecture for Mobile and Fixed Network Environment (OSAM), Initial Release, Volume 2 - Impact of Mobility, December 1996

[7] ITU-T Recommendation Z.120, Criteria for the use and applicability of formal description techniques- Message Sequence Chart (MSC), WTSC, Helsinki, March 1-2, 1993

[8] T. La Porta, M. Veeraraghavan, P. Treventi and R. Ramjee: "Distributed Call Processing for Personal Communications Services", IEEE Communications Magazine, Vol. 2, No. 3, June 1995

[9] ACTS- DOLMEN - AC036, Intermediate Deliverable AC036/FUB/APD2/v1, Performance Evaluation Framework in an Open Service Architecture Environment, April 1997

[10] M. Veeraraghavan, T. La Porta, R. Ramjee: "A Distributed Control Strategy for Wireless ATM Networks" ACM/Baltzer Wireless Networks, Vol.1, No 3, October, 1995

Using CORBA to Support Terminal Mobility

Mika Liljeberg, Kimmo Raatikainen
University of Helsinki
Department of Computer Science
P.O.Box 26 (Teollisuuskatu 23)
FIN-00014 University of Helsinki
Finland

Mike Evans, Stephen Furnell
University of Plymouth
Network Research Group
School of Electronic
Plymouth
United Kingdom

Nicolas Maumon
SEMA Group Telecom
7 & 9 Rue Hélène Boucher
78280 Guyancourt
France

Eric Veldkamp, Berny Wind
Royal PTT Netherlands (KPN)
KPN Research
2260 AK Leidschendam
The Netherlands

Sebastiano Trigila
Fondazione Ugo Bordoni
via Baldassare Castiglione 59
00142 Roma, Italy

Abstract

Nomadicity is quite a new challenge for computing and communication technologies. Some of the key issues in the field are, what is the impact of nomadicity on client-server interaction, how to support terminal mobility, and how to cope with the unique performance characteristics of wireless access. The first issue concentrates on the question of whether the computational viewpoint client-server interaction mechanisms for fixed environments remain valid in mobile environments. The other two issues are related to the engineering aspect of dealing with mobile routing and the unpredictable performance and reliability of wireless networks, generally orders of magnitude below those of wired networks. In this paper, we show a novel way to deal with these issues in a CORBA based distributed processing environment.

Object technology—CORBA, in particular—is already mature, or is at least reaching maturity. Today CORBA provides a software execution and development environment that simplifies distributed computing and application integration. The main stream of CORBA-based solutions is targeted to LAN-based applications relying on (quite) fast and reliable connections. In this paper we will show that the CORBA 2.0 specification also provides the means to support nomadic computing. We will demonstrate how the CORBA 2.0 Interoperability Architecture, together with some CORBAServices (Common Object Service Specifications), can be used to provide seamless support to terminal mobility and communication through slow wireless connections. The corner stones of our solution are mediated bridges, and an Environment Specific Inter-ORB Protocol (ESIOP) tailored for wireless networks.

1 Introduction

The ACTS project DOLMEN is developing an Open Service Architecture for Mobile and fixed environments, called OSAM [5]. One aspect that is looked at in more detail is terminal mobility. In that area, continuous terminal mobility is seen as the most challenging issue as it places the highest requirements on the architecture.

The scope of this paper is the impact of (continuous) terminal mobility on client server interaction mechanisms over a wireless access and how this can be supported by a CORBA [4] compliant environment.

From the computational viewpoint the client server model is used to structure the interactions between objects. The mechanism used by a client to invoke an operation offered by a server comprises two steps:

- Retrieval of a reference to an instance of the interface that gives access to the desired operation.
- Invocation of the operation across the interface, using the obtained reference, provided that a valid reference has been obtained.

In that respect, terminal mobility implicitly means frequent changes of object references; in particular those of nomadic objects because the mobile terminal and the objects contained therein are continuously on the move. This reveals the main problems of wireless access from the computational point of view:

0-8186-8335-X/98 $10.00 © 1998 IEEE

Can the basic interaction mechanism, as outlined before, be used when mobile terminals are involved? Do mobile terminals impose additional constraints on the validity of interface references?

The answers to the questions are yes and no, respectively. This paper outlines a solution in which the impact of terminal mobility on the basic client server interaction mechanism is tackled in the engineering view point by using CORBA bridging techniques.

The following requirements are imposed on the support of mobility from the computational point of view:

1. The DPE should support object interaction according to the conventional mechanism, over a wireless access network, from terminal to network and vice versa.

2. Under the assumption that a reference changes as a result of mobility, even though it points to the same interface instance, the DPE should be able to detect those changes, and to issue a new valid reference for the interface instance transparently to the client.

3. The DPE should notify interested (client) objects of the reference change.

The rest of the paper is organised as follows. In Section 2 we introduce the basic concept of CORBA, that is *bridging*, used in the DOLMEN solution for supporting terminal mobility on the DPE level. In Section 3 we describe how the bridging is used in the DOLMEN solution to solve problems due to terminal mobility. Section 4 is devoted to bridge handover and loss of signalling connection. Optimisation of the CORBA Inter ORB Protocol for wireless communication paths is addressed in Section 5. Other issues—including relocatable references, Location Register and address resolution of objects, Global Naming Service—of terminal mobility in CORBA are discussed in Section 6, whereas. Finally, in Sections 7 and 8 we discuss our implementation experiences and present a summary.

2 Basic Concepts

The CORBA 2.0 architecture supports "out-of-the-box" interoperability between different CORBA implementations by means of the *General Inter-ORB Protocol* (GIOP) and *Internet Inter-ORB Protocol* (IIOP). GIOP specifies a standard transfer syntax (low-level data representation) and a set of message formats for Inter-ORB communication. GIOP can be viewed as an abstract protocol, as by itself, it cannot provide complete interoperability; for actual implementation, a mapping must be defined between it and the transport used. IIOP defines such a mapping of GIOP onto the TCP/IP protocol used in the Internet. A CORBA 2.0 platform must implement both GIOP and IIOP in order to claim CORBA Interoperability conformance. However, CORBA seeks to interoperate with a wide and diverse range of different environments, and a single set of interfaces is not adequate in all circumstances. Hence, CORBA also makes provisions for Environment Specific Inter-ORB Protocols (ESIOPs). ESIOPs allow ORBs to interoperate with other objects which run in a completely different environment. For example, the DCE-CIOP (DCE Common Inter-ORB Protocol) allows CORBA objects to interoperate with objects located in an OSF DCE environment.

In the CORBA 2.0 architecture [4], *bridging* is one of the cornerstones of building interoperability support between different communication environments. A common use for an interoperability bridge is to act as a gateway between a CORBA domain and a non-CORBA environment, and translate between IIOP and a particular ESIOP designed for that environment. While ESIOPs may be optimised for particular environments, all ESIOP specifications are expected to conform to the general ORB interoperability architecture conventions to enable easy bridging.

On the architectural level, the term bridging simply means mapping between different CORBA compliant domains. In more concrete terms, a bridge resides between two domains and translates each message related to an object invocation across the domain border into a format understood by the destination domain. The translation may involve a mapping between object references, operation names, parameters, values, types, security credentials and the like. In the CORBA 2.0 specification *mediated bridging* means that the elements of an interaction relevant to a domain are first transformed at the domain boundary from the internal form of the original domain to a common form of representation, then transmitted over a communication media, and finally transformed from the common form to the internal form of the new domain.

3 Building an Interoperability Bridge for a Wireless Access Network

The concept of bridging seems very appealing in its potential to interconnect mobile terminals to the fixed network. In particular, a mediated bridge would appear to be particularly well suited to the task. Implementing two half-bridges, one residing in a mobile terminal and the other in a well-known access point within each mobility domain in the fixed network, allows us to introduce an efficient light-weight Inter-ORB protocol for use over the wireless access network. This approach also allows us to address terminal mobility, performance, and reliability issues.

Figure 1 shows how mobile terminals can be connected to a fixed network domain with mediated bridging. The wireless access domain and part of the core network domain is divided into mobility domains. The core network part of each mobility domain instantiates a set of mobile-specific support services, including one or more *Fixed DPE Bridges* (FDBRs) that serve as access points to the fixed network. The rest of the core network domain serves fixed terminals and acts as a backbone network. Each mobile terminal has its own ORB that provides object services to the applications running on the terminal. Invocations of objects outside the local access domain are directed to the *Mobile DPE Bridge* (MDBR) on the mobile terminal.

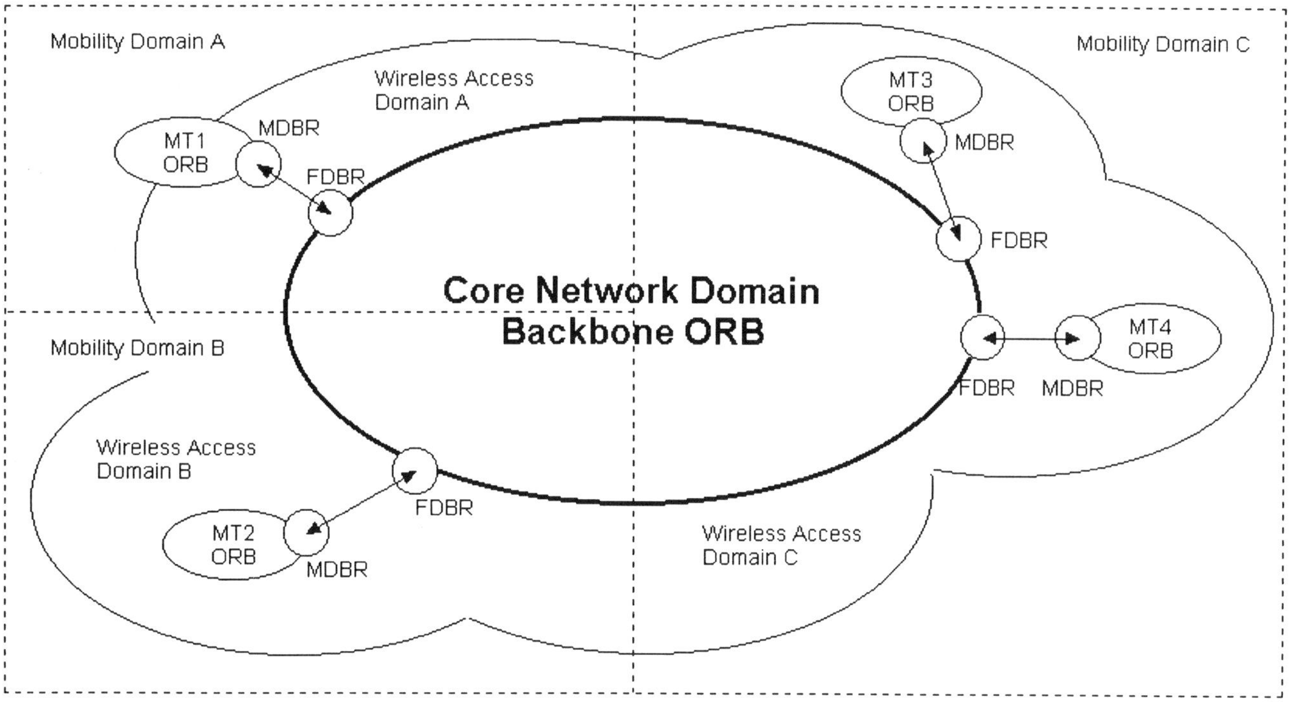

Figure 1: Bridging Mobile Terminals to a Fixed Network

The MDBR forwards the invocation to the FDBR, which then invokes the desired object. The FDBR acts as the representative of the mobile terminal within the fixed network, invoking operations in other objects on behalf of the mobile terminal. The FDBR also accepts invocation requests for objects located on the mobile terminal from objects within the core network. The FDBR forwards an invocation request to the MDBR, which then invokes the actual object and returns the response through the FDBR.

4 Bridge Handover and Loss of Signaling Connection

When a mobile terminal is in contact with the core network (see Figure 1), a physical signalling connection (a dedicated signalling channel) exists between the two bridges. When the terminal moves to another mobility domain, this signalling connection must be released, a new FDBR within the new domain must be contacted, and a new signalling connection must be created. This procedure is referred to as a bridge handover.

However, more than the switching of physical signalling channels is involved in a bridge handover. As already explained, the location register plays a central role in addressing objects with relocatable object references. During a bridge handover, the Location Register is updated and the new FDBR is registered as the current access point of the mobile terminal. This allows future invocation requests to be routed to the mobile terminal through the correct FDBR.

Object invocations that are in progress during a bridge handover, however, present some complications. One of the basic requirements for the mobility bridges is that they must hide the effects of mobility from client and server objects. This implies that the invocations in progress at the time of handover must be reliably and correctly completed despite the momentary break in connectivity that is inherent in a handover.

This reliability is achieved by buffering invocation related messages in the old FDBR until the mobile terminal has successfully connected to a new FDBR. A forwarding connection, termed a tunnel connection, is then set up between the old and new FDBRs and the buffered messages are forwarded to the new FDBR. Because IIOP is a connection-oriented protocol and an invocation reply is always sent over the same connection from which the request arrived, it is not possible to re-route an invocation replies. Therefore, the tunnel connection is be maintained until replies for all pending invocations have been delivered their destination through the old FDBR.

In a wireless environment it is possible to encounter a rapid fade-out that causes a sudden loss of signalling connection before the mobile terminal has the chance to attempt a handover to a base station with a better signal strength.In these cases, it is likely that some messages are lost. Since CORBA communication requires a reliable message service, the bridges must perform recovery after an unexpected loss and subsequent re-establishment of the signalling connection. The LW-IOP protocol provides the means for such a recovery: each LW-IOP message must be acknowledged by the receiver before it can be discarded by

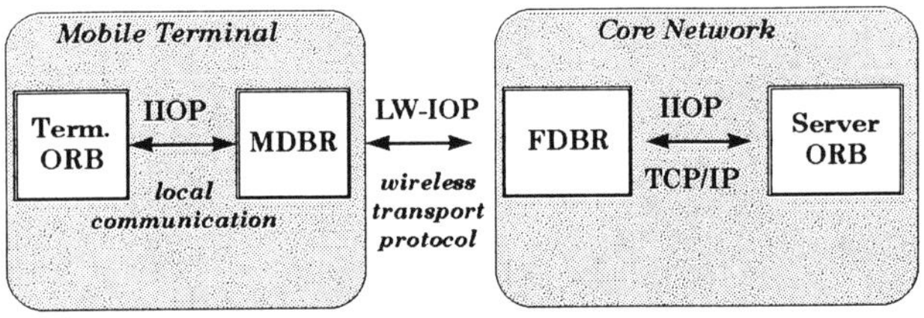

Figure 2: Protocols in Invocations over a Wireless Network

the sender. In the event of a communication error, any unacknowledged messages can be re-sent after the communication channel has been re-established. Recovery from a loss of signalling connection also often entails a bridge handover, since the new connection may be established through a different base station.

5 Light-Weight Inter-ORB Protocol

Since all object invocations between a mobile terminal and the core network pass through MDBR and FDBR, as depicted in Figure 2, the two bridges effectively form a closed interoperability domain. Thus, it is possible to transparently implement an ESIOP between the two bridges. We have taken advantage of this and designed a special *Light-Weight Inter-ORB Protocol* (LW-IOP) for wireless access networks with low-bandwidth signalling channels. The following basic requirements were identified for the LW-IOP protocol:

1. LW-IOP should be functionally compatible with GIOP.
2. Protocol message representation should be as efficient as possible.
3. Redundant information in discrete protocol messages should be minimised.

The rationale for LW-IOP protocol becomes clear when one considers the design of the GIOP protocol in the context of low-bandwidth wireless access networks. GIOP was designed to be very simple to implement, and to make minimal assumptions about the underlying transport connection. The protocol itself is stateless, consisting of simple request-reply interactions. However, result of this simplicity is that each message carries a relatively large amount of redundant information that could be communicated once and then left out of subsequent messages. The data representation format is also somewhat space-inefficient.

The LW-IOP protocol is based on the GIOP protocol in the sense that the exact same functionality is supported. However, the set of messages and their representations are more efficient. In addition, caching and compression techniques have been utilised in order to reduce the number of octets sent over the wireless access network.

GIOP specifies a total of seven message types (Request, Reply, CancelRequest, LocateRequest, LocateReply, CloseConnection and MessageError). The existing structures for these messages have been transformed into a set of light-weight equivalents.

Many GIOP message elements are allocated storage which is both unnecessary and inefficient for a wireless implementation. For example, request identities are allocated 32 bits, thus allowing for the unlikely scenario of 4,294,967,296 concurrent requests. LW-IOP request ids permit 16,384 concurrent requests, and thus require only 14 bits. Another example is the message size fields, which LW-IOP reduces from 32 bits to 16 bits. If a client wants to send a message longer than the maximum message size on the wireless link, then LW-IOP fragments and rebuilds the message.

In message headers and bodies the basic type 'unsigned long' is frequently used, e.g. to indicate the length of a string. Since these values are rather small in general, a representation with a variable size is used, in 1 to 5 bytes instead of systematically 4 bytes. In most cases, the numbers are represented in 1 or 2 bytes. In addition, the Common Data Representation (CDR), which is part of GIOP, inserts a lot of padding bytes for alignments of data. These are not present in LW-IOP.

In GIOP, all messages are preceded by a standard header containing a protocol identifier, version number, byte order flag and data indicating the type and size of the message. Of this, only the message type and size strictly need to be sent with each message and, as such, LW-IOP abstracts the version number into a new OpenConnection message, which is only sent at the initiation of an association, whereas other elements are discarded because they are irrelevant in LW-IOP. The byte ordering is always big endian. The remaining message type and size elements are optimised to further reduce the size of the standard message header.

Coded representations of textual and binary strings, which are very usual in GIP, are employed to avoid the need to always transmit the full information. Two approaches are used here. Firstly, certain strings are predefined which are known to be used on a frequent basis (e.g. the name of the core network domain). Secondly, a string history is maintained through caching on each side of the link. Thus when the transmitting side wishes to resend a string that it still maintains in its cache, it simply transmits a cache reference, enabling the recipient to extract the appropriate full details from its own cache. A mechanism is used to delete the less often transmitted strings when the cache is going to be full, while maintaining cache consistency between both ends of the links. All data transmitted between caches, i.e. cache references and delete notifications, are optimised in their Light Weight Representation (LWR). A cached string can be replaced by a cache reference, which is 1–3 bytes depending on the cache size. The mechanism of maintaining cache consistency can be embedded into marshalling algorithms.

Aside from the restricted bandwidth, the other main limitation of the wireless environment is the relative unreliability of the transport medium. It is conceivable that connections will drop out or become intermittent, causing messages to be lost. As such, the LW-IOP incorporates a reliability safeguard, with client and server ends assigning a unique message sequence number as part of the message header. These are then used by an LW-IOP-specific Acknowledge message, sent by client and server, to indicate that messages have been received and may be removed from the buffer. These messages provide a mechanism for re-synchronising the transmission in the event of the signalling connection having to be recreated, as each side will know the messages that require resending.

6 Other Issues in Terminal Mobility

6.1 Relocatable and Migrable Object References

The concepts of *Relocatable Object*, *Relocatable Object Group Identifier*, and *Relocatable Object Reference*, are introduced here to compensate for some lacks in the CORBA objects for the support of features such as mobility and migration. In brief, a Relocatable Object (RO) is a CORBA interface instance which implements a TINA engineering Computational Objects (eCO) interface instance. CORBA permits a system to enhance the basic concepts towards RO concepts, while staying compliant to CORBA.

Both migrable and mobile objects are kind of relocatable objects. The main difference is that when a mobile object moves, it stays in the mobile terminal, whereas when a migrable object migrates, it changes terminal. A fixed client will perceive a mobile object as being relocatable, because when the terminal moves, the client sees it through another bridge. The common aspect between migrable and mobile objects resides in the fact that both their references become out of date after a relocation, since the route to invoke them changes. We only concentrate here on mobile objects.

It is necessary to assign an identifier to a mobile object so that it is always possible to keep track of it. The identifier, Relocatable Object Group Identifier (ROI), is composed of information which will never change during the life cycle of the object. Theus, the ROI does normally not contain any information on the object instance location, with the exception of elements of location information which dot not change during the life cycle. A service is needed to bind the ROI to the current location of the object instance. Such a service, called Location Register (LR) has been introduced in [2]. The LR binds each ROI to the variable elements of location information. These elements and the ROI are the main components of a Relocatable Object Reference (ROR). Clients will use the ROR to always point to the same object instance and invoke it, even if the location changes.

In distributed applications all the objects instances do not need to be relocatable. The DPE must offer distribution transparency to the application and to the service control architecture. A client, at the application level, does not have to know whether or not an object instance is relocatable in order to invoke it. A client application will invoke non-relocatable and relocatable object instances exactly in the same way. Conversely, a server, at the application level, will always know whether or not an object it owns is to be instantiated as a relocatable object. A server application instantiates an object and generates its reference in different ways depending on whether or not the object is relocatable. This allows the server application to attribute ROIs to the Relocatable Objects it creates. When a mobile object reference enters the fixed network through the bridge, the bridge behaves like a fixed server which would instantiated the object. The bridge translates (or creates) the reference by embedding an ROI into it. Consequently, IIOP mobile object references that are introduced in the fixed network domain by the bridge are also RORs.

An ROR is made by adding an ROI to the reference of the object instance which currently constitutes the RO, with the constraint that the format of the ROR remains CORBA compliant. The only possible place for the ROI in an object reference is in the object key since only the object key is transmitted in an object invocation. If the ROI would not be in the object key, the server could not know whether or not an invoked object instance is relocatable. This would imply that the relocation transparency could not be provided.

In our implementation the Global Terminal Identifier (GTID) is used as an ROI. The GTID is inserted in the object key of the IIOP object reference. Mobile objects are perceived as being relocatable only in the fixed network, and hence only the mobile terminal needs to be identified, not the exact object location into the mobile terminal.

6.2 Location Register and Object Address Resolution

The basic problem in mobility is that, in order to enable objects in the core network to invoke objects in the mobile terminal, the location of the mobile terminal must be known at all times. To be precise, the bridge through which the mobile terminal can currently be accessed, must always be known. The LR is a database that maintains this information within the core network. The address of the current primary bridge of the mobile terminal can always be retrieved from the LR if the GTID of the mobile terminal is known.

The LR in our prototype implementation contains a persistent flat list of couples (GTID; FDBR TCP/IP address). The TCP/IP address is the one of the current FDBR through which a fixed client can invoke mobile object instances on the terminal. Thus, the LR permits a per terminal registration. If object instances on mobile terminal would be migrable, a per object registration would be required in the LR. In this case the LR should have a greater storage capacity, and should be able to process more frequent invocations than for a per terminal registration.

The LR offers an IDL interface that has at least the basic operations to store and remove couples of (GTID; FDBR address) and to get an FDBR address for a given GTID. The LR should be distributed, and tailored for frequent updates and read accesses. The Global Naming Service (see Section 6.3) and each FDBR will have to acquire the object reference of the

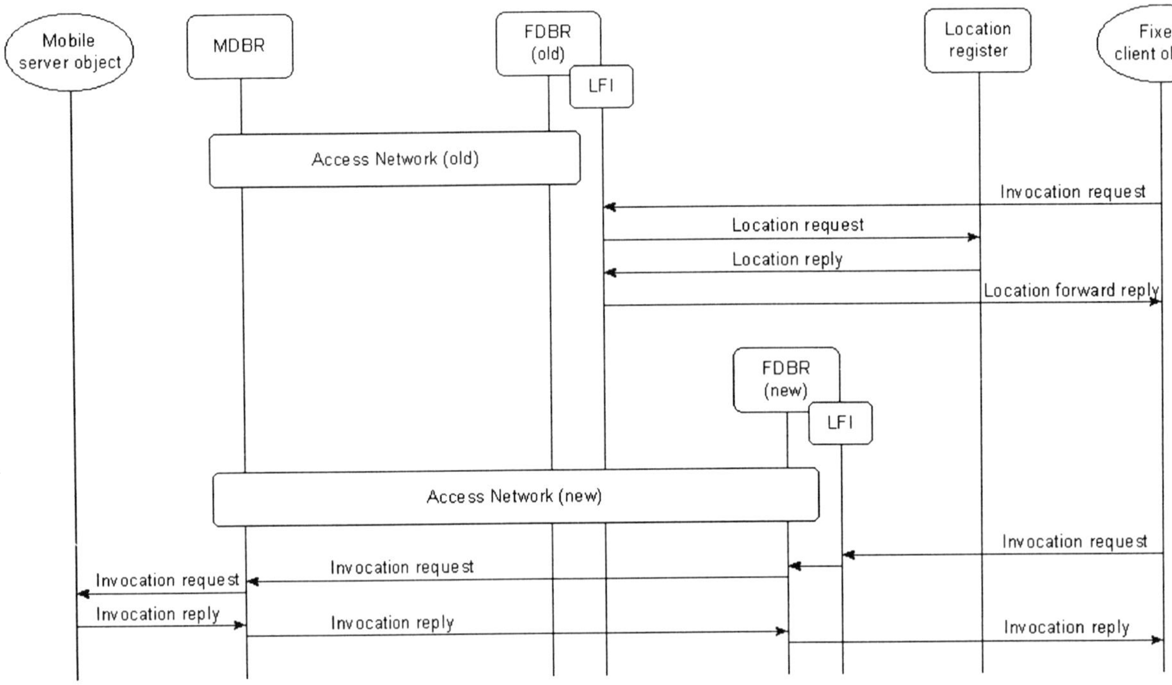

Figure 3: Message Sequence Chart of Reference Update

LR interface in their configuration step. This reference should be modified as rarely as possible.

In world-wide implementations the LR must be distributed. Most probably there will be one LR per access provider domain. This also affects the contents of LR. It should also be noted that in the home domain of the user there must be a mapping from a permanent UserId to the current access provider domain so that initial access initiated from the core network would be possible.

On the DPE level there are two cases of address resolution that need to be taken into account. Mobile-originated object invocation is the simple case: An object on the mobile terminal invokes an object within the fixed kTN network. Since the invoked object is behind the wireless access network, the invocation request is intercepted by MDBR and forwarded to FDBR. Since the FDBR is part of the core network domain, it can invoke the operation directly on the invoked object.

The mobile-terminated object invocation is the hard one of the two cases. As usual, in order to invoke an operation on an object, the invoking object must first determine the object reference of the target object. We require that the object reference of an object located on a mobile terminal includes a separate terminal identity (the GTID) and address. Therefore, the address, when resolved, enables the complete route from the fixed client to the object to be determined. In particular, the address is composed of the local address of the object in the mobile terminal and of the address of the FDBR in the fixed network which is currently bridging the mobile terminal.

As the FDBR address in the reference may become out of date after a handover, FDBRs can receive invocations for objects hosted on mobile terminals that they are no more bridging. In order to enable the invocations also after handover, the FDBR includes a functionality called *Location Forward Indicator* (LFI), which helps the DPE to offer relocation transparency. The LFI makes full use of the LOCATION_FORWARD mechanism offered by CORBA in GIOP. The principle is explained below.

When a fixed client invokes a mobile object through an old FDBR which is no more bridging the mobile terminal, the LFI in the old FDBR replies to the client proxy with a LOCATION_FORWARD status and with the up-to-date reference of the target object. This reference contains the address of the new FDBR which is currently bridging the mobile terminal. At reception of a Request, the LFI has to (an example is shown in Figure 3):

1. Extract the GTID from the object key, which is transported in the header of the request. The object key is an element of an object reference.
2. Verify if the FDBR is currently bridging the mobile terminal which hosts the target object, using the GTID.
3. If yes, the LFI allows the FDBR to transmit the request towards the mobile terminal.
4. If not, the LFI interrogates the LR (Location Register), by providing it the GTID to get the address

of the new FDBR, in the fixed network, which is currently bridging the mobile terminal.

5. The LFI rebuilds a new object reference, which includes the address of the new FDBR.

6. The LFI replies to the client with a LOCATION_FORWARD status and the new reference.

7. The client replaces its old reference by the new one, and reissues the request through the new FDBR. Since all this is performed by the client proxy, it is transparent for the client application.

The only constraint of this method is that FDBRs should always remain alive to be able to answer to clients which have old references. This constraint is acceptable since FDBRs are permanently active.

6.3 Global Naming Service

The need to have a global naming service on the DPE has been expressed, in order to name all the object instances in a centralised way. The scope of a global name is unlimited on the whole DPE, which is deployed over mobile and fixed network environment. A given global name is understandable everywhere on the DPE, without any translation which could depend on a particular context. Global naming is one of the enhancements brought to the CORBA platform to constitute the DPE.

The DPE is required to support distributed objects belonging to different stakeholders, and to hide the heterogeneity of the network. A stakeholder (e.g. a service provider or a retailer) is here an organisation who deploys objects over the DPE, and expresses the need that they can be accessed from everywhere on the DPE by using a name of a unique format (i.e. a global name). Thus, a stakeholder will publish the name of one of its objects to other stakeholders. To achieve this, the DPE should provide an internal support service to name the objects, called *Global Naming Service* (GNS). This service is defined here using concepts defined in RM-ODP and TINA, such as domain, policy etc.

The names are organised hierarchically in a tree. To be global, the tree must be unique. Each node or leaf in the tree has a simple name, except for the root which has no name. Each node represents a naming domain, which applies its own naming policy, thanks to a naming domain manager object. Each node is bound to a naming domain manager object reference. Each leaf is bound to any object reference.

Each node of the first level in the tree represents a stakeholder. Each stakeholder organises its domain (or sub-tree) as it wants, provided that it can resolve a name (i.e. return the bound object reference) on behalf of a foreign requester. A given policy can permit, for example, to give several names to a same object interface.

The root is an authority which has its own naming policy which must be accepted by all the stakeholders, and applies only at the first level. Whom belongs the root GNS is an open issue. The root naming policy is defined as follows:

- a stakeholder which wants to create a new naming domain, with a given name, officially requests the root GNS to accept the name and the creation of the domain (at the first level of the tree). In exchange, the stakeholder must provide the location of its own naming domain manager, and must officially notify the GNS of any change in location.

- a same stakeholder naming domain cannot have several names (for reason of simplicity).

- a new stakeholder domain name is unambiguous, and has never been used in the past.

- a stakeholder can request to change the name of one of its naming domains.

- a stakeholder can remove a naming domain.

- the subordinates of the root GNS are uniquely stakeholder naming domains.

Every stakeholder takes care of its part of the GNS in the stakeholder domain of its own. Other stakeholders have access to a particular stakeholder naming domain if they have a federation agreement with that stakeholder. Access can be provided by explicit replication of naming data or by referencing to the naming domain of other stakeholders. The GNS root will be quite stable. Therefore, it can be replicated into naming domain of each stakeholder.

Regarding mobility, the GNS should to deliver up-to-date references for mobile objects, although the system tolerates the use of out of date ones. Several alternatives have been identified for the implementation of the GNS.

Alternatives using a dedicated naming sub-tree for names of mobiles objects have been eliminated. They constrain users to know if an object is mobile or not to name it, since the name must be stored in the dedicated naming sub-tree, in the GNS. In the other alternatives, names of mobile and fixed objects can be mixed in the same tree. Three sub-alternatives have also been identified so that the GNS delivers up-to-date references:

1. The GNS checks if the object reference is up-to-date or out of date before delivering it to the client, by sending a LocateRequest message towards the referenced object. This solution has the drawback that when the reference is up-to-date, the LocateRequest message would be forwarded to the mobile terminal, and thus would increase traffic on the wireless link.

2. The GNS is notified by the LR each time the mobile terminal becomes attached to a new FDBR. This alternative has the drawback that the GNS would have to search in the whole naming tree all the occurrences of object references which contain the identifier (GTID) of the concerned mobile terminal. The GNS would also have to update them.

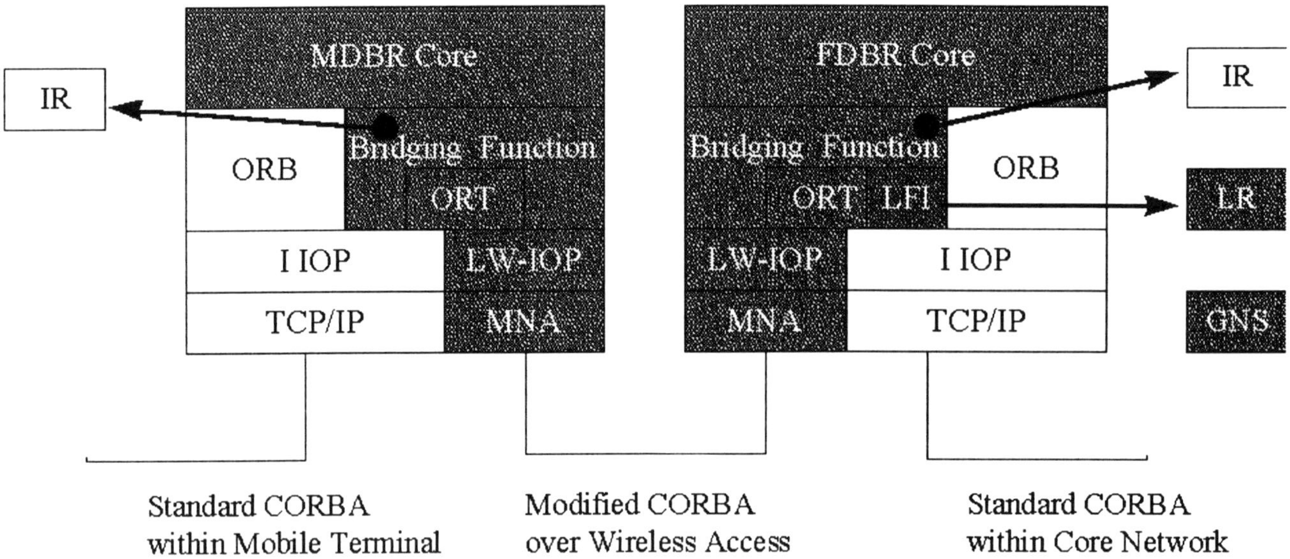

Figure 4: Software Modules in the DOLMEN Mobility Bridges

3. The GNS checks the validity of an object reference only if the stored reference is too old (expiration timeout). In that case GNS invokes the LR in order to get the TCP/IP address of the current primary FDBR. If necessary, the new TCP/IP FDBR address is set in the object reference. The latter is then registered again in the naming tree and delivered to the client.

The IDL interface of the CORBA Naming Service can be used for the GNS.

7 Implementation Experiences

We have implemented the MDBR and FDBR bridges using two commercial CORBA implementations as the basis: COOL ORB r4.1 [1] on the mobile terminal and Orbix 2.2 [3] within the core network. Figure 4 depicts the major software modules in our implementation.

Our initial plan was to implement a request-level bridge as described by the CORBA 2.0 standard. In practise, we encountered difficulties in using the Dynamic Skeleton Interface (DSI) for general purpose bridging. The CORBA standard seems more focused on application-specific bridging, where the semantics of the bridged interfaces are known to the bridge, rather than general purpose bridging. For instance, a separate dynamic skeleton must be registered for each different interface type, causing the ORB to create a proxy object for that interface type. The DSI language mapping also does not provide the means to iterate through the message contents or to manipulate the message contents in a machine independent format. As a result, a general purpose request-level bridge must contain a large amount of platform dependent code and is therefore not portable across different platforms.

Since the option of implementing an inline bridge was not available with commercial CORBA platforms, we decided to implement MDBR and FDBR as gateways that perform the tasks of a CORBA bridge on the Inter ORB Protocol level. Despite the additional burden of writing an IIOP protocol engine, this turned out to be a fortuitous choice. Instead of creating a proxy object in the bridge for each bridged interface, we were able to arrange routing on a message by message basis, only keeping track of the current location of each mobile terminal but not specific objects. The absence of proxy objects simplified the design of a bridge handover considerably, as we did not have to worry about how to rebuild the proxy objects in the new FDBR after a handover. In addition, a large part of the bridge implementation was common to both MDBR and FDBR. The other considered approaches would have involved less reuse of program code.

Our experiences in implementing a generic bridge lead us to believe that the CORBA 2.0 interoperability standard is not yet mature enough for building general purpose bridges. A generic bridge is required to modify object references that are passed as invocation parameters. This forces the bridge to unmarshal all invocation requests and responses, translate any references, and then remarshal the messages. An external interface repository is required to store the necessary knowledge about each interface. All this amounts to a great amount of overhead that limits the scalability of generic bridges. It is considered that extensions to the IIOP protocol are needed for general purpose bridges to support object reference translation without forcing a bridge to unmarshal and remarshal each message completely. Efficiency is especially important with multi-domain bridging, where a single invocation request might travel through multiple bridges before reaching the server object.

When performing a protocol conversion between IIOP and an environment-specific inter-ORB protocol, access to low-level data representation encoding and decoding operations is essential. These routines exist in some form or other in every ORB implementation, but currently the CORBA standard defines no API that could be used to access these routines.

Finally, the dynamic skeleton interface does not provide access to all features of the underlying inter-ORB protocols. For instance, it is not possible to return a LOCATION_FORWARD status through the DSI to indicate the migration of a dynamic object implementation. The DSI language mapping should also be extended to define a platform independent representation of invocation parameters. The current representation is highly dependent on the hardware platform, operating system, and even on differences between compilers of the same language.

8 Summary

In this paper we have considered CORBA based object communication in the context of mobile networks and terminal mobility. We have shown that it is possible utilise the interoperability mechanisms introduced in the CORBA 2.0 specification to support continuous terminal mobility in object communication. The object interaction model is not affected as the presented mechanism is completely transparent to client and server objects.

Our solution is based on the concept of *interoperability bridges* described in the CORBA 2.0 architecture. We have designed and built a prototype of two such bridges. The *Fixed DPE Bridge* (FDBR) serves as a DPE access point for mobile terminals. The *Mobile DPE Bridge* (MDBR) connects the local ORB domain of a mobile terminal to the core network ORB domain by interacting with an FDBR over the wireless access network. Together, MDBR and FDBR perform location management functions and DPE handovers, enabling terminal mobility on the DPE level.

We have defined a relocatable object reference format that together with the bridges and a special *Location Register* allows us to support relocatable object referencing domains. The location register maintains the current mapping between a relocatable referencing domain and the bridge currently in charge of delivering object invocations to and from that domain. Bridges update this mapping when a change of location is detected. We have shown that, the mobile terminal being an example of a relocatable referencing domain, this innovative technique can easily be applied to reliable object addressing of mobile objects in a way that is completely transparent to the communicating objects themselves.

We have also enhanced the reliability and performance of object communication in the wireless environment by introducing a *Light-Weight Inter-ORB Protocol* (LW-IOP) between the MDBR and FDBR. The LW-IOP protocol defines efficient message formats and a compressed data representation for object communication. In our experience, LW-IOP significantly reduces the bandwidth consumed over the constrained wireless medium.

Acknowledgements

This paper is based on the work carried out in the EC/ACTS project DOLMEN (AC036) that is a TINA Auxiliary Project. The authors want to thank all their co-workers in the project for fruitful co-operation.

References

[1] Cool ORB. http://www.chorus.com/Products/Cool/, 1997.

[2] EC/ACTS Project DOLMEN (AC036). *Definition of an Enhanced Distributed Processing Platform for DOLMEN*, Public Deliverable MPD2, December 1996.

[3] Orbix. http://www.iona.com/Products/Orbix/, 1997.

[4] Object Management Group, New York. *The Common Object Request Broker: Architecture and Specification*, Updated Version 2.0, July 1996.

[5] S. Trigila. An architecture for integration of fixed and mobile services: The DOLMEN approach. In *ACTS Workshop on Mobile Multimedia Services*. Signapore, September 1997.

Session 3

Experience Reports I

Richard Soley, OMG

A TINA Trial
Interworking Experience with the Legacy Telephone System

Tim Eckardt
DTAG/GMD-FOKUS,
Kaiserin-Augusta-Allee 31
D-10589 Berlin, Germany
Eckardt@fokus.gmd.de

Patrick Kielhöfer
DTAG/GMD-FOKUS,
Kaiserin-Augusta-Allee 31
D-10589 Berlin, Germany
Kielhoefer@fokus.gmd.de

Didier Guy
France Telecom CNET, Technopole
Anticipa, 22307 Lannion Cedex, France
Didier.Guy@lannion.cnet.fr

Victor Pérébaskine
France Telecom CNET, Technopole
Anticipa, 22307 Lannion Cedex, France
Victor.Perebaskine@lannion.cnet.fr

Abdul Akram
Sprint, 9300 Metcalf Avenue
Overland Park, KS 66212, USA
Abdul.Akram@mail.sprint.com

Abstract

This paper describes a large scale experimental implementation of multi-operator services on a multi-retailer TINA platform. Of the set of the services that are being developed, the call completion service is particularly high-lighted, showing a possible application of the TINA concepts to legacy telephone systems.

1 Introduction

The TINA consortium (TINA-C) was founded in 1993. Its goal is to specify, before the end of 1997, an overall architecture for the next generation of telecommunication networks. The main characteristics of the resulting overall architecture are as follows :

- the proposed architecture relies on ODP reference model [1], normalized at ITU-T,
- CORBA specifications [2] (and therefore CORBA compliant products) have been adopted as a possible instance of the DPE infrastructure [3],
- a service architecture [4] has been defined in order to provide a common framework for service construction and deployment, and to abstract away the exact nature of the underlying network infrastructure,
- a business model has been developed that identifies reference points (e.g. Ret-RP, RtR-RP, 3Pty-RP) between the software systems operated by the different actors of the telecommunication market (e.g. clients, retailers, third party service providers).

Various projects, such as EURESCOM P.508 [5], have already studied the TINA architecture along with the possible migration path from existing network and service architectures, but for most of them no experimentation was made at the scale of the targeted networks.

Today, even if there will be no industrial TINA products by the end of 1997, it is clear that the success of TINA strongly depends on the development of TINA products by various participating industrial parties. Currently different projects try and provide industrial prototypes of either the TINA DPE or service architecture. The ReTINA project [6] aims at building a flexible ORB using a binding approach [7]. The basic idea of the EURESCOM P.715 project [8] is to build an experimental TINA based environment with the view to study the construction and the deployment of telecommunication services at a pan European scale.

0-8186-8335-X/98 $10.00 © 1998 IEEE

2 The DTAG, FT, Sprint and Global One TINA Trial objectives

The TINA Trial (TTT) gathers a set of projects (currently three) which aim at a proof of concept for the TINA architectural concepts. In this context the TTT project[1] leaded by DTAG, FT, Sprint and Global One focuses on TINA Service Architecture [4] and reference points [9] (c.f. Figure 1). Its purpose is mainly twofold: specify, design and experiment an open and programmable service architecture for public or private network and residential terminals (SOHO); validate the TINA reference points (at least the Ret-RP between consumer and retailer, and the RtR-RP between retailers). However, this project will take advantage of the emerging switch technology (programmable switches through CTI interface) to extend the scope of the TINA service architecture to corporate networks and introduce a new business case: flexible and programmable services for corporate networks. This project also intends to provide the TINA consortium with feedback concerning the specification of reference points.

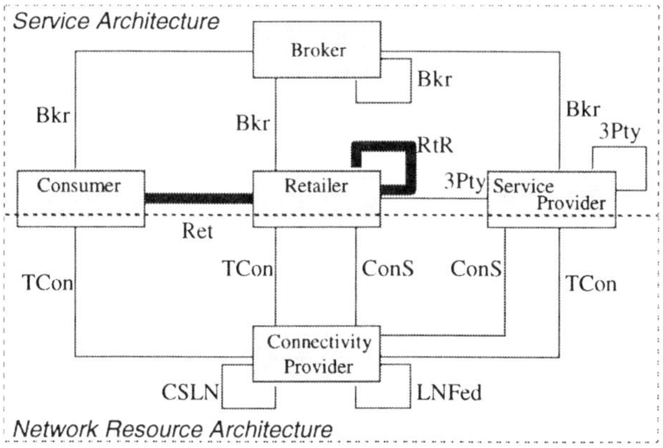

Figure 1. TINA business model and reference points.

At first stage, this project will use the existing CORBA technologies in order to deploy TINA computational objects at a worldwide scale, on top of retailer specific platforms (c.f. Figure 2). When available, this project may also be influenced by more enhanced - and more suited to telecommunication constraints - TINA middleware technologies, such as ReTINA DPE.

[1] Please note, in the sequel, when mentioning TTT, the TTT of DTAG, France Télécom, Sprint and Global One is meant.

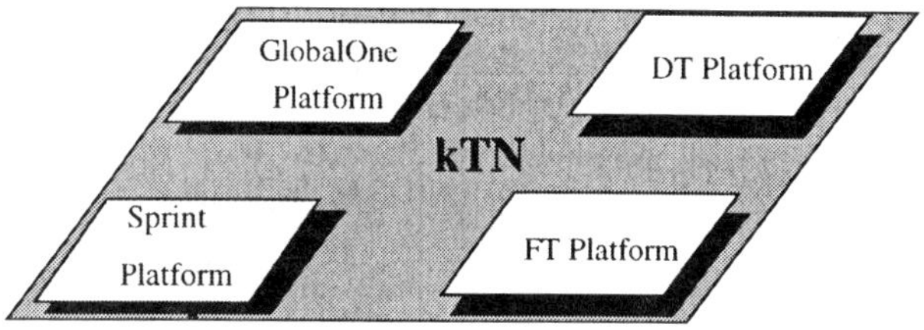

Figure 2. The TTT platform.

So as to fulfill its goals, the project itself has been divided into three experimentation scenarios, based on three different services. All of those scenarios will run on top of the same infrastructure and will share a set of computational objects.

The first service deals with call completion and user mobility. The second one will provide a uniform access to customer services: the objective of this service is to provide a customer with the possibility to access the services it has registered to from anywhere in the world operated by Global One for the price of a local call (or at least for the price of a call to the closest point of presence). No assumption is made about the customer's terminal software (different customers from different retailers may use different software). Furthermore, the visited retailer ought to be able to interact with the customer's home retailer in order to carry out the demanded service. The third service which is to be demonstrated will provide a set of group communication services including video conference and joint document edition facilities.

3 General Experimentation Description

This section, along with the rest of this paper, focuses on the first demonstration scenario, namely *call completion with user mobility*. The rest of this section overviews the intended service capabilities, including a brief service scenario and the major functional blocks required.

3.1 Demonstration Service Perspective

The objective of this service is to make sure that all calls to a customer can be completed, wherever this customer is (within the scope of a global service provider like Global One). That is, a roaming corporate user can be reached via her personal number (directly addressing a user instead of a terminal) at any location, at any phone - provided this user has previously registered to a terminal/phone. With the help of a user profile management application, the user of this service has also the ability to define a personal schedule which enables an automatic schedule-based registration. Lastly, the user can monitor and control the terminating call handling by choosing some simple rules like "don't call me after 20h00", "if it's my wife, reach me any time".

At network level, calls will be handled using programmable switches or PBXs. Those network elements will request the support of TINA system-related capabilities for personal mobility-enhanced call completion.

For the service demonstration experiment, the TINA systems at each partners site interact via a common kernel Transport Network (kTN) for global provision of call completion with personal mobility. Here, the kTN acts as a kind of signaling network allowing the different TINA systems to interwork as needed to support roaming corporate customers of an International Virtual Private Network (IVPN) service.

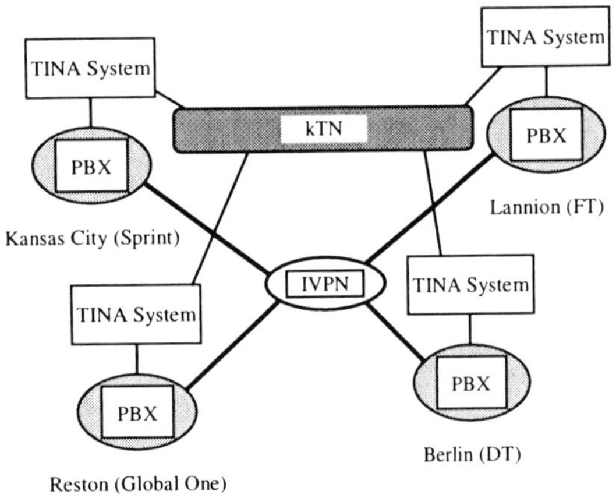

Figure 3. Global installation for the demonstration service.

As delineated in Figure 3, each site (i.e., Kansas, Paris, Berlin, Reston, etc.) consists of:

- a PBX or programmable switch that is connected to all PBXs/Switches at the partner sites either *via* an IVPN or *via* the public network,
- a TINA System comprising Access Session components (*Initial Agent* and *named User Agents*) and User Profile Management UAPs dedicated to the registration of the user at terminals,
- a link between the TINA systems and the legacy system world (ISDN IVPN) allowing the TINA System to control the PBX/Switch for a proper call completion,
- access to a kTN interconnecting all the TINA systems.

3.2 Demonstration Scenario

The demonstration scenario would consist, for instance, of a roaming user/customer who manually registers to one of the terminals in the demo room. There are three ways of registering a user at a terminal (i.e., phone):

(i) registration using a dedicated registration User Application (UAP) residing on a TINA end-user system.

(ii) registration at terminals by dialing a certain number at the terminal to be associated with the user (corresponding to the UPT Access Code (UPTAC) defined in [10]).

(iii) registration at terminals by TINA system-initiated activities (e.g., registration by schedule comprised within the namedUA).

A call within the demo scenario can be initiated in a number of different ways, including by dialing the personal corporate number (a sort of UPT scenario) of the mobile user, or by connecting a PC to the local or remote TINA system, browsing the corporate directory and selecting the icon representing the callee.

3.3 Demonstration Service Functional Blocks

Each site consists of the following functional blocks: (i) the PBX/Switch, (ii) the TINA System which handle access sessions and profile management, and (iii) an Interworking Unit (IWU) providing the "glue" between the TINA world and the legacy world of ISDN PBXs/Switches (cf. Figure 4).

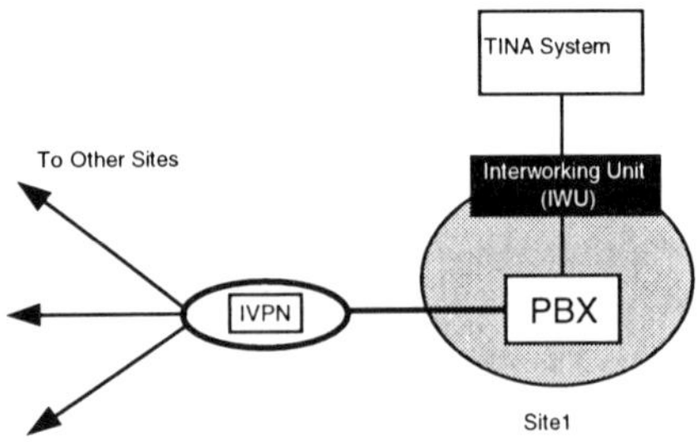

Figure 4. Major Functional Blocks at each Site.

3.4 TINA System

The TINA system consists of reusable components related to (i) TINA Access Session and (ii) user profile management (e.g. a registration management application allowing the user to register at terminals or locations via a PC- or workstation-based GUI).

3.4.1 Access Session Components

The Access Session Components required are Initial Agents (IA), used to request the set up of an Access Session with the provider domain, and Named User Agents (namedUAs). A namedUA represents a single user within the provider's domain. It manages a user's preferences on service access and usage, thereby providing support for personal mobility. The namedUA considers resource configuration information of the users system including knowledge on terminals and their network access points.

The namedUA within the TINA system is used to handle calls/invitations sent to the called party in order to perform a call completion even if the called user is not at her home office, currently not available, or busy. An invitation operation requested at the namedUA of the invited/called party will usually (if no exception is raised) return with indications for:

a) a refusal to accept the invitation/call (i.e., call blocking),
b) a forwarding of the invitation to a different, alternative recipient (i.e., call forwarding),
c) or (in the usual case of accepting an invitation/call) in the selection of the terminal where the user has registered.

Note, that for the call completion step of the demonstration scenario, there is no access session established. The caller is not necessarily a user of a TINA system (the call completion service is offered to the potential callees). Should the caller be a user of a TINA system, an adaptation would be needed in order to be able to set up an access session from a POTS telephone (e.g. interactive voice system). This functionality has been left for further developments.

3.4.2 User profile management

User profile management (UPM) is a key service component of advanced telecommunication services which provides the customer with individual control over the behavior of the subscribed service(s). The TINA-C concepts of User Applications (UAPs) running on intelligent end-user systems and the concepts of *ancillary services* (i.e., 'helper' services that provide management capabilities supporting the subscribed telecommunication or information services) allow to realize sophisticated user profile management applications and services. In this regard, particularly the Worldwide Web (WWW) and Java represent the technology of choice when trying to implement flexible and platform independent UPM capabilities. Java applets (comprising CORBA objects as well as an ORB) communicate via IIOP with a remote TINA-C access node of the retailer, which in turn takes care of the modifications within the TINA-C provider domain (e.g., user profile).

Currently, at least two user profile management UAPs are needed for this demonstration service: a user registration management application and a reachability management application. The user registration management application allows the end-user to register at terminals or at locations (i.e., more general targets) for incoming invitations. Subsequently, invitations send to the registering user's UA will be delivered at the registered terminal (or at any terminal within a registered location). A complementary functionality, that is *de-register*, is needed to cancel any type of registration. A future option could be to allow a kind of pre-scheduled registration on the basis of the entries in the end-user's preferred calendar application.

The second type of user profile management UAP is a kind of reachability management application. This application allows the end-user to set up an individual invitation handling policy (i.e. preferences for a personalized invitation handling) kept inside the namedUA service component. When receiving an invitation request from an SSM, the UA within the provider domain processes that invitation request according to the preferences for a personalized invitation handling specified by the end-user. The resulting behavior of the UA when receiving an invitation request could be either a selective blocking, a forwarding to a different UA or an explicit delivery of the invitation to the registered terminal/location. The application of personalized invitation handling concepts has been studied in the TINA-C auxiliary project "PCS in TINA" [11]. The reachability management UAP should allow the user to examine and modify his/her individual invitation handling policy (i.e., to set up rules such as "all calls on Nov. 22 1997 to be forwarded to Ms. Moneypenny").

3.4.3 Interworking Unit (IWU)

The Interworking Unit (IWU) is a *mediator* between the TINA system and the PBX/switch (the latter representing the ISDN/IVPN legacy system). Legacy system service sessions (i.e., calls) as well as legacy terminals are adapted by the IWU to TINA-compliant behavior as expected at TINA reference points.

From the perspective of the PBX/switch, the adaptation function implemented by the IWU acts as a telephony server that manages the monitoring of incoming calls directed to personal numbers. For this sake, it accepts events issued from the PBX/Switch and sends commands to the PBX/Switch in order to control routing, connection set up, and call completion.

From the perspective of the TINA world, the IWU acts as an end-user system[2]. Actually, this end-user system is associated to the invited party (the party associated with the personal number the call is directed to). As the inviting party belongs to a legacy system, it cannot be associated to any TINA conformant system. Thus, the inviting party cannot be associated to a PA, which means that an adaptation is needed in order to step into the TINA system. That is, a call to a personal number triggers an "invitation-request" issued by the IWU and sent to the namedUA of the callee. The response from the namedUA indicates to the IWU how to proceed with call completion.

3.4.4 Discussion

Functionally, what is needed in order to implement such a service, is twofold: a detection point for incoming calls to personal numbers, and a PBX/Switch control point, in order to carry out call completion.

These two functions have already been implemented by public network operators, based on IN architecture (implemented as IN Service Nodes). This paper describes an alternative solution based on TINA service architecture, namely TINA Service Nodes.

One of the advantages the TINA approach related to the call completion service has over traditional IN architecture is that it allows the distribution of the service logic, based on CORBA technology, should this technology show enough reliability and performance so as to fulfill telecommunication needs.

Furthermore, we expect to take advantage of the TINA service architecture concepts to distribute the desired control over several network elements (multiple points of control). This evolution should allow us to control two or three PBXs/Switches possibly involved in the completion of a single call, in order to optimize network resources (this topic is left for further studies).

[2] Part of the IWU (i.e., the SSM and SF, cf. section 4) behaves like a different retailer domain sending invitations to UAs in the TINA system. This aspect will be elaborated when RtR Reference Point specifications become available.

At last, the use of standardized reference points should allow the interconnection of different TINA systems, from several stakeholders (in our context, DT, FT and Sprint), in order to deliver a service at a pan-alliance scale.

4 IWU Design

This section is intended to provide a short glance at the IWU design.

The Figure 5 shows the target software solution, including the three basic components: PBX, IWU, and TINA system. Only the interfaces and interactions between the IWU component and the TINA System component are subject to specification. The internal structure of the IWU component is not mandatory and rather exemplary or descriptive. It is given to clarify the behavior of the Provider Agent (PA) interfaces, the PA Factory (PAF) interfaces, the Service Session Manager (SSM) interfaces, or the Service Factory (SF) interfaces.

4.1 Registration

4.1.1 Functional description

Registration means association between the personal number of a user and a specific terminal (i.e. a phone line or set). Any call directed to the personal number mentioned above can be completed to the registered terminal.

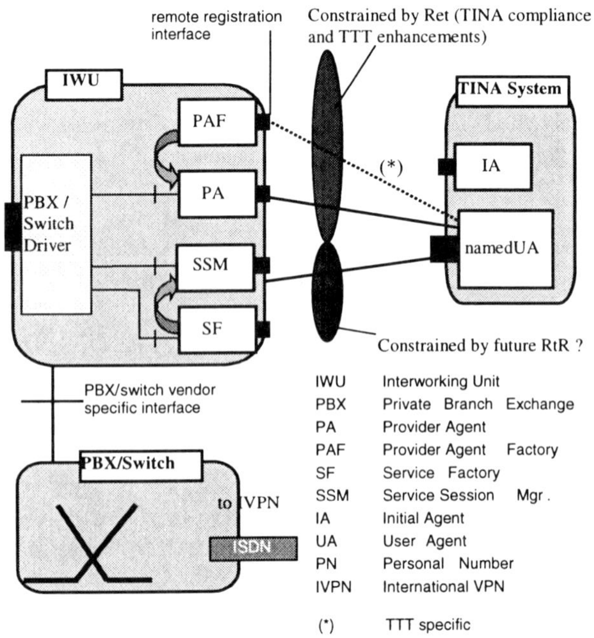

Figure 5. IWU design.

In the IWU, a successful registration will result in the creation of a new PA computational object, dedicated to this user's registration. This PA holds the information about the registered terminal (e.g., a E164 phone number). A reference to this object is stored by the namedUA of the user, for call completion purpose.

4.1.2 Message/event sequence

The figure below depicts a registration procedure, issued from a TINA enhanced terminal.

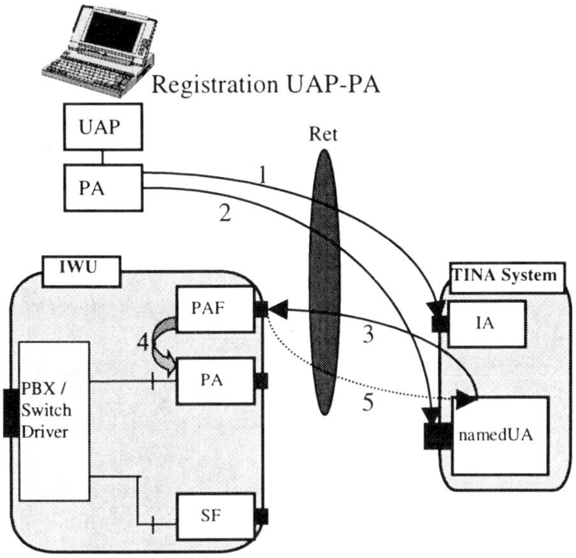

Figure 6. Registration sequence.

[Step 1] The user establishes an access session to the TINA system, using a 'registration UAP-PA' tuple, according to TINA specifications. As an output of this step, the registration-PA holds a reference to the namedUA of the user.

[Step 2] The user can specify the terminal he/she wants to be registered at using a dedicated GUI, which will invoke a 'register' operation, *via* the registration-PA, on the namedUA, passing the registered terminal data as a parameter (typically, a E164 phone number).

[Step 3] In turn, the namedUA will invoke a 'register' operation on the PAF of the IWU, passing the registered terminal data as a parameter.

[Step 4-5] The PAF creates a new PA, dedicated to the terminal the user is registered at. The new PA will store data about the terminal, in order to be able to complete a call to this terminal (typically, a E164 phone number). The new PA reference is then passed back to the namedUA, as a result of step 3 invocation. This reference is stored by the namedUA for later use.

Note that a user can de-register from a terminal using the same UAP-PA tuple.

4.2 Call completion

For a clear understanding of the design of the IWU, we will consider a simple case, were a single system is used: this means that no interaction will occur between the systems of the different retailers potentially involved in the service.

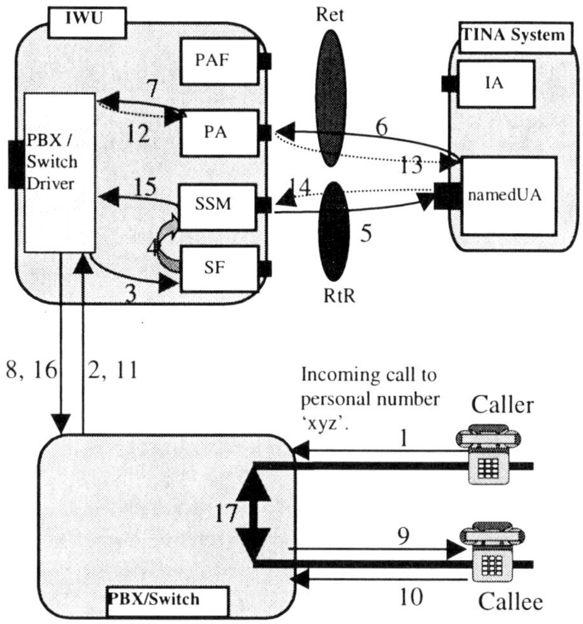

Figure 7. Call completion sequence.

[Steps 1-2] The caller dials 'xyz', which is the personal number of a user who has subscribed to the call completion service. This call is routed by the network to the PBX/Switch. The PBX/Switch, which has been configured to monitor the specific numbers associated to personal numbers, places the termination of the call under the control of the IWU.

[Steps 3-4] The (PBX/Switch) driver invokes from the Service Factory of the call completion service the creation of the Service Session Manager associated with the incoming call, passing all the data needed to complete the call to it. The SF then creates the SSM.

[Step 5] After having invoked a TTT specific operation on the IA in order to find the namedUA of the user whose personal number is 'xyz', the SSM invokes an 'inviteUser' on it (as defined in our first specification of RtR reference point), indicating that the invitation is for the call completion service.

[Step 6] Relying on a specific call management logic, as explained in the beginning of this paper, the namedUA finds the reference of a PA (in the IWU) representing a user's registration, and invokes on it an 'inviteUserOutsideAccessSession' request, as specified by Ret.

[Steps 7-8-9] The PA requests from the driver the set up of a call to the registered terminal it is associated with. The driver forwards this request to the PBX/Switch, and a call to the terminal is set up (the phone is ringing!).

[Steps 10-11-12] When the callee picks up the phone, this event is forwarded to the driver, which gives a 'success' answer to request number 7, indicating that the outgoing call is correctly established.

[Step 13] The PA answers 'success' to the invitation (request number 6), indicating that the invitation has been accepted by the invited party (the callee).

[Step 14] The namedUA answers 'success' to the invitation (request number 5), indicating that the invitation has been accepted by the called party.

[Steps 15-16-17] The SSM request from the driver the 'join' between the incoming and outgoing calls. This request is then forwarded to the PBX/Switch, and lastly the call is completed. In the current implementation, no explicit 'joinSessionWithInvitation' request will be made.

4.3 Discussion

4.3.1 Reusability

A major goal of the design of the IWU has been to allow a TINA compliant namedUA component to provide personal mobility to a legacy service (IVPN via ISDN) as well as (by the same instance of namedUA!) to any type of TINA-compliant multiparty telecommunications service. Therefore, the IWU interacts only via a subset of Ret reference point operations (as well as RTR reference point operations when specified) with the TINA system. The customer of this demonstration service could be a consumer of the enhanced IVP service and at the same time a consumer of pure TINA telecommunications services. Despite the fact that both services were designed and operated completely independent from one another, both could interact with the same instance of namedUA representing a certain corporate customer for instance.

4.3.2 Adaptation role

Again, it is important to note that PA, SSM and SF are only *adapters* providing TINA interfaces towards the TINA System but internally interact within the IWU in a way not defined by the TINA SA.

The PAF and PA interfaces of the IWU (i.e., the `i_ConsumerRegister` interface and the `i_ConsumerInitial` interface) provide the TINA system with a TINA-compliant view on non-TINA compliant terminals (i.e., POTS/ISDN phones). The interaction between these interfaces on one hand and the IA and UAs on the other hand are constrained by the TINA Ret reference point (or a TTT specialization of it).

The SF and SSM interfaces of the IWU provide the TINA system with a view of a service session issuing invitation requests. Note that the interactions between these interfaces on one hand and the UAs of the TINA system on the other hand might be constrained by a future TINA RtR reference point (or a TTT specialization of it). When RtR reference point specifications become available, the SF and SSM interfaces of the IWU can be re-designed according to this inter-domain reference point allowing the IWU to invite users of multiple different TINA retailers. Thus, this part of the IWU would behave (from the TINA perspective) like another retailer interacting with TINA system via the RtR reference point.

Note that since the caller is not necessarily a user of the TINA system, it is impossible to set up any access session before starting the call completion service, as it should be the case according to TINA specifications. However, if the caller could be identified as a TINA system user, a more TINA compliant behavior could be implemented.

4.3.3 Communication Session Management

In this demo service experiment, no TINA-compliant CSM has been specified by the partners. If needed (for example to provide third party call control to TINA Service Session Management - SSM - components) this functionality will be added. A scenario requiring this type of third party control could be to enable a user to connect a PC to a TINA system, browse a directory and initiate a call by selecting an icon that represents the callee.

5 Conclusion

In this paper, we report experience from a service prototype which provides personal mobility on legacy service (IVPN via ISDN). This prototype has been designed in order to be compliant with TINA Service Architecture, and the TINA Ret reference point. In addition, this work also considers interactions between two retailers through the RtR reference point.

Although the Ret reference point had to be extended (i.e. additional data types, operations and interfaces), the TINA Service Architecture was left unchanged: Neither an existing computational object of the Service Architecture had to be modified, nor a specialized to be added in order to provide the here described service. The TINA compliance had been achieved by adapting the objects of the IWU (i.e. the PAF, PA, SF and SF), in order for them to look and act like TINA-objects from the TINA point of view.

Comparing to the traditional IN architecture, the TINA approach has the advantage to allow the distribution of the service logic (based on the CORBA technology) related to the call completion service. Another advantage is the possibility to distribute the desired control over several network elements, thus having multiple points of control.

This project also takes advantage of the emerging switch technology (programmable switches through CTI or other interface) to demonstrate that the scope of the TINA service architecture extends to corporate networks (as well as to public network service nodes), and introduces a new business case: flexible and programmable services for corporate networks.

6 References

[1] Object Management Group, *The Common Object Request Broker Architecture: Architecture and Specifications*, Revision 2.0. OMG , July 1995.

[2] ITU-T I ISO/IEC Recommendation X.903 I International Standard 10746-3, *ODP Reference Model: Architecture*. January 1995.

[3] TINA-C, *DPE Architecture*. December 1994. http://www.tinac.com/deliverable/deliverable.htm .

[4] TINA-C, *Service Architecture 5.0*. June 1997. http://www.tinac.com/deliverable/deliverable.htm .

[5] EURESCOM P508, PIR 8.2, *Migration Path toward TINA (Refinement)*. To be published.

[6] The ReTINA Project. http ://www.chorus.com/Documentation/retina.html

[7] F. Dangtran, V. Pérébaskine, J.-B. Stefani, B. Crawford, A. Kramer, D. Otway. *Binding and Streams, the ReTINA approach*. Proceedings of the TINA 96 Conference. September 1996.

[8] EURESCOM P715. http://www.eurescom.de/public/projects/all/pr715.htm .

[9] TINA-C, *TINA Business Model and Reference Points 4.0*. May 1997. http://www.tinac.com/deliverable/deliverable.htm .

[10] ITU-T Recommendation F.851, *Operations And Quality Of Service Universal Personal Telecommunication - Universal Personal Telecommunication (UPT) – Service Description (Service Set 1)*,. February 1995

[11] T. Eckardt et al., *Personal Communications Support in the TINA Service Architecture - A new TINA-C Auxiliary Project*, TINA '96 Conference, September 3 -5, 1996, Heidelberg, Germany

7 Acronyms

CO	Computational Object
CORBA	Common Object Request Broker Architecture
DPE	Distributed Processing Environment
GUI	Graphical User Interface
IA	Initial Agent
IVPN	International Virtual Private Network
IWU	Inter Working Unit
kTN	kernel Transport Network
ODP	Open Distributed Processing
PA	Provider Agent
PAF	Provider Agent Factory
PBX	Private Branch Exchange
POTS	Plain Old Telephony Service
SA	Service Architecture
SF	Service Factory
SSM	Service Session Manager
TTT	The TINA Trial
UA	User Agent
UAP	User Application
UPT	Universal Personal Telecommunication
UPTAC	UPT Access Code

Service Creation on a TINA Platform: an Experience Report

Lorenz Lehmann
Swisscom
Corporate Technology
Switzerland
Lorenz.Lehmann@Swisscom.com

Mauro Cadorin
Swisscom
Corporate Technology
Switzerland

Claude Eric Würgler
Swisscom
Corporate Technology
Switzerland
ClaudeEric.Wuergler@Swisscom.com

Abstract

Tomorrow's telecommunications market in an open competitive environment will increase the pressure on operators to differentiate themselves from their competitors. This will mainly be done on the service level, thus increasing the importance of a flexible service architecture allowing for fast and cost effective development of new services. Furthermore we need a technology and provider independent architecture to provide new services on top of our heterogeneous legacy networks.

The problem was already known in pure narrowband networks and most operators introduced the concept of a centralised service management platform in the network, the IN (Intelligent Network). But the IN architecture was designed for simple narrowband services and proved to be strongly technology and provider dependent. Therefore a new architecture has to be introduced to create new complex services in a fast and cost effective manner. The most promising service architecture at the moment is TINA.

This paper presents the experience gained during the implementation and integration of different services on a TINA platform. Special attention is given to the time needed for the delivery of the services and the hurdles encountered during the implementation.

1. Introduction

Tomorrow's telecommunications market in an open competitive environment will increase the pressure on operators to differentiate themselves from their competitors on two levels: provision of pure network connections and provision of services.

- The main differentiating factor for transport network connections will be the pricing structure. This will put enormous pressure on the cost of network connections and will increase the need for efficient Network Management tools. However, the profit margin on the provision of network connections will heavily decrease.
- Bigger profit margins can be expected on the provision of user tailored services. The main differentiating factor in this domain is fast deployment of personalised services.

The key for operators to hold on existing and to gain future markets is the introduction of a Service Architecture which allows for a fast development of user tailored services in a cost effective manner. Therefore most operators introduced the concept of a centralised service management platform in the network, the traditional IN (Intelligent Network).

But quick service creation, flexible adaptation of services and reuse of service components is a goal which was only partially met with the introduction of IN. This is mainly due to the fact that the current IN solutions are:

- network specific:
  different solutions are used for the management of wireline and wireless networks.
- technology dependent:
  IN is designed for simple narrowband voice services.
- vendor specific:
  the IN is heavily coupled with proprietary switching equipment. Reaching interoperability of different IN platforms is time consuming and requires the co-operation of all involved vendors.
- operator specific:
  many operator specific (country specific) options are still in use, due to the different legislatorial restrictions and the different historical background.

Furthermore, the traditional IN architecture is not designed for complex multimedia services and heterogeneous transport networks and has to be extended or replaced in the future.

The most promising architecture for the evolution of network service management is TINA, which combines

0-8186-8335-X/98 $10.00 © 1998 IEEE

Information Technology and the traditional telecommunications environment.

It is designed to manage all kinds of services in a technology independent manner and is therefore well adapted for the support of a wide range of services provided in a multi-vendor equipment environment.

But although the first phase of the TINA-Consortium is ending this year and a consistent set of specifications is provided, many issues are still open. TINA is still far from being a serious alternative for operators. In order to justify big investments, the feasibility and advantages of TINA have to be demonstrated, the TINA-C specifications have to be completed through feedback from experiments and - probably most important - ways have to be found to migrate from today's legacy systems into tomorrow's broadband multimedia network.

In order to demonstrate the feasibility and to study the advantages and bottlenecks of TINA, the Unisource partners R&D laboratories have started the SPOT project (Service Pilot On TINA). In this project operators (KPN Research, Telia Research and Swisscom Corporate Technology) work together with equipment manufacturers (Alcatel and Ericsson) on a collaborative basis, focusing on service development and integration and platform interworking. Feedback is expected from pilot users involved in an international field trial with a dozen SPOT terminals based on off the shelf PC equipment in four countries.

This report concentrates on the first point mentioned above, namely the service development and service integration.

In section 2 of this paper a short overview of the used TINA Service Platform compared to actual TINA-C baseline documents is given.

Section 3 presents the implemented services and shows how they reuse the same Service Generic Components and Supporting Services, thus reducing the overall complexity of the system.

Section 4 presents the hardware and software environment used in the project.

Section 5 shows the results of the service creation in terms of development time and discusses the encountered hardware and software hurdles.

In section 6 possible future steps of the project are discussed and section 7 summarises the results of the project.

2. Service platform overview

As can be seen in figure 1, TINA provides a clear separation into different architectural levels, thus de-coupling the services from the connection management and allowing for the reuse of service generic components within different services.

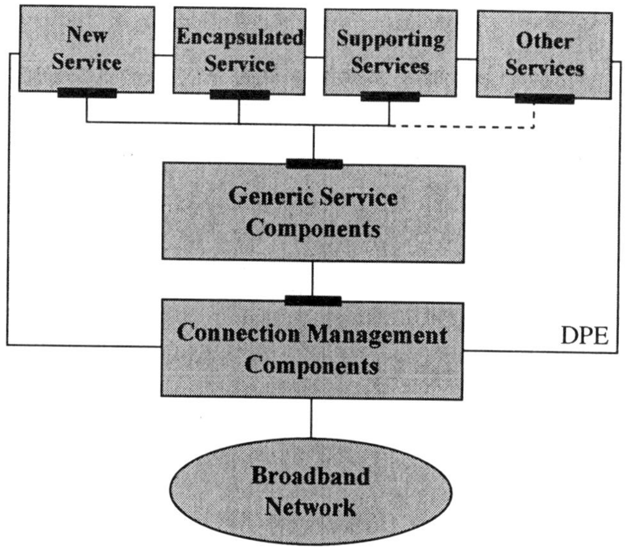

Figure 1: Architectural separation in TINA

The services use different capabilities offered by the underlying service platform and by other services. This results in an optimal reuse of generic components and keeps the service specific components as simple as possible.

The underlying platform (i.e. generic service components and connectivity server) builds a framework for the provision of TINA-C compliant services. The generic service components offer a service independent interface and support generic mechanisms like authentication, subscription and session management. The Connectivity Server manages the network resources and supports federation mechanisms to allow the management of connections across multiple provider domains.

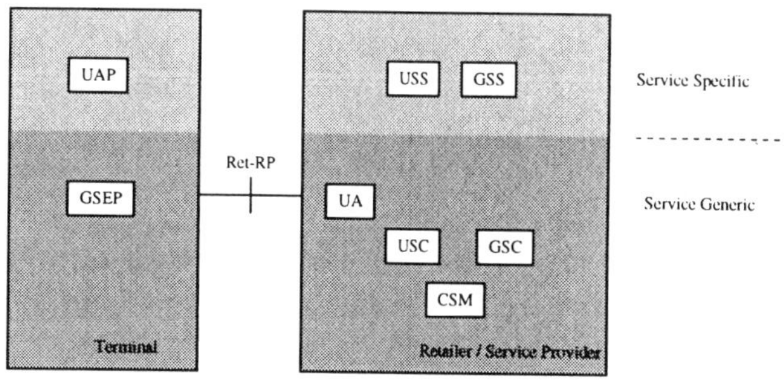

Figure 2: Computational Objects

Figure 2 shows the Computational Objects implemented in the project. The Retailer/Service provider domains are combined and the service specific code in the domain is integrated into the USS and GSS components.

As no stable Reference Point specifications existed when the project started, the IDLs were developed in parallel with the generic and service specific components. The work was based on the TINA-C Service Architecture version 4.0 and therefore cannot (yet) provide feedback to the current specifications of Reference Points. Anyway a lot of the work done in this project was fed into the working groups for the different Reference Points.

3. Implemented services

In order to prove the flexibility of the architecture, the reusability of service components and the speed of service development, three different types of services have been developed:

- One new service, created from scratch: Desktop Video Conference (DVC)
- One existing service, encapsulated into TINA components and using the generic service components: DeskTop Presentation (DTP) based on MS PowerPoint
- Two supporting services which are used within the stand-alone services: Accounting Service (Acc) and Address Book Service (AdB)

These services are currently tested in an international trial. Feedback from pilot users is expected during the last quarter of this year.

All these services were developed in the following order: service specification (enterprise viewpoint, stage 1 description), design (computational and information viewpoint, stage 2 description), IDL definition, implementation and local test, integration test, deployment, user feedback.

This section describes the services implemented on the TINA-platform and gives approximate values for the time needed to complete the different steps.

3.1 New service: DVC

Service overview, GUI:

The DVC service implemented allows for a high-quality multiparty video conference. Currently up to four participants are supported in a conference, each displayed in a separate window. The throughput of 2Mbps per video channel assures high-quality images.

The DVC session is at all times owned by one participant, initially by the party which requested the creation of the session. The owner controls the session and can accept or reject other participants, change the quality of service or change the billing scheme.

As depicted in Figure 3 an end-user must first provide the necessary configuration data for starting a DVC session.

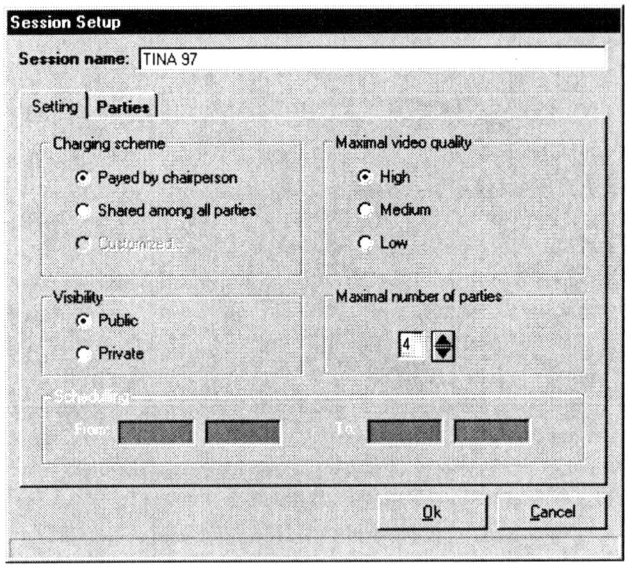

Figure 3: Graphical User Interface for the DVC session set-up

High level COs overview:

As depicted in figure 4, the DVC service is composed of 6 high-level computational objects (COs), which base on the Generic Service Components and are thus independent of the underlying network infrastructure.

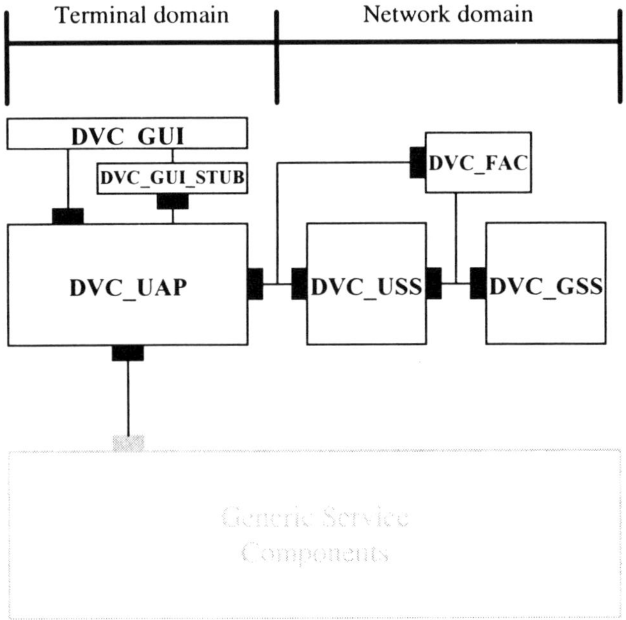

Figure 4: DVC service specific components

The functionality of those objects is the following:

DVC_GUI (Graphical User Interface):

The DVC_GUI is implemented with MS VisualBasic and translates the end-user actions into invocations on the DVC_UAP interfaces.

DVC_GUI_STUB:

The DVC_GUI_STUB is an intermediate object which eases the communication between the DVC_UAP and the DVC_GUI. It acts as the gateway from the CORBA world towards the non-CORBA world. When the DVC_UAP wants to invoke an operation on the DVC_GUI, it has to invoke this operation on the DVC_GUI_STUB, which forwards it to the DVC_GUI.

DVC_UAP (User Application):

The DVC-UAP implements the DVC-specific code in the terminal which is not related to the Graphical User Interface.

DVC_USS (User Service Segment):

The DVC_USS enables the interaction between the DVC_UAP and the DVC_GSS.

DVC_GSS (Global Service Segment):

The DVC_GSS provides centralised management of all active DVC sessions.

DVC_Fac (Factory):

The DVC_Fac enables the DVC_UAP to create and delete DVC_USS COs and passes the DVC_USS interface reference to the DVC_GSS.

3.2 Encapsulated service: DTP

Service overview, GUI:

Using the Desktop Presentation (DTP) Service, an end-user can present a slide-show to other end-users using Microsoft PowerPoint.

To do so the user who wants to show a presentation has to start a DTP session and invite other users into the session. The file containing the presentation is transferred to the invited users via ATM and is automatically opened with Microsoft PowerPoint. Now the presenter is able to control the sequence of the slides at all other end-users' terminals using the controls provided by the DTP Graphical User Interface (see Figure 5).

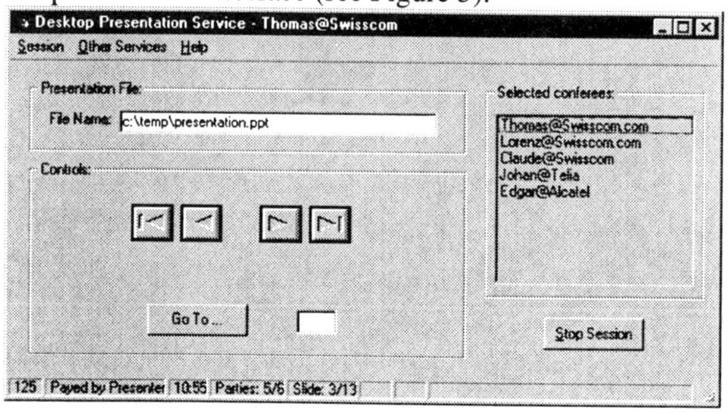

Figure 5: DTP Graphical User Interface

This implementation shows that it is possible to wrap legacy applications, as long as the APIs are known. The service does obviously only work if the participants in the presentation have Microsoft PowerPoint properly installed on their PC.

High level COs overview:

The Desktop Presentation Service is constructed using the same Computational Objects (COs) as the DVC Service (e.g. DTP_UAP and DTP_GSS). Additionally the DTP_GUI_Stub facilitates an interface with PowerPoint, as it is not possible to send OLE-Automation commands (used for the external control of PowerPoint) directly from CORBA (see Figure 6).

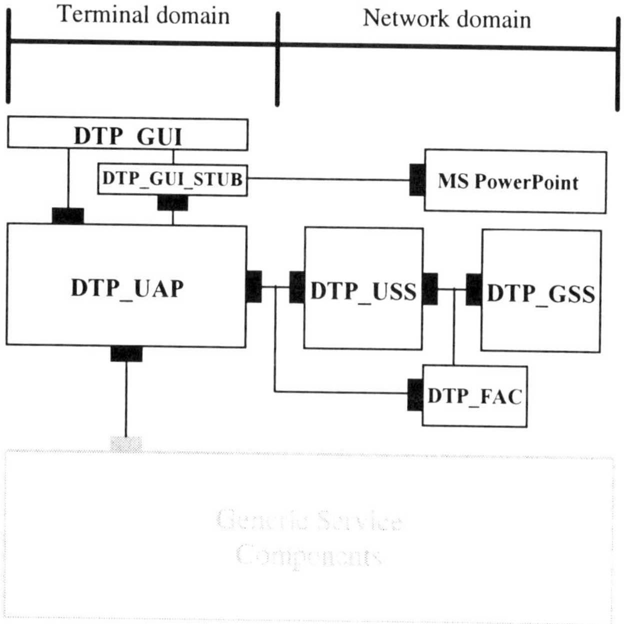

Figure 6: DTP service specific components

3.3 Supporting services

Supporting Services were implemented using the same Generic Service Components. They are independent of the DVC but can be used in the DVC or any other service for value added functions. They don't require explicit subscription, i.e. every end-user who has subscribed to a retailer is automatically allowed to use the administrative services. Two supporting services are currently implemented in SPOT:

1. Address Book

Service overview, GUI:

The Address Book service allows the end-user to manage a list of aliases. Every alias represents a group of user-identifications. The user can work with these aliases to invite other parties into a service session. The Address Book service also provides a complete list of all users currently subscribed to the retailer.

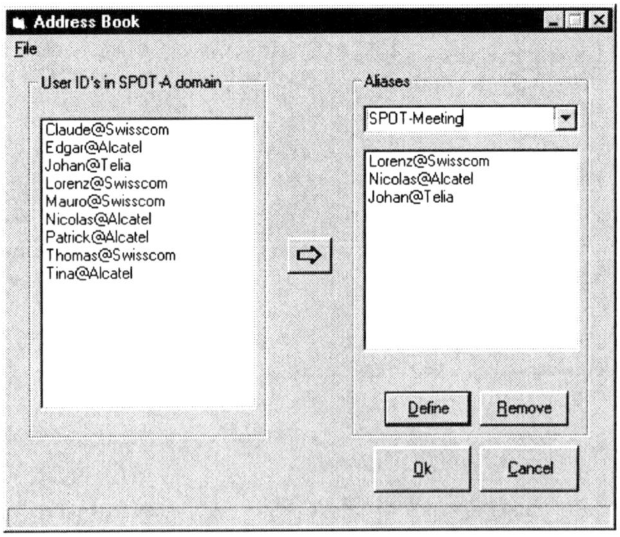

Figure 7: Address Book Graphical User Interface

High level COs overview:

Figure 8 presents an overview of the address book specific components. The DVC_GUI is indicated as one of the potential clients of the address book. The same goes for the DTP_GUI.

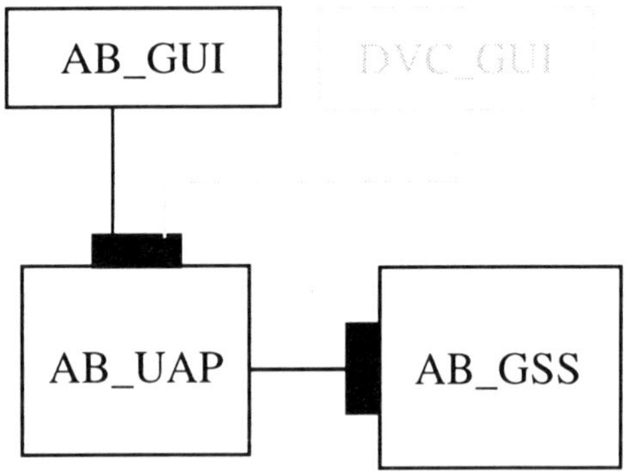

Figure 8: Address Book specific components

The functionality of those components is the following:

AB_GUI:

The AB_GUI offers the end-user the ability to manage his personal list of aliases.

AB_UAP:

The AB_UAP implements the Address Book specific code in the terminal domain which is independent of the graphical user interface. It offers an interface to all service GUIs that allows them to retrieve the address book information. But only the AB_GUI can create a new address book or modify an existing one.

AB_GSS:

This component keeps track of the personal address book for each end-user. In this object the address book information is stored.

2. **Accounting**

Service overview, GUI:

The Accounting service informs the user about the actual retailer's accounting scheme and keeps track of the user's personal bill. The accounting scheme considers duration of a session, number of parties in a session, quality of service and time of the day (see Figure 9).

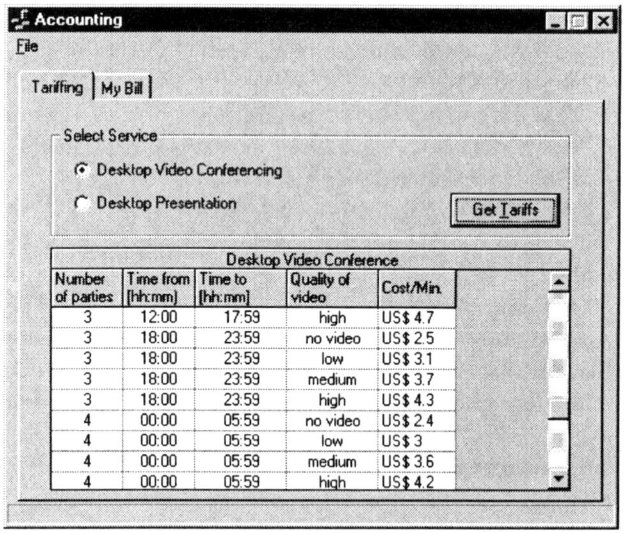

Figure 9: Accounting Graphical User Interface

High level COs overview:

Figure 10 shows an overview of the accounting specific components. The DVC_GSS is indicated as one of the potential clients of the accounting service.

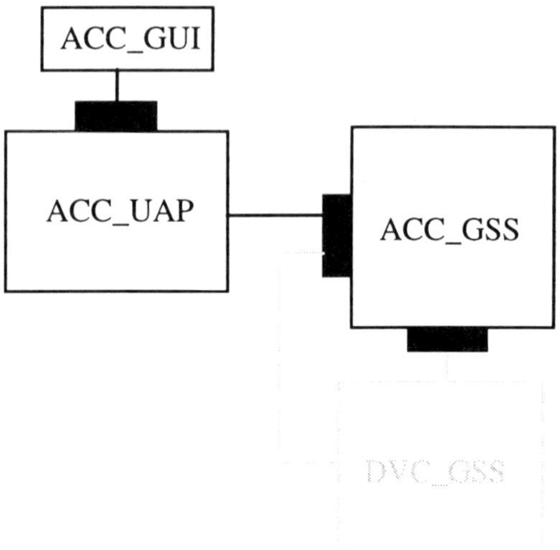

Figure 10: Accounting specific components

The functionality of these components is the following:
ACC_GUI:
The ACC_GUI offers the end-user the ability to retrieve the tariff table and billing information of the different services.
ACC_UAP:
The ACC_UAP is the contact point for the GUI on the terminal side for the access of tariff and billing information.
ACC_GSS:
The ACC_GSS stores the tariff tables of all different services offered by the retailer. It offers different clients such as DVC_GSS and DTP_GSS interfaces to get the tariffs or to issue requests to update a certain user's bill.

4. Hardware and Software environment

4.1 Deployment environment

The demonstration services presented at the TINA World Wide Demo at Telecom 95 impressed many potential customers. However, these services were implemented using expensive high-end terminal equipment (workstations), which is not acceptable for the mass-market. Therefore high priority was given to have PCs as user terminals. The user shall be able to work in his usual environment.

To simplify the transport network, FTTH (Fibre to the home) was chosen, which allows to carry the data streams directly over ATM. It was not the goal of this project to bother with the access, therefore the access was kept as simple as possible. In a real environment, the access network would rather be xDSL, ISDN or IP.

The kTN was implemented as IP over ATM, which allows to carry signalling and data over the same fibre access. The data streams were implemented as dynamically controlled ATM-PVCs.

Figure 11 presents the ATM transport network separated from the information network, carrying ATM to the home.

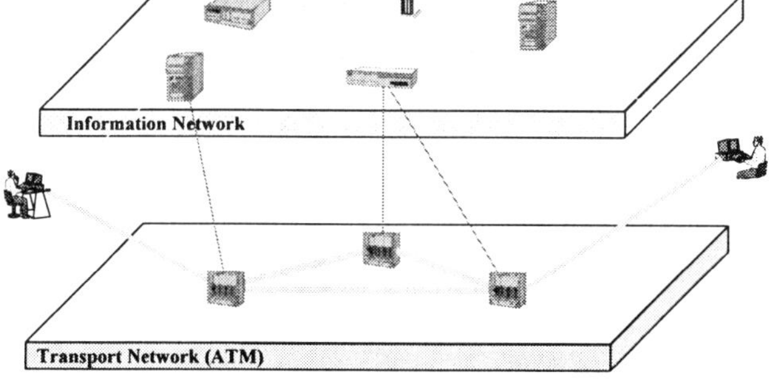

Figure 11: Architectural separation of information network and transport network

4.2 Terminal

As stated before, one basic goal of the project was to provide the services in the usual office environment and therefore to provide the services on a cost-effective PC-based terminal.

Consequently the DVC service was implemented on Pentium PCs running Microsoft Windows NT 4.0, equipped with high-performance audio, video and ATM boards. Each PC is able to mix and simultaneously display multiple incoming audio-visual streams, thus eliminating the need for specialised conferencing bridges in the network. Visual Basic and Visual C++ were used for the implementation of the Computational Objects residing in the terminals.

The user terminals have one, and only one, physical network connection, that is, a fibre to an ATM network. This means that signalling (TINA Kernel Transport Network) and user data (video, audio, and data streams) use the same physical network connection and ATM network, but different Virtual Channels with different Qualities of Service.

4.3 Network

UNIX workstations (Sparc 20) are used as network servers, offering the high computing capacities needed to handle the management of the different services. The server based COs were implemented using the Sun Sparc C++ compiler.

For the ATM connection we used standard FORE ATM OC3 boards installed in the workstations and a legacy Alcatel ATM public backbone Switch. The ATM Switch is controlled by the TINA connection management. To be able to do so a resource adapter object had to be implemented, but NO changes had to be made on the switch itself.

4.4 Middleware

The communication between objects was handled by a middleware compliant to the CORBA 2.0 standard. We have chosen Orbix to be conform to other internal projects and applications. Furthermore Orbix was the most mature and widely used product at that time.

The project started with Orbix version 2.0.1, which was then upgraded to version 2.2 due to compatibility problems with OLE.

5. Results

5.1 Service deployment

As described in the introduction of the paper, speed of service development and integration will be the main differentiating factors in the telecommunication market. This paragraph gives approximate values for the manpower needed to develop and integrate various services and thus shows that TINA indeed allows for the quick deployment of new services.

It has to be noted here that the resulting services are still prototypes and would need further tuning to be released as products. Additional time would be needed for this. Furthermore the whole marketing aspects of the deployment of new services was not taken into account. On the other hand much time was lost because the technology is still not stable and not completely understood.

Desktop Video Conference

The development of the DVC service can be divided into four phases:

- The work on the external service specification (describing in detail the functions and the GUI of the DVC service) took about 1.5 person-months (PM).
- The service platform design specifies the interfaces (IDLs), the operations offered by each Computational Object and the sequence of operations (scenarios) caused by each user interaction with the DVC_GUI: About 1 PM was used for this work.
- The implementation and testing of the stand-alone DVC service components took about 3 PM. The clean IDL design allowed a parallel development of the different objects.
- The integration and testing of the DVC specific components with the Generic Service Components took almost 2 PM. This number shows that still many problems occurred when integrating the (stand-alone tested) service on the platform. The time needed for integration still tends to be underestimated.

DeskTop Presentation

The development of the DTP service required approximately 6 PM.

- Describing the functions of the DTP Graphical User Interface took about 1 PM.
- The specification of the service platform design (i.e. IDLs, operations, etc.) took about 1 PM.
- The implementation and testing of the stand-alone DTP service took about 3 PM.
- The integration and testing of the DTP specific components with the Generic Service Components took almost 1 PM.

Administrative services

The development of the two administrative services required for each 1 PM. The time for the definition and specification of the services was quite short as there was no co-ordination needed with the common components.

Overview

Table 1 shows an overview of the time needed to develop the different services. It is quite obvious that the Desktop Video Conferencing service used the most time for all stages. This has two reasons:

- The DVC service was the first service implemented on this TINA platform. It was developed together with the generic service platform, the general structure of the platform was not stable.
- The DVC service is quite complex and was developed from scratch (based on commercial hardware).

The development of the DeskTop Presentation service was much shorter because it could be designed on top of a stable platform and a lot of the service description and code could be reused from the DVC service. Furthermore the DTP service is based on the existing MS PowerPoint application which only had to be wrapped.

Finally the administrative services (Address Book and Accounting) are of limited complexity and could therefore be implemented quite quickly.

time in man days	DVC	DTP	Acc	AdB
Description (Stage 1)	34	20	2	2
Specification (Stage 2)	17	18	3	3
Implementation	107	61	15	10
Integration/Test	27	21	3	5
Total time	185	120	23	20

Table 1: Invested time for the different development stages

5.2 Encountered problems

Software:

The software problems encountered during implementation and integration of the services can be divided into three categories:

- unclear, interpretable specifications:

  As in all international R&D projects, we encountered

some problems due to a different understanding of the service specifications. But these could be kept within limits because of the carefully specified IDLs and well co-ordinated scenarios.

- technical problems:
  Various products were used in this project where first a sufficient know-how had to be built.
- gaps and bugs in the used Software products:
  Most of the encountered problems were due either to missing functionality or bugs in the used Software products.
  Many problems occurred (as expected) with the Orbix-OLE interface, due to lack of compatibility. One example is the type "sequence", which represents an unbounded list of any type of elements in CORBA and is particularly difficult to be represented in VisualBasic.
  Moreover the Orbix server can be multithreaded, but the OLE automation server can not. This leads to many problems when operations are invoked asynchronously.
  Another problem was that the C++ code written for a Sun Solaris was not directly portable to a PC with Windows NT. Indeed, the code mapping from CORBA IDL to C++ is different from Sun to PC. This leads to a severe problems when the type "any" is used. This type allows to define a variable of a generic type, the real type can be given during the execution. If an Orbix PC Client invokes an operation with a parameter of type any on a Orbix Sun Server, this will cause a system error as the C++ representation of a variable of type any is not the same for the two systems. In order to avoid this error, the stub generated by Orbix on the PC or on the Sun must be adapted manually.
  Furthermore the Microsoft C++ Compiler was not capable of handling strings longer than 2048 characters. This lead to problems when IDLs with interfaces offering many operations were defined. This problem has been solved with a workaround by inheriting part of the operations from a subclass.

Hardware:

The project was meant to offer a high quality multiparty video conference on cost-effective PC-based terminals. Unfortunately, the goal of working on a "cost-effective" terminal could not be reached completely. In order to get good results, we had to use high-end PCs, equipped with 48MB of RAM and expensive prototype boards.

The need of 48MB of memory was mainly due to the fact that Windows NT, the Orbix daemon, the TINA generic components and all developed services were running on the PC at the same time. Windows NT all by itself already used 25MB of RAM, while the Orbix daemon merely used 1MB. Another 16MB was then used to have all service running at the same time.

The requirements for the Audio, Video and ATM boards were very high, as the goal was to offer a high quality multiparty video conference on PC. Therefore the acquisition of those boards turned out as one of the main problems of the project. There were no video boards on the market that met our requirements. Indeed the actual market for video boards on PC focuses on applications using limited bandwidth networks, such as N-ISDN. Therefore the actual implementation uses prototype boards which allow for four video streams at 2 Mbps each.

6. Future evolution, issues to be solved

The future work in this project will be done in two different directions:

- **Homogeneous TINA environment**: The SPOT platform as such is far from being complete and will be further improved. Possible work domains are: performance tuning, creation of new and extension of old services, service federation, adaptation to and verification of the "new" stable set of TINA-C baseline documents.
  An important domain to be investigated is the use of lightweight terminals and personal network based agents. The graphical user interface and applications are dynamically downloaded from the network, thus minimising the installation and management effort for the end-user. The web technology is now generally accepted as a de facto standard also in the telecommunication environment and customers are used to the look and feel of web browsers. Therefore future work will be based on Internet front-end technology.
- **Heterogeneous environment**: In a real telecommunications environment TINA networks have to deal with legacy systems and with alternative broadband multimedia solutions. Therefore it is important to study **and implement** a migration path from the different legacy systems to TINA networks as well as their interworking.

7. Conclusions

Speed of service development, reusability of SW components and flexibility of service management are key factors in today's telecommunication environment.

Even though the used platform and services are limited in scope and not fully compliant to the current, stable TINA-C specifications, the SPOT project proved that TINA is not only able to meet these needs, but also to provide a framework for the deployment of broadband multimedia services in the future liberalised telecommunications market. TINA specifically allows for quick deployment of new services and the encapsulation of legacy hardware and software.

Main problems in the provision of TINA services are the lack of maturity and the cost of the underlying infrastructure.

Acknowledgements

The authors would like to thank all the contributors to the SPOT-project for the fruitful co-operation and Thomas Kijftenbelt for his work on the DTP service.

References

[1] TINA Service Architecture, Deliverable from TINA-C, http://www.tinac.com

[2] TINA Business Model and Reference Points. Deliverable from TINA-C, http://www.tinac.com

[3] J. Bengtsson, P. Hellemans, L. Lehmann, N. Mercouroff. Implementation of Services for Computer Supported Co-operative Work on TINA: The SPOT Project. Proceedings of ISS'97, Toronto, Canada, September 1997.

[4] James Rumbaugh, Michael Blaha, William Premerlani, Frederick Eddy, William Lorensen. Object-Oriented Modelling and Design, Prentice Hall 1991.

[5] Robert Orfali, Dan Harkey, Jeri Edwards. The Essential Distributed Objects Survival Guide. John Wiley & Sons 1996.

Application of TINA-C Computing and Service Architecture Concepts to the Development of an Advanced Information Brokerage Service in the Context of EC

Juan I. Asensio[†], José I. Moreno, Víctor A. Villagrá, Jesús Redondo
Dep. of Telematic Systems Engineering
Technical University of Madrid (DIT-UPM), Spain
{jasensio, jmoreno, villagra, jrv}@dit.upm.es

[†]*Visiting Researcher from Univ. Valladolid, Spain*

Alexandre Nolle
France Télécom-CNET, France
alexandre.nolle@cnet.francetelecom.fr

Ionel Tothezan
SEMA GROUP TELIS Telecom, France
TOTHEZAN@pmail.sqy.tel.telis.fr

George T. Karetsos
Telecommunications Laboratory
Dep. of Elect. Engineering & Comp. Science
National Technical Univ. of Athens, Greece
karetsos@softlab.ece.ntua.gr

Abstract

The concept of Electronic Brokerage is based on the use of new information technologies so as to provide a service capable of facilitating and organising the relationship between the Supply and the Demand in an "electronic marketplace" environment. This new service supported by the so-called "Electronic Brokers", will allow service providers to publish and advertise their offers and, at the same time, will help end-users to access the offered services and information in an easy and efficient way.

This paper presents an architecture for this electronic brokerage service that is currently being developed within the scope of the ABS[1] (Architecture for Information Brokerage Service) ACTS European project. This architecture is based on concepts and principles contained in the TINA-C computing and service architecture and in the OMG EC-DTF Reference Model

[1]This work is partially sponsored by the Commission of the European Union under the project ABS, project number AC206 of the ACTS programme. The ABS consortium is integrated by France Telecom-CNET (France), DeTeBerkom (Germany), Telecom Finland (Finland), Portugal Telecom-CET (Portugal), Telis (France), VTT (Finland), NTUA (Greece), DIT-UPM (Spain), Degriftour (France) and PoP (Germany). The views presented here do not necessarily represent those of the ABS consortium.

1. Introduction

The ABS brokerage service, devoted to match Supply and Demand in a preliminary phase of commercial transactions, is foreseen as an advanced distributed information service in the context of the electronic commerce intended to be deployed over a distributed processing environment. Due to those features, the ABS consortium has decided to develop an architecture of brokerage services taken into account results from TINA-C architectures [1] and the OMG EC-DTF (Electronic Commerce Domain Task Force) Reference Model [2].

The ABS brokerage service architecture is related to TINA-C in several aspects. First of all, TINA-C Computing Architecture principles are used in order to get a complete model of the brokerage service (applying the five ODP [3] prescriptive viewpoints). Secondly, the obtained enterprise viewpoint is, in most of the aspects, very similar to the TINA Business Model [4]. Third, stakeholders identified in the ABS brokerage service enterprise model follow the User/Provider paradigm introduced in the TINA-C Service Architecture [5] and, in this way, the ABS brokerage service architecture can reuse the set of service components derived from the TINA-C Service Architecture. And fourth, the ABS brokerage service architecture has been designed with the objective of its adaptation to a TINA-compliant DPE (Distributed Processing Environment), like the one that is subject of the ACTS RETINA project.

0-8186-8335-X/98 $10.00 © 1998 IEEE

With respect to the OMG EC-DTF Reference Model, the ABS brokerage service as a whole constitutes one of the identified EC-DTF facilities: the brokerage facility. This demonstrates the key role that the brokerage service will play in the future development of the Electronic Commerce. But, furthermore, different parts of the ABS brokerage architecture offer services and facilities that are also included in the EC-DTF Reference Model: service invocation interface, service catalogue, service session facility, management facility etc.

This paper will explain how the ABS architecture is being developed under the above influences and what are the main results obtained so far. It will also try to demonstrate both the importance of the brokerage service and the suitability of the approach adopted by ABS for the definition and design of its brokerage service architecture.

The paper will be structured as follows: in section 2 the electronic brokerage service is presented: its objectives, its added-values, its functionality etc. as part of a more general mediation process which will also be defined. Section 3 and 4 are devoted to the description of ABS brokerage service business and information models respectively, as a result of the application of the TINA Computing Architecture principles. Section 5 will be devoted to a general description of the ABS computational model (the key aspect of the ABS service architecture) and its relation with the TINA Service Architecture components and the OMG EC-DTF Reference Model. Finally, section 6 concludes the paper presenting the main conclusions and the work that has still to be done within the ABS project.

2. The brokerage service in the context of the mediation process

Trading is based on the exchange of goods and services. When customers need something, a demand is created in the market. On the other hand, if there are providers offering goods and services, there is also a supply. Providers and customers (or suppliers and demanders) must meet each other in order to proceed with the trading process. Some kind of platform is needed so as to support that meeting of customers and providers, and their subsequent interaction. Years ago, that platform consisted of the physical marketplace but nowadays a new "electronic marketplace" has emerged enabling trading with no physical travelling requirements.

The trading process in the electronic marketplace requires the support of a set of mediation services. A mediation service can be defined as that provided by an existing or intended system by which individuals and organisations obtain what they need through the exchange of information, products and value with each other. Mediation services (grouped under the so-called *Mediation Platforms*) support trading by helping during the negotiation phase and supplying transactional and logistic services. The concept of mediation services can be illustrated by the content of figure 1.

The brokerage service is one particular mediation service. This service is used during the first stage of the relationship between customers and providers: it informs a customer about a set of providers that might fulfil its demand and helps him in finding which of them would be the most suitable. At the same time, it helps the providers in putting their offers closer to the customers. In other words, it brings the Supply and Demand together in order to proceed with the trading process.

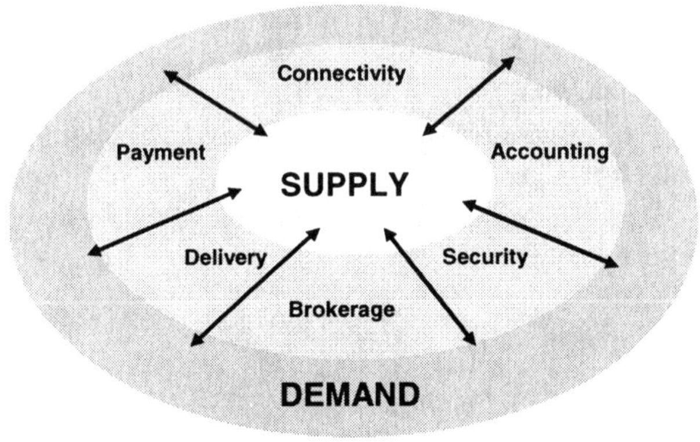

Figure 1. Mediation Services facilitates the interaction between Demand and Supply in the context of the electronic marketplace.

In this context of mediation platforms and brokerage services, the ABS appears with the main goal of specifying and designing an architecture of brokerage services suitable for the development of electronic brokerage systems able to offer a set of added values that can fulfil their users' requirements. These added values can be summarised as follows:

- **Low search costs for both customers and providers**: by means of the brokerage service, customer or demander will not have to access several places and systems to find the best fitting provider or supplier. Simultaneously, those providers will have opportunities to reach new kinds of customers. Brokers can maintain databases of customers' preferences, and reduce search costs by selectively routing information from providers to consumers.
- **Structuring of information sources**: efficiency in matching offers and demands will be based on their adequate structuring in the electronic broker. The obtained structure will enable customers to navigate

through available offers. In this way, customers would be able to participate more directly in the matching process.

- **Combination of information**: the electronic broker will combine different information sources in order to satisfy the customer demand in a better way. The broker might even gather product information from sources other than product providers as, for instance, independent evaluators and other customers.
- **Presentation of information**: if the electronic broker finds lots of information sources that might fulfil a customer request, the results should be presented to that customer in the way that best facilitates its decision-making process.
- **Homogeneous access to heterogeneous information sources**: the customer must perceive a homogeneous view of the whole set of providers that exist in the electronic marketplace.
- **Reliability of content**: the electronic broker will keep information consistent and up-to-date. It will also take care of legal aspects of the maintained information about actors in the electronic marketplace.
- **Confidentiality**: when agreed, the electronic broker will not reveal the identity of the actors involved in the Supply-Demand matching process.

In order to support the above set of added values, the electronic broker (the system supporting the brokerage service) will have to incorporate a precise set of functionalities, also shown in figure 2:

1. **Customer demand definition**: the electronic broker will provide some mechanism by means of which customers can express their requests. Those requests will include both the desired good/service and some kind of preference criteria that might be used to discriminate and organise the potential set of results. The broker might persistently store some of those preference criteria in order to be applied, by default, to all the requests of a particular customer.
2. **Directory and information retrieval functions:** the electronic broker must maintain databases of description of goods and services offered by providers. Those databases should be structured in such a way so as to allow customer the searching and navigation through their contents. To some extent, the customer should perceive the brokerage service as some kind of homogeneous yellow pages directory.
3. **Offer registration:** this functionality is devoted to allow the providers to give the electronic broker information about the contents that they maintain. At the same time, by means of this functionality they will be able to supply information about administrative aspects and about the way the broker can access their domains in order to obtain precise information about particular offers.
4. **Information processing function:** this functionality covers all the necessary information processing between the request of a customer and the offer presentation generated by the broker. The

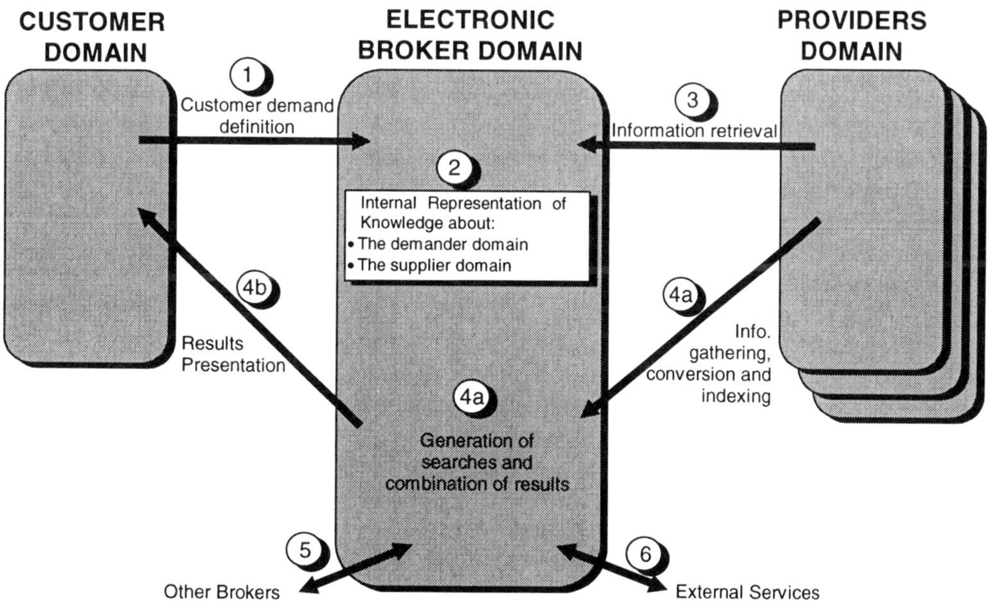

Figure 2. ABS broker functionalities.

broker must perform two subsets of information processing activities. First of all, the broker has to create, structure and maintain its databases. In order to achieve this goal, it will periodically gather information from the providers' domains, convert the retrieved information into a defined broker internal format and index all that information in the databases. The second subset of information processing activities is related to the handling of customer demands: the broker will have to receive the customers' requests, evaluate and filter them, generate the required searches in the databases (and, if necessary, in the providers' domains) receive the responses and, finally, present the results to the customer in the most suitable way.

5. **Federation function:** an electronic broker should be able to interoperate with other brokers in order to obtain better and more precise responses to customer requests. Nevertheless, the broker must hide to the customers the involvement of other brokers in the processing of their demands.
6. **Interfaces to external functionalities:** the electronic broker must be able to use other external mediation services in order to be integrated in an overall mediation platform capable of supporting electronic commerce transactions.

Once the brokerage service, together with its environment, added values and functionalities, have been defined, the following natural step is the definition and design of an architecture of systems capable of supporting this service.

As it has already been said in the introduction, the ABS consortium has decided to reuse the concepts and principles contained in the TINA-C computing architecture for the development of the brokerage service architecture. Thus, the application of the five viewpoints prescribed in the TINA-C computing architecture (which are those defined in the ITU-T Reference Model of Open Distributed Processing) will lead ABS to a complete description, design and even implementation of electronic brokerage systems.

The models obtained after using the viewpoints will:

- Clarify the set of actors that participate in brokerage service environments (enterprise viewpoint) refining those already introduced (customer, providers, and brokers),
- Give more details about the information storing and processing activities in the broker (information viewpoint),
- Describe how the functionalities of the electronic broker are supported by distributed components (computational viewpoint),
- Study what infrastructure is needed so as to support the distribution of functional components (engineering viewpoint),
- Guess what technologies are suitable for supporting the whole system (technology viewpoint).

The ABS consortium has already applied the first three viewpoints (enterprise, information and computational viewpoint) to obtain a first version of prototypes of electronic brokers. The rest of the paper is devoted to the description of the results from the application of those viewpoints.

3. The ABS brokerage service enterprise model

The ABS brokerage service enterprise model [6,7], which is obtained from the analysis of the service from an enterprise viewpoint, describes the brokerage from the perspective of what actors are involved in its operation and usage (their roles, their requirements their obligations...) and what policies are applied to their behaviour.

The ABS project has already obtained the enterprise model of the electronic brokerage service in which 6 types of actors and their requirements have been identified (they all are shown in figure 3):

- **User of brokerage service**: is that entity (human being, machine or application) that uses the brokerage service in order to satisfy its own desires. The *user of brokerage service* can be a *subscriber* or even an anonymous *end-user*, depending on the contractual relationship with the *brokerage service provider* (which will be described afterwards). The *users of brokerage service* demand a precise quality of service, a particular level of security, added values that differentiate a particular *brokerage service provider*, and simplicity of use.
- **Brokerage Service Provider**: is that entity that offers information to the *users of the brokerage service* about services and contents maintained by third parties. To some extent, the *brokerage service provider* can be considered as an extension of the *ODP trader* [8] as it translates the concept of matching services offered by distributed objects with client's requests to the idea of matching services and goods offered by suppliers with demanders' requests. The *brokerage service provider* requires a certain level of security and information about the characteristics and preferences of the *users of brokerage service*.
- **Service Provider**: is that entity that provides telecommunication services, information services or both, as well as application services. These services can be divided in two main groups: services that

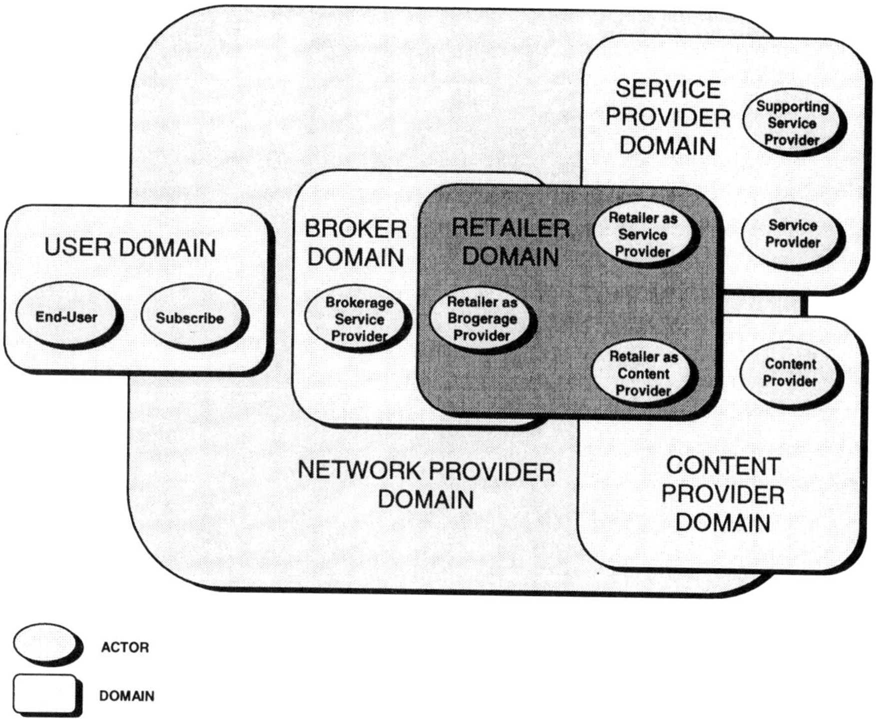

Figure 3. Actors and domains in the ABS brokerage service enterprise model.

might be the subject of the brokerage service (in the sense that they can fulfil some demands) such us VoD, videoconference, etc., and supporting (mediation) services as, for instance, security, payment, etc. The requirements of this kind of actors are very similar to those of the *users of the brokerage service*, i.e., security, quality of service, added-values and simplicity of use.

- **Content Provider**: is that entity that offers tradable goods or information to *customers* or *end-users* by means of the brokerage service. The requirements of this actor are the same than those of the *service providers*.
- **Retailer**: this entity supplies services or contents to users *of the brokerage service* but taking into account that these are services and contents bought from third parties. The *retailer* performs its activity thanks to contractual agreements with *service* and *content providers*. The *retailer* incorporates brokerage functions in order to help *user* to find the offer that can satisfy their requirements in the best way. The requirements of the *retailer* can be considered as a combination of those of all the previously described actors. It requires flexibility (it must be able to get adapted to almost any possible business area), quality of service, security, added-values and simplicity of use.
- **Network Provider**: is that entity that provides all necessary networking functions to other actors. It has been considered as a different actor, instead of having been considered as a *service provider* because its service is not intended to be offered by means of the brokerage service and because it is not an specific mediation service.

The ABS brokerage service enterprise model, considers four different roles that the identified actors can play:

- **Provider**: an actor with this role interfaces with another with the *customer* role so as to agree in the provision of a service. This agreement will be usually governed by a contract between both parties.
- **Customer**: actors in the customer role interfaces with a provider so as to obtain a particular service.
- **Operator**: in the operator role, an actor is involved in the operation, management and maintenance of the resources (owned by the actor) needed to provide a particular service.
- **User**: this role is very similar to the *provider* role. The distinction resides in the fact that the *user*

cannot negotiate the characteristics of the desired service. All those characteristics will be fixed since the sign of the contract.

The roles that the actors of the ABS brokerage service enterprise model can play, are summarised in table 1 [6,7].

Table 1. ABS brokerage service actors and roles.

ACTOR	ROLE: Provider	Customer	Operator	User
Brokerage service user		X		X
Brokerage service provider	X	X	X	
Service provider	X	X	X	
Content provider	X			
Retailer	X	X	X	
Network provider	X		X	

The last aspect of the brokerage service enterprise model that should be described is the identification of *domains*. A domain, from the enterprise viewpoint, can be defined as the grouping of several actors that interact so as to achieve a set of objectives. There can be *administrative domains* (created due to geographic, politic, organisational, etc. criteria) or *functional domains* (groups of actors with similar functionalities).

From an enterprise viewpoint, *administrative domains* can be considered of greater importance because they can better reflect the business objectives intended to be achieved by means of the use of the system under study.

In the ABS brokerage service enterprise model, the following *administrative domains* (also shown in figure 3) have been identified:

- **User domain:** it groups *users of brokerage service* of all types.
- **Broker domain:** it is the domain associated with the *brokerage service provider*.
- **Service provider domain:** it groups the two type of *service providers*: those that offer services that can be offered by means of the brokerage service and those that offer supporting services.
- **Content provider domain:** it is the administrative domain associated with the *content providers*.
- **Retailer domain:** it is the administrative domain associated with the *retailer* actor. The superposition of this domain with others, in figure 3, tries to show how the *retailer* may behave in different ways depending on the circumstances.

It is very important to underline the similarity between the obtained ABS brokerage service enterprise model and the TINA-C Business Model defined in [4]. Actors in the User Domain might adopt the TINA-C Consumer role, actors in the Broker Domain might match the TINA-C Broker role, actors in the Retailer Domain might behave as TINA-C Retailers, actors in the Service and Content Provider Domains might adopt the TINA-C Service Provider role and, finally, Network Providers would behave as TINA-C Connectivity Providers. This matching between both models, and the fact that interactions between different stakeholders in the ABS brokerage service enterprise model are governed by the user/provider paradigm, make the ABS brokerage service suitable to be integrated under the "umbrella" of the TINA-C service architecture. This integration will be explained later on.

4. The ABS brokerage service information model

The information model, obtained from the application of TINA computing architecture information viewpoint focuses on semantic aspects of the service, trying to describe what information is needed to support the brokerage tasks and how that information is processed.

In section 2, the information processing, information retrieval and directory functions (figure 2) were introduced as required functionalities devoted to the support of the added values offered by the ABS brokerage service architecture.

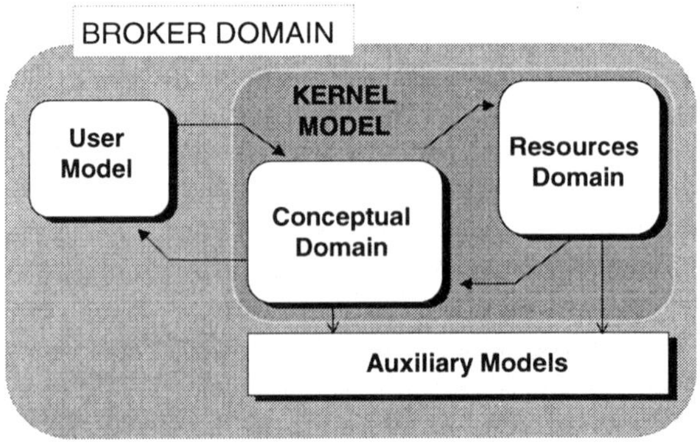

Figure 4. ABS Information Sub-models and their relationship.

The information model must reflect all the aspects related to the handling of information associated to those functionalities. Using this constraint, the information model has been divided in several submodels as shown in figure 4.

- **User information model**: contains information about the users of the service. This information is intended to facilitate the query and navigation processes and to customise the service for each user. The key aspect of this submodel is the *user profile*. Attitudes, skills, preferences, etc. of each user are the typical information to be included in its *user profile*.

- **Conceptual domain**: this submodels contains information about the different business areas that are being supported by the brokerage service. The main aspect of this conceptual domain is the so-called *conceptual network*. It can be defined as a complex multirelational structure of nodes and relations among nodes. Each node represent a concept of the real world, and a set of concepts, and their relations, can be projected over the so-called *viewpoint* which constitutes an abstract representation of a particular business area. Figure 5 shows our vision of the conceptual network.

  In that figure we can appreciate the Conceptual Network (CN) on the top. The concepts of the CN are grouped according to different business areas. Each group of concepts (each business area) constitutes a *viewpoint*, which is composed of a set of *shadow objects* (SO). A SO could then be defined as the projection of a concept over a particular *viewpoint*. The same concept from the CN could be projected over different *viewpoints* thus creating different SOs. So, for instance, Madrid might be a concept that could be projected over the "European Flights Viewpoint" as a possible destination or could be projected over the "Tourism in Spain Viewpoint" as one potential place to visit. SOs derived from the same concept are related by the so-called *bridges*.

  The way the *conceptual network* has been designed has two very important advantages:

  1. **It is a general model**: it is very easy to add new business areas. The ABS electronic broker might be able to cope with almost any business area.
  2. **It is a flexible model**: by means of the *viewpoints*, the user can navigate through different business domains. Furthermore, by means of the *bridges*, the user is able to go from one business area to another very easily. Of course, apart from the navigation capabilities, the model also supports random searches. The user will be able to switch from navigation to searching and the opposite whenever he wants to.

- **Resource domain:** a resource is a composed entity that encapsulates all necessary informational items from the providers' domain. It includes *content descriptions* (*what* the providers offer), *query capabilities* (*how* the broker can ask the providers' sites for some needed information) and *access descriptions* (indicating how the broker can *access* the providers' domain to make a query).

  Resources are "attached" to SOs by means of *queries*. In other words the broker does not store all the information related to a particular SO (a *concept* in a particular *viewpoint*) but it stores a *query* that

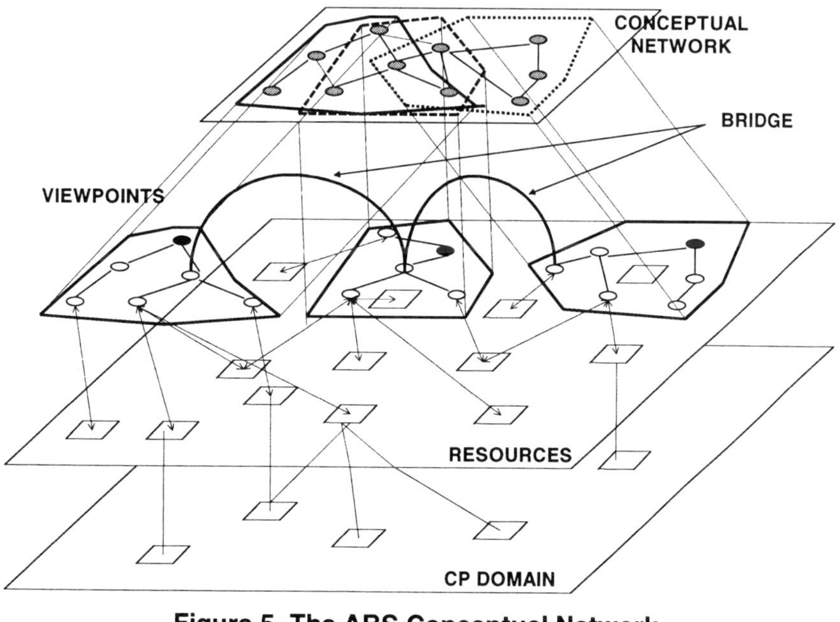

Figure 5. The ABS Conceptual Network.

will be subsequently posed to one or more providers' query interfaces. When a provider's information sources are queried in this way, they are supposed to return the information that was needed.

- **Auxiliary Model:** contains information about external mediation service providers.

5. The ABS brokerage service computational model

Figure 6 shows the ABS Brokerage service functional blocks derived from the application of the computational viewpoint and their relation to the TINA-C Service Architecture and the OMG EC-DTF Reference Model. This computational model can be considered as the "core" of the ABS brokerage service architecture. As it can be appreciate the ABS service architecture will reuse the session components contained in the TINA-C Service Architecture and, on

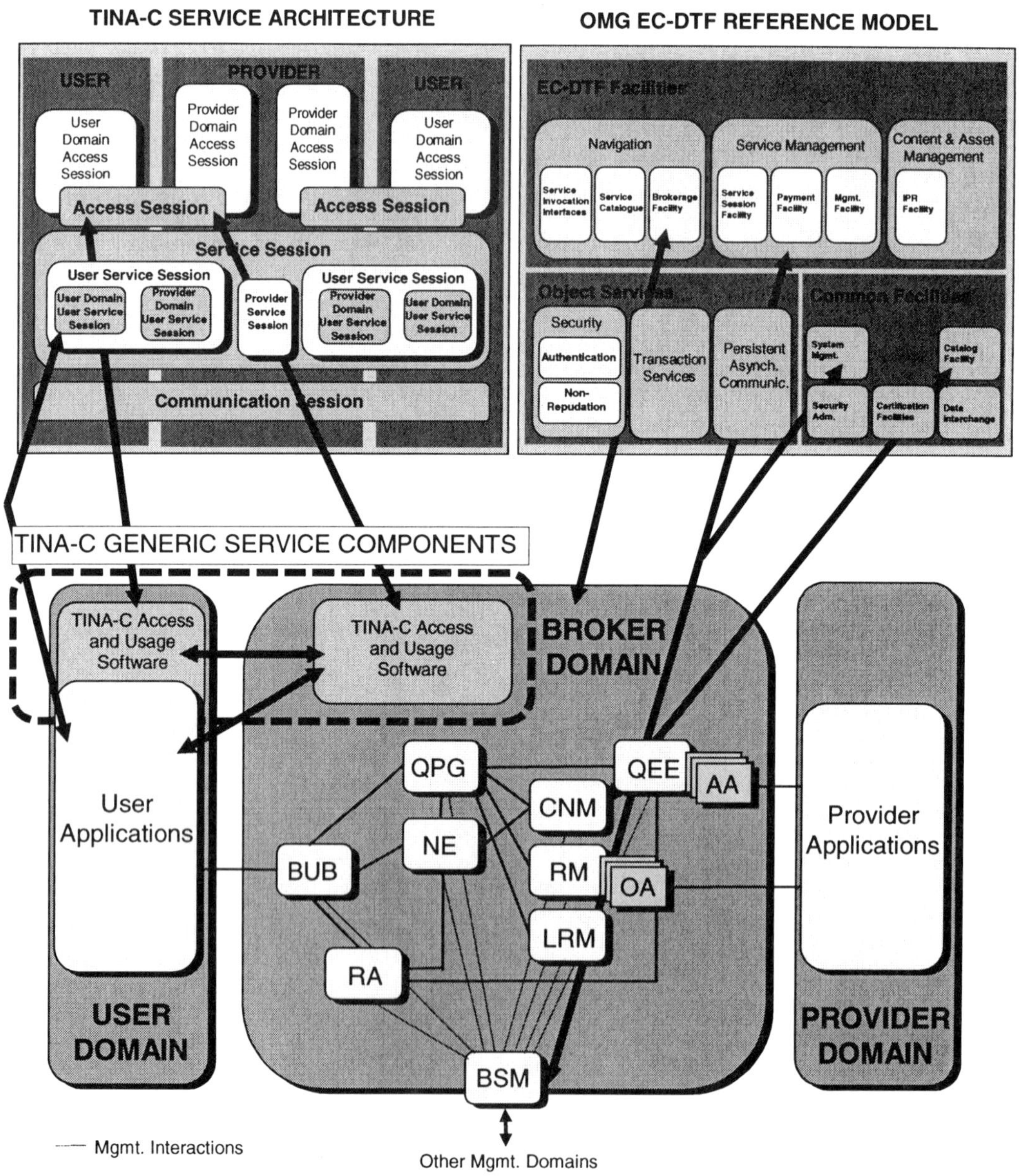

Figure 6. ABS brokerage service functional blocks and their relation to TINA and OMG EC-DTF.

the other hand, some components of the ABS service architecture can be matched to EC-DTF facilities in the reference model for EC. Obviously, the ABS brokerage service computational model also incorporates some concepts developed within the project, the most important of which is the *conceptual network*.

Let us briefly describe the main blocks of the ABS service architecture for the broker domain and how they support the functionalities identified in section 2:

- **CNM** (Conceptual Network Manager): is responsible for managing the conceptual network and the viewpoints. This could be considered as the key block of the architecture. It supports, together with the QPG, the *directory function* and part of the *information processing function*.
- **RM** (Resource Manager) and **OA** (Offer Agents): the RM manages offers description (information about Providers and about locally stored data). The OAs provide *Offer Registration* in the RM and, optionally, in the LRM.
- **LRM** (Local Resource Manager): is responsible for the local databases management. It is, to some extent, a *cache* of the information maintained by the Providers.
- **BUB** (Broker User Block): represents the users in the broker's domain. It is the broker's domain end-point of an access session with a user. It is accessible from the user's domain, regardless of the domain's location. It is also responsible for deciding if the user demand is navigation or query, and to ask for the execution of the demand to the corresponding blocks. It might be considered as the functional block that supports the *customer demand definition function* (and some of the tasks belonging to the *information processing function*).
- **QPG** (Query Plan Generator): is responsible for building an execution plan corresponding to a user query. The QPG block should identify the concept and the viewpoint corresponding to the query by interacting with the CNM, ask the LRM to know if locally stored data is consistent or if it needs an access to the Provider database, and build an execution plan. Together with the CNM supports the *directory function* and part of the *information processing function*.
- **QEE** (Query Execution Engine) and **AA** (Access Agents): the QEE is responsible for the execution of the plan generated by the Query Plan Generator. The Access Agents are responsible for information retrieval on the Provider and local (broker) domain. The QEE should execute the plan on the provider domain and on the local (broker) domain (the plan may consist of information retrieval or updating of the locally stored databases) and structure the results. Both QEE and AAs support the *information retrieval function* and part of the *information processing function*.
- **RA** (Result Assistant): is responsible for first presenting treatment of the results. It supports that part of the *information processing function* that deals with the formatting and presentation of results to the users.
- **NE** (Navigation Engine): it is responsible for the treatment of the navigation operations requested by the users. It gives support to some of the *information processing functions*.
- **BSM** (Broker Service Manager): it acts as the front-end for brokerage service management and broker application management. It is intended to monitor and control the behaviour of the components of the electronic broker and, at the same time, to enable the management of the ABS broker from other management frameworks (acting as a management gateway). By means of this gateway functionality, the service and application management will be integrated with network and middleware management.

Table 2. ABS computational blocks and their support for ABS brokerage service functionalities.

FUNCTIONAL BLOCK	FUNCTIONALITY				
	Cust. demand definition	Directory	Information retrieval	Offer registration	Information processing
CNM		X			X
RM & OAs				X	
LRM				X	
BUB	X				X
QPG		X			X
QEE & AAs			X		X
RA					X
NE					X

Table 2 summarises how computational functional blocks support the different ABS brokerage service functionalities identified in section 2. The *federation function* and the *interfaces to external functionalities* do not appear, as they have not been included in this first version of the ABS brokerage service architecture. For the second version that will be developed during 1998, these two aspects will be taken into consideration.

6. Conclusions and Future Work

In this paper the Brokerage Service has been introduced as an advanced information service with a key role in the development of the future Electronic Commerce. Its inclusion in the OMG EC-DTF Reference Model demonstrates that importance.

TINA-C Computing Architecture concepts and principles have been applied by the ABS project in the development of an architecture of brokerage services. Those concepts and principles, based on those contained in the Reference Model of ODP, have been of a great value in the performance of ABS development tasks. Some results of the work of ABS have also been described: the enterprise model has clarified the set of actors that appear in brokerage service scenarios, the information model (and its central aspect, the *conceptual network*) describes the way of structuring and storing the brokerage service information so as to offer a flexible and powerful service, and the computational model has identified the functional blocks needed in order to support the functionalities intended to be provided by the ABS brokerage service.

Currently, the ABS consortium is involved in the implementation of the first broker prototype which is a simplification of the global architecture and that will be validated in a set of trials to be held in October 1997 in Portugal, France and Germany. The implementation is based in the IONA's JAVA implementation of CORBA (OrbixWeb 2.01) in order to take advantage of distributed processing capabilities of CORBA and the flexibility, portability, and Object-Orientation of JAVA. The second version, planned for mid-1998, will incorporate more sophisticated features like security, dynamic search and the use of specific facilities provided by external service providers.

7. References

[1] TINA Consortium. "Overall Concepts and Principles of TINA". Version 1.0. February, 1995.

[2] Object Management Group. "Electronic Commerce DTF Reference Model". *OMG document ec/96-09-03*. September, 1996.

[3] ITU-T Recommendation X.903. "Information technology - Open distributed processing - Reference model: Architecture". 1995.

[4] TINA Consortium. "TINA Business Model and Reference Points". Version 4.0. May, 1997.

[5] TINA Consortium. "Service Architecture". Version 4.0. December, 1996.

[6] ABS Consortium. "Broker Business Model". *Public Deliverable D23*. Ocotober, 1996.

[7] I. Tothezan, E. Athanassiou, P. Alzon and G. T. Karetsos. "Enterprise Modelling of Information Brokerage and Retailer Services". *First international enterprise distributed object computing workshop (EDOC'97).* Marriott Resort, Gold Coast, Australia. October 1997.

[8] Draft International Standard ISO/IEC 13234-1, "Information technology - Open Distributed Processing - ODP trading function - Part 1: Specification". 1995.

Session 4

Internet II

Emmanuel Darmois, ALCATEL

TINTIN
the TINA to INTERNET Project

Filip Vandermeulen*, Jean-Marc Reynders+, Laurent Missa+, Piet Demeester*

* Department of Information Technology (INTEC)
University of Ghent, Sint Pietersnieuwstraat 41, 9000 Ghent (Belgium)
(Phone: + 32 9 264 33 43, Email: {fvdmeul, piet.demeester}@intec.rug.ac.be)

+ Technology Management division / Network and Services Management Integration
BELGACOM, 177 Bld E. Jacqmain, Belgacom Towers 9T31, 1030 Brussels (Belgium)
(Phone: + 32 2 202 81 79, Email: {jean-marc.reynders, laurent.missa}@is.belgacom.be)

Abstract

Interworking with existing systems is a key issue for the success of the telecommunication information networking architecture (TINA). In this paper, The TINA service architecture and the stream binding mechanism have been investigated and ported on the Internet. Based on these concepts, a chat service combining data and voice has been implemented to demonstrate TINA-Internet interworking scenarios. The field trial results are outlined and further migration steps are also investigated.

1. Introduction

The telecommunication scene is a fast moving target with the deregulation and the global opening of the market. To cope with this changing environment, the PNOs need a network architecture being able to rapidly adapt to market fluctuations and for flexible enough to rapidly create, introduce and deploy new services on the network according to the market needs.

When TINA comes on the telecommunication market, it will offer a platform able to tackle all these constraints in a coherent framework. But at that time, other ad-hoc solutions, each addressing a specific domain, will already be in place. For instance, the Telecommunication Management Network (TMN) which provides management in the FCAPS domains is being widely accepted by the PNOs; the intelligent network, which successfully separated the intelligence of a service from the network already provides advanced telecommunication services. But, one of the most promising domains in the future telecommunication landscape is the interactive multimedia market with promises of rapid and colossal revenues. Solutions based on enabling technologies such as ADSL, that drastically increase the end-user capacity, are boosting the market of advanced multimedia services.

In this context of multimedia, the Internet appears to be a living laboratory for the development of new multimedia applications. It is clear that the success of Internet is mainly bound to the Web. The generic interface of a Web browser offers an intuitive and platform independent access to information and the concept of HyperText, further extended with the HyperMedia allows a seamless and intuitive navigation through this information stored in a world-wide distributed database of web servers. Even though intrinsic problems such as lack of Quality of Service remain, there are currently trials for distribution of Video on Demand and telephony on the Internet. Moreover, the World Wide Web appears more and more as an integration medium for network-centric applications that are platform and technology independent, using the open standards of the Internet. In this context, it is then indisputable that the Internet will soon offer a wide range of commercial multimedia services in the network.

Nevertheless, up to now, these services are mostly asymmetric services such as data retrieval, and do not provide much support for session control on a peer to peer basis. It can be noted also that there is no real support for subscription and authentication resulting in lack of security and untrusted service behaviour. It is then certain that TINA, with its advanced management and service architecture, has huge added values with regard to current Internet services. On the other hand PNOs are not willing to trash all the investments they are doing in TMN, IN and Internet. To be accepted, it is of paramount importance for

0-8186-8335-X/98 $10.00 © 1998 IEEE

TINA to propose a strategy and scenarios for the integration of existing legacy systems.

The goal of the TINa To INternet project (TINTIN) is to propose an architecture and scenarios for the interworking of Internet and TINA. Migration and interworking strategies will be developed and validated through prototypes and field trials. The project is scheduled in three phases, the first phase will come up with a lab version of a TINA architecture offering limited Internet services (Web-Chat, FTP). The second phase enhances the prototype with advanced interactive multimedia services and will address the management of an ATM based network by TINA. In this phase we will study how new protocols which are actually developed at the IETF (RTP, RSVP, PIM, flowspecs,...) could be incorporated in the TINA/ATM connection management. The third phase will come out onto a pre-commercial field trial where the prototype is deployed on an IP/ATM/ADSL high speed network. Also interworking scenes of TINA connection-management with ATM-emulated LANs (LANEs) will be studied. During the second and the third phase our research will also address the problem of how TINA should manage multidomain-multiprovider landscapes in the Internet environment.

The rest of the paper is structured as follow. Chapter 2 outlines, the current status and trends of the Internet. Chapter 3 addresses the TINA service architecture as we applied it for some Internet applications. Chapter 4 presents some implementation issues and chapter 5 describes the basic application used to validate the architecture. Further work is summarised in Chapter 6 and a conclusion is given in Chapter 7

2. The current trends and state of the Internet

Why is the Internet so successful? At the network layer, the Internet can be seen as an architecture made of nodes that exchange datagrams based on the IP protocol. Is this the key of the success of Internet? We think that the answer is "no". Good architectural grounds are not the key to success, though they can help. For instance, the Microsoft Windows operating system had to wait until 1995 to provide multitasking. Other operating system architecture such as the UNIX had had this feature for nearly 30 years, working perfectly. This didn't prevent Windows to colonise 90% of desktop computers in the world. The difference is that early MS Windows, though conceptually less advanced than other architectures, provided the user with cheap, friendly and numerous applications.

The second factor is the number of users. To attract attention, an architecture must provide services or applications. If these applications are attractive enough, more users will come to this system, increasing the number of potential users for new service. Third parties, being companies or private users, will find some interest to publish new services, and applications to these new potential users, attracting by the way more users and the wheel goes on[1].

To start this process of "potential users" and "service providers", the system must reach a first critical mass in terms of number of users. Below this point, the system has a major risk to be a flaw. Above this point, the success is almost guaranteed. The conditional form of the two previous phrases indicates that these statements are not rigorously demonstrated but are rather common sense.

For the Internet, it has been exactly the same process. For nearly 30 years, the Internet has been the playground of military people and scientists. It's only very recently that the Internet arose on the public market place, even playing the role of a social revolution[2]. The exponential process is very explicitly depicted on the following figure (1). From 1993, an exponential growth of the number of host can be noted .

What is interesting now is to look at the increase of the number of web sites. As depicted on the next figure (1) According to what has been said before, it looks that Internet reached the critical mass of potential users around 1994. At that time, the World Wide Web appeared on the Internet and provided the user with a friendly, easy to use interface. The WWW was then the trigger of the exponential growth of the Internet. Enough information ,and a consistent way to access it, was there to attract users. These users were themselves feeding the Web with information, attracting more and more users, etc[3].

---

[1] This is exactly what happened in France with the "Minitel". Minitel does not provide highly graphical pages, with animation's, etc. But it provides the users with a multitude of services, making it part of the daily life of the French people.

[2] By the way, it is amazing the number of people thinking that Internet is a brand new baby of the 90ties technology!

[3] Nowadays it is said that every 23 seconds a new PC is connected to the web.

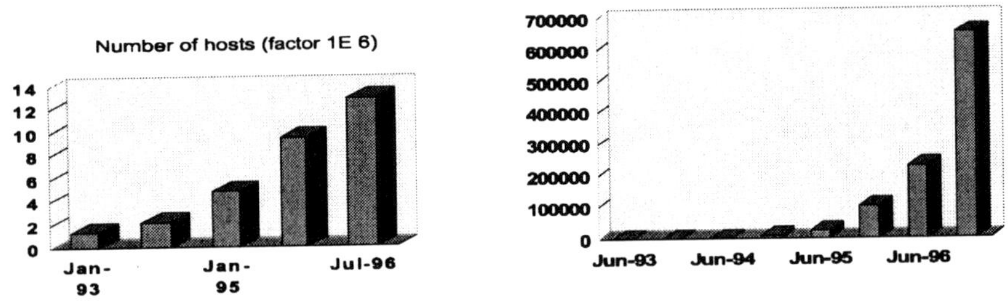

Figure 1 : Left: the growth of Internet host volume Right : the growth of website volume

The lesson of the Internet is that it is not the architecture that makes a success. TINA can be based on firm architectural grounds ([4],[5],[6],[7],[8]), introducing session concepts, mobility, service and business architecture, etc. However, it will only succeed if it reaches the critical mass of users, information and applications.

It is then of prime and paramount importance for TINA to reuse, somehow, the information contained in the Internet, as seamlessly as possible, and possibly bringing added values to Internet in a gradual way (see [3]). We state and believe that using the existing Internet infrastructure can act as an excellent research laboratory for rapid introduction and testing of TINA service platforms adapted to Internet semantics.

3. The TINA Service Architecture applied to the Internet

In this section we will outline a visionary framework sketching how the Internet could evolve from it's current state to a tinarised environment offering a broad range of services varying from the current Internet services to a more enhanced multimedia multiparty service assortment (video and voice conferencing, chat services, whiteboard, joint document editing, teletraining, and many more). In this framework, the latter class of services would take benefit from the TINA Service Architecture (TINA-SA) [7] for the service-level management aspects, such as a session oriented service control. As such some aspects as security, authentication and integrity of trustworthy service usage can be introduced in the Internet.

3.1 The business model

When one speaks of introducing multimedia service packages, and later on QoS metrics to support the reliable delivery of these services, it is obvious that some business roles and business interests will come into existence. In Figure 2 on the left side a summarising sketch of the TINA business model is shown. Several stakeholders are distinguished and the relationships between them are basically the *use* relationship and the *contract* relationship. With use is meant that the using stakeholder uses the services of the used stakeholder on a rather temporarily basis. With contract is meant that the involved stakeholders have a longer term agreement on a prepaid basis. On the right side of the figure we sketched the business model we applied to introduce TINA services in

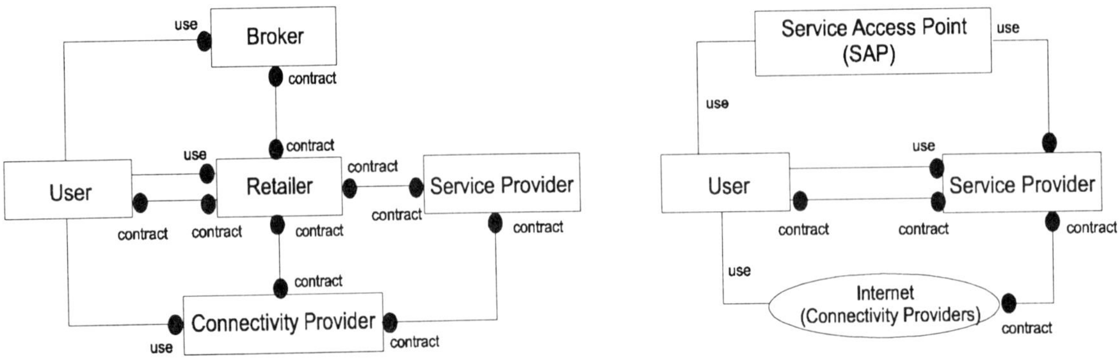

Figure 2 : Business Model for the interworking of TINA with Internet

(left: TINA model Right Internet model for a first phase)

the Internet in an initial phase of TINA Internet interworking. The Internet business model is a subset of the TINA model, but we had to face some mappings. At first, in the TINA model the service provider is a third party subcontractor of the retailer who will offer a service properties or to get a list of providers offering particular service offers. Behind the covers of the "Tina search " button trading mechanisms for finding service providers and services would be necessary.

The retailer in the TINA sense of the word would be assortment of one or more provider. In such a position the retailer can apply some service composition on the services of different providers to offer an aggregated service. In the Internet however, a service provider will, in a first phase, mainly live and act on itself. Therefore the end user will have a direct interaction with the service provider based on use and contract relationships.

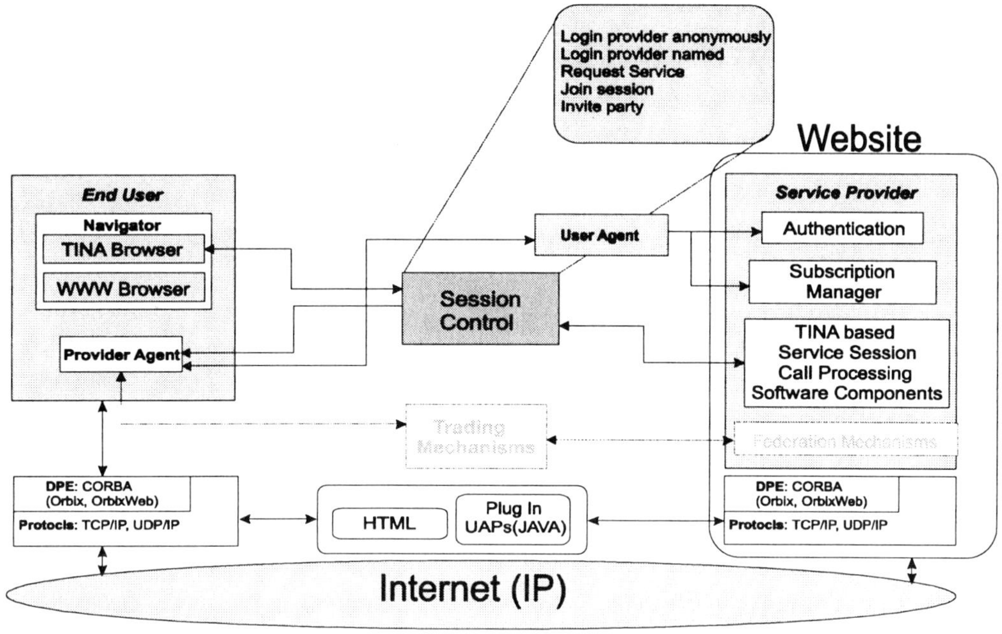

Figure 3 : Overview of the technology model for the first phase of TINTIN

A second difference is that the broker in the TINA model is regarded as an entity with a business role, supplying trading functions and search engines for providers and services. In our Internet business model we regarded the TINA broker as some kind of a common TINA Service Access Point (SAP). This TINA-SAP works similarly as the website *home.netscape.com* which is contacted by pressing the "Net Search" button of our WWW browser, and which delivers us search engines and the hot websites of the day. Clicking some "Tina Search" button or typing the URL *www.tinatscape.com* would bring the user to a website that downloads an HTML page carrying some hyperlinks to service providers and an applet that implements a Graphical User Interface(GUI) to interact with the search engine of the TINA-SAP server. This GUI offers a menu to search for a service with certain introduced in some later stage of the TINA-Internet migration.

3.2 The technology model

The technology model (*Figure 3*) of phase 1 of the TINTIN project relies on the *existing* protocols of the Internet (TCP/IP and UDP/IP). On top of this protocol stack a middleware platform (DPE: CORBA: Orbix [9] and OrbixWeb [10]) is supported that enables the distributed execution environment of the TINA service architecture.

The TINA service platform allows the smooth introduction of new services in a multi provider landscape. On top of the CORBA platform the different roles and stakeholders are indicated: the enduser, the broker (trader) and the TINA enhanced website. The session control logic is the key concept of the usage of multiparty services.

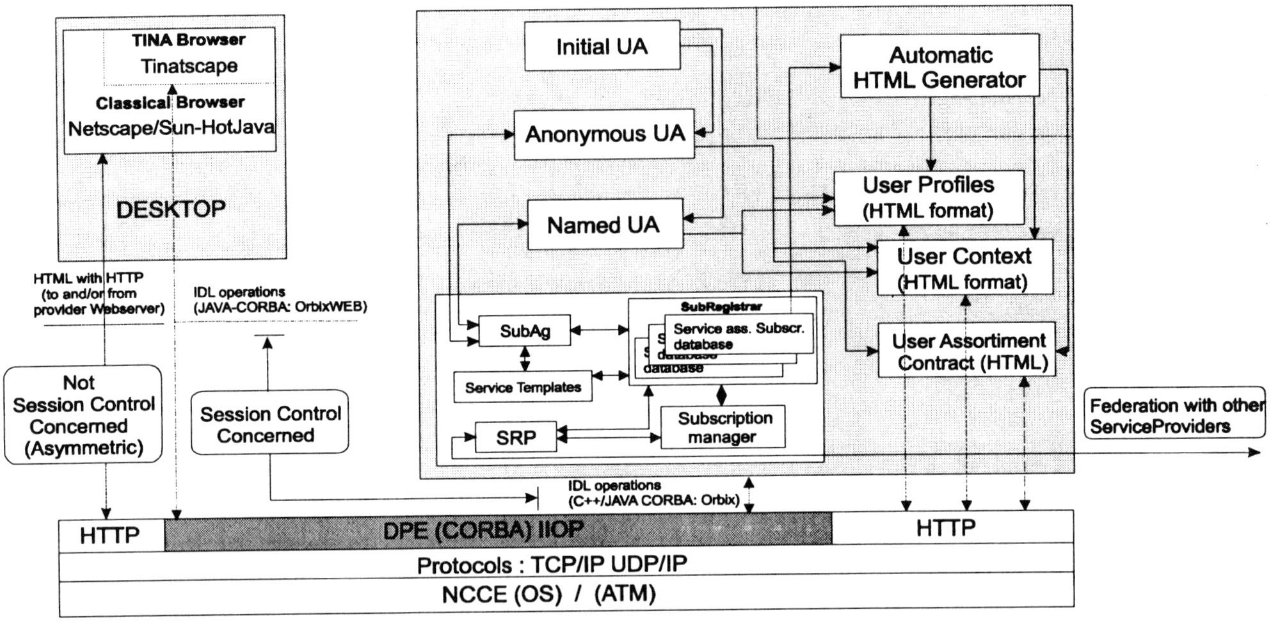

Figure 4 : Modelling of a TINA enhanced access management system

3.3 The End user system and the provider access management

The desktop of the user is deployed with an ordinary web browser (see Figure 4) enhanced with e.g. persistent JAVA classes implementing the computational objects for provider agents and access sessions. When using SUNs HotJava browser a button "TINA search" can easily be added to access the TINA world. Pressing this button causes the downloading of a GUI to access the TINA-SAP. This GUI offers the functionality to contact a provider or to get lists of providers. With a technology as OrbixWeb the events (logging in to a provider, searching a TINA service, getting lists of providers) coming from the TINA GUI applet are handled behind the covers of the browser via the CORBA based DPE. The TINA GUI can be regarded as a TINA browser.

The access management system of a service provider is in a certain sense hybrid because it interfaces via the CORBA DPE with the IIOP protocol and via the HTTP protocol. The initial agent is the entrance point of the provider which is contacted with the provider agent (PA). Via the initial agent the user can log in from his TINA browser anonymously are named, and accordingly a UA_{named} or a $UA_{anonymous}$ is created:

The user logs in as a named user: In this case a named UA is created based on a complete user identification provided by the provider agent. The user will be accounted according to the services he uses and the modalities and features (e.g.: QoS) attached to the service usage. This means that the bill the user will get presented will contain a list of all the sessions in which he was involved and the price he has to pay for them. The advantage of this is that the user can subscribe on the fly to whatever service, but the identity of the user, the services he uses, and the sessions in which he was involved are disclosed to the provider.

The user logs in as an anonymous user: In this case still a trusted relationship has to be established between user and provider, but without disclosing the identity of the user to the provider. This is realised by a so-called user alias. The provider configures packages of services which are the subject of a prepaid contract. When the user subscribes online from his terminal for an *anonymous contract* he gets a registration form in which he can configure his contract (range of services with quality and feature levels, duration of contract, terminal system) and give his user alias. During the set-up of this contract the user will have to provide his identity, but this identity is only used to make up an aggregate bill for the package. Once the bill is made the identity is removed and the provider ends up with an association of the alias and the agreed profile of the service contract. As such the user can make use of the services in an anonymous way. Some functionality is foreseen to allow the user to change his alias. When the user cancels the registration form he will be able to subscribe to some kind of default service package for which no accounting is necessary.

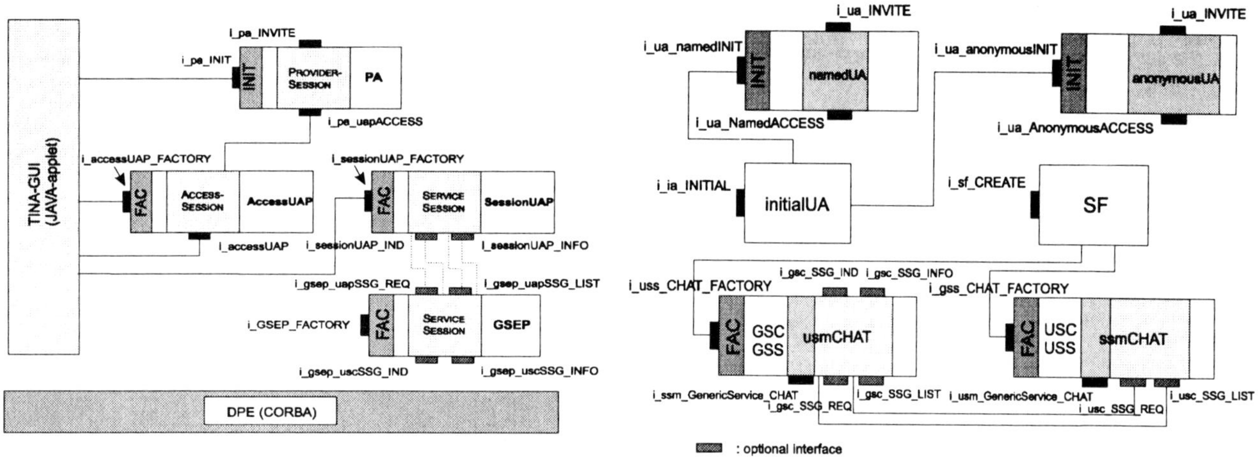

Figure 5: left : The end user terminal: the TINA application environment right : the service components of the provider domain for e.g. a chat service

On getting a UA also a Subscription Agent (SubAg) is created that allows the user to browse in the services assortment of the provider. Information on services can be downloaded in HTML format. The SubAg is also the contact point for getting HTML formatted information on longer term user service assortment contracts or for changing user service profiles and contexts. Events concerned with contacting (PA versus Initial UA) and logging in to a provider (UA: named or anonymous) are handled by the DPE, but exchange of registration forms, contracts and profiles are performed by HTML and HTTP. These HTML formats are automatically generated by a HTML generator.

4. Computational mappings

4.1 The end user terminal: user application (UAP) and provider agent (PA)

The end user TINA application environment (figure 5, left side) consists of different factories that manage the creation of the appropriate components for access sessions and service sessions. The factories for PA and Access Sessions are persistent or could just as the TINA GUI be downloaded by the TINA-SAP in the form of JAVA classes that work on top of CORBA via e.g. OrbixWeb[4]. When the user asks from his TINA-GUI the access to the services of a certain service provider, then an access-session and an associated provider agent are created and the object reference to the initial user agent of the provider is returned by the TINA-SAP to the PA. This is the scenario for *contacting a provider*. For every access-session with a new provider a new PA and access session are created.

On *requesting a service* from a service provider a service session factory JAVA class is downloaded by the provider together with an appropriate GUI. These components can be 'plugged in' the other JAVA components that came from the TINA-SAP.

4.2 The session processing components: USM, SSM and SF

The service factory (SF) offers a generic interface `i_sf_CREATE` from which the creation of the service components of a particular service (determined by the serviceId) can be requested. The SF contains for that purpose the references to the factories of each of the offered services. In figure 5 (right side) the factory interfaces of the USM and SSM are showed.

An important aspect of the concept of the USM and SSM components of a certain service type is that they are customisable to a certain extent. Every USM and SSM will provide the possibilities to create dynamically the interfaces for a Service Session Graph. This means that the opportunity to use an SSG will exists and can be negotiated at session creation. However applications *should* not be obliged to use the SSG as is described by TINA. Applications can also rely on other mechanisms to manage multiparty service behaviour. The concept of *interfaces on demand* has been realised by a *generic interface* that the USM and SSM of every service have to offer. For the 'chat service' for example, this interface on the USM is called `i_usmGenericInterface_CHAT` and `i_ssmGenericInterface_CHAT` on the SSM. These interfaces offer retrieval operations for the interfaces of the SSG in the case the use of a SSG has been agreed upon. They also provide the operations necessary for retrieving service specific interfaces. The interfaces for the SSG allow the managing of it. Four types of SSG interfaces are distinguished:

[4] Also the integration of Netscape and Visigenic's ORB could be a technological solution

i. **Request interfaces (`REQ`):** This interface is used to request modifications to the SSG: adding parties, adding stream bindings, adding stream interfaces, etc.

ii. **List interfaces (`LIST`):** This interface is used to request enumeration of objects in the SSG such as getting a list of parties, getting a list of stream flow end points , getting a list of stream bindings.

iii. **Indication interfaces (`IND`):** This interface is used to negotiate on alterations of the SSG between members of the session. The SSM sends via this interface to every party an announcement of a pending change. The parties can hereupon deliver an acceptation or a rejection. After collecting all the answers a votingrule is applied to decide whether the change will be executed or not

iv. **Info interfaces (`INFO`):** This interface is used to notify each party in a session about a change that has been executed in the SSG.

The `GenericService` interfaces will also be a key concept for multidomain management aspects.

4.3 The trading and federation components

Trading is not a small subsystem and is in fact one of the key issues in designing and operating a distributed system. Though our trading approach could be subject of a future publication, we will outline some of the basic issues of the implemented trader which is based on the concept of an ODP-trader [2].

A trader basically is a computational object that maps a service offer, expressed in terms of a name and optional information, with an object reference. In a first step we assumed only one organizational domain and consequently also one trader, which contains all the service offers. With respect to the discussed computational objects, these service offers are mainly the UA's of different potential users and the references of the actual and future Service Factories. The trader consists of a Directory Service Agent (DSA) with his interior Trader Database (TDB) and the Federation Service Agent (FSA). Exporting objects can register object references along with their properties and name via the `export` interface and the references can be imported via the `import` interface. The export interface supports the operations `TraderModify`, `TraderWithdraw` and `TraderExport`. The import interface supports the operations `DirectorySearch`, `DirectorySelect`, `DirectoryList` and `DirectoryExport`.

The trader database (TDB) is organised as follows: For every service type (based on the type of the interface of the service object) a binary tree is maintained of all the object references from that service offer. These binary trees are maintained in a hash table which is accessed based on the interface name which is used as the service identifier. In the binary tree itself the service offers are sorted by a comparison function defined on their names and/or properties and/or values of these properties. The TDB is persistently stored on disk, and at regular intervals (i.e. after each 10 updates) the TDB is written to disk.

The FSA is already included because in future, different

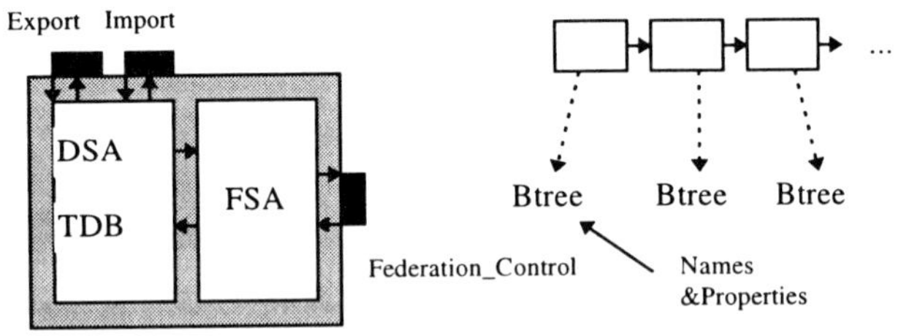

Figure 6: Trader and Federation related components

traders will be federated across several administrative or organisational domains. Several traders will then form a hierarchical network, and a search to service offers which are not successfully completed in the local trader, must then be propagated through this network by the FSA sub-modules. The precise internal structure of the FSA however, remains for further study.

5. Case Study : Chat Application

In this section we will sketch some scenarios explaining how we introduced and tinarised a chat service exemplifying how Internet and TINA meet each other in the realm of phase 1 of our project [13]. This means that we only made use of the existing Internet protocols (TCP/IP) which we used as the KTN for the CORBA based DPE (Orbix and OrbixWeb), and for the transport of streams. We deployed the chat service in two ways: For the primary version we used operational CORBA interfaces for the exchange of the chatted string data. In the enhanced version we used UNIX sockets and a UNIX PCM-voicecoding process for exchanging audible voice streams. As such we deployed a test version of a simple TINA based chat service on the real Internet.

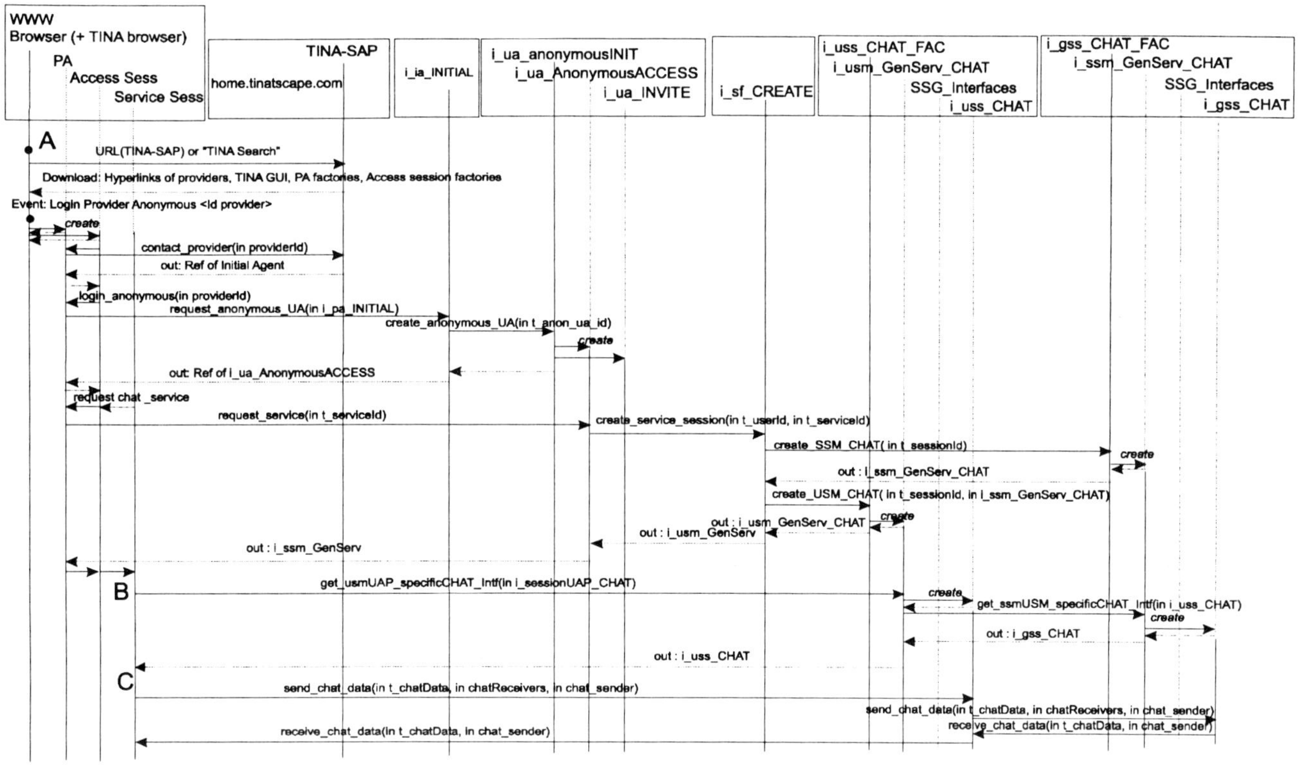

Figure 7 : Event trace diagram: Contacting the TINA-SAP, logging in to a provider, requesting a chat service and using it.

5.1 The primary chat service

The service specific interfaces of the User Service Segment USS(USM) and Global Service Segment GSS(SSM) of the chat components look as follows:

```
interface i_uss_CHAT {
  void send_chat_data (
       in t_chatData chat_data,
       in t_userNameList list_of_chatReceivers,
       in t_userName chat_sender);
  void receive_chat_data (
       in t_chatData chat_data,
       in t_userName chat_sender)
}
```

The interface `i_gss_CHAT` contains an analogue send operation and the interface `i_sessionUAP_CHAT` contains an analogue receive operation. In the event trace diagram of

Figure 7 : *Event trace diagram: Contacting the TINA-SAP,*
logging in to a provider, requesting a chat service and using it.

the scenario's of requesting for a chat service and using it are stated, the scenario works as follows:

1. Phase A: Contact the TINA-SAP (broker) with it's URL or by pressing the "TINA Search" button. The TINA GUI applet together with factory classes for PA and access sessions are downloaded to the enduser's WWW browser.

2. The end user indicates from the downloaded TINA browser that he/she wants to contact a certain provider anonymously. The result of this is the creation of a PA and an access session for this provider. From this PA a request is invoked on the TINA-SAP to contact this provider. Via some trading mechanism the broker searches the reference to the Initial Agent (IA) of the provider returns it to the PA.

3. The PA requests on the IA the creation of an anonymous UA for logging in anonymously to the provider. The $UA_{anonymous}$ is created and a reference to it is returned. The provider will also download (not shown in the figure) a service session factory (together with GUI components) and a list of default available services (e.g. in HTML format). If the user would like to use an advanced service he/she could ask for a registration form to subscribe to the service anonymously and provide an alias.

4. The user invokes the use of a default chat service. The session UAP for this default service could already be present in the in step 3 downloaded service session factory or it is downloaded now. The generic service interfaces on the USM and SSM are created and the reference to the chat generic service interface on the USM is returned to the session UAP of the user.

5. Phase B: when the user presses on the "initiate" button of the chat service GUI the session UAP requests the creation of the chat specific service interfaces. The service interface `i_uss_CHAT` of the USS is returned to the user.

6. The user invites one or more other users. This step is not shown in the figure.

7. Phase C: the user sends text string data to one or more other users. This request is passed via the USM to the SSM. The SSM will then invoke with this chat data a receive operation on the USMs of the users for whom the message is destined.

5.2 The enhanced chat service

The concept of stream is not useful in the case of a chat application since there is no need for continuous flow of information. The operational interface are sufficient for the required functionality. To validate the concept of stream, we've implemented an audio session capability embodied by the "audio talk" button in the bottom of the user interface. A typical scenario would be as follow. During a chat session, user A decides to start a voice conversation with an other user B. To make it so, A clicks on the name B in the "connected people" list and pushes the "audio talk" button. Upon clicking on this button, a request for a stream binding creation is sent to the session manager. This later takes case of setting up the streams. The following section expand the concept of stream as it has been implemented for the audio chat.

6. TINA streams on TCP

The concept of separation of call and communication is based on the idea that, before using resources, two end-users should first negotiate on the connection they really need. This appears to be a very sensible idea once Quality of Service is involved. That's why two levels were introduced to separate the call control for end to end negotiation and the connection set-up. TINA translates the call control into the service session and the connection control is embodied by the communication session and the whole connection management architecture. Up to now, that separation does not exist in Internet, that's why we introduced the concept of streams in the Internet framework to abstract the communication session and socket related concepts from the application. The streams provide the abstraction needed to conceive an Internet application using, during the same service session, several communication sessions dealing with different technologies. This is a first step towards the Quality of Service in the Internet but also towards a convergence between PNO's network and the Internet.

6.1 Mapping for Stream bindings

To use the concept of stream in the Internet framework, we have to provide a technology mapping for all the streams entities involved in the stream binding process at both service and communication level. The connectivity level, which is under the responsibility of the connectivity provider, won't be addressed since we will keep the Internet native routing. TCP is used at the transport layer since a connection oriented transport layer appears very convenient. TCP is very close to the concept of stream : it provides a path between two ports that are registered in their respective terminal by the applications.

In the TINA stream concept, the latest object to be technology dependent is the Network Termination Point (NWTP) [5]. We decide to map the NWTP directly on the TCP port concept, these terms will be used indifferently.

The Network Flow End Point (NFEP) references an addressable Network Termination Point (NWTP). The NFEP is a software representation of the NWTP that is manipulated by the Terminal Communication Session Manager (TCSM). The NFEP is bi-directional and is well suited to represent the TCP port at the communication level. The socket is a concept that is found in Internet (BSD socket) but also in ATM (Fore socket interface). It can be stated that the socket API is a technology independent interface for the transport layer. For these reasons, we choose to introduce the sockets at the communication level. In the Internet, a socket descriptor (SD) represents basically a TCP path which is a quadruple (IP machine1, TCP machine1, IP machine2, TCP machine2). With this concept, it is then easy to translate a single point to point NFC in terms of TCP paths.

We can then map a point to multipoint NFC into several TCP paths. The advantage of this mapping is that several socket descriptors can be attached to one single TCP port provided that the quadruple is different for each socket. In other words, provided that they all reach different NFEPs, which is sensible. The root NFEP of a point to multipoint NFC is then an object containing a number of socket descriptor being equal to the multiplicity of the point to multipoint NFC.

The SFEP can either be a source or a sink. We chose to map each instance of a SFEP in a threaded object containing a method overloading the stream operator/ class (C++ / JAVA mapping). In that way, the stream can directly be used by the application in a easy and consistent way, keeping intact the semantic of the stream. During the stream binding process unidirectional SFEPs are associated to NFEPs. The "SFEP" thread is provided with a table with the list of sockets involved in the potential point to multipoint connection. When writing on a root source of a point to multipoint connection, the thread copies the data on each SD, resulting in a totally transparent point to multipoint SFC.

Within the terminal, the Terminal Flow Connection (TFC) ensures the connection between the NFEP and the SFEP. No mapping is really needed since it is the socket API that ensures the "connection" between the NFEP (TCP port) and the SFEP. This "connection" is materialised by the buffer associated to the socket attached to the TCP port. Indeed, the application can directly read/ write data from/ to the NFC from/ to the buffer.

A stream interface is composed of many SFEPs. Hence, we can then say that the stream interface is composed of SFEP threaded objects processing the data in reading or writing. In this way, we believe that the whole mapping is in line with the session/ communication concepts described in the TINA specification [4] and the overall idea of streams [5]. Figure 8 illustrates this mapping.

6.2 Implementation of the streams

The link between two interfaces (stream or operational) is called a channel [5] [14]. The channel is composed of three parts: the stub, the binder and the protocol adapter. The stub support the representation of the interface to the user application; the binder maintains the information about the channel such as a data buffer. In our case, there is no protocol adapter since TCP is used at each node.

Architecture of the Stream Interface Management.

When it comes to operational interfaces, invoking a method on a remote object is completely transparent thanks to the stub and the skeleton generated from the IDL description of the interface. In our case, the stream interface is created dynamically and use an explicit binding instead of an implicit one. It is therefore not possible to use the same concept of "stream interface compiler". Nevertheless we would like to provide as much as possible reusable components hiding the complexity and the mechanisms involved in the stream interface.

We decided to have a specific objects managing the life cycle of the Stream Interface : the SIM (Stream Interface Manager) which is hidden in the stubs of the UAP. This object is able to create a Stream Interface, grouping SFEPs and managing the life-cycle of the pool of threads associated to the SFEPs (creation, deletion, etc.) which are the internal representation of the Source/ Sink SFEP. The SIM is also responsible for abstracting the SFEPs. For instance while sending information on the source SFEP of a point to multipoint connection, the SFEP manager copies the information on each SD involved in the connection. The Stream Interface has relevance only to the application that asks for a stream binding. hence, all these concepts are translated into JAVA classes.

If the terminal posses specific hardware, this could result

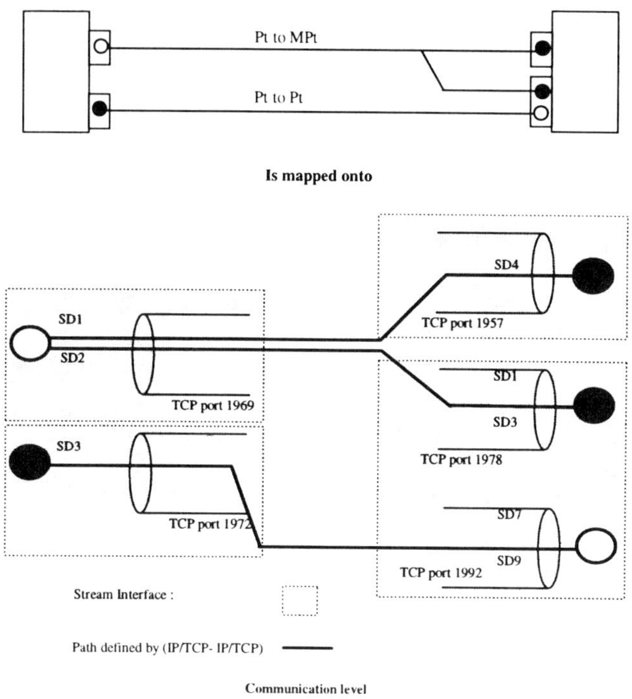

Figure 8 mapping for the stream interface

in specific NWTP for this terminal. For instance, getting the audio_in stream is different for a SUN workstation or for a PC based computing node. The idea is to have a specific computational object whose responsibility would be to map the NWTP onto NFEP and this is indeed the TCSM. The TCSM is responsible for managing the TCP ports, that is to find a free port, open it, close it and, if necessary, reserve it. The TCSM is also responsible for managing the SFCs that is to create the SDs (mapping of the NFCs), attach it to a TCP port (mapping of NWTP), and perform the life-cycle of both.

The SIM package

Java provides the a logical grouping of classes into a package. We decided to group all stream related classes into a *tinaStream* package so that any other Java program wishing to use the TINA stream concept will just have to import this package.

The classes of the stream package and the relationship between these classes is depicted on the following OMT diagram :

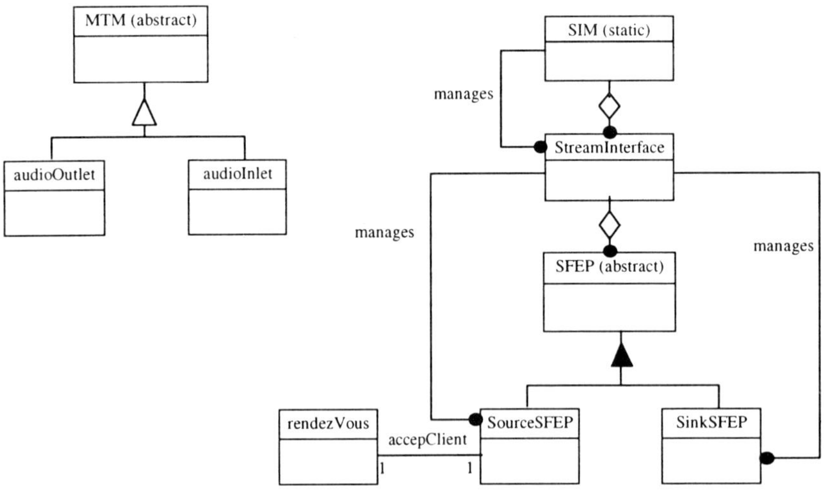

Figure 9 relationship between the tinaStream package classes

- The stream interface related classes

The Stream Interface Manager is a static class in the Java sense of the term. This means that one doesn't need to create an instance of the class to use it, since there is only one SIM per object (that manages all the interfaces).

The SFEP class is in fact an abstract class that extends (Java terminology to say that the class derives from...) the Thread class. From the SFEP derives the source and sink SFEPs which implement the logic of the TINA Stream Flow End Points. The rendezVous class extends also the Thread class and is used by the SourceSFEP to accept leafs (Sink clients). The details are explained in the strip down of the code.

- The multimedia (MTM) related classes

the MTM classes is an abstract (generic) class that extends the Thread class. The audioInlet is a class representing an audio input (mike, cd-player, ...) and that deals with platform specific constraints. It would be elegant to derive the platform specific classes from the audioInlet class, for instance by having a SUNaudioInlet or a PCaudioInlet, etc. but this is not useful in our case. The audioOutlet represents an audio output, for instance a speaker. Here again, it would be nice to have specialised classes extending the audioOutlet with, for instance, tone control, balance control, etc. the MTM could be also the root for other multimedia classes such as video inlet, etc.

Together with the stream concept, the MTM is a powerful tool for the Java programmer. We are now going to run quickly through the tinaStream package.

Asking for a binding.

The TINA stream concepts provides the user with a high level of abstraction regarding the connectivity the UAP needs. For the UAP, the only relevant concepts are the ones of parties to be connected, directionality and QoS. All the management is done on the retailer side and in the stream interface stubs. Though the concept could have been easily extended to multipoint to multipoint connections, we only implemented the point to multipoint since it is the most widely used within common applications.

The corner stone of the SFEP classes and MTM classes is the way they are going to communicate the data flow to the application. That's why we chose the PipedInputStream and PipedOutputStream Java classes to export/import the data flow processed by the SFEP. Hence, the SourceSFEP posses a PipedInputStream in which data (bytes) can be written and the SinkSFEP posses a PipedOutputStream where bytes can be read from.

Put simply, a piped input stream is the receiving end a communications pipe. They key of these classes is the connect method by which an input pipe is glued to an output pipe. This is very useful to allow two threads to communicate by having one thread send data through a piped output stream and having the other thread read the data through a piped input stream. And since the SFEPs, as well as the MTM objects, derive from the Thread super class, the pipedstream are ideally suited.

Once the SFEPs and MTM objects are set-up, the former by the retailer domain and the later by the application itself (in conjunction with the TCSM), the User Application (UAP) is provided with a bunch of PipedOutputStream and PipedInputStream instances. All the UAP of party A has to do to let party B hear his voice, once the stream binding has been performed, is to connect its audio inlet on the SFEP source of the stream binding. This is done by a simple call :

```
speaker.outStream.connect(chatAudioSI.
                          si(index).pos);
```

where speaker is an instance of an audioInlet.

Stream binding seen from the retailer side.

One must remember that the stream binding is asked by one party but is set up by the SSM. So, the first thing the SSM must do is contact all the parties which are to be involved in the stream binding to see if they agree or not in participating in the SB. The **joinStreamBinding** method serves this purpose. Typically, the implementation comprises a pop-up window prompting the user to join the stream binding. This is like the ringing of the phone, one has the choice to hook up or refuse the communication. There can be also other information required before joining the stream binding. For instance the user could be asked to provide a preference on the quality of service (QoS) of the connection (usually, the user pays according to the QoS and may prefer a lower quality when it is not strictly required).

According to the service profile, the SSM may decide not to pursue the stream binding process if not all the parties accept to partake in the stream binding. This is how it has been implemented but this is not a strict rule.

When the conditions to pursue the stream binding are fulfilled, the SSM contacts again the parties to ask them to register Stream Flow End Point that would suit the stream binding. To make it so, the SSM invokes the **registerSFEP** method on the user application. The method returns a audioSI structure that contains the identifier of a source and a sink. Note that it is not useful to pass the reference of the SFEP itself. It is indeed the user application that sets up the SFEP and that will use them. The SSM only manages and co-ordinates the operations on the SFEPs, therefore, only an identifier is passed to it. It is worth noting also that, for the sake of the implementation, the stream binding is bi-directional by default. Moreover the type of connection is also set by default to the audio type.

Put simply, the root SFEP is mapped onto a TCP server source and the leafs onto TCP client. To be able to make the connection, the information on the source must be provided to the client. This information is simply composed of the server hostname and the TCP port number. These information are returned in a structure by the **GetInfoSource**. The argument is the identifier of the SFEP performing the role of root.

This information is then forwarded to the TCP client (SFEPleaf) with the **SFEPBound** method. The arguments are the SFEPleaf identifier and the info on the server it has to connect to. The invocation on this call forces the *connect* to the server.

The root can now be advised that the connectivity has been established with the leafs with the void **SFEPinit** whose argument is the identifier of the sink SFEP. Note that the implementation of the SFEP - TCP server of TCP client - is not related to the nature of the SFEP - root or sink- since, once the connection is established, the TCP server can read/ write as well as the client. We found the mapping root=TCP source very convenient since the most common connections are point to point and point to multipoint.

7. Further Migration steps

In the next phase of our project we will study how the new multimedia related protocols being developed at the IETF for the Integrated Services Internet (ISI) such as RTP, RSVP, PIM, and the flow specification protocol could be interwork with the TINA architecture. In addition we are working on migration and interworking steps for IP and ATM. An ATM switch adapter has been developed based on AToMMIB [11] to translate connection setup commands of the TINA connection management into SNMP protocol commands for the agent in the ATM switch. As such we believe in a migration path that starts from TINA-IP and evolves to TINA-IP/ATM.

8. Conclusion

On developing the strategies for TINA Internet interworking the essential goal we had in mind was to introduce TINA concepts seamlessly in the Internet. An Internet user's usage plane is then regarded as a subset of the TINA usage plane, but a user subscribing to TINA services should not be aware of the fact that these services are based on a TINA architecture. A second aspect is that the service architecture had to be immediately deployable on the present Internet without excessive costs and efforts. We advocated a smooth introduction of TINA concepts rather than TINA itself. In the realm of this we still believe that also for a good architecture as TINA, easy access and a high quantity of available information is a decisive factor for its acceptance.

The experience we obtained by carrying out the interworking study of TINA with Internet is that TINA offers the fundamental and conceptual framework for global interoperability with the Internet. Analysing the benefits of TINA migration paths we postulate that from the service point of view TINA could gradually introduce a business model into the Internet via a very smooth way which is guaranteed by the DPE solution. The architecture could bring both the Internet and PNO's in a more favourable position to interwork and to take advantages of each other's position: PNO's could take benefit from the popular IT model of the Internet i.o.t. deliver services, while the Internet could take benifit of advanced networking infrastructure such as an ATM backbone. A service that could flourish in this dual world for example is a VPN service on an ATM network with an outsourced administrator management interface via the Internet. Because of the architectural concepts themselves TINA would bring the ultimate solution for rapid service provisioning in a virtual information society of which the Internet is an integral part.

References

[1] C. S. Hong, H. Abe, K. Dai and Y. Matsushita, "Realising Distributed Multimedia Networking Services On DPE", *Proc. GLOBECOM 95. pp. 607-613*

[2] C. Popien, B. Meyer, "Federating ODP Traders: An X.500 Approach ", *Proc. ICC 93. pp. 313-317*

[3] EURESCOM Project P508, "Evolution, Migration Paths and Interworking to TINA", February 1996

[4] TINA-C, "Overall Concepts anfd Principles of TINA", February 17, 1995

[5] TINA-C, "Engineering Modelling Concepts (DPE Architecture)", December 94, 1994

[6] TINA-C, "Computational Modelling Concepts vers. 2.0", February 1995

[7] TINA-C, "Definition of Service Architecture vers. 4.0", October 28, 1996

[8] TINA-C, "Network Resource Architecture vers. 3.0", February 10, 1997

[9] IONA Technologies, "OrbixWeb Programming Guide", November 1996

[10] IONA Technologies, "Orbix Programming Guide", release 2.0, November 1995

[11] RFC 1695, "Definitions of Managed Objects for ATM Management Version 8.0 using SMIv2", August 1994

[12] ITU-T, Recommendation X.903, "Reference Model for Open Distributed Processing", February 1995

[13] Jean-Marc Reynders, Christel Knops, "Interworking between CORBA distributed applications and the World Wide Web", Euromedia 96, December 1996.

[14] Fredéric Dang Tran, Victor Perebaskine, Jean-Bernard Stefani, "Binding and Streams: the ReTINA approach", submission to the TINA 96 conference, March 15 1996

Managing TINA Streams that use Internet Transport

Hyun Cheol Kim
ETRI, South Korea

Frank Steegmans
Alcatel Telecom, Belgium

Narayanan Natarajan
Bellcore, USA

Jarno Rajahalme
Nokia, USA

Francis Wellesplein 1
B-2018 Antwerpen, Belgium
Telephone: +32-3-240 7957
email: steegmaf@rc.bel.alcatel.be (fste@tinac.com)

Abstract

The current TINA Network Resource Architecture provides a flexible framework for integrated management and usage of various connection oriented transport networks.

By introducing the concept of Layer Network Binding at the Layer Network level this framework can also support connection-less transport networks. Interfaces allowing the manipulation of TCP and UDP endpoints in the terminal, are specified so that IP Connectivity Providers can control the IP network part of TINA Streams and associate it with the terminal part.

The combination of these interfaces with the Layer Network Binding concept facilitates the usage of basic and advanced internet concepts such as IP uni-cast and multi-cast for TINA Streams. Furthermore, it allows Connectivity Providers to transparently use any combination of transport network technologies, including IP, inside their network to meet the network quality of service requirements for TINA Streams.

1. Introduction

The *TINA Network Resource Architecture* (NRA) is the TINA component architecture that specifies concepts and principles for the design of network control functions (i.e., functions that provide connection setup and control) and network management functions (fault, configuration, accounting, performance, and security management) for a TINA network [1]. An important aspect of the NRA is that it defines an end-to-end connectivity service for transporting information between TINA applications resident in Customer Premises Equipment (CPEs) attached to a TINA network. This connectivity service is based on the concepts of *stream flow connections* and *communication sessions*. A stream flow connection is a connectivity (or binding) between two or more stream flow end points, where each stream flow end point is either a source or a sink for an information stream (typically associated with some continuous medium) that has a specific format and quality of service (QoS) characteristics including bandwidth and timing. A communication session is a context in which one or more stream flow connections can be set up and managed as a unit.

The physical structure of a TINA network that supports stream flow connections has two main components: a *connectivity layer network* and communication resources in the CPEs (which may range from simple audio and video devices to a general purpose computing system) attached to the connectivity layer network. A connectivity layer network is a transport network made up of components of different network technology, such as ATM, Frame Relay, N-ISDN, wireless, SDH, etc.

A primary goal of NRA has been to define the interface for communication session management function hiding from the clients of this function (service level applications) the technology specific details of the network used for transporting streams including details of interworking between different technologies. This goal has been achieved in the NRA (see [1]) by defining the communication session management interface in a generic manner using the concept of *logical connection graph*. A logical connection graph describes the connectivity between TINA applications in terms of aggregates of stream flow connections and the characteristics of the end points of the stream flow connections.

The current version of the NRA [1] has specified in detail the mechanisms for communication management for the network structure where the components of the connectivity layer network are different types of connection oriented transport networks (such as ATM, Frame Relay, N-ISDN, and POTS). In such a network structure, setup of a stream flow connection involves set up of different levels of connections in the network:

- a connection across the connectivity layer network, called a *network flow connection*
- a connection across a *layer network* which is a portion of the connectivity layer network based on a specific technology; such a connection is called a *trail*
- a connection across a subnetwork within a layer network; such a connection is called a *subnetwork connection*

The NRA has specified the functional components (i.e. computational objects and interfaces) required for communication session management in such a network structure. (See Section 2 for a brief overview of these components). Management of communication sessions over connectionless networks, of which the Internet is a prime example, was not addressed. Given the wide prevalence of Internet, this is a serious limitation. This paper proposes an approach for TINA communication session management using internet transport[1]. This paper is based on work in progress in the TINA-C Core Team on this topic.

In a nutshell, the proposed approach modifies the current TINA connection management architecture without requiring changes in the communication session management interface. Thus, network transparency is preserved for service level applications. The concept of layer network, which has been invariably used in telecommunications networking literature to represent only connection oriented networks, is generalized to represent also connectionless networks. A layer network supports *layer network bindings*. A layer network binding represents a resource that transports information across a layer network. For a connection oriented layer network, the concept of layer network binding is further specialized into the concept of trail which is already supported in the NRA.

1. In this paper, the term 'internet' denotes the transport technology. Therefore all descriptions are applicable for any network that uses internet protocols including the 'Internet'.

Internet transport is modelled as a connectionless layer network capable of supporting different types of layer network bindings. The different types of layer network bindings supported by internet vary in the topologies supported and QoS guarantees provided. Establishment and management of internet layer network bindings are realized using existing and emerging internet protocols (such as UDP, TCP. RTP, IGMP, and RSVP) without requiring any modifications to them. As elaborated in Section 3, management of internet layer network bindings is accomplished using layer network level components of the TINA Connection Management Architecture. Internet specific details are transparent above this level.

The remainder of this paper is organized as follows. Section 2 presents an overview of the current TINA connection management architecture to set the context for Section 3. Section 3 describes in detail the proposed approach including changes to the current connection management architecture. Section 4 outlines how multicast IP addresses can be used while Section 5 gives an example scenario. Section 6 concludes the paper by summarizing the main results and highlighting the merits of the proposed approach.

2. Current TINA Connection Management Architecture

Figure 1 illustrates the current TINA Connection Management Architecture that is described in detail in [1]. A brief description of the various levels of connection management is given below

- *Communication Session Level*: The function provided at this level is the setup and management of communication sessions and stream flow connections (SFCs). Setting up a stream flow connection involves setting up a network part (called a *network flow connection or NFC*) in the connectivity layer network and a terminal part (called a *terminal flow connection*) in each CPE that contains an (application level) end point of the stream flow connection. The object *Communication Session Manager Factory* (CSMF) provides an operation for setting up a communication session. At communication session setup time, one or more stream flow connections may also be set up. When a communication session setup is requested, the CSMF object creates a *Communication Session Manager* (CSM) object and delegates the setup request to the CSM object. The CSM provides operations for the management of an individual communication session including addition, removal, and modification of stream flow connections that are contained in the

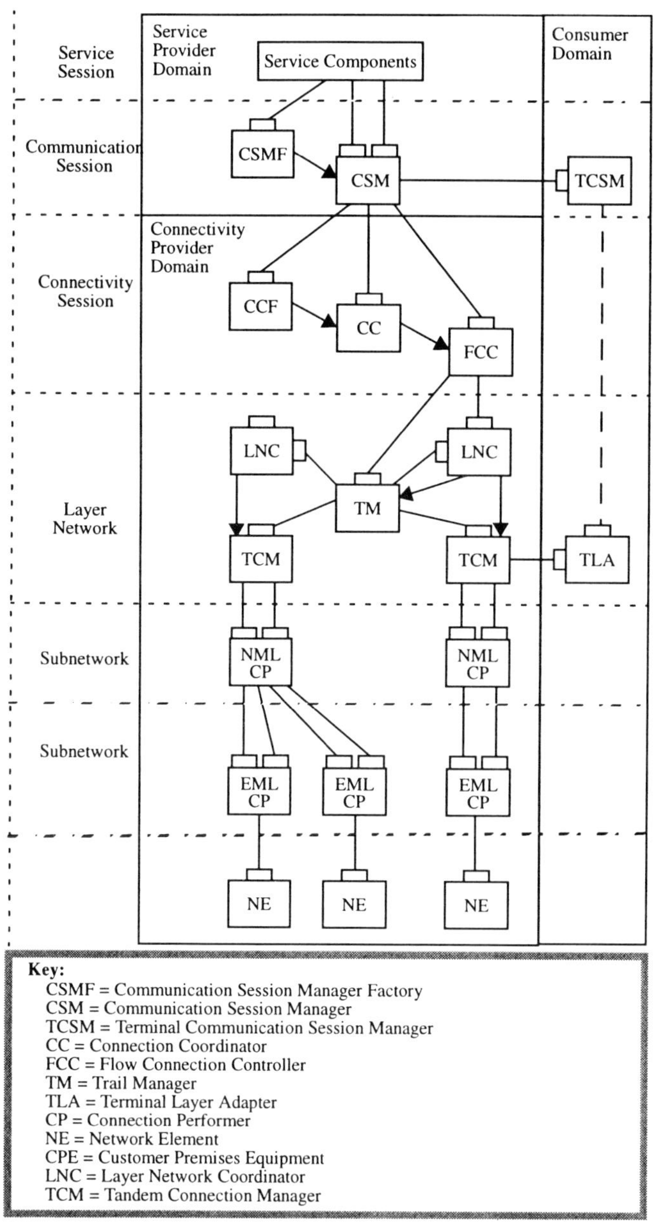

Figure 1. TINA Connection Management Architecture

communication session. The object *Terminal Communication Session Manager* (TCSM) exists in a CPE and manages the terminal part of all communication sessions in the CPE including the management of terminal flow connections which interconnect the application level end-points of the stream flow connection and the network level end-points.

- *Connectivity Session Level*: The function provided at this level is the setup and management of *connectivity sessions* and network flow connections. A connectivity session is a context in which one or more network flow connections can be established and managed as a unit. Setting up a network flow connection involves setting up one or more trails in the layer networks that make up the connectivity layer network. More than one trail needs to be created to realize a network flow connection if the end points of the network flow connection (called *Network Flow End Points*) are on different layer networks. The object *Connection Coordinator Factory* (CCF) provides an operation for setting up a connectivity session. At connectivity session setup time, one or more network flow connections may also be set up. When a connectivity session setup is requested, the CCF object creates a *Connection Coordinator (CC)* object and delegates the setup request to the CC object. The CC provides operations for the management of an individual connectivity session including addition, removal, and modification of network flow connections that are contained in the connectivity session. When a network flow connection setup is requested, the CC creates a *Flow Connection Controller* (FCC) object and delegates the setup request to that object. A FCC object provides management operations for the network flow connection associated with it including addition of branches, removal of branches, and network flow connection release.
- *Layer Network Level*: The function provided at this level is the setup and management of trails in a layer network, which is the part of the connectivity layer network that is based on one network technology. In the case where different portions of the layer network are under the control of different administrations, each such portion is called a *layer network domain* (LND). The portion of a trail within a LND is called a *tandem connection*. Within each LND, there is a *Layer Network Coordinator* (LNC) object that provides two functions. To the FCC object in the same administration, it provides trail setup functions. With the LNC object in a neighbor LND, it has a peer-to-peer relationship. It requests the latter to perform tandem connection management operations, and also performs tandem connection management operations when requested by the latter. When the LNC is requested to setup a trail, it creates a *Trail Manager* (TM) object and delegates the setup request to it. Similarly, when the LNC is requested to setup a tandem connection, it creates a *Tandem Connection Manager* (TCM) object and delegates the request to that object. A TM provides all trail management operations for the trail associated with it. Similarly, a TCM provides all management operations for the tandem connection associated with it. To setup a boundary tandem connection of a trail, the

LNC interworks with an object in the corresponding CPE, called the *Terminal Layer Adapter* (TLA) that manages the channels on access links that interconnect the CPE with a switch contained in a layer network.

- *Subnetwork Level*: The function provided at this level is the setup and management of subnetwork connections. Topologically, a layer network domain consists of a top level subnetwork that can be decomposed into two or more *subnetworks* interconnected by *links*. This decomposition may occur in many levels until each subnetwork maps to an individual network element (NE). Associated with each subnetwork is a *Connection Performer* (CP) object that manages subnetwork connections across the subnetwork. Reflecting the multiple levels of subnetwork partitioning, there is a hierarchy of CP objects. Figure 1 illustrates a two level hierarchy composed of a *Network Management Layer* (NML) CP that manages a connections across a subnetwork made up of several NEs and a *Element Management Layer* (EML) CP that manages subnetwork connections (cross-connections) in an individual NE.

3. Establishing Stream Flow Connections using Internet

The TINA Connection Management Architecture (CMA) as described in Section 2 is quite modular where network technology specific aspects are handled in the layer network level and the levels below. Thus, the most logical place to introduce a new network technology in the CMA is the layer network level since this is also the level that distinguishes the different connection oriented network technologies.

The layer network level in the current TINA CMA handles only connection-oriented networks using concepts, such as trails and subnetwork connections, based on similar concepts defined in ITU-T standards G.805 [2] and M.3100 [3]. Since these concepts are not applicable to connectionless networks such as internet, these concepts have to be generalized to accommodate connectionless networks. The manner in which this is done by reusing much of the current CMA components is described below.

3.1 Description of the new architecture

Figure 2 depicts the part of the TINA CMA model that has been modified. The modified components have been shaded in the figure. In the current CMA, (see Figure 1), the Flow Connection Controller (FCC) uses the Layer Network Coordinator (LNC) interface to setup a trail and the Trail Manager (TM) interface for managing a specific trail. These interfaces used by the FCC need to be the same for connectionless and connection oriented networks since the FCC needs to use these networks transparently. The current interfaces that are based on trails are unsuitable since the trail concept does not exist in connectionless networks. The solution to this problem is the introduction of a new concept called *Layer Network Binding (LNB)*.

The LNB is the generalization of the concept of trail and represents an end-to-end connectivity across a layer network (connection-oriented or connectionless layer network). The end points of a LNB are called *Layer Network Access Points* (LNAP). In the case of internet, a LNAP represents either a UDP port or a TCP port depending on the internet transport protocol used for the stream flow. From a resource modelling perspective, a LNB has the following attributes: characteristic information (such as ATM VP, ATM VC, Frame Relay, UDP, TCP, etc.); topology (point-to-point, point-to-multipoint, etc.), QoS parameters that can include bandwidth and timing. A trail is then a specialization of LNB for connection-oriented networks.

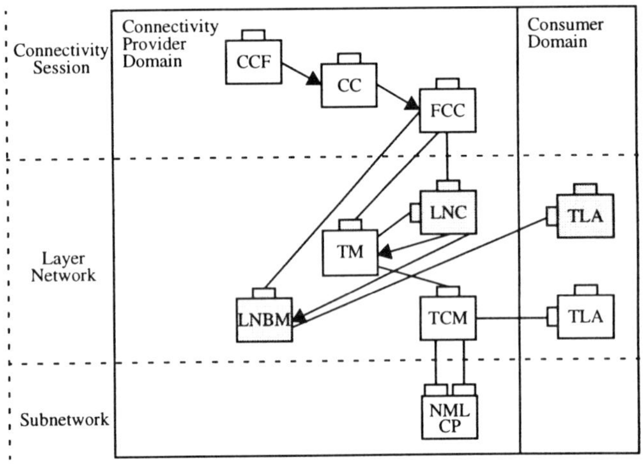

Figure 2. Modified part of the TINA Connection Management Architecture

The current LNC interface (used by FCC) is generalized to support creation of LNBs. The FCC identifies the end points of a LNB using NFEP Pools (as in the current CMA). In the case of internet, a NFEP Pool is a set of IP addresses. When it receives a LNB setup request, the LNC creates depending on whether it manages a connection-oriented or (exclusive) connectionless layer network, a TM or a *Layer Network Binding Manager* (LNBM). The TM created then manages the specific trail. Similarly, the LNBM created manages the specific LNB. To provide transparency to FCC, both LNBM and TM support an identical interface (type), called the LNB interface, that provides the following LNB management operations:

- Add/Remove end points
- Activate/Deactivate end points
- Delete LNB
- Modify_QoS of end points

After it creates a TM or a LNBM as a result of a SetupLNB, the LNC returns to the FCC a reference to the LNB interface of the created component.

The other CMA component that requires modification is the Terminal Layer Adapter (TLA). In a CPE, one TLA exists for each layer network to which the CPE is attached, and the interfaces of a TLA are specific to the corresponding layer network. Thus, to handle internet transport, a TLA exists in each CPE. This TLA encapsulates the internet protocol details and offers interfaces for LNB management, one for each internet transport protocol supported in the CPE. To setup a LNB using a specific internet transport protocol, the LNBM uses the appropriate interface offered by the TLAs in the CPEs that are participants to the binding.

The rest of this section describes the setup of a LNB in the internet and the LNBM-TLA interactions needed for this setup. Two cases are considered: UDP and TCP. Each case is described below.

3.2 TCP/IP TLA interface

This section describes the setup of a LNB using the TCP transport. The TLA for the internet layer network offers an interface that has the following operations:

- *CreateTCPep* takes an optional (IP) end point description as input. This description identifies the set of IP addresses (network interfaces) from which the TLA can select one for creating an end point for the LNB. If the IP end point description is omitted, the TLA selects one of the IP addresses associated with the CPE. After creating[2] the end-point it returns the fully resolved end point description[3] (e.g. adds selected port).
- *Accept* takes optionally a local and a remote end-point descriptions as input. It creates the local end point if it does not exist yet. The local end point will accept TCP connection requests only from the remote end point if it is specified. If no remote end point is specified it will accept connection requests from anywhere. It returns the fully resolved local end point description.
- *Connect* takes an optional local and a remote end point description as input. If the local end point does not exist, it creates one. It connects to the remote end-point and returns the resolved local end-point.
- *Disconnect* takes a local end point description as parameter and disconnects the connection if it exists.
- *DeleteTCPep* takes a local end point description as parameter and deletes it and the associated connection, if any.

Figure 3 shows the interactions between the LNBM and the TLAs for the setup of a point-to-point LNB using TCP.

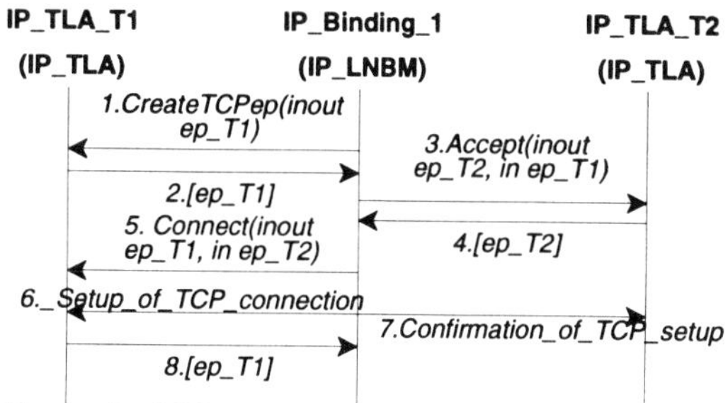

Figure 3. LNBM-TLA Interactions for a TCP/IP LNB

3.3 UDP/IP TLA interface

The TLA interface for the UDP transport supports both point-to-point and multipoint bindings. The TLA interface for UDP provides the following operations which can be applied consistently to both point-to-point and multipoint bindings:

- *CreateUDPep* has the same semantics as CreateTCPep, the only difference is that the end point is a UDP end point.
- *AddSource* takes an optional local and a remote end point description as input. If the local end point does not exist, it creates one. Then, it adds the remote end-point to the list of accepted sources for the local end point, starts a receiving process if this is the first receiving end point and returns the fully resolved local end point description.

2. All operations that can create end-points must also have a correlation identifier as 'in' parameter. This is a reference to the terminal part of the stream flow connection. For simplicity it is omitted in the operation descriptions.
3. The specification of the end-point description is out of the scope of this paper, but it must be able to contain full or partially resolved references to IP end-points. A URL like approach might be suitable (e.g. TCP://ip_address:port).

- *AddDestination* takes an optional local and a remote end point description as input. If the local end point does not exist, it creates one. Then, it adds the remote end point to the list of destinations for the local end point and returns the fully resolved local end point description.
- *RemoveSource* takes a local and a remote end point description as input. It removes the remote end point from the list of accepted sources for the local end point and stops the receiving process if this is the last receiving end point.
- *RemoveDestination* takes a local and a remote end point description as input. It removes the remote end point from the list of destinations for the local end point.
- *DeleteUDPep* takes an end point description as input. It stops the receiving process if it is running and deletes the end point.

Figure 4 illustrates the LNBM-TLA interactions needed for the setup and release of a point-to-point LNB using UDP/IP

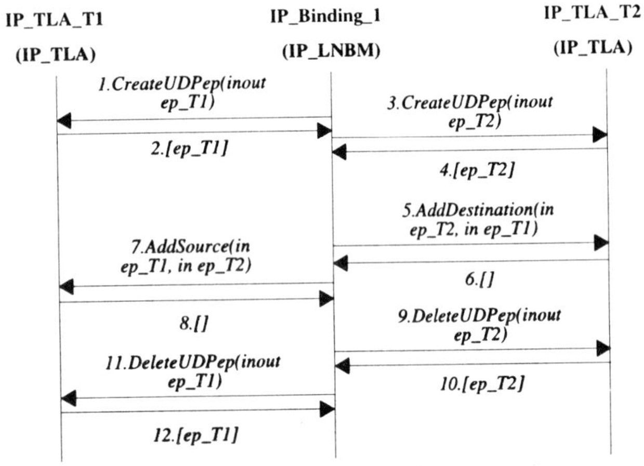

Figure 4. LNBM-TLA Interactions for a point-to-point UDP LNB

4. Using IP multicast groups for a point-to-multi-point UDP LNB

Although the UDP TLA interface described in the previous section can handle multi-point to multi-point information distribution, it will only be used for point-to-point or point-to-multipoint since this is what the Layer Network Binding interface supports.

To establish a point-to-multipoint Layer Network Binding, the Layer Network Binding Manager can use uni-cast UDP IP by adding multiple destinations to the source end point. As a result the UDP end point will send multiple copies of the same datagram to each uni-cast address. Because of the copying, this method uses the network resources very inefficient. However, it is applicable to any IP network.

IP multi-cast has been designed to optimize the usage of network resources especially for the case in which datagrams need to be send to multiple destinations. IP multicasting delivers IP datagrams to a "host group" identified by a single IP destination address without delivering a separate copy to each destination. When all the hosts are within one network segment, only the involved hosts need to be specially configured. If hosts reside in multiple network segments, interconnecting routers also need to be configured for the multicast group.

In the TINA NRA, the Layer Network Binding Manager is responsible for the management of a Layer Network Binding in one administrative domain. Therefore, it is well placed to manage the well-known administratively assigned IP multicast address for a multicast group. TINA foresees the 'federation' concept for inter-administrative domain communication setup, which nicely fits the needs of multicast router management. However, this paper only intends to point out the power of applying TINA to the internet and not to explore all the possibilities in detail.

A LNBM that manages a IP multicast address can easily use this for a point-to-multipoint UDP LNB as demonstrated by Figure 5. Figure 5 illustrates the LNBM-TLA interactions needed for the setup of a point-to-multipoint LNB using IP multicasting..

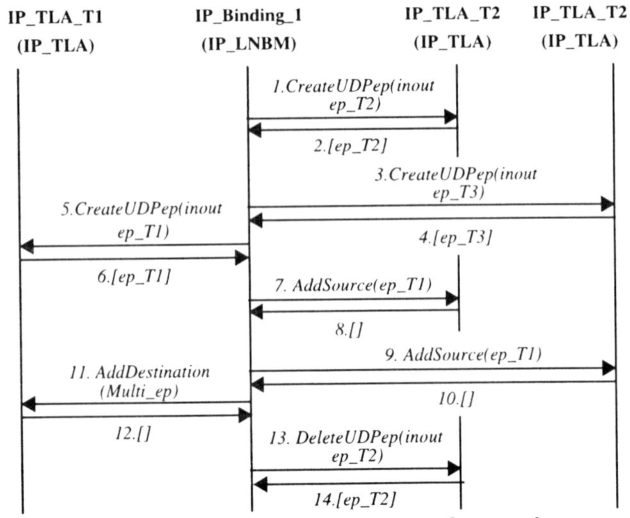

Figure 5. LNBM-TLA Interactions for a point-to-multipoint UDP LNB

5. Illustrative example

This section describes a scenario of a TINA stream connection setup that uses both IP and ATM networks for the transport of information. In this example the service uses only one connectivity provider with one administrative domain. This connectivity provider manages a connectivity layer network that consists of two layer networks, namely one for Internet and another for ATM (Figure 6).

After the Communication Session Manager (CSM) in the service provider domain receives the request for setting up a Stream Flow Connection (SFC) - this always happens in the context of a communication session -, it will negotiate the codec and session protocol part of the associated Terminal Flow Connections (TFCs) with the Terminal Communication Session Manager (TCSM) based on the required quality of service specifications. For this purpose the terminal will announce its capability set to the Communication Session Manager. The capability set describes a network or graph of capabilities that the terminal can use to accommodate a certain communication with required quality of service [4]. The final result of this negotiation process is the establishment of the codec and session protocol part of the Terminal Flow Connection. Consequently, two Terminal Flow Connections will be partially established in the example scenario.

The Communication Session Manager determines the network requirements for the Stream Flow Connections based on the network capabilities upon which the selected codecs and session protocols depend. This information was conveyed in the previously exchanged capability sets. From this information it compiles a set of Network Flow Connections (NFCs) that interconnect Network Flow End Points (NFEPs). A Network Flow End Point represents both the Network Flow Connection end point and the Terminal Flow Connection end point. In the example scenario, the service provider's Communication Session Manager will request the connectivity provider's Connection Coordinator (CC) to setup one Network Flow Connection from terminal A to terminal B. Network Flow End Point A represents a IP network termination point while Network Flow End Point B represents a ATM network termination point.

The connectivity provider's Connection Coordinator determines that both Network Flow End Points point to network termination points in different Layer Networks. As a result the Connection Coordinator locates an interworking unit that is capable of handling the interworking between the two layer networks and requests an interworking connection[4]. In this example, the Connection Coordinator

4. Note that this is a new idea that does not have TINA Consortium consensus yet. It will be introduced in the next release of the Network Component Specification [5].

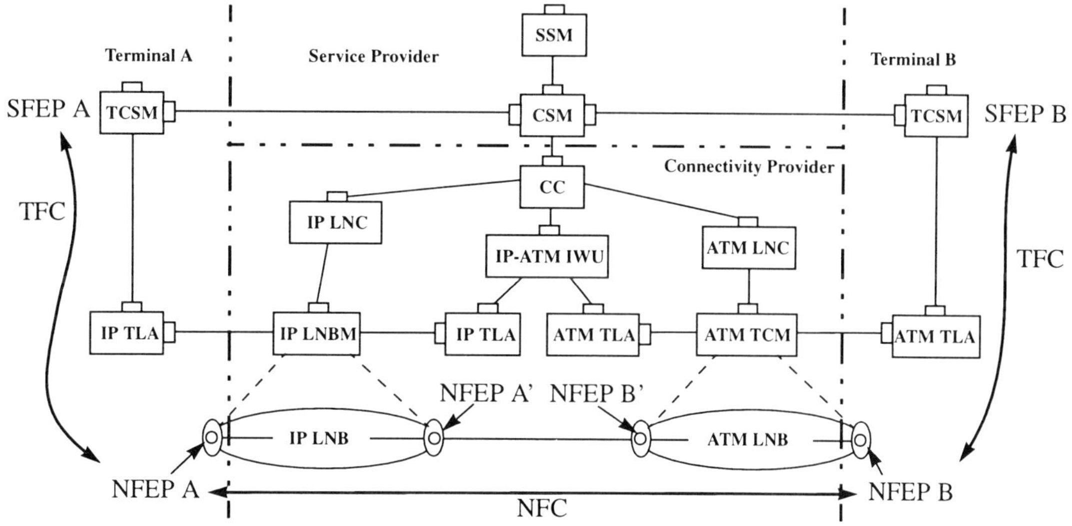

Figure 6. Example: a Stream Flow Connection using both ATM and IP as transport technology

will pass the two Network Flow End Points to the IP-ATM interworking unit. After successfully setting up the interworking connection, the interworking unit will return Network Flow End Point A' pointing to the IP network termination point and Network Flow End Point B' pointing to the ATM network termination point of the 'interworking' connection. Subsequently, the Connection Coordinator will request the setup of a Layer Network Binding (LNB) to the IP Layer Network Coordinator (LNC) between Network Flow End Point A and A' and to the ATM Layer Network Coordinator between Network Flow End Point B' and B. Because this paper focuses on internet and IP, it will not further discuss the ATM part of the Network Flow Connection.

The IP Layer Network Coordinator will instantiate a Layer Network Binding Manager (LNBM) that manages the Layer Network Binding. The Layer Network Binding will derive the Terminal Layer Adapter (TLA), network protocol and addressing information from the Network Flow End Points. In this example, it will derive the Terminal Layer Adapter interface reference and IP address of terminal A as well as the protocol that needs to be TCP, from Network Flow End Point A. The same kind of information will be derived from Network Flow End Point A', but than for the interworking unit.

The Layer Network Binding Manager will use the in Section 3.2 described interface to configure the end points at both sides. As a result the Layer Network Binding is established and the Terminal Flow Connections are completed.

6. Conclusion

This paper presented an approach for the management of TINA streams that use internet transport. The proposed approach uses the existing TINA CMA and suggests a minor modification to the CMA for using internet transport, namely the Layer Network Binding. The integration of internet transport with other types of transport networks in the CMA enables a rational selection of specific layer networks for stream flow connections based on requested Quality of Service. For example, the following selections are possible:

- Time critical connections in a communication session can be supported by isochronous networks while the others are supported via the Internet.
- Usage of internet LANs as access networks to several core networks such as ATM, SDH, ISDN or even PTNS.
- Interconnection of applications on terminals connected via different technologies (e.g. GSM, Internet, N-ISDN).

Another interesting application of TINA in the Internet environment is the possibility to introduce coordinated policing, which is a prerequisite for accounting. This capability might accelerate the introduction of RSVP in the public Internet. A variation on this theme in combination with the introduction of administrative domains could make it possible to dynamically manage (corporate) firewalls (e.g., for user based setup of tunnels between firewalls).

Future work will investigate the above applications, and will also address internet resource management (FCAPS).

In conclusion, it needs to be emphasized that the ability to introduce internet transport under the scope of the CMA without requiring fundamental changes to the CMA, vindicates the modularity of the CMA.

7. References

[1] *Network Resource Architecture* Version 3.0, NRA_v3.0_97_02_10, TINA-C, February 1997.

[2] ITU-T Recommendation G.805, *Architectures of Transport Networks,* June 1995.

[3] ITU-T Recommendation M.3100, *Generic Network Information Model*, 1992.

[4] *Quality of Service Negotiation in TINA*, Jarno Rajahalme, Telma Mota, Frank Steegmans, Per F. Hansen, Fernanda Fonseca, TINA '97 Conference, November 1997

[5] *Network Component Specification*, Version 2.2 Draft, TINA-C, to be released December 1997

PalaTINAt: A TINA Based Control Platform for Telecommunication and Information Services

Stéphane Pensivy
Tel : + 33 2 96 05 34 09

Vincent Gouraud
Tel : + 33 2 96 05 34 85

Gilles Lecorgne
Tel : + 33 2 96 05 34 38

France Telecom – BD/CNET/DAC/ARP
2, avenue Pierre Marzin
22307 Lannion Cedex
France
{stephane.pensivy, vincent.gouraud, gilles.lecorgne}@cnet.francetelecom.fr
Fax : +33 2 96 05 37 84

Abstract

Today's network infrastructures are heterogeneous and target well-defined networks and services. This multiplicity of solutions prevents from reusing existing functions or services already deployed in the network. The need of a common control network infrastructure, enabling reusability of existing functions, and by doing so, reducing the design and development time is obvious. This paper presents PalaTINAt, a TINA based control infrastructure expanding three major domains: public network operator, corporate private networks and residential terminals.

1 Introduction

With deregulation, the telecommunication market is becoming more competitive than ever. It is crucial for an operator to provide cost efficient and time-to-market services in order to suit the requirement of end-users becoming more and more exacting. On the other hand, the existing network infrastructure is not designed in such a way that we can easily reuse existing services or functions already deployed in the network. For instance, numerous mobile telephony services are functionally the same as fixed telephony services[1], some narrow-band services are functionally identical as broadband services. The lack of a common control infrastructure and the variety of signalling protocols used today as well as the static and centralised approach of the service logic (like in IN for example), doesn't help in getting this common control infrastructure.

The TINA architecture brings numerous concepts useful to fulfil the requirements mentioned above:

- A kernel Transport Network that can (in theory) rely on different protocol stacks (TCP/IP, SSNo7)
- Service logic location transparency
- An object oriented approach that can facilitate the reusability of existing components, functions
- An easy deployment of Computational Objects (COs)
- A high level of abstraction to define the control logic (an Object Definition Language)
- Well defined reference points (Ret) that facilitates interworking

This paper intends to show how the TINA architecture could be used to provide this common control infrastructure required to face the coming competitive market. The prototype called PalaTINAt, a TINA based control platform for network mediation services

[1] But relying on a different control infrastructure

0-8186-8335-X/98 $10.00 © 1998 IEEE

developed by the CNET is introduced as validated results on testing this universal control network.

2 Overview of existing network control infrastructures

To get access to a service, the end-user first needs to get access to the network. To that end, the notion of access session, such as defined by the TINA Service Architecture [2], fits naturally to supply different kinds of services like authentication, accounting, network access control, bandwidth negotiation, load balancing, authorisation (to third party services or service providers)... We call these services network mediation services with regard to application mediation services which are rather related to the application (subscription, customisation, service management,...). Considering these network mediation services, it is crucial for a telecommunication operator to control the access to its networks, to reject or accept the requests as fast as possible in order not to waste resources (on access node) for a service which might won't be delivered at the end.

It might be also very convenient to factorise the network mediation services with application mediation services. For instance, if the Network Access Provider (NAP) supplies also the services, it's more convenient, for the end-user to be authenticated only once. It is also very important, for performance purposes, to be able to deploy software components (even dynamically) where they are solicited and, on the other hand, to reuse existing software components already deployed in the network. The need of a flexible software architecture enabling the functional entities (software components) to interact together irrespective of their location and to be deployed quickly and smoothly where and when required, is then crucial.

Traditionally, with POTS, the information used for control purpose (ringing, off-hook,...) and information used for the service itself (voice for telephony service) was not clearly separated. The control information was carried in the band. The main drawback of this solution was a very poor semantics for control information and, as a consequence, for the control capabilities.

Numerous and heterogeneous control networks....

The idea to overcome this limitation was to clearly distinguish the control and transport data information. As we have traditionally the notion of transport network (to carry voice, data,...), the notion of control network is necessary. The purpose of a control network is to carry only control information and then, to clearly separate (at least logically), the transport information (i.e. information used for the service itself) from the control one.

With the advent of ISDN, the control information is logically separated (outside the band) from the transport information. In N-ISDN the control information is sent over the D channel with Q.931 for access signaling, logically distinct from the B-channels. In B-ISDN, a dedicated VC might be used to carry control information (Q.2931). Signalling System No7 is another example of well-known dedicated control network, used as network signaling relying on MTP (Message Transfer Part) for numerous services like ISDN or IN services with a worth mentioning difference: this network is even physically different from the transport network.

The TINA architecture relies on a logically dedicated control network, the kernel Transport Network (kTN). An important feature of the TINA kTN is that it can rely on different protocol stacks. Today, if we consider Corba [3] as a first approximation of the TINA Distributed Processing Environment (DPE) laying on the kTN, it relies on TCP/IP but as outlined in [7], nothing precludes the usage of a SS7 protocol stack. The TINA kTN provides a high level abstraction to a service designer who only deals with computational objects interaction no matter where these COs are located (locally or distributed) and what the underlying transport protocol stack actually is (TCP/IP, SCCP/MTP,...).

With the advent of these control networks, it's then possible to rapidly enhance the semantic of the control information to provide new value added services for customers.

Numerous and heterogeneous control infrastructures...

Different control infrastructures exist today to provide different services. An example is the intelligent Network, relying on an infrastructure based on Service Switching Point (SSP), Service Control Point (SCP), Service Data Point (SDP) and Service Management Point (SMP)[2]. On the other hand, mobile services might rely on a Network Subsystem (NSS) that consist of a Mobile Services Switching Center (MSC), a Home Location Register

---

[2] Different control networks might be used in IN service delivery. For example SS7 and X.25.

(HLR) and a Visitor Location Register (VLR), but also, with the concept of Virtual Home Environment (VHE), on a MSC interacting with a CAMEL server. Information services control infrastructure might rely on a Network Access Server (NAS) connected to an Internet Access Provider (IAP). These control infrastructures use the control network (or kTN) to carry the control information. The Figure 1 shows how these control infrastructures are clearly separated, with different processing nodes not interconnected between each others.

This lack of reusability does not lead to cost effective solutions (slow and expensive process).

Numerous and heterogeneous service control...

Here again, different solutions exist for service control. Intelligent Network (IN) services rely on a Signalling System No 7 (SS7) control network with INAP interface; ISDN services also rely on a SS7 control network but with an ISUP interface, with the interaction problems we know [11], and mobile network rely on Camel Application Part (CAP) interface. Some information services[3] rely on a TCP/IP kTN, with HTTP protocol for Web services, SMTP for e-mail services or SNMP for management services. All these services use their own service control approach and are not designed to interact. As depicted in figure below, they all require network mediation services but they all have their own and proprietary implementation to fulfil the same functionality.

The TINA Architecture relies on high level of abstraction for service control. This is provided by the Object Definition Language, an object oriented language independent of any programming language and of the control network finally used for service control. This language facilitates the interactions between service components because it doesn't target any specific environment, machine or programming language.

[3] One worth mentioning difference is that in these services, the kTN and the TN are the same.

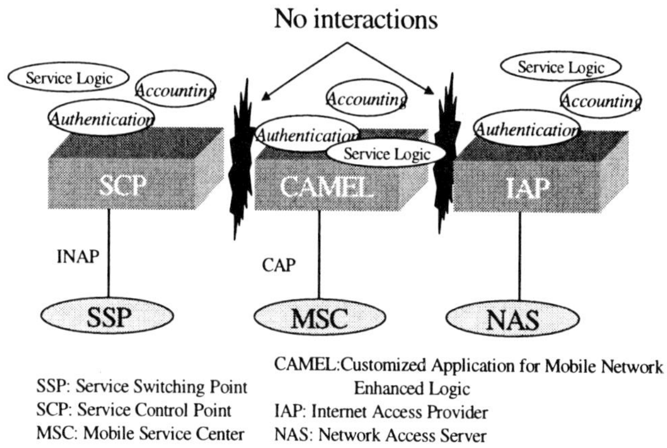

Figure 1. Example of current control network infrastructures

The figure above is an example of the drawbacks of today's solutions where the same services are deployed for different network without any possibility to reuse existing piece of code because of targeting different machine, environment and networks.

In the TINA service architecture, there is a clear separation between the access session, the service session and the communication session. Through a single access session, the end-user accesses several services, no matter the transport network required. Indeed, as the service and communication sessions are clearly separated, it is easy for a designer to focus on the service logic with a total abstraction of the underlying transport network that will be used to supply the service. This is a major difference regarding existing systems that target specific transport networks.

These TINA features mentioned above are the main input of PalaTINAt, a TINA based control platform expanding three major domains:

1. Public network operator
2. Corporate private Network
3. Residential terminal

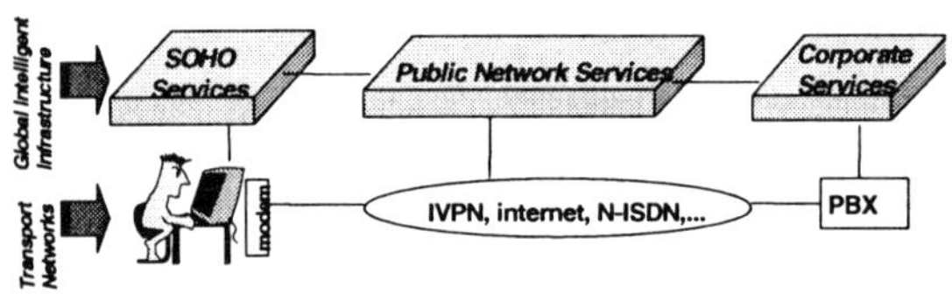

Figure 2.Objectives of PalaTINAt

In the following sections, we try to show through PalaTINAt how these features are applied and what the added valued is.

3 The PalaTINAt architecture

The PalaTINAt architecture considers the access session, service session and communication session as defined by TINA. According to the TINA business model [1], the following business roles are considered:

- Consumer
- Retailer
- Third Party Service Provider
- Connectivity Provider

As depicted in the Figure 3, the access part of the TINA service architecture (the main focus of our prototype) was implemented with various configurations with regard to engineering and technology choices. Technology because different kinds of terminal are used: phone sets and computers. Engineering because different nodes are considered: full DPE nodes, and non-full DPE nodes, leading to constraints when deploying computational object on the nodes.

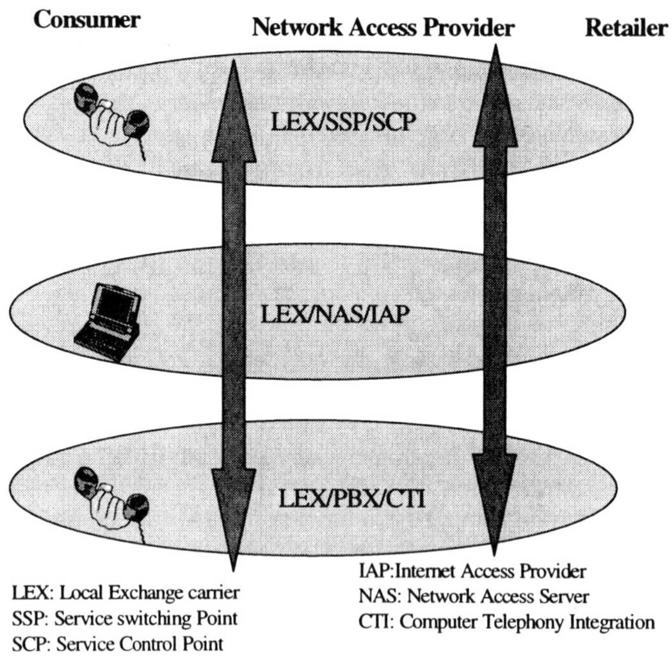

Figure 3. Access scenarios in PalaTINAt

The consumer gets access to his retailer or service provider through a network access provider using legacy applications and protocols. The terminal is either a dumb (phone set) or an intelligent (PC) one.

The overall architecture of PalaTINAt is depicted in Figure 4. A service is achieved through a set of service components interacting with each other. These interactions rely on the underlying Distributed Processing Environment (DPE), namely CORBA [3] implementations (waiting for a full TINA DPE), on top of a TCP/IP kernel Transport Network (kTN). The DPE used is Orbix 2.2 MT, an IONA [6] product, on the server side, and OrbixWeb on client side. The servers are developed in C++ whereas the clients are developed in Java.

The backbone kTN (control network) relies on IIOP (Internet Inter ORB Protocol) messages over TCP/IP connections. Point to Point Protocol (PPP) might be used to get access to the network. There is a clear separation between the kTN used to control the service and the Transport Network used to transport the information required by the service (voice, data, video,...), except, in our prototype, for Web services where the kTN and the TN are the same (TCP/IP).

The Transport Network is controlled from the application through a gateway, which translates Corba requests into switch calls (either CSTA or INAP-like calls).

Different kinds of terminals are used. Intelligent terminals where the intelligence (a CORBA object, namely the Provider Agent) is deployed in the terminal itself; dumb terminals where part of the intelligence is deployed in the network.

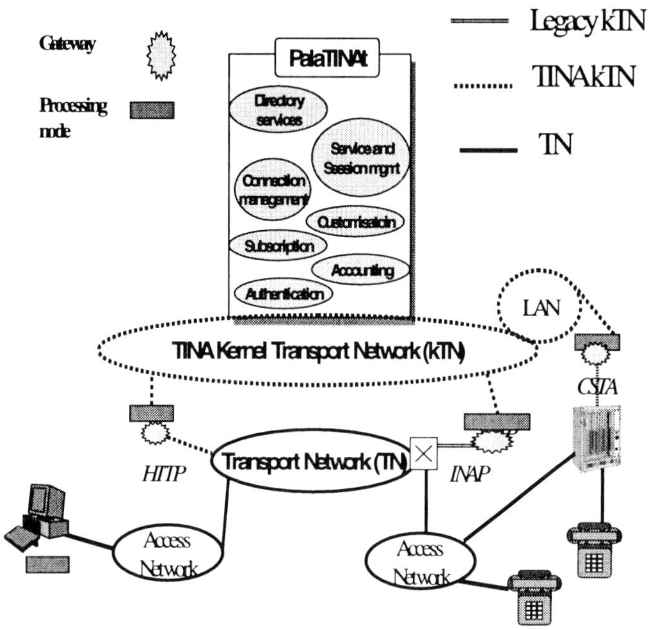

Figure 4.Overview of PalaTINAt infrastructure

The above considerations (taking into account legacy terminals, applications, protocols and networks) led to study in details the engineering of our applications. Indeed, according to RM-ODP [4] [5], the TINA architecture specifies the business model, the information model and the computational model. The next step, purposely left to the designer of a TINA application corresponds to the engineering viewpoint defined by RM-ODP.

The considerations are twofold:

1. We first need to access to the TINA kTN (either directly or though some gateways)
2. The node connected to the TINA kTN either fulfil the TINA requirements to run service components (a full DPE), either some functionalities of these service components are deployed in the network on a node providing a (or access to a) full DPE.

The next section explains in details how we took into account these engineering considerations.

4 The access part

One of the aims of PalaTINAt was to apply TINA principles to control the access to services without making any assumptions about the terminals, the physical networks, and the control network used. The first step of the access session is to get access to the TINA network mediation services in order to control the services. Obviously, this step is quite different with a terminal like a phone set, a computer with no DPE or with a computer with a full DPE. Indeed, if we consider intelligent or enhanced terminals, it might be possible to deploy the TINA Network mediation services on the terminal, whereas with dumb terminals, we have to interact with legacy control networks and service control.

This chapter shows the different possibilities in PalaTINAt to control access to services in a TINA way without reconsidering existing terminal and control networks.

4.1 Getting access to network mediation services through a TINA kTN

This section explains how, from a terminal that has a TINA kTN (TCP/IP stack), it is possible to get access to network mediation services.

Traditionally, the telecommunication terminals were relatively dumb. But new kinds of terminals are emerging at the network edges. A good definition of these terminals is that they are « intelligent » enough to offer easy access to mediation services (graphical interfaces thanks to a convenient browser and graphical screens), good support of numerous protocols (thanks to convenient pieces of software like TCP/IP stack, sockets API,...). A good and popular example of such equipment is the home PC.

Our approach to get access to the network mediation services was to rely on off the shelf and common equipment and software to suit the largest amount of terminals. With such considerations in mind, the basic terminal configuration we selected was: common browser (Netscape, Internet explorer, Mosaic...) and a TCP/IP stack; a configuration that can be found on most of the equipments on the market. Our wish to open our platform to such terminals and control networks led us to consider TCP/IP as an existing control network to open to. This was achieved on a PSTN link (we also experimented successfully an access through an N-ISDN) by simply adding on the end-user side and retailer side a PPP stack (Point to Point Protocol) and a modem to allow transport of TCP/IP on a point to point link.

Note that in our experiment we use a Network Access Server (NAS) in the network access provider domain (in our experiment a Cisco 2522). This adds authentication, identification and accounting functionalities via TACACS or RADIUS protocols to the classical PPP functionality. We can regret the lack of gateways between such protocols and Corba.

This kTN connectivity between the end-user's terminal and the network access point is not enough to get access to network mediation services. Indeed, the common control infrastructure relies on TINA service components interacting with each other. According the TINA principles, this interaction assumes a full DPE on each node hosting TINA COs. But, as described below, it might not be the case.

4.1.1 Study case: a full DPE on the end-user's terminal

This section gives an example of intelligent terminal, namely a PC or a workstation with enough capabilities to host a DPE. This DPE is either resident, either dynamically downloaded. The latter case is possible with Java. The stubs are generated with OrbixWeb and the applet (the PA) is dynamically downloaded along with the access to the DPE. In this scenario we have a TINA kTN (TCP/IP) in the user's terminal and there is no need of gateways. This scenario shows that the common control infrastructure is able to take advantage of the latest capabilities of the terminals, more and more intelligent, by deploying the intelligence at the network edges.

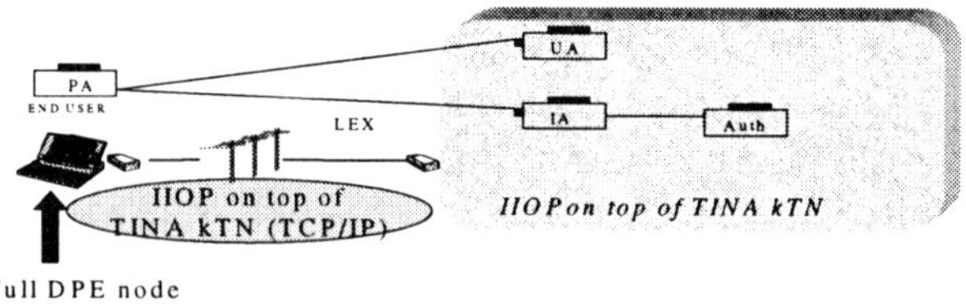

Figure 5. Access to PalaTINAt from an intelligent terminal

4.1.2 Study case: no DPE on the end-user's terminal

Considering such terminal (no DPE) it was obvious that the user's terminal couldn't support the deployment of full TINA components (PA for example) because they are supposed to run on a full DPE. We had to develop gateways between the processing environment and control infrastructure used in the user's terminal (HTTP/TCP/IP) and the TINA DPE (IIOP/TCP/IP). The semantic of this gateway is to get access to the TINA common control infrastructure. Thanks to the AnsaWeb tool [9] this gateway was easily implemented. This tool generates a CGI script from an IDL description of a CORBA object. This script allows the invocation from the client side (HTTP compliant) of any CORBA object on the retailer side.

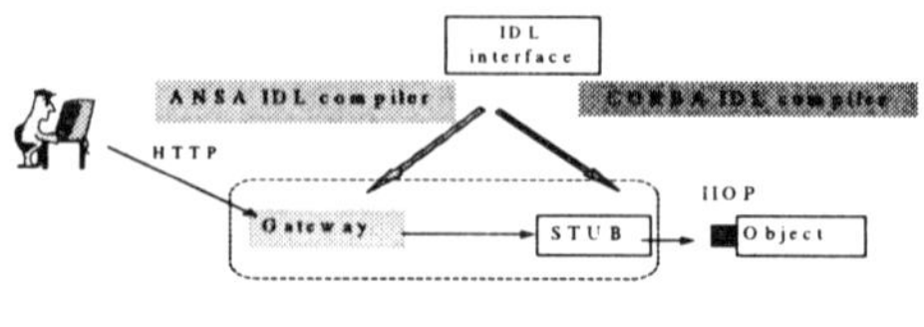

Figure 6. Integration of different service controls

As there is no DPE in the terminal, some DPE functionalities have to be considered in the network. According to the TINA service architecture, the PA is deployed in the consumer domain. In our experiment, some functionalities of the PA are fulfilled in the terminal (e.g. the piece of software delivered by an Internet Access Provider), some others in the network. For instance, the location transparency normally provided by a DPE has been developed and deployed in the network. The main problem we had to confront with is due to the « stateless » nature of HTTP protocol. Indeed, once the client is connected to the network access provider (some functions of the dedicated PA exist in the network) each request from this client should be directed to the right PA. This can be achieved by adding to each client's request, an identifier (Cookie, Session identifier written in the URL...) which references the right TINA object. As a matter of fact, the role of this « access session identifier » is the equivalent of the PA IOR (Interoperable Object Reference) in the HTTP domain and the gateway acts as a request broker to find the right PA (and through the PA, the right User Agent).

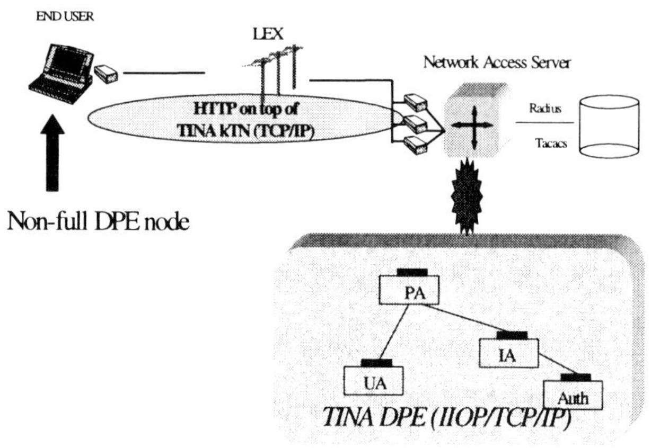

Figure 7. Access to PalaTINAt from a dumb computer without existing network connectivity

4.2 Getting access to network mediation services from legacy kTN

This section presents some engineering solutions to get access to the TINA network mediation services from legacy kTN.

The kind of terminal considered in this section is a phone set. As this node doesn't provide any DPE environment, neither TINA kTN some gateways have to be considered. Two gateways are necessary in this scenario:

1. The gateway from the legacy kTN to the TINA kTN

2. The gateway from the legacy control infrastructure to the TINA control infrastructure and service logic

Thanks to the network equipment we use (a switch and a PBX) the first gateway is natively provided and gives access from the legacy kTN to TCP/IP (TINA kTN) and vice-versa.

The next step is to get access to the TINA DPE. Two kinds of legacy service control were identified in PalaTINAt. One based on an ECMA CSTA control with an Alcatel 4400 PBX; the second one based on INAP-like protocol with a Redcom switch. Through these gateways, we get access to the TINA kTN and service components. A more detailed scenario using the CSTA gateway is explained below.

The CSTA (Computer Supported Telecommunication Applications) protocol is currently used in the Tina Test Trial [8]. The PBX is directly controlled by a CTI link. An Inter Working Unit (IWU) handles the incoming calls. The callee is identified by a personal number (an alias), which is translated into a TINA user_id. There is a great of flexibility for CO deployment, either on the IWU itself, closer to the caller, either elsewhere in the network. This is a worth-mentioning added value of the TINA architecture. An important characteristic when integrating telephony services in a TINA way is that the access session functionality is partially fulfilled by the POTS. Furthermore, it would be too restrictive to consider only TINA caller. Then, the caller UA is not considered at all in the current version. Only the callee is known to the TINA system and has a dedicated User Agent. Through this User Agent, the end-user updates his personal profile to register at the terminal he wants to be reached at. Nothing precludes considering the caller as a TINA user and this will be done in next releases. According to the TINA service architecture [2], an invitation request is sent to the invited PA. Upon registration, a PA is dynamically created on the right IWU. This shows the flexibility of the TINA architecture to dynamically create objects where and when needed.

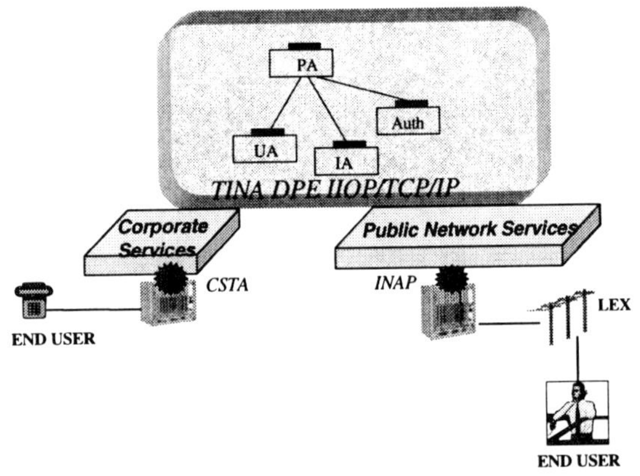

Figure 8. Access to PalaTINAt from a phone

These scenarios (access from computers and phones), although using their legacy control networks and service control interfaces, get access to the same control network (the TINA kTN) and use the same software components to fulfil some functions. For instance, for a particular end user, this is the same User Agent, which is referred to in all scenarios. A lot of value added services (combined services) can then be envisaged.

4.3 Discussion

The access session interface specifications considered in PalaTINAt conform to the Ret reference point specifications as defined by TINA-C. The added value of being compliant to this reference point is that PalaTINAt is accessible from any end-user's domain, and can access to any other network access provider and retailer, as long as it conforms to Ret. This was validated in the context of the TINA Test Trial with DT, FT, Global One and Sprint. Without sharing any piece of software, but by closely working together to get a common understanding on the semantic of the interfaces and operations, all the partners have access to the others platforms just by being compliant to Ret. Furthermore, there was no need to agree upon programming language, machine or environment. Heterogeneous configurations are used by each partner (different ORBs, operating system, programming languages) without any interaction problems. Through this trial, the common control infrastructure is geographically extended easily and seamlessly, simply by following the same architecture principles, those defined by TINA. But, with such a large-scale control network, we might face performance and congestion problems. Of course, the technology choice for the kTN (on top of ISDN, ATM, ...) impacts directly the performances. But this is not enough. If the engineering choices are not carefully considered (where and when to deploy COs), it might impact directly the performances of an unnecessary overloaded control network. The granularity of the clusters might also impact the performances. For example, the performances won't be the same if the capsule corresponds to a domain of couples of UA, hundreds UA or thousands UA. Indeed, we might end-up to an overloaded capsule by too much simultaneous requests, with a great impact on the performances. Furthermore, in the latter case for instance, it might not be realistic to have all the objects activated together in the capsule. The usage of databases might then be very useful. But the integration between databases and DPE have also to be designed carefully, especially when we want to dynamically fetch objects from the database, a generally too slow process.

5 The usage part

This section gives an overview of the services currently deployed on top of PalaTINAt. Thanks to the common control infrastructure explained in the previous sections, it's easier to integrate different kinds of services (telecommunication and information services).

A user accesses the services through an access session. Based on the user's profile, a list of services is proposed to him. The services are stored in different databases, object oriented and relational ones. Once the user has selected a service, the right Service Factory is called in order to instantiate the service specific components.

5.1 Information services

PalaTINAt allows subscribers to access to the World Wide Web. The rationale of this service is to improve existing Web services by giving a customised and more friendly environment to the user, to improve the QoS by dynamically routing the request to the right server by taking into account the servers overhead or user's profile, to control the usage of web services, based on the service and user profiles, and to bill the usage of the service based on different criteria.

The user can get access to the service from either a dumb (no DPE, no Java virtual machine) or an intelligent terminal. If he does not have the network connectivity, he can get access to the kTN through the POTN or ISDN over a PPP connection, as explained in the previous sections.

From this access session, the user can either browse the Web (through a gateway using the libwww [10]) or access to TINA services (directory services,...). The user can update his personal profile (ancillary usage) to register at the terminal he wants to be reached at. This registration will be used later on for telephony services (call set-up).

One worth mentioning characteristic of this service is that the kernel Transport and the Transport Network are the same.

5.2 Telephony services

The purpose of this service is to allow a TINA user to be reached at the terminal he is registered at. This terminal might be a pager, a phone, or a fax. The end-user might register at a terminal through Web services as explained in the previous section.

The caller can call from any phone set (phone booth, home, office). The caller doesn't need to be known from the TINA system. The callee terminal identifier (the phone number in this scenario) is dynamically assigned based on the registration repository updated from the previous scenario. The transport network is controlled by CSTA and INAP calls depending on the gateway being used. This control is achieved through IDL interfaces provided by a CSM.

5.3 Information services and network services

Through the common control infrastructure supplied by the TINA kTN, the combination of services (information and telecommunication) is facilitated. One example considered in PalaTINAt allows a user to browse a directory service on the Web. When he clicks on the icon representing the user we wants to talk with, the call is established between the two users. The caller's terminal identifier is dynamically fetched from his personal profile (through the UA) and the callee terminal identifier is the one considered by the directory. Like the previous scenario this service is very easy to design because both information and telecommunication services rely on a common control layer and service components (the UA for instance).

6 Conclusions

The TINA service architecture is now becoming stable. If the burden of implementing such architecture the first time is not negligible this architecture allowed us to reuse existing software pieces, namely the ones used for network mediation services, from one service to another, improving by this means the design and implementation time and consequently the cost. The engineering problems we met during the implementation of this prototype showed us how important it is for a TINA application to consider these aspects carefully. It's what we are striving to look at as well by trying to define engineering choices and tuning. Through these examples we tried to show how we could reuse a common control infrastructure for network mediation services. The TINA architecture and its underlying middleware are good candidates for such an infrastructure. Indeed, with the DPE and the distribution and location transparency it provides, we have the flexibility to deploy services, if necessary dynamically where and when required. This is worth mentioning regarding the more centralised and static approach that exists today for IN services.

Through such a common control infrastructure we can imagine a lot of new value added services for the consumer combining different kinds of services (telecommunication services and information services), different kinds of networks, control networks and legacy applications with no significant extra cost.

It's the reason why it's now crucial for TINA to go beyond specifications by proving concretely its attractiveness through real implementations, prototypes and products.

7 References

[1] "The TINA Business Model and Reference Points", Version 4.0 – TINA-C Deliverable – http://www.tinac.com

[2] "The TINA Service Architecture" , Version 4.0 – TINA-C Deliverable - http ://www.tinac.com

[3] CORBA – OMG – http://www.omg.org

[4] ISO 10746-2 RM-ODP Part 2 "Descriptive model"

[5] ISO 10746-3 RM-ODP Part 3 "Prescriptive model"

[6] Orbix - IONA - http://www.iona.com

[7] "Introduction of distributed computing Middleware in Intelligent Networks", White Paper, Eurescom Project P508

[8] "A TINA Trial. Interworking experience with the legacy telephone system", TINA'97, Santiago November 17-20 1997

[9] ANSA – APM – http://www.ansa.co.uk

[10] The World Wide Web Consortium – http://www.w3.org

[11] "Open Networking: is it technically possible?", R. Kung, , Proceedings of ISS '95, Berlin April 23-28 1995, pp 212-216.

Session 5

Network Management

Kazumi Endo, Fujitsu

The VITAL Network Resource Architecture

Juan Pavón, Alcatel CRC, Spain
Telma Mota, Portugal Telecom/CET, Portugal
George Pavlou, University College London, UK
Lisbeth Veillat, Universidad Politécnica de Madrid, Spain
Panagiotis Palavos, Strabo, Greece

Abstract

TINA aims to provide an architecture based on distributed computing technologies that will enable telecommunications networks to support flexible introduction and control of new advanced services. The provided services and the supporting network and computing infrastructure will be managed in an integrated fashion. Network management aspects in TINA are addressed by its Network Resource Architecture (NRA), which has so far concentrated on Connection and Configuration management, with Accounting, Fault and Performance management aspects to follow.

VITAL is a European collaborative research project that aims to validate a long-term telecommunication architecture, based on TINA, through the design, realisation and trials of a large scale system. This paper presents the VITAL work related to the realisation of the network resource architecture in the areas of resource configuration, connection and accounting management. The problems encountered are highlighted and the architectural solutions adopted are presented, together with examples of operation that illustrate better the design of the VITAL NRA system.

1. Introduction

TINA has defined a Network Resource Architecture (NRA) [1] that covers management areas of Connection, Configuration, Fault, and Accounting Management. Some of these areas have been defined in a great level of detail (e.g. Connection Management) while some others have only started being designed (e.g. Accounting Management) and even some are not yet defined at all (e.g. Performance Management).

VITAL (Validation of Integrated Telecommunication Architecture for the Long-term) is a European collaborative research project of the ACTS (Advanced Communication Technologies and Services) programme that aims essentially to validate a long term Open Distributed Telecommunication Architecture (ODTA) based on TINA. The concepts and specifications developed by TINA have been followed in areas where a good definition of the relevant models and objects existed. In other areas, VITAL followed its own approach.

This paper intends to present a summary of the work related to VITAL Network Resource Architecture [7]. The problems encountered and the solutions adopted by VITAL are presented. In addition, an example is given in order to illustrate the design of the VITAL telecommunication software system. Some TINA concepts are introduced in the paper but the reader should be familiar with the basic aspects of TINA Network Resource Architecture and its relevant terminology.

In summary, this paper has the following structure. Section 2 discusses the status of TINA network resource architecture. Section 3 presents the design and realisation of VITAL network resource architecture. In both sections exist separate subsections for resource configuration, connection and accounting management. Section 4 presents scenarios concerning the use of VITAL NRA for ATM technology networks. Finally, the paper closes with a summary and conclusions regarding the validation of TINA NRA in VITAL.

0-8186-8335-X/98 $10.00 © 1998 IEEE

2. Current Status in TINA

2.1 Resource Configuration Management

While the functionality of Resource Configuration Management in TINA has been identified for quite some time, the current state of the relevant specifications and the overall architecture are pretty immature. The latest specification [1] represents a complete change compared to the previous one [10]. VITAL specification has taken into account input from all TINA specifications prior to [1] and has introduced additional concepts and models. VITAL approach is documented in [11] and will be fed back to TINA-C for the next version of the Network Resource Architecture [1].

The main requirements of RCM for managing network, service and computing resources are the following. RCM should maintain an inventory of resources under its influence and manage their relationships in the form of a *resource map*; it should ensure the map is updated with newly installed or deleted resources; it should allow the activation, deactivation, reservation and release of resources; and it should support automatic resource installation through the DPE wherever possible.

A key aspect of RCM is the maintenance of resource related information. Since the resource information is object-oriented, a major requirement is a computational construct offering a database-like access mechanism. This should provide access to objects representing the resources and should allow to navigate their relationships in a flexible fashion. Such a generic access interface is provided by OSI System Management (OSI-SM) and we have mirrored its capabilities in a generic computational interface for the resource map. The adoption of this approach means we can exploit TINA Network Resource Information Model (NRIM) specifications by mapping them directly to information objects handled by the resource map. In addition, the use of a OSI-SM-like computational interface implies that resource maps can be easily federated due to the hierarchical nature of the relevant information model. While this type of interface is powerful and *generic*, clients of the resource map may provide simpler, task-oriented interfaces, targeted to particular activities.

A key architectural addition to TINA RCM is the consideration of the management resources themselves, i.e. the computational objects offering TINA management services, in addition to the network, service and computing resources. For example, the Connection Management Configurator (CMC) is a computational entity that sets up and *manages* the Connection Management computational objects, based on topological information. Such an object is a manager for management resources and should belong to a new domain defined in VITAL, the Management Configuration Management (MCM). Subsequently, the other RCM domain considered in VITAL is the Network Topology Configuration Management (NTCM) domain which deals with the management of static network resources. In summary, the VITAL RCM modelling has started from existing concepts in TINA but modified and extended them to create a design suitable for a real prototype. VITAL RCM system is described in section 3.1.

2.2 Connection Management

Since the early times of TINA-C activities, Connection Management has deserved particular attention. Considered the core of TINA Network Resource Architecture, it was intentionally decoupled from the resource configuration management area. While this area deals with management of static resources, Connection Management addresses the management of dynamically allocated connectivity resources. It has been defined and specified in a level of detail in [3,4]. The *graph* information model was first proposed to represent connectivity but latter on it has proven to be too 'heavyweight' and the terminology not always adequate. Therefore, TINA has proposed another information model, based on the first one but simpler and more appropriate from the terminology point of view. However, some important mechanisms have not yet been included in the new Network Resource Architecture [1].

Currently, Connection Management in TINA is satisfactorily defined regarding the general mechanisms required for setting up and releasing connections, though unfortunately, it still contains a number of deficiencies. Some of them have already been resolved, for example the large amount of operations required to provide simple functionality, but some others still remain open. An overall definition on how to handle Quality of Service (QoS) in a technology independent manner, is an example of the still incompleteness of TINA. QoS generic negotiation mechanisms need to be introduced at different levels of the CMA, as well as at the service management level [13]. Another open issue, is the mapping between stream flows, which are by nature unidirectional, and network flows that might be bi-directional. The model proposed by TINA does not explicitly preclude this mapping but does not explain either what is the most suitable approach regarding this issue. The mapping has to be flexible enough to allow different possibilities of supporting one or more stream flow connections over a single network flow connection (e.g. ATM VP). As an alternative, it maybe useful to send a stream flow over two different network flow connections, for security reasons for example. In addition, the structure of the Endpoints (Stream Flow End Points, Network Flow End Points and Reference End Points) has not yet been clearly defined. The use of Reference End Points (RFEPs and ANFEPs) is ambiguous as is the whole procedure of resolving the Stream Flow End Points by associating Network Flow End Points belonging to different layer networks. Furthermore, scheduling, routing and performance are still some of the open aspects which make federation not very well defined. There is very little on accounting for connectivity resources and finally, the most important limitation in current TINA specifications, which is pertinent not only to the Network Resource Architecture, is the lack of a consistent way of defining object behaviour.

For an implementation and validation project like VITAL it is very important to find a common understanding of the behaviour of TINA computational objects. A compromise between a very generic but complex approach and a more specific, simple but comprehensive solution has been followed. In summary, VITAL did not intend to solve all the open issues but it has identified some important inconsistencies and unspecified areas, proposed some solutions and provided valuable input to TINA Network Component Specification.

2.3 Accounting Management

Accounting Management in TINA is not yet considered stable enough and complete. The dynamic, flexible and distributed nature of TINA presents a challenge regarding accounting functionality and features. The current work undertaken in TINA, presented mainly in [1][8], presents the requirements and principles for a flexible, distributed, reliable, multi-domain accounting management architecture. The work is mainly focused on resource level accounting from an information viewpoint.

TINA specifies the decomposition of the accounting management activities in four cycles, namely Usage Metering, Classification, Tariffing and Billing. The usage metering cycle is the first step and the basis for all the other cycles. Its purpose is the generation, collection and logging of resource usage information. In the classification cycle collected information is classified and categorised. In the tariffing cycle the charges are calculated taking input from the classification cycle and the tariff structure. Finally, in the billing cycle the bills are produced, stored and forwarded to customers according to the billing schedule.

TINA accounting management architecture defines an accounting management context as the object that specifies the details of accounting management to be performed in each stage of the accounting cycle. The introduction of the accounting management context concept facilitates the dynamic, flexible, distributed and reliable aspects of the architecture.

Unfortunately, despite the fact that TINA accounting management principles and architecture are attractive, they are not yet directly implementable. In addition, the current specifications are mainly focused at the information viewpoint while there is a lack of computational specifications. VITAL considered the principles and architecture presented in [1][8] but extended and refined them, defining a generic but implementable accounting management model, specifying the accountable resources and defining the metrics applied to them.

3. VITAL Network Resource Architecture

VITAL Network Resource Architecture has been applied to the management of ATM-VC and ATM-VP layer networks. The basic components of this architecture are depicted in Figure 1, and described in the following sections.

3.1 Resource Configuration Management

The computational design of VITAL RCM follows TINA documents prior to NRA 3.0 [1] with some deviations and novel aspects documented in [11]. A key deviation with respect to TINA is the division of VITAL RCM system into two subsystems, Management Configuration management (MCM), managing management resources, and Network Topology Configuration Management (NTCM), managing network topology resources.

Both subsystems have a subsystem manager computational object; MC in the case of MCM and NTMan in the case of NTCM (Figure 1), responsible for the integrity of the respective subsystem. The MCM comprises the Connection Management Configurator (CMC) computational objects within one Connectivity Provider domain. There exists one CMC instance for every layer network within that domain, responsible for the configuration of the computational objects in the Connection Management system. This configuration is tightly coupled to the topology of the layer network, provided by the NTCM subsystem, but distinct in the sense that the CMC manages Management Resources (the CM Computational Objects), whereas the NTCM subsystem manages Network Topology Resources.

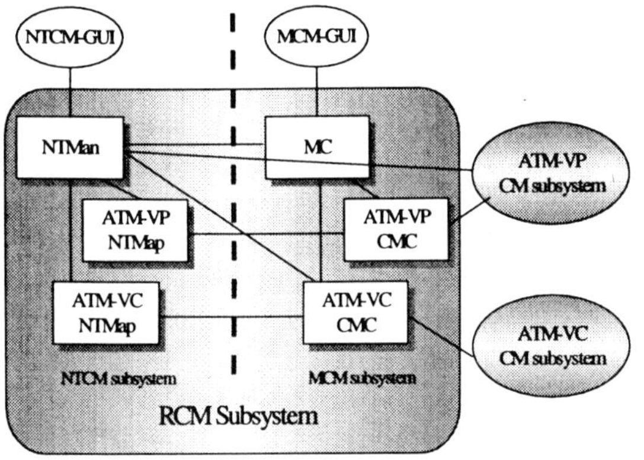

Legend:

CM	Connection Management	MCM	Management Configuration Management
CMC	Connection Management Configurator	NTCM	Network Topology Configuration Management
GUI	Graphical User Interface		
MC	Management Configurator	NTMan	Network Topology Manager

Figure 1: VITAL Network Resource Architecture

The NTCM comprises instances of the Network Topology Map (NTMap), one per layer network, within each Connectivity Provider Domain, in the same way to the CMCs in the MCM subsystem. The NTMap manages the topology map of one layer network. There exist two layer networks in VITAL, ATM Virtual Path and ATM Virtual Channel, in a server-client relationship. The NTMap is influenced by TMN methodologies and specifications and provides a generic, CMIS-like interface that allows to navigate the information objects that constitute the map in a very flexible fashion. The NTMan provides task-oriented, higher abstraction interface, with operations tailored to the needs of its client objects. The NTMan checks also the validity of clients' requests and thus maintains the integrity of the maps. This is particularly useful in the case of server-client networks, where a change in one layer / map potentially means changes to another layer and map, respectively.

VITAL RCM considers two cases of configuration: initial and dynamic reconfiguration. The initial configuration is a total configuration of the Connection Management system, carried out immediately after the bootstrapping of the system and reflecting the initial state of the network. The dynamic reconfiguration reflects any subsequent changes made to the map, with the consequence of reconfiguring partly the Connection Management system. A scenario for dynamically adding or removing a ATM VPC and the necessary reconfiguration are presented in section 4.1.

3.2 Connection Management

The set-up of a connection and negotiation process in VITAL

The computational model implemented in VITAL follows the one proposed by TINA. TINA Business Model and Reference Points defined in [5] were also taken into account. The proper behaviour has been added and a minimal set of operations and parameters have been implemented in order to coherently set-up and release connections.

Whenever a connectivity session is required, the Communication Session Manager (CSM) is contacted in the retailer domain. As a result of the stream binding realised at the service level, a set of connections is required in order to accomplish the service session demands. These connections are explicitly requested to the CSM. This computational object is responsible for negotiating with the terminals involved, the common information needed to run a session. Both, the type of codification of the stream flows (e.g. a_law for audio, MPEG for video) and the transport protocol are negotiated between the CSM and the Terminal CSM (TCSM) objects. At the consumer domain, the TCSM object manages the mapping between the codec information and the transport protocols. In VITAL second prototype, a simple mapping has been considered; this means that no multiple choices are offered to the CSM for it to decide which should be taken depending on the terminal capabilities. Basically, it has been assumed that all users will have the protocol determined by the source Stream Flow End Point (SFEP) and the connectivity providers will always have the needed technology (in case of VITAL, ATM). Therefore, the mapping is being done in the terminal in a very simple but efficient manner (this is foreseen to be enhanced for the next prototype). As a consequence of this approach, the CSM does not have to take care of specific technology parameters. The QoS information is decided at the service level, the codec's are negotiated at the CSM level and once agreed, the terminals will do the mapping between the service level QoS and ATM specific parameters. This information is then directly passed to the Layer Network Coordinator (LNC), a computational object which is by nature technology dependent.

The package composed by the LNC and subordinate objects (i.e. the CPs) is always technology specific; at least one instance of each of these objects must exist for each technology domain. The object that takes care of the specific information in the consumer domain is the TLA (Terminal Layer Adapter). There is one TLA per layer network. The Connection Co-ordinator (CC) computational object is the contact point at the connectivity provider domain. The current CC version in VITAL has been kept simple since only one technology is involved (i.e. ATM) but the need of having the CC controlling the adapters (peer-to-peer and client-server) and the definition of some kind of federation at the CC level have been identified. A CC is meant to be instantiated at each connectivity provider domain. When more than one connectivity provider is involved in providing a connection and each of them owns a different technology, the existent federation mechanisms defined at the level of the LNC do not apply any more. Since LNCs are technology specific they are supposed to federate only within the same technology domain.

The Federation Mechanism in VITAL

The VITAL demonstrator will be deployed over an ATM network by four network operators (Belgacom, CSELT, Telefónica, and Portugal Telecom). In order to make services available and interoperable across all domains, there is a need to support the establishment of connections through different network domains.

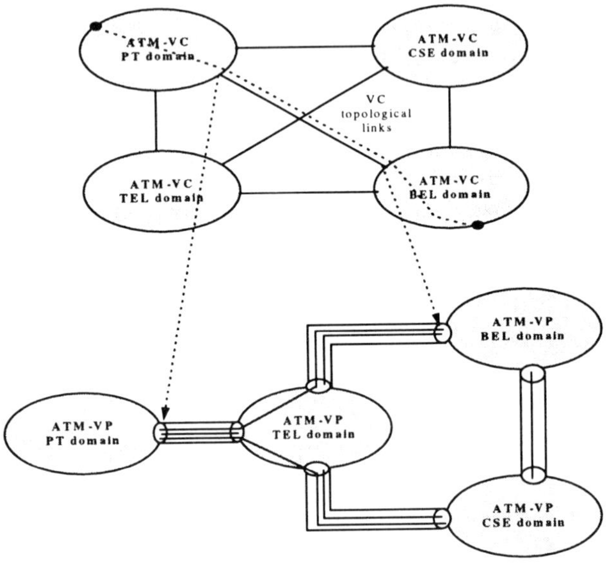

Figure 2: Federation at VP layer network supports federation at VC layer network

In spite of federation relationships can be defined at any layer network, there may be implications upon the choice. For instance, considering the configuration of ATM VC layer network shown in Figure 2, ATM VC topological links are established using trails in the ATM-VP layer network across all domains. If it is possible to set up a VP trail from Portugal Telecom (PT) to Belgacom (BEL) domain, it would be possible to define a topological link in the VC layer network between the corresponding domains of these operators. This situation implies that a request for a VC trail from a point in PT domain to a point in BEL domain goes transparently to Telefonica (TEL) domain, as shown in the figure. However, if there is not such VP trail, then the VC trail is established involving also TEL domain in the VC layer network. These considerations are useful when defining federation relationships between network operators, and are tightly coupled to the implications from the enterprise viewpoint.

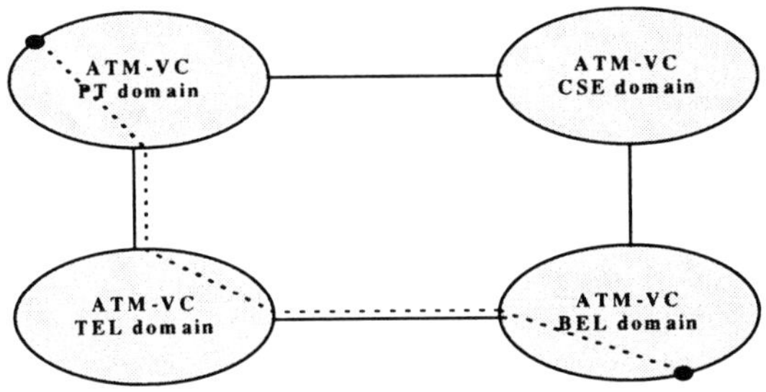

Figure 3: Different federation relationships

Federation relationships are supported by the LNC computational object. Each LNC offers the interfaces to create, control and destroy tandem connections through the corresponding domain. In principle, a LNC can be requested to establish a trail which starts or finishes in other domains. The LNC analyses which part corresponds to its own domain, and requests its corresponding NML-CP to establish and set-up that part, and to neighbour LNCs (those which are reachable through a topological link) for the remaining parts if any. Routing is made, using the network address of the end points, normally depending on the prefix of the given addresses.

Current specifications in VITAL consider that federation relationships may change, so LNC routing tables can be modified through the configuration interface, by specifying the name of the neighbour domain (this is used to get the reference of the neighbour LNC federation interface), the topological links between domains, and the list of network addresses which are reachable through the neighbour domain (this can be specified as a group by defining their prefixes).

Operations and parameters defined at the interface offered to the CC are also considered in the federation interface, with the addition of parameters which define a negotiation policy between the domains. These parameters determine, for instance, how accounting is to be performed, or whether monitoring information on the tandem connection is needed.

Federation interfaces can be applied also to avoid a deep hierarchy of CP computational objects, as it provides a mechanism for horizontal decomposition of the control of the network (see Figure 4).

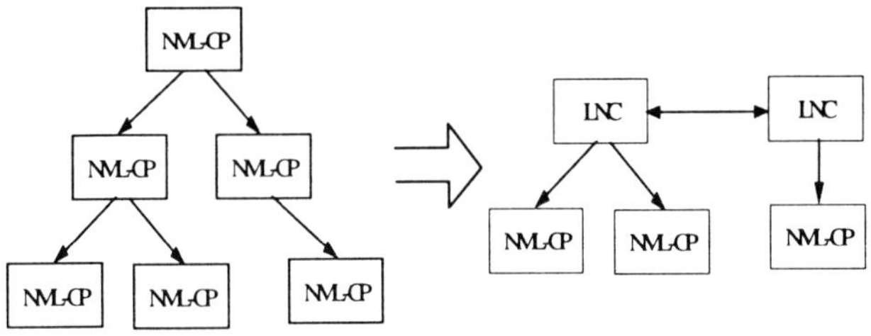

Figure 4: Decentralisation of connection management component hierarchy using LNC federation

3.3 Accounting Management

The accounting management model specified in VITAL follows as closely as possible the principles set in TINA accounting management architecture. For the second trial of VITAL the main focus was on the specification and implementation of the resource level metering cycle. This task is concerned with the definition of the accountable objects and the metrics applied to them, as well as with the definition of an implementable information and computational model applicable to both Retailer and Connectivity Provider domains for usage metering.

An accountable object is the modelling concept that represents a real resource which is subject of accounting management. The mapping between a real resource and an accountable object is not necessarily one to one. A real resource can be represented by more that one accountable object, and an accountable object may represent more than one real resource. The accountable objects can represent resources that belong to several TINA layers and to several aspects of TINA architecture. These resources can be service level, resource level, network, management and computing resources.

The possible network resources that are subject of accounting management can be the resources defined in TINA NRIM [9]. Such resources are, Subnetwork, LayerNetwork, Topological Link, Link Termination Point, NWTTP, NWCTP, NWTPPool, Connection, Subnetwork Connection, Tandem Connection, Edge and Trail. In addition to the above resources which are technology independent, it is possible to define a series of technology dependent accountable resources that correspond to the specific technology of the layer network.

Since the Connectivity Provider offers connectivity services to its clients, the network resources that constitute the connectivity fragment of TINA NRIM (Connection, Subnetwork Connection, Tandem Connection, Edge and Trail) appear to be the most suitable ones to be subject to accounting management. In VITAL, the Trail has been selected to be the accountable resource and the Trail Manager (TM) to be the accountable object that represents this resource.

The information and computational models for accounting management in VITAL are defined in a generic manner, so that it can be performed on both service and network resources. In principle, accounting management in VITAL is event driven. The accountable objects generate accountable events, each time the state of the accountable resource changes. In VITAL the following Accountable Events have been defined: registration, request, accept and complete. The Registration, Accept and Complete events model changes in the state of the service accountable resources. The Request event happens each time a change occurs in the state of the Trail that effects the usage of the service. Such a change can be noticed when one or more branches are added to or deleted from a Trail, or when one or more branches of the Trail became active or inactive.

The accounting events contain also usage metering information. This information is forwarded from the accountable object (TM in case of VITAL) to the Usage Metering Data object (UMData). The UMData object is responsible for metering the usage of the accountable objects (TM). It is also in charge of controlling the temporary logging of the accounting events and management of the logging records. In VITAL there is an instance of the UMData computational object per active Trail. Accounting events modelling was influenced from X.742 *Usage Metering Package.* Accounting event blocks contain information about the type of the event, the party that is related to the event, additional unit information and a time-stamp.

Upon the termination of the life-cycle of an accountable object, its corresponding UMData computational object reports the usage data to the Usage Metering Log (UMLog) computational object which stores the usage information in order to keep it persistent at the Connectivity Provider domain.

The metering and storage of accounting information in the connectivity provider domain is depicted in Figure 5.

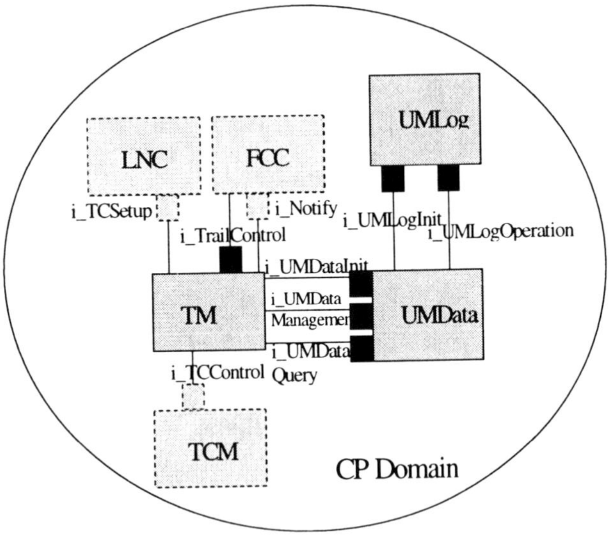

Figure 5 Connectivity Provider Usage Metering and Storage

4. Scenarios applied to ATM Networks

The following scenarios illustrate the interactions among the computational objects described above. Two steps are considered. First, the configuration of the resources needed to set-up and release connections. Secondly, the procedures for providing the connectivity required by session level objects. In order to make clear the control of connections a simple network configuration will be considered where only one connectivity provider domain is involved (see Figure 6).

Note that in VITAL second trial the four operators act as both retailers and connectivity providers. Therefore the access part of ConS Reference Point (CSM-CC) [2] has not been considered.

4.1 Configuration scenario

In this scenario we consider the dynamic reconfiguration of connection management subsystems when adding or removing a new ATM Virtual Path Connection (VPC). As explained in section 3.1, the system is initially configured to reflect the initial topology of the network, both at the Virtual Channel (VC) and Virtual Path (VP) layers, which have a client-server relationship. The initial VPCs have been configured according to network planning activities. From then on, new VPCs may be added or existing ones may be removed by the network operator according to changing requirements. The steps involved in this process are the following, referring to the computational components of Figure 1.

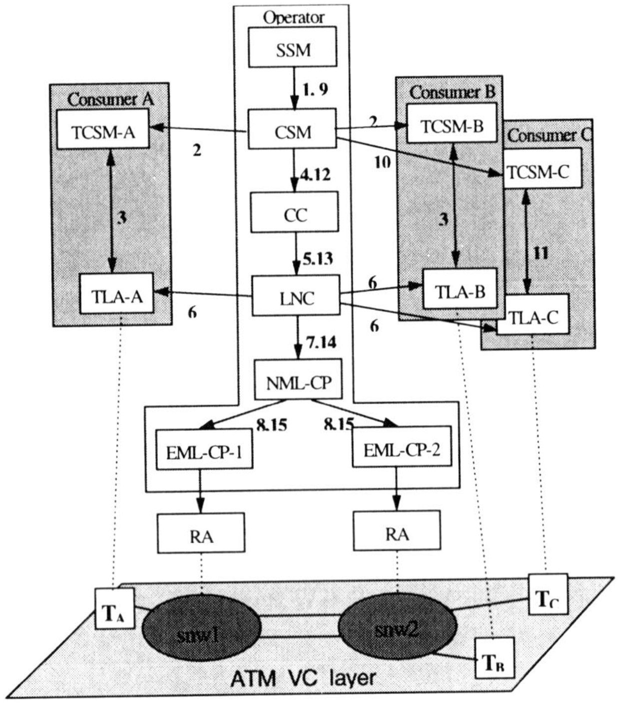

Legend:
CSM - Terminal Communication Session
CC - Connection Co-ordinator
LNC - Layer Network Co-ordinator
CP -Connection Performer
NML - Network Management Layer
EML - Element Management Layer
TCSM - Terminal CSM
TLA - Terminal Layer Adapter
snw - Subnetwork
T - Terminal
RA - Resource Adapter

Figure 6: Configuration for scenario

- The NTCM-GUI retrieves and displays the current network topology from the VP and VC NTMap objects through the NTMan. The network operator may then request the addition / removal of a new VC topological link or VP trail through the graphical interface.
- The NTCM-GUI passes the request to the NTMan which checks its validity by accessing information in the VP and VC NTMap.
- In the case of a valid request, the NTMan asks the actual request implementation to the LNC of the VP Connection Management system. The latter implements it by passing it through the CPs and RAs to the actual ATM cross-connects.
- Upon successful completion, NTMan updates the VC NTMap with the topological link change (a VPC manifests itself as a topological link in the VC layer). The NTMan subsequently notifies the VC CMC about changes on the VC layer network topology.
- The VC CMC reconfigures the VC CM subsystem to take into account the topological change. From then

on, VCC requests to it will be served according to the new topology of the VC layer network.

4.2 Set-up connection scenario

The aim of this scenario compared to Connection Management in TINA is to show how computational objects in the Consumer Domain (TCSM and TLA) negotiate connection related parameters (e.g. QoS, VCI) with those located in the Connectivity Provider Domain (CSM and LNC). It is also important to highlight that the higher level components (i.e. SSM, CSM and CC)) are not aware of technology specific information (e.g., VCI values); this is handled by the Consumer Domain computational objects and the low level components (i.e. LNC and CPs). The scenario presented in this section is more focused on high level mechanisms rather then on technology specific ones. In addition, federation is not further described here due to its complexity. Details concerning this issue can be found in [12].

1• • The chosen scenario is initiated by SSM, which requests CSM the creation of a communication session. The stream flow end points to be interconnected are given to the CSM together with the TCSM references.

2• • The CSM then contacts each TCSM object located in each terminal in order to know the characteristics of the correspondent network flow end points (e.g. network address, layer network, media flow codification and network related protocols). By knowing this information, CSM can then perform the negotiation between the different terminals. Currently, in VITAL a simple negotiation process has been implemented. CSM looks for commonalties among the parameters provided by the different terminals and in order to take the final decision gives high priority to the choice of the source. Similarly, only a simple mapping between stream flow end-points and network flow end-points has been defined. This mechanisms will be enhanced in the next trial.

3• • While each TCSM is a repository of the existent stream flow end-points, a TLA keeps track on the technology dependent information, which in the case of VITAL is ATM specific. The layer network type, the network address and the transport protocols are then requested by TCSM to TLA. Latter, when the VCI is chosen by the LNC in co-operation with the CP objects, the TLA helps the TCSM to resolve the mapping between stream flow end points and network flow end points.

4• • The CC is a less complex object since the access part has not been considered in the current VITAL trial and only the ATM VC layer is handled. In the future, CC will choose the connectivity provider and the layer network based on constraints or/and requirements arising from consumer domains.

5• • The LNC receives a request from the CC for the provision of a trail. The CC sends the network addresses of the terminals, the references of the associated TLA objects, and the correlation identifier.

6• • The parameters mentioned above allow the LNC to contact the TLA objects and negotiate the VPI/VCI values, together with the connection parameters (e.g., quality of service and bandwidth), which are supported by the terminal for that connection. If a terminal was attached to a network point in the scope of other connectivity provider domain, the LNC would delegate the negotiation with the corresponding TLA to a federated LNC. The LNC is prepared to suggest some values of VPI/VCI and quality of service to the correspondent TLA. Currently in VITAL, the TLA is suggesting such values to the LNC since the mapping of QoS is totally done in the terminal and no constraints from the consumers have been considered yet. Nevertheless, the negotiation mechanism already exists for future enhancements.

7• • The LNC then requests to its top NML-CP object to create a subnetwork connection among the identified network termination points. Each NML-CP deals with routing functionality in its own domain.

8• • Each NML-CP determines the lower CP objects (finally, at Element Management Level, EML-CP) to request the connectivity for the subnetworks in the domain. EML-CP objects keep control on the use and reservation of the resources associated to a certain subnetwork, e.g., termination points and bandwidth associated to the subnetwork.

Up to this moment, resources have been reserved but they have not effectively been allocated in the real switches. This happens when there is a request to CSM to activate the connection graph (or some specific part of it). This request goes from the CSM down through the CC, LNC, NML-CP, and, finally, EML-CP. A Resource Adapter (RA) class of objects is used for interacting with the real switches. RA objects provide to each EML-CP independence over vendor's specific interfaces supported by the switches. This allows the use of the same CP objects over different switches, as it is the case in VITAL, where each operator has its own.

Once the connections in the switches are established and set-up, multimedia flows can be transferred between the different terminals involved in the communication session.

Add a Branch

Operations 9 to 15 enable to add a new branch to a point-to-multipoint connection. The CSM must re-negotiate the terminal capabilities with the parties involved. Similarly to the other terminals, the technology dependent parameters (including QoS) are negotiated at the LNC-TLA level. The mapping between QoS parameters concerning services and those network related is done within the terminal.

5. Conclusions

The validation effort of the TINA architecture performed in the VITAL project has driven to the identification of under-specified components behaviour and interfaces. In order to develop working services on a real multi-operator ATM network, TINA specifications have been refined and extended when needed. This work is currently being considered by TINA-C core team and will be reflected in final TINA-C specifications, in concrete Network Resource Architecture [1], Connection Management Specifications [4], Accounting Management [8], and Network Resource Configuration [10].

The experience achieved by mid-97 has been centred around Resource Configuration Management, Connection Management and Accounting Management. It is interesting to note that specifications developed in VITAL for Resource Configuration Management and Accounting Management are not only applicable to Network Resources, but also to the Service components of the TINA architecture, because of the generality of the initial information and computational models. These have been specialised for the network part, and next year will be applied for the service part. Another important aspect developed in VITAL is the way application stream flows and QoS requirements are mapped to network connections, and the implications of the architecture in the terminal side.

Acknowledgements

This paper describes work undertaken in the context of the ACTS AC003 VITAL project which is partially funded by the Commission of the European Union.

We would also like to specifically acknowledge design work undertaken by our colleagues in the VITAL project related to the VITAL NRA. In particular we would like to thank David Griffin of University College London, UK, and Juan Carlos Garcia Lopez of Telefonica, Spain.

6. References

[1] *TINA Network Resource Architecture*, version 3.0, TINA document no. TB_v3.0_97_02_10, TINA-C, Feb. 1997.

[2] *The ConS Reference Point*, version 1.0, February 97

[3] *Connection Management Architecture*, document no. TB_JJB.005_1.5.94, March 95

[4] *Connection Management Specifications*, document no. TP_NAD.001_1.2_95, March 95

[5] *TINA Business Model and Reference Points*, version 4.0, Approved Public Version, May 97

[6] *Network Resource Information Model Specification*, TINA Baseline Draft, May 96

[7] *Connection Management Specifications for VITAL 2nd phase*, VITAL internal draft document, May 97

[8] *Accounting Management Architecture*, version 1.3 (Draft), Engineering Note, February 1996.

[9] *Network Resource Information Model*, Document No. TB_LRS.011_2.0_94 TINA-C, December 1994.

[10] *Network Resource Configuration Management*, version 2.0, October 1996.

[11] G.Pavlou, D.Griffin, "*Realising TMN-like Management Services in TINA*", to appear in the Journal of Network and Systems Management Special Issue on TINA, 4th quarter 1997.

[12] J. Pavón, E. Montes, "*Network Provider Domain Federation*", TINA 97.

[13] J.Rajahalme, T.Mota, F.Steegmans, P.F.Hansen, F.Fonseca, "*Quality of Service Negotiation in TINA*", TINA 97.

Distributed Management System for IP Network - Implementation and Performance Evaluation

Nobutaka Nakamura
NTT MultimediaNetwork Labs.
1-1, Hikari-no-oka, Yokosuka
Kanagawa, 239 Japan
nobu@nttmhs.tnl.ntt.co.jp

Akira Murashige
NTT MultimediaNetwork Labs.
1-1, Hikari-no-oka, Yokosuka
Kanagawa, 239 Japan
mrsg@nttmhs.tnl.ntt.co.jp

Yasuharu Yamada
NTT Multimedia Network Labs
3-9-11, MIdorimachi, Musashino
Tokyo, 180 Japan
yamada.yasuharu@na.tnl.ntt.co.jp

Abstract

The network management systems for IP networks should be able to manage a large-scale network because the number of network elements increases. Currently there are many applications of the IP network management system. These applications are been using by many users and are readily available for LANs. Though it is expected to manage the large-scale network using SNMP, it is difficult to make a network management system using SNMP on a single computer. An effective way to solve the problem is to construct a network management system using a distributed processing environment.

We have developed a new distributed IP network management system. The system consists of several sub-systems, which enables the processing load for IP network management to be reduced. In the system evaluation, the system was implemented on DCE, a client-server environment, and it was found that the overhead time of the distributed processing is small in comparison with the overall response time.

1. Introduction

IP network that is popular in LAN grew into a large-scale network according to increase of Internet users. Therefore, the number of elements in IP network has been increasing recently. Network management systems have to be made more powerful to manage all these elements. Managing a large-scale network requires the use of management functions, such as OSI management [1], and protocols, such as CMIP [2]. However, almost all network elements support SNMP [3], and users expect to be able to manage a large-scale network using SNMP. The distributed processing [4] of network management tasks is one way to solve this problem [5]. A distributed processing environment, such as TINA [6], is effective for network management [7]. In a distributed environment as TINA, it is especially important to carefully select the management functions required by users because an ill-designed distributed network management system will greatly increase the overhead of distributed processing, resulting in reduced processing efficiency.

This paper describes the implementation and evaluates the performance of a new distributed IP network management system. The proposed system consists of several sub-systems, which enables the processing load to be reduced in the IP network management. Each sub-system comprises of several modules; these modules cooperate with the intra-sub-system's module through inter-communication on a single computer and with other sub-system's modules through remote procedure calls in a Distributed Computing Environment (DCE) [8].

2. IP Network Management Function Grouping

We examined the management functions of the existing IP network management applications to clarify the bottlenecks in these functions when a large-scale network is managed by the application. The applications examined are almost all commercial IP network management products that have a major market share.

Table 1 shows the functions for IP network management and the function units that comprise each function. For example, "Configuration Data Management" should be expanded to manage the state of the network element according to increase of elements and "Discovering Agent" should be expanded to make it easier to find the agent because the managed area expands with increasing network scale. The table shows the function units whose performance has to be boosted (marked "O") as the network scale

Table 1. IP network management functions and their grouping

IP network management functions	Function units	Higher performance needed with lager network scale	Function groups
Configuration Management	Configuration Data Management	O	Configuration Management
	Database Engine	O	
	Discovering Agent	O	
	SNMP Operation	-	
Agent Monitor	Configuration Data Management	O	Agent Monitor Management
	Monitoring MIB Value	O	
	SNMP Operation	-	
Remote Agent Monitor	Remote Monitor	O	RMON Management
	SNMP Operation	-	
Summarization Function	Data Collection	O	Summarization Management
	Statistics	-	
Event Control	Event Control	-	Operation Management
	SNMP Operation	-	
Log Control	Log Control	-	
	SNMP Operation	-	
MIB Browser	SNMP Operation	-	
	MIB Dictionary	-	
GUI	GUI	-	

becomes larger. The performance of the functions that include one or more marked function units should be expanded and improved in proportion to network scale.

Taking these points into consideration, we adopted the following three-part design policy.

(1) The functions with one or more marked function units are made up into one group and remaining functions are combined into another group, i.e. Operation Management.

(2) As network elements are added, a set of functions is duplicated on a new computer to manage them.

(3) From the viewpoint of software development, function units should be dealt with independently of software units.

3. Implementation

3.1 Selection of Distributed Environment

In general, there are three main ways to achieve distributed processing:

(1) A distributed operating system messaging between the many processes that construct the functions [9].

(2) A client-server environment messaging between servers and clients that have divided functions.

(3) A distributed object environment messaging with an Object Request Broker (ORB) between objects.

Method (1) was popular a few years ago and methods (2) and (3) are popular now. With method (2), the load of distributed processing is small because the client-server relationship is determined in software design. Method (3) enables dynamic configuration among components, and eases system expansion. As the distributed environment for this method, CORBA [10] has been standardized by OMG and DPE [11] was proposed for TINA.

We adopted method (2) because of its ability to reduce overhead on the distributed environment and selected the OSF/DCE, the standard for the client-server environment.

3.2 Sub-System Design Approach

This section discusses the communication method employed in our system.

Each of the function units shown in Table 1 is implemented as a software module and each of the groups is implemented as a sub-system that is a unit for installation. To avoid duplication of function units in the different sub-systems, which would cause data inconsistency, a client-server mechanism is adopted. In this mechanism, a module that calls modules in other sub-systems is defined as a client module (CM), and a called module is defined as a server module (SM). For example, Configuration Management is implemented as an SM in a sub-system, and other sub-systems utilize this module from the corresponding CM within the sub-system. A module that dedicates itself to tasks within the sub-system is called a common module (CoM).

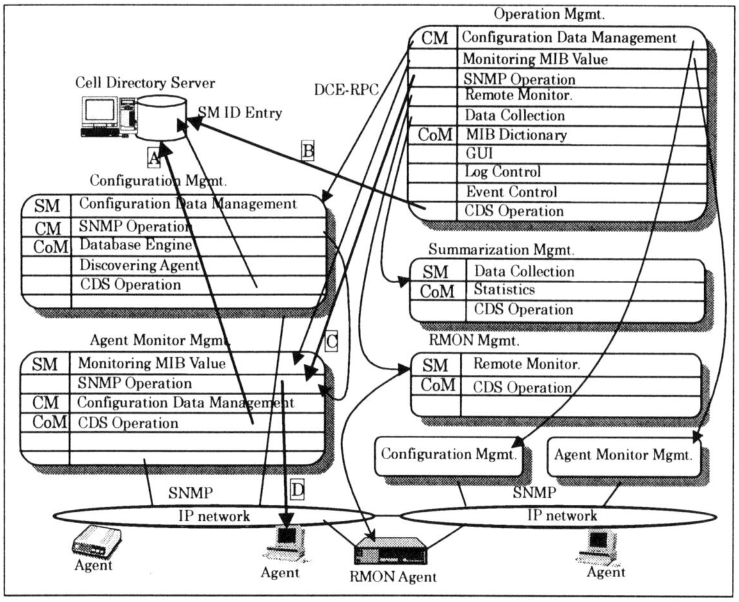

Figure 1. System structure

Remote Procedure Call (RPC) [12], which is provided by DCE, is used for communication between SMs and CMs. For communication between modules within a sub-system, a simpler communication method is preferable, because all modules in a sub-system are running on the same computer and there is no need to use a distributed communication method. Therefore, a pipe mechanism is adopted for in-subsystem-module communication.

A distributed IP network management system that provides the function units shown in Table 1 was constructed. The system is outlined in Fig. 1.

3.3 Design of Directory System for IP Network Management

To transfer CM operation requests to the managed network element (SNMP agent) , the Cell Directory Server (CDS) of DCE is used as follows: The SM's ID entry of the subsystem that manages the SNMP agent is registered into CDS before the management operations are executed (A in Fig. 1). Next, CM obtains the SM's ID entry using the agent ID from the CDS (B). Then CM actually sends the request to the SM (C) using RPC, and finally the operation to the agent is converted to SNMP by the SM (D).

The CDS entry data on searching is shown in Table 2. The subsystem group is the set of subsystem managing a whole network. The subnetwork group is the set of networks managed by the system. The agent group is the set of agents managed by the system. These entries are registered as the result of detection by "Discover Agent".

Table 2. Data entry on Cell Directory Server

Category	Entry name	Member value
Subsystem group	Grp_CFmg	CFmg_[IP address] (Configuration subsystem entry name)
	Grp_IPmg	IPmg_[IP address] (Agent Monitor subsystem entry name)
	Grp_RMmg	RMmg_[IP address] (RMON subsystem entry name)
	Grp_SMmg	SMmg_[IP address] (Summarization subsytem entry name)
Sub-network group	Grp_CFbyNW	CFmgbyNW_[NET_address]
	Grp_IPbyNW	IpmgbyNW_[NET_address]
	Grp_RMbyNW	RMmgbyNW_[NET_address]
	Grp_SMbyNW	SMmgbyNW_[NET_address]
Agent group	Grp_CFmgbyAg _[NET address]	CFmgbyAG_[IP address]
	Grp_IPmgbyAg _[NET address]	IPmgbyAG_[IP address]
	Grp_RMmgbyA g_[NET address]	RMmgbyAG_[IP address]
	Grp_SMmgbyA g_[NET address]	SMmgbyAG_[IP address]
Subsystem name	CFmg_[IP address]	Subsystem Interface end-point
	IPmg_[IP address]	Subsystem Interface end-point
	RMmg_[IP address]	Subsystem Interface end-point
	SMmg_[IP address]	Subsystem Interface end-point
Managed sub-network name	CFmgbyNW_[N ET_address]	CFmg_[IP address] (Configuration subsystem entry name)
	IpmgbyNW_[N ET_address]	IPmg_[IP address] (Agent Monitor subsystem entry name)
	RMmgbyNW_[NET_address]	RMmg_[IP address] (RMON subsystem entry name)
	SMmgbyNW_[N ET_address]	SMmg_[IP address] (Summarization subsytem entry name)
Managed agent name	CFmgbyAG_[IP address]	CFmg_[IP address] (Configuration subsystem entry name)
	IPmgbyAG_[IP address]	IPmg_[IP address] (Agent Monitor subsystem entry name)
	RMmgbyAG_[I P address]	RMmg_[IP address] (RMON subsystem entry name)
	SMmgbyAG_[IP address]	SMmg_[IP address] (Summarization subsytem entry name)

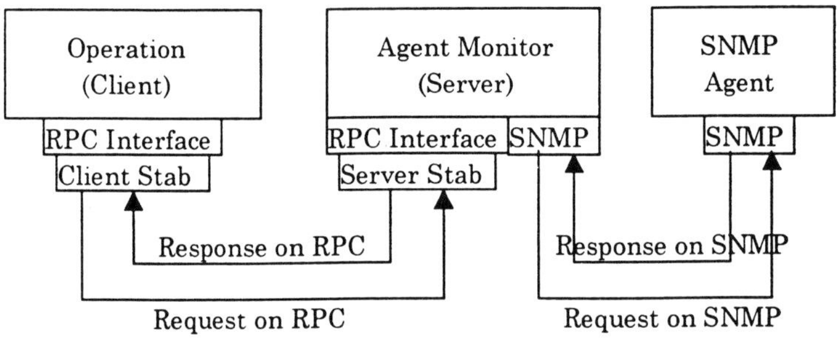

Figure 2. SNMP protocol proxy

The subsystem name category indicates the SM's RPC end-point list. Managed subnetwork name entries are the names of the subsystems. Likewise, managed agent name entries are the names of subsystems. The CM searches the SM of the subsystem using these names.

3.4 Design of SNMP Access Service

SNMP access service provides the SNMP v1 operation proxy as a module in the "Agent Monitor" subsystem and its services are implemented using RPC. The interface between CM and SM is defined by IDL. IDL includes the following parameters: agent address, error-status, error-index, and variable-binding-list. The SNMP access service processing is illustrated in Fig. 2. The CM's request defined by IDL is sent to the "Agent Monitor" subsystem's SM, an SNMP access service module. The SM receives the request and transfers the request into the SNMP protocol data unit (PDU). Then the SM sends the SNMP request to the agent indicated in the IDL parameter. The SM takes out error-status, error-index, and variable-binding-list from SNMP response PDU and sends these data to the CM on RPC.

The SNMP service consists of Get, Get-next, Get-bulk, Set, and Cancel. Get-bulk and Cancel are original services that are not SNMP v1 operations. Get-bulk is used to collect a large amount of data, such as a table, in a single operation. The SNMP module translates a Get-bulk request into several Get-next SNMP PDUs and sends them to the SNMP agent. This mechanism can reduce communication loads between the SNMP module and client module. Cancel is used to interrupt Get, Get-next, Get-bulk, and Set operations. Cancel is especially useful to interrupt a Get-bulk operation that takes too much time.

4. Performance Evaluation

4.1 Evaluation Condition

DCE is available on Windows NT as well as UNIX. We tried to put the network management system on a Windows-based DCE, because this makes low cost implementation possible. Windows-based DCE can provide CDS service and security service in the same environment. We used Windows NT3.51 and a Pentium machine (166 MHz, 64 Mbyts memory). As a managed agent, we used the SNMP Agent of Windows NT on a PC.

4.2 Performance Evaluation

We measured the response time corresponding to the GET operation, as this is one of the most important factors for evaluating the performance of the distributed environment. The operation (T) consists of searching for the CDS entry (T1), the communication from the CM to the SM (T2) and the SNMP operation of the SM (T3), as shown in Fig 3. The processing time for one GET operation is approximately 170 ms. For the existing IP network management application, the processing time is about 50 ms. This difference occurs in the processing time of T1 and T2. Because T1 searching and T2 communication are the factors that dominate the system performance, we measured them in order to clarify the details of T.

4.2.1 CDS Entry Searching

Figure 4 shows the response time of the GET operation taking the number of CDS entries as the parameter. T is

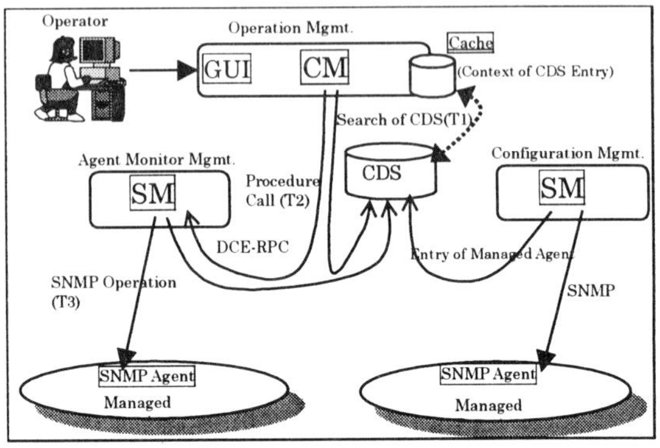

Figure 3. Evaluation system

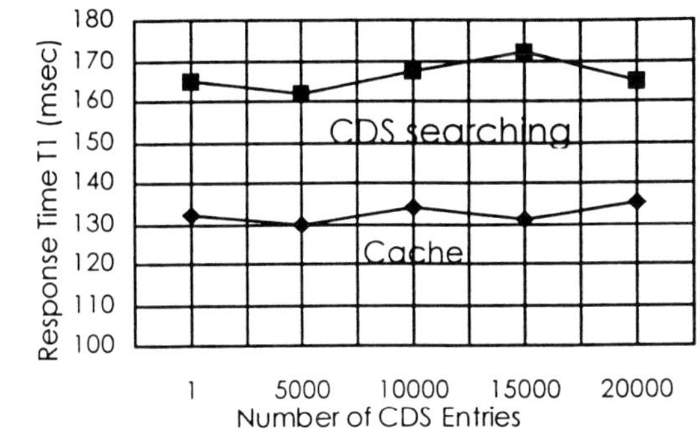

Figure 4. System response time for number of CDS entries

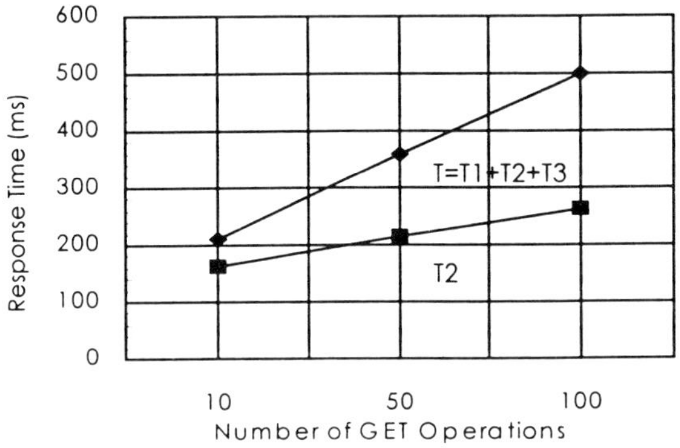

Figure 5. System response time for GET operations

always approximately 170ms for up to 20,000 entries. If the cache is used, the searching is 35ms faster. 20,000 entries have shown in figure 4 means the management of a few thousand agents. CDS is available for general directory service. It was found that the searching overhead is independent for the few thousand agents as a result. It is also available to search higher on PC environment.

4.2.2 Communication between CM and SM

Figure 5 shows the system response time (T) and the communicating time (T2) taking the number of GET operations as the parameter. Because T1 is constant for a given number of CDS entries, the difference in the rate of increase between T and T2 can be regarded as the increase in the rate of the SNMP operation response time (T3). The difference increases as the number of GET operations increases. The increase rate of T2 is smaller than that of the T as a whole. This is because SNMP operations time T3 effects T more than T2. It was found that the communication overhead in the distributed environment is smaller than it is in the essential SNMP operation.

5. Conclusion

The growth of the IP network, which is popular in LANs, has increased the operation processing load of the network management system. Therefore, the distributed processing has become necessary. However users are continuing management based on SNMP and desire low cost and simplicity in the distributed management environment.

We have proposed a distributed network management system that makes it possible to distribute the processing load in IP network management. The proposed system was implemented on DCE, a client-server environment. In the system evaluation, it was found that the overhead time of the distributed processing is small in comparison with the overall response time. As a result, a network management system that targets the management functions of the existing network management application could be constructed in distributed processing environment using PCs.

CORBA is now available in a PC environment. As part of our future research, we plan to evaluate the overhead of communication between objects on CORBA and compare client-server communication with our distributed environment.

Reference

[1] ITU-T X.701 (01/92), "Information technology - Open Systems Interconnection - Systems Management Overview," 1992.

[2] ITU-T X.721, "Information Technology – Open Systems Interconnection – Common management Information Protocol Specification – Part 1," 1991.

[3] J. D. Case, M. Fedor, M. L. Schoffstall, and C. Davin, "Simple Network Management Protocol (SNMP)," RFC 1157, 1990.

[4] D. Bowen, "Open distributed processing," *Computer Networks and ISDN Systems*, Vol. 23, pp195-201, 1991.

[5] Simon Znaty, Guy Genilloud, "Networked System: Introducing Network Management Models into ODP," *Proc. GLOBECOM '95*, pp121-126, 1995.

[6] TINA-C, "Overall Concepts and Principle of TINA," Document No. TB-MDC.018_1.0_94, December 1994.

[7] Qinzheng Kong, Graham Chen, "Integrating CORBA and TMN Environments," *Proc. NOMS '96*, pp86-96, 1996.

[8] W. Rosenberry et at., "Understanding DCE," *O'Reilly & Associate, Inc.*, 1994.

[9] A.S. Tanenbaum and R. van Renesse, "Distributed Operating Systems," *ACM Computing Surveys*, Vol.17, Vol.4, pp419-470, 1985.

[10] OMG, "The Common Object Request Broker, Architecture and Specification," OMG Document, No.91.12.1, 1991.

[11] Yagi, H., et al., "TINA-C Service Components," *Proc. TINA95 Integrating Telecommunications Distributed Computing – from Concept to Reality*, February 1995.

[12] John Bloomer, "Power Programming with RPC, O'Reilly & Associates", Inc., 1992.

Operator-Oriented Approach for the Inter-work of Service- and Network Management

Yoshitsugu TSUCHIYA and Sakae CHIKARA
NTT Multimedia Networks Laboratories
3-9-11 Midori-Cho Musashino-Shi, Tokyo 180 JAPAN
Telephone: +81 422 59 2930 Facsimile: +81 422 60 7372
Email: tsuchiya.yoshitsugu@na.tnl.ntt.co.jp
chikara.sakae@na.tnl.ntt.co.jp

Abstract

In order to gain flexibility of management systems, inter-work among the service management components, network management components, and operators is necessary. To fulfill this requirement, we propose a module-based implementation framework to enhance the implementation aspect of the TINA Service Architecture.

We propose a module-based implementation framework using the mapping relation of user interfaces and management functions that mediate the user interface in low-cost customers/operators environment and TINA service components. To support the inter-work between the modules, we introduce a inter-working mechanism using management context description that is based of the storage of data format to be shared by the modules.

1 Introduction

To satisfy the complex and various demands of the customers today, telecommunication networks must provide their services rapidly and flexibly, with the support of efficient service and network management systems.

Regarding the demand of the customer to control part of the network resources, management systems should support collaboration among customers, operators, and the management functions. In addition, For the quick response to the change of operation scenarios, tasks of operators should include not only the usage and maintenance, but also the customization and supervision of the system. Part of these jobs can be shared by the customers, depending on the network providers' policy. Thus, the management systems should support the environment for the customers and the operators to change system configurations.

To meet these requirements, a demand-driven service management model is studied in TINA Service Architecture[4]. However, to realize actual management systems, the framework lacks the implementation aspect focusing on the operator-side customization of the system.

In this paper, we aim at realizing a management system that is able to relate management function with operators' management procedure, with the concept of exchanging and storing the management context schema.

In the following section, we focus on three requirements on distributed network management: flexibility for the change of operation scenarios, inter-work of service- and network management, and the lightweight customer environment. To fulfill these requirements, section 3 introduces the relevant concept from the TINA Service Architecture. Section 4 presents the proposed module-based implementation framework, for the flexible customization of management information, network management functions, and user interfaces. Section 5 proposes an implementation framework using the concept of management context schema in order to support the module inter-working. Section 6 deals with advantages regarding the application of our approach to an actual service management scenario.

2 Requirements on distributed network management

To meet the customers' demand on network services, we think that network management systems should fulfill the requirements described in section 2.1 to 2.3.

2.1 Quick response to the change of operation scenarios

Through the service or network operation in practice, modification of operation scenario is likely to occur. This may be caused by revision of service definitions, replacement of software components, renewal of network equipment, or exceptional customer requests. In such cases, system developers have to supervise the management system by modifying the relevant part. Supervising in this sense corresponds to the manual invocation of the function for the purpose of system customization. For the quick response to the change, management systems should provide environment for the supervision not only for the system developers but also for the operators.

2.2 Inter-work of service- and network

0-8186-8335-X/98 $10.00 © 1998 IEEE

management

We analyze that the inflexibility of the present management systems stems from the lack of inter-work between service- and network management. Revision of service specification may require change of information items on network management. Similarly, renewal of network management functions may require change of information used in the service management. To share the information of the modified part, we need a framework to support the inter-work among management functions.

Inter-working framework is also necessary to realize collaboration of operators and management systems. Since management cannot be completed without manual operations, network management systems must be capable to relate operators' intention with the information handled inside the system.

2.3 Lightweight customer environment

To enhance the availability of the TINA services and the customizability of management systems, access environment for the customers and operators should be lightweight. Since improved performance of the customers' equipment and the current technology such as WWW and Java provides flexibility of the systems design, we should utilize these technologies that require minimum cost for the access environment.

3 The TINA service management model

For the flexibility and inter-work of management systems as required in section 2.1 and 2.2, a service management model using the concept of management context is discussed in the TINA Service Architecture. The model aims at the service-oriented customization of the network management function.

3.1 Demand-driven service instantiation[4]

To realize the customization of management operation, it is effective to separate software component that may change according to the revision of service or development phase as modules and the other as generic component. Flexibility can be obtained by replacement of the modules.

For the module-based service composition, service instantiation using the concept of management context is introduced in TINA Service Architecture. According to TINA, service components are created and instantiated by the service factory component. Service factory creates instance of objects and configures them by giving the necessary initial parameters. The data that are used before the creation of a object instance and that must be passed to the instance are described as management context. Management context is a set of information to be passed to the object instance, which includes type and status of the data, reference of object instances that process the data. Reference for the instances are described as management items. According to the management context description, service factory picks up the relevant set of services. These objects are dynamically attached to service components when they are instantiated at service factory. This demand-oriented service instantiation is called the Pick and Mix approach[4].

3.2 Requirements on the model

Although the above approach can be used as a guideline for module-based systems design, we need to further determine the implementation framework focusing on the support for inter-module communications. In TINA/ CORBA systems, inter-work among modules is realized using the concept of Interface Definition Language (IDL) [5]. Using the IDL, system developers can develop subsystems in reference to the interface definition. However, as required in section 2.1, since not only the system developers but also the operators should participate in the system supervision, management systems should provide a mechanism to support operator-side customization.

Moreover, advancement of the WWW and applet technology allows us more dynamic instantiation of the modules. As required in section 2.3, we need a framework utilizing this technology in order to bind the TINA-based module structure of the system with low-cost customers/operators environment.

Hence, in the following section, we propose an implementation framework focusing on the module-based customization of management systems.

4 Module-based implementation framework

In this section we propose an interface module concept in order to satisfy the requirement for the flexibility of management systems. Information model of our proposal is shown in OMT diagram in Figure 1.

4.1 Interface module concept

Burden of the operators that work on variable tasks can be reduced by similarity of operation procedures. It is thus desirable to design generic components of the systems to be operated with common schema that can be shared with different subsystems[3]. Example of the common operation schema can be based on the business objects. If we are to model the managed objects, it may be based on the common logical view specified in the network information model[1][2].

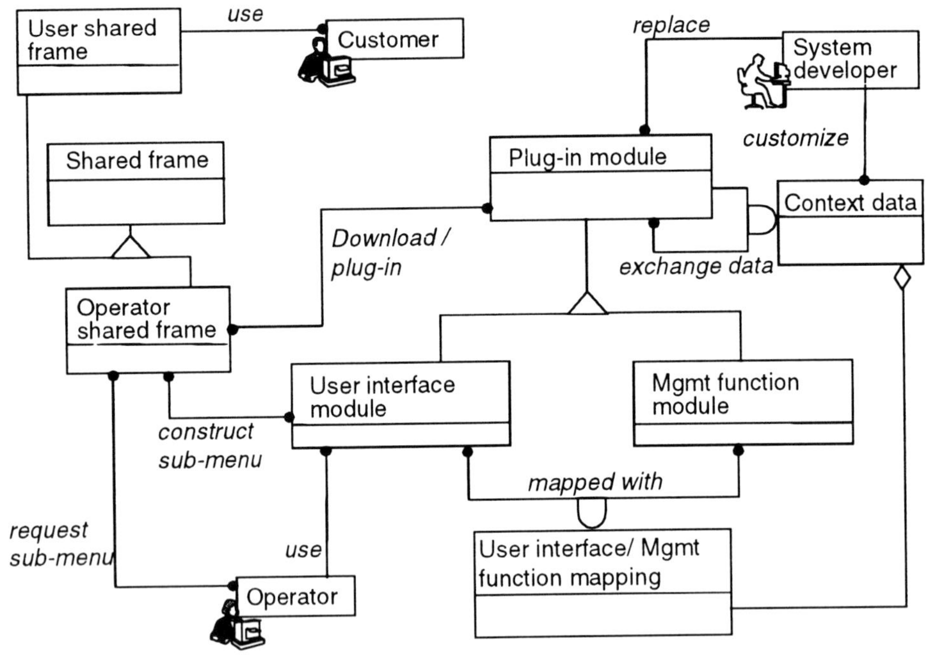

Figure 1 OMT diagram for the proposed interface module concept.

When a minor change of the system scenario or introduction of a new management tool is needed, systems have to be easily customized. Some part of the interface should be deployed as modules and replaced if necessary. For this purpose, we define function of plug-in module (PIM), a module that can be replaced and that does not have common operation schema.

4.2 User interface module and management function module

It is effective for the inter-work of service- and network management to manage the relationship between user interface part and the back-end management functions. We realize the module structure of management system according to the mapping relation of user interface and management function. This information will be stored by the management system to form the operation scenarios.

Management function module represents implementation of management information, management interface, and information processing procedure. It can be used as adapters to link user interface with TINA services. User interface module represents resources for providing interaction between operators and the management function modules. As described in Figure 1, both are modeled as subclasses of PIM.

Operators access the added management function modules with user interface modules, which help sharing the management knowledge among operators. They contain information that helps the operators to operate the new management function modules, by guiding them to relate the new operation scheme with the current management knowledge.

4.3 Shared frame function[3]

To realize the plug-in of user interface module requested by the operators, management systems have to employ mechanisms to download user interface modules to the operators' environment. Since this cannot be realized without the human control, we need an environment that represents relationships between user interface modules and management function modules, to relate operators' image with the information used inside the system. We define group of functions that are necessary to provide such environment as shared frame.

Shared frame assembles user interface modules to provide user interface for the operators. It functions as menu-structured GUI frame that attaches user interface modules with the currently used user interfaces, by adding the attached module into its sub-menu. Shared frame also has function of temporal data storage and data access control to assist information sharing for the operators' cooperative work.

5 Framework for the inter-work of service- and network management

In this section, we apply the proposed module-based architecture to the TINA service management model. We propose an implementation framework using the concept

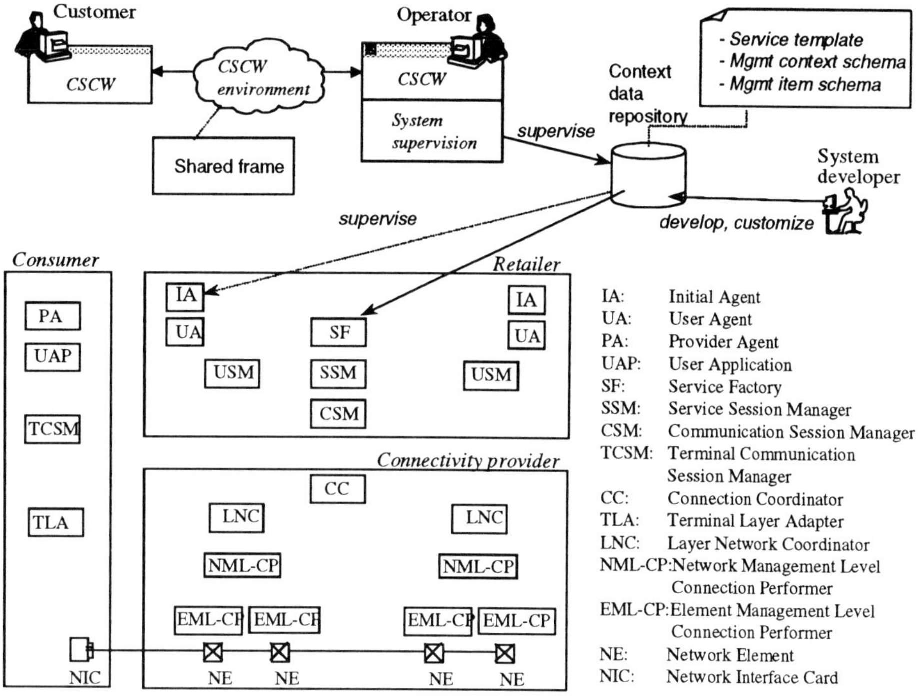

Figure 2 Mapping of the proposed approach to the TINA service component model.

of management context schema in order to support the module inter-working.

Figure 2 shows the mapping of our model to TINA service component model. Our aim is to implement lightweight operator/customer environment that can supervise and customize the service- and management components.

5.1 Context data repository

To facilitate inter-module communication, we need to share the information on the usage of data that are exchanged by the modules.

One approach for this requirement is the usage of type repository function that stores type information inside the management system. Type information includes data type, procedure, and reference for the modules that process the data. This approach however has a drawback since it requires extra cost on the operator/customer equipment to prepare for DPE-based environment.

Our proposal targets operator-side system customization and the implementation of inter-module communication mechanism in a lightweight environment. On this standpoint, we propose a description of management context schema that includes data type mapping, reference for the relevant PIM. Proposed schema does not include information procedure. Schema is described in a customizable format and stored by context data repository.

Our proposal should co-operate with IDL mechanism. IDL definition of each distributed object can be used for the linkage with other objects in the system development phase. Proposed mechanism will be used in the system supervision and customization phase, for the plug-in of developed components to link with the other component that are already in use.

5.2 Example of management context description

We have implemented a prototype system to test the feasibility of our proposal. An example of context schema description based on our approach is shown in Figure 3. Each description in the figure corresponds to service template, service profile, and management context schema. These are defined in a format similar to the C++ type definition, and stored in the system to define necessary information to set up a "Terminal Registration" service.

Service template description contains the definition of data types that are used in the service, to specify how the data should be used in other modules. In this example, "OwnerID" and "terminalType" are defined for the "Terminal Registration" service. Service template also contains reference to the PIMs. Variable "form" and "toolkit" contains reference to the implementation of user interface module "TerminalRegistration_i" and management function module "ChatMediator_i" respectively.

Service profile description lists up the reference to the data that must be filled in for the completion of the service. It also determines the initial value of the data if necessary. In this example, data "owner" and "TType" are mapped with the values that are set by "TerminalRegistration_i" module, while the data "adr" is left blank to be processed in other modules.

```
//Service Template
subscription TerminalRegistration {
    title = "Terminal Registration";
    id = 104;
    response = callback;
    form = TerminalRegistration_i;
    toolkit = ChatMediator_i;

    long ownerID;
    string terminalType;
}
```

```
//Service Profile
serviceprofile TerminalRegistration {
    long owner = ownerID;
    string TType = terminalType;
    string adr;
}
```

```
//Management Item Schema
mgritem AddressAssign {
    title = "Address Assignment";
    toolkit = AddressAssignProcessor_i;

    long ownerID;
    string terminalType;
    string address;
}


//Management Context Schema
mgrcontext TerminalRegistration {
    title = "Terminal Registration";

    mgritem AddressAssign AddressA {
       ownerID = owner;
       terminalType = TType;
       address = adr;
    }
}
```

Figure 3 Example of management context description.

Management context description includes reference to the PIMs as management items, that must be called during the operation. It also maps the data type defined in the service template with those handled in the PIMs. In this example, management item "AddressAssign" will be instantiated as "AddressAssignProcessor_i", and data types of "ownerID", "TerminalType", and "adr" are shared with this module.

This mechanism enables easier reconstruction of the system and module inter-working, by declaring the reference to the PIM with the types of data exchanged by these modules.

Another advantage of the proposal is that it supports the data consistency inside the management system. Management items can be used as reference for accessing PIMs. The modules share information by exchanging this reference. This requires less complex procedure in keeping consistency of data of each management function, compared to the way of directly sending data.

In addition, information used for network management is organized with management context to represent type of services. This allows the operators and the management system to efficiently exchange their tasks with each other.

6 Applying to service management example

Using the module-based system structure introduced in section 5, we propose a usage of the mechanism to realize plug-in of management functions and user interfaces. Figure 4 shows the instantiation of modules according to the information stored in the context data repository and the plug-in of user interface modules into the shared frame GUI.

6.1 Job progress management view

Service management such as operators' job progress management can be done by managing the status data included in the management context description. Management system can interpret management context description and indicate the status of the jobs in the main menu of the shared frame. Reference for the management items are listed in this view as required tasks for the operators.

6.2 System customization using the plug-in mechanism

Description of management context that maps management function modules to relevant user interface modules helps the system developers to design the management Usage of repository enables plug-in on the subsystems. With the concept of management context schema, we can separate the details of management function from the service-related information. Thus, information necessary for the collaboration between management system and operators can be customized for each type of services. Proposed framework enables customization of services component by modifying relevant management items in the management context description. On the contrary, customization of network management function is possible, without giving influence to service specifications.

Management systems can be re-built by replacing the PIMs. Number (1) to (4) in Figure 4 corresponds to the following items which describe the sequence of the plug-in of user interface modules and management function modules.

(1). When replacement of a service or network management function is necessary, System developers add a new PIM, i.e., a pair of management function module and a user interface module. To enable data exchange between the PIM and the currently used components, they modify the relevant part of management context description and update the repository. Management context description is used as information to map management function with user interface, since it includes user interface modules as management items.

(2). A PIM attached to the shared frame parses the modified part of the management context schema, and lists the reference for the other PIMs in its menu. As operators proceed their jobs using the shared frame main menu, they request the plug-in of the added user interface module via this reference.

(3). Triggered by the operators' request, service factory, together with context manager, searches management context description to extract information necessary for the instantiation and initial configuration of the PIM. Initial configuration includes the downloading of the user interface modules to the operators' environment, setup of management function module to interact with the other modules, preparation of information exchange between the user interface module and the management function module.

(4). Shared frame receives the reference for the user interface module and re-build its sub-menu. Thus, operators can manipulate the managed resources via the user interface module which is added to this menu.

6.3 Module inter-working

Inter-work between the service management components and those for network management will be supported by the exchange of the stored management context information. In the example shown in Figure 4, interaction between management function module A and B is realized by referring to the data type store in the context data repository.

6.4 Example of operation scenario

We illustrate an example of service- and network management according to the proposed concept. Figure 4 illustrates an example of a flow-through operation for an order in which a customer requests operators to register the user terminal. Necessary information for this operation is defined in the context data presented in figure 3.

Order form is submitted to the operator as subscription data. This information is shared among operators as service management data using the shared frame function. Operator's jobs include checking of subscription data, job progress management, and configuration of managed resources. In this example, operators tasks are "Terminal Registration" and "Address Assignment".

Operation schema that depends on a specific technology can be developed as management items. In this example, "Address Assignment" task is separated from

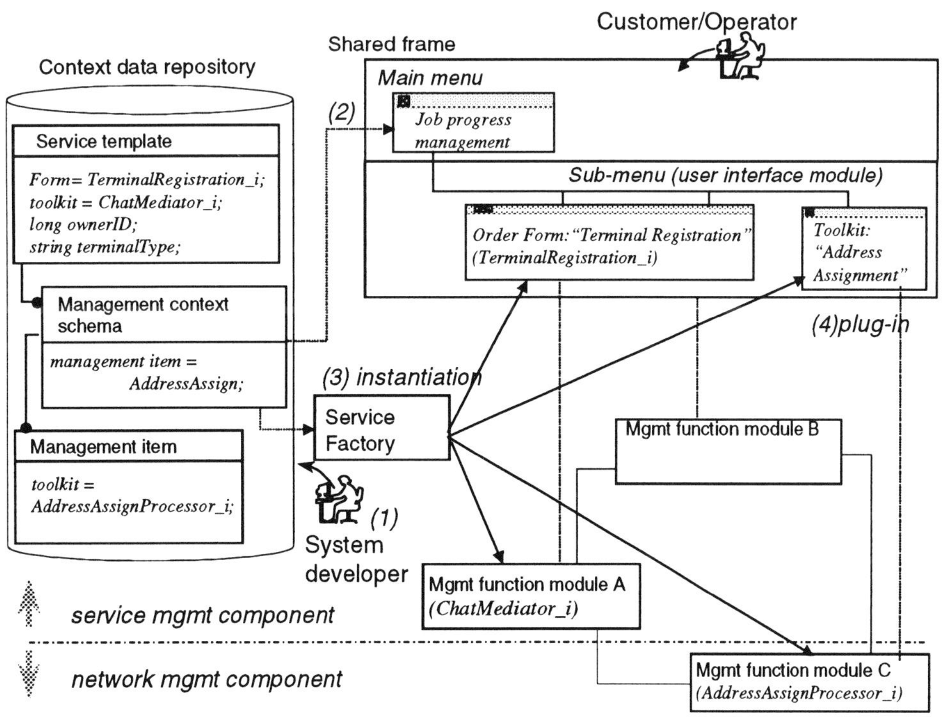

Figure 4 Mechanism for the module instantiation and plug-in.

the "Terminal Registration" task, for the reason that it may depend on a specific transport technology.

7 Implementing lightweight access environment

We present an example to apply current implementation technology to the proposed module-based management structure. We suppose that operators/customers can access management system using WWW browser.

7.1 Downloading user interface modules

We utilize download/upload mechanism to support information sharing. Customers download order forms that are created according to the service template, in which the customers put necessary information and upload toward the provider domain as subscription data.

Plug-in of user interface modules can be supported by network-downloadable applets. User interface modules will be downloaded and executed in the operator's environment. Information such as subscription data, service profiles, management status, etc. can be conveyed through the network as HTML document with Java Applets.

Shared frame function can be implemented as a part of user application component. Function of the shared frame to assemble user interface modules and to provide GUI for the operators can be realized in operator's environment using HTML hyper-link mechanism with attachment of those applets.

7.2 Communication with TINA services

To facilitate communications between management function modules, user interface modules, and shared frame function that are distributed through the network, these functions are implemented as CORBA objects. Interaction between shared frame and user interface module is realized by communication protocols such as IIOP, which facilitates interoperability between Java- and CORBA objects.

Management function module will be used to link the operator's user interface with backend TINA services. It can be executed either in the operator's environment or as a part of TINA service component. An example of the usage of management function module in the operators' environment is to contain ORB stub for the applet-based user interface modules to communicate with CORBA servers.

7.3 Sharing mangement context schema

When the PIMs are downloaded to the operators' environment, service factory gives the initial value according to the descriptions stored in the context data repository. Necessary data for the initial configuration can be passed through as HTML or as input parameters for Java applets.

For sharing the data type information among modules, choice of using DPE-oriented technology such as type repository will be feasible, if standard operator's terminals equipment support full DPE-based environment in the near future.

8 Conclusion

In this paper, we proposed an implementation framework using the mapping relation of management function module and user interface module. We applied this concept to the service management model discussed in TINA. The proposed framework helps us to relate operators' manipulation schema with management information structure inside the system. This realizes interwork between operators, service management functions, and network management functions.

To realize lightweight access environment, we proposed the concept of context data repository that stores information necessary for the inter-module communication. This facilitates efficient supervision of the management system by the system developers as well as by the operators. This is necessary for the quick response to the change of operation scenarios.

To evaluate the feasibility of our approach, we have developed a prototype system based on the architecture. Detailed tuning of the system considering the performance issue is left for further study. Moreover, we regard that further analysis on the actual business operation in practice should be done, for the effective application of our proposal to the systems in use.

Acknowledgement

The authors would like to thank Mr. Gen Kagawa of Toyo Information Systems Co., Ltd., for his fruitful discussions on the implementation of our prototype system.

References

[1] ATM Forum, M4 Interface Requirements and Logical MIB: ATM Network View, 1996.

[2] ITU-T Recommendation G.853-01, Common Elements of the Information Viewpoint for the Management of the Transport Network, 1996.

[3] Y. Tsuchiya and S. Chikara, "A Study on User Interface for Distributed Cooperative Network Management", Proceedings of the 1997 IEICE General Conference B-7-45 (written in Japanese)

[4] TINA-C, TINA Service Architecture Version 5.0, 1997.

[5] Object Management Group, Common Object Request Broker Architecture and specification Revision 2.0, 1996.

Session 6

Security

Alvaro Campos, PUC-Chile

Dynamic Role Creation from Role Class Hierarchy
- Security Management of Service Session in Dynamic Service Environment -

Takeo Hamada

Fujitsu Laboratories Ltd.

TINA-C c/o Bellcore, NVC-1C234
331 Newman Springs Rd., Red Bank, NJ 07701, U.S.A.
TEL: +1 732 758 2715, FAX: +1 732 758 2865
e-mail: hamada@tinac.com

Abstract: Dynamic role creation scheme from role class hierarchy is proposed. In combination with security space representation, algebraic operation on roles creates a new, composite role. Strongly-roled system and role mapping between federated domains are presented. By using dynamic role creation and role mapping, complex security management in dynamic service environment with federation and composition becomes possible.

Keywords: Security Management, Role-based Control, Role Class Hierarchy, Dynamic Role Creation, Security Space, Strongly-roled System, Role Mapping, Service Management.

1. Introduction

It would be fair to say that most urgent issues of TINA security lie at the service layer, where much of the flexibility, dynamism as well as complexity of TINA come from. In particular, recent development of the federation and composition concept in the service architecture opened a huge potential for flexible development of composite services in TINA, but at the same time it introduced new roles, and new security issues along with them. On the other hand, the service session graph (SSG) and the service transaction have been available as generic session/service management tool [1], but they were not designed with the role-based security framework in their requisition. Although service session graphs have been successfully applied to multimedia broadband services ([2][3]), roles are fixed mostly for stream binding, and security management has not been incorporated as an integral part of it.

In this paper, we propose a role-based security management framework for service session based on dynamic role creation and security space representation. For a composed or federated service, a new role may need to be created or added to the existing roles, as roles such as VOD-user are service specific and dependent on the services being composed or federated. Likewise a role of a federated service may have to be restricted to a reduced set of access privileges, due to the degree of trust between federated domains. In either case, a new role has to be created or restricted on demand as a result of cross-domain service configuration.

This paper presents a systematic approach toward this goal, using a role-hierarchy and algebraic operations on roles. The paper also proposes the concepts of strongly-roled systems and role-mapping between federated domains, with which complexity of security management in federation and composition can be made manageable.

2. Issues of TINA Security Framework

Although it has been widely recognized that security is one of the most critical issues in TINA and in distributed computing in general, the issue is far from being resolved. In a sense, security involves every aspect of TINA. The followings are brief summary of our understanding of the current security issues in TINA.

2.1 Authentication Framework

Security starts from authentication, and TINA needs a public authentication framework. Standardization activities (X.509, PKIX) are under way, and progress of smart card technology will enable reliable authentication. Authentication framework is utilized in the access part of the service architecture.

2.2 Complex Interactions between Objects

Reference point specifications (Ret: retailer reference point [18], Cons: connectivity service reference point [19]) are illustrated using event sequence diagrams, which practically serve as a behavior model of the reference points. Although the sequence itself is only given as an example, and thus does not cover the entire behavior of the reference point, which involves complex interactions between multiple interface objects. It offers an unique and new challenge to the telecom software development from security point of view, how reference point interactions can be assured to be security-safe. Work is currently in progress as a part of TINA security framework [4].

0-8186-8335-X/98 $10.00 © 1998 IEEE

2.3 QoP and Fairness

There are various levels of quality of protection (QoP), from basic ones (data integrity, data confidentiality), to more advanced ones (non-repudiation, protection against denial of service), and it is still a matter of question, to which degree TINA should be responsible for the different levels of QoPs. For example, the denial of service relates to the fairness of resource allocation, which can be considered as a part of general QoS problem. In principle, we believe that the basic QoPs should be covered, and in fact they are already part of DPE security. For example, security issues of communication service (TN: Transport Network), signalling network (kTN: kernel Transport Network) security belong to this group. The more advanced QoPs should be covered by negotiable/dynamic part of service session.

2.4 DPE Security

Much work has been done at OMG [5], X-Open [8], and supplementary frameworks are available using ECMA [6], SESAME [9] standards. Through analyses of reference point specifications (Ret [18], Cons [19]), it was found that CSI level 2 interoperability (CSI [7]) is necessary in TINA [4].

2.5 Resource Architecture Security

Network resource objects are owned by the communication service provider, and they have been seen almost exclusively from the management/control point of view. The development of service architecture, in particular federation/composition combined with 3rd party call control, resource reservation, intelligent agent-based call control etc. would require a modified view. Separation of security interests of concerned parties combined with the role-based access control [10][12] will be necessary. Basic computational tools such as layering (a la TMN), management domain ([16]), object grouping ([15]) will be used in combination with the role-based access control.

2.6 Privacy and Anonymity

Security interests of an individual user and those of a provider are not the same, and they do conflict typically at privacy and anonymity. For example, it would be desirable from the security point of view if all the user activities can be logged, but that is likely to violate the privacy of the user. Anonymity may appear in obviously two different reasons; the user does not want to disclose its identity, or the user is not willing to pay. In the current TINA service architecture, only the latter is considered [1].

2.7 Self-Testable Component

TINA components such as Provider Agent (PA) and Terminal Communication Session Manager (TCSM) reside in User domain, still they are expected to perform management or monitoring functions, complementary to those belonging to Retailer domain's, such as Communication Session Manager (CSM) [21]. For example, TCSMs in User domain coordinates with CSM in Retailer domain to set-up a end-to-end stream binding. At the same time, TCSMs are capable of performance monitoring, such that they can report QoS problems which may be observed in the User domains. For this to be possible, all the components must be self-testable, such that the components themselves and the events generated by the components are authentic, and cannot be forged by the user.

2.8 Enterprise Level Issues

There are a diverse set of issues from enterprise security management policy (and its management) to firewall management, from QoP guarantee to billing adjustment policy etc., which may be collectively called enterprise level security issues. TINA addresses, however, only those relevant to the service architecture. As Security of FCAPS,[1] the service transaction and the security management context [1] covers user-provider aspects of the problem.

Not to mention, the above list of issues are by no means separate nor independent. For example, object grouping is a DPE concept but can be most effectively used in the resource architecture, authentication and fairness and privacy and bill-ability of the service are intricately related to each other.

3. Security Management in Service Session

Service session is performed by a set of service components such as SSM and USM, which are separate from those service components in other service sessions. Clearly, a service session itself represents a separate security interests from those of other service sessions, even though they may be used by the same session members.

In TINA service architecture, authentication occurs in access session, and the rest of the security management occur mostly within service session. Each service session is characterized by a session key in symmetric systems, or a pair of keys in asymmetric systems. In either case, session keys are used to separate security interests of one service session (and communication sessions spawned from the service session) from those of other service sessions.

For example, an SSM of a service session can be protected from operation invocations from objects in another service session by using a certificate generated from a characteristic session key, even though interface references of the SSM may be accidentally revealed to the other service session. Other

1. Fault, Configuration, Accounting, Performance, and Security.

protection methods such as message encryption and integrity check naturally follows and derived from the characteristic session key, by using standard encryption techniques.

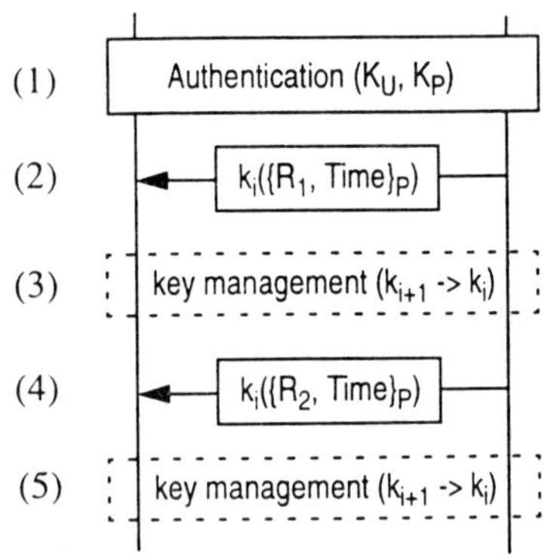

Figure 3-1. An Example of Secure Service Session

Figure 3-1 illustrates an example of secure service session.

1. The session starts with authentication. The authentication should be mutual, as the user may connect to different, more than one providers. The details of the authentication are subject to the protocol being used. It is assumed in the figure that two asymmetric keys (K_U, K_P) are used for authentication. Authentication keys need to be cryptographically strong, such that it is eligible for long-term use.
2. Role certificate[2] for Role R_1 of the user is given by the provider. The role is timestamped (*Time*) and digitally signed by the provider ($\{\}_P$), which is then encrypted by the current session key (k_i).
3. Session keys often need to be replaced by new keys, as they do not always fit for continuous, long-term use. The service session, as a result, is periodically interspersed by key management sequences, which replace session keys.
4. A new role certificate (R_2) is given by the provider.
5. The current session key is replaced by a new one.

Although roles can change and session keys can be replaced, and even authentication keys may be updated, there is a continuation throughout the session, which we would call a common security interest between the two party. When there is a common security interest shared by the two parties, we call it a *security space*. Roles can be mutually and correctly understood, only when the two parties (and the roles) are in the same security space.

2. Role certificate proves the role of the certificate holder, which is digitally signed by the provider. Unlike capability, however, access privileges of the role may not be contained in the certificate itself. More details are explained in Section 4.5.

Multi-party service session in general, however, requires finer separation of security interests, simply because each party's security interests are not the same.

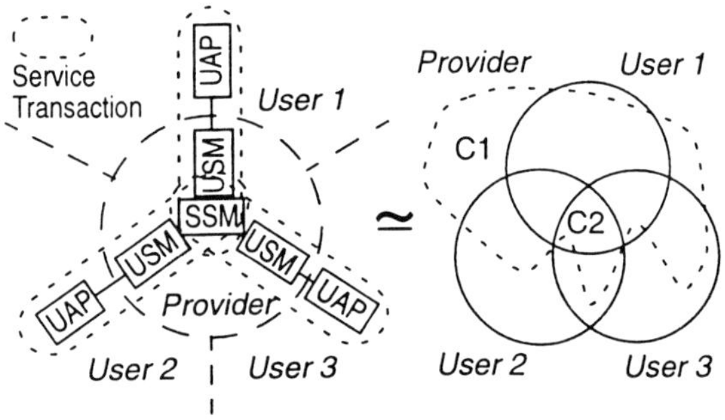

Figure 3-2. Service Session, Service Transaction and Security Space Representation

Figure 3-2 illustrates relationship between service session and service transaction, and their corresponding representations in security space using Venn's diagram. In security space, interests of a stakeholder is represented by a closed area. For example, security interests of User 1 is represented by a real circle, and that of Provider is represented by a dotted closed curve in Figure 3-2. C1 is a characteristic of the security space exclusive to the provider, and C2 is a characteristic of a security space common to all the stakeholders.

$$C1 \in Char(Provider \cap \overline{User1} \cap \overline{User2} \cap \overline{User3})$$

$$C2 \in Char(Provider \cap User1 \cap User2 \cap User3)$$

Apparently, two security spaces can be distinguished only by choosing different characteristics. In this example, the space C1 may correspond to resource objects which are only accessible by the provider, and the space C2 may correspond to a communication service session using a common stream encryption key given by the service session. By characteristics, we mean either session keys or certificates, which are cryptographically strong, practically unforgeable, and are used to protect and separate security interests. Security space can be further divided into finer granularity by applying well-known partitioning principles to the security space. For example, a hierarchical construction of object groups naturally lead to a hierarchical construction of security spaces, by assigning a different characteristic to each level of grouping hierarchy (for more details, see Section 5.3).

4. Role-Based Control in Service Session

4.1 Role-Based Access Control

Role-based access control has been proposed [10][12][13] to help reduce the complexity of system administration of large networked systems. The permissions are associated with a role, rather than an individual or a group of individuals; a role is believed to be more stable over time than individual entities. Our motivation for introducing role-based access control into service session management is essentially the same; we'd like to reduce the complexity of security management of service session. At the same time, it is desirable that the impact of introducing a new role can be analyzed in the planning stage of service lifecycle.

The latter feature is particularly useful in TINA service management framework, where new services are introduced and old services are withdrawn under a control of service life cycle management. The feature is also a natural fit for the user subscription management in TINA. When a user subscribes a service, the user becomes eligible for a role in a concerned service session, i.e. a role certificate can be issued to the user, such that interfaces associated with the service type will be accessible to the user.

4.2 Service Session Management in TINA

Two kinds of roles can be recognized in TINA service architecture. The first kind is generic and statically assigned within session architecture, and roles such as User, Provider, Peer, Party, etc. belong to this kind. The second kind of roles can be more service specific, and are dynamically added-on to a service session and to its participants by using session relations (SRs), which are represented internally in the service session graph (SSG) of the service session. We consider that the latter kind of roles are more important in dynamic service environment, as new roles and new security spaces are to be added through federation and composition.

When a role is mapped onto a computational model, the role controls access privileges of the objects and interfaces in the session security space.

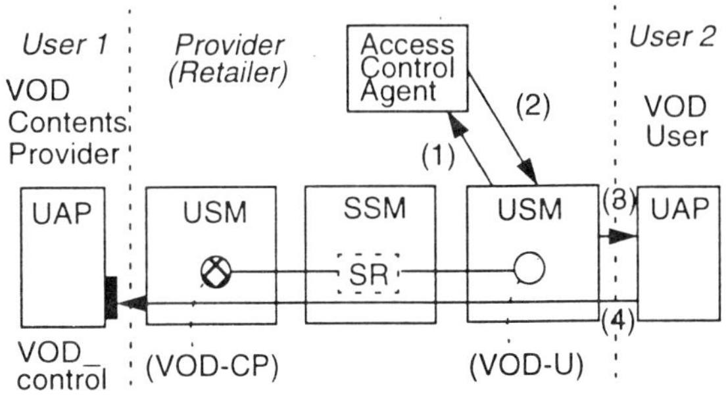

Figure 4-1. Role-based Control of Composed VOD Service Session

Figure 4-1 illustrates a mapping example of roles and role-based control of composed VOD service session. In this mapping, the SSM retains the session relation, which is represented by a set of tuples of roles, interfaces and associated access privileges. When the VOD service is composed, an *owner* of the service session creates the SR and the corresponding roles in the SSM and the USMs, respectively. In the figure, User 1 plays the role of VOD Contents Provider (VOD-CP) and User 2 plays the role of VOD User (VOD-U).

1. VOD-U is able to obtain its role certificate (a characteristic of the added VOD service security space) with an assistance from Access Control Agent.
2. The Access Control Agent returns a role certificate to the requesting USM
3. The USM passes the certificate to the end-user UAP (VOD User).
4. VOD User is able to activate VOD_control interface of the VOD-CP.

It should be noted that the VOD service is virtually added to the service session security space, when its corresponding USM assumed VOD-CP role.

4.3 Access Control Agent and Owner Role

In Section 4.2, we illustrated a centralized view of role management in TINA service session, in which role creation and issue of certificates at access control agent are exclusively controlled by the retailer. Other scenarios, however, are possible within generic TINA service architecture, as owner role can be taken by a party member (e. g. VOD contents provider), which is entitled to create or give a new role to a session participant.

Owner of a service session is entitled to create or modify session relations and their associated roles [1]. This assumption is fairly strong, as it allows the owner to arbitrarily assign a new role to a user, or to deprive an existing role from the user. Practically, the assumption can make sense only when session participants can retain the right whether to accept or refuse the role operation issued by the owner,[3] as it happens in invitation sequences. The owner concept, however, opens possibility of distributed role management.

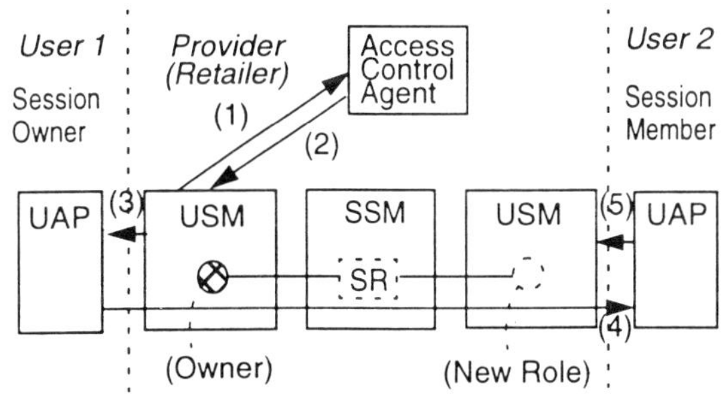

Figure 4-2. Role Assignment Initiated by Owner

Figure 4-2 illustrates role assignment initiated by an owner of the service session. We assume User 1 is the owner, whereas User 2 is a session member, which User 1 is assigning a new role.

3. Therefore, the meaning of 'owner' is quite different from the one used in domain-based theory [14].

1. The owner's USM initially holds the owner role. The USM requests an owner role certificate to the access control agent.
2. The access control agent positively responds with an owner role certificate, which is returned to the USM.
3. The certificate is returned to User 1 UAP.
4. The User 1 UAP signs the combined new role assignment request with the owner certificate, which is sent to User 2 UAP, to which the new role is to be assigned.
5. If the User 2 UAP recognizes that both the owner certificate and the sign by User 1 are valid, and if the User 2 UAP agrees with the new role assignment, it responds positively to its USM, such that a new role can be created within the USM.

In the above sequence, a new role is assigned to the User 2 UAP. When an existing role is to be taken away from the User 2 UAP, the new role assignment issued by the session owner needs only to be replaced by a role revocation assignment[4] by the owner. In contrast to the sequence of Section 4.2, the role management function is controlled by the session owner. Roles stored in USMs are not essential in this case, as the retailer (thus the access control agent) is only aware of the owner role, and other roles are no longer relevant to the retailer.

There are pros and cons for the two models; in the first model (centralized model), roles are managed by the retailer, thus it requires a more involved role management functionality in the retailer domain. As such, roles are mostly *public*, in the sense that roles must be recognized and used by all the users who might interact with the retailer. In the second model (distributed model), roles are mainly managed within user domains, thus roles can be kept *private*, some roles remain valid only within a small group of session participants. It may be said that the first model is less scalable than the second one. In the following part of this paper, however, we assume a centralized model, since it applies better to the generic TINA business model, where services can be federated and composed in arbitrary combination (Section 7.1).

4.4 Roles and Access Privileges

By describing access privileges in terms of roles, not by individual users, role-based access control separates access privilege management from user management, possibly reducing the overall complexity of security management. In the original presentation by Sandhu et al. [10], only the accessing entities have roles, but not the objects being accessed. In multi-party object-oriented communication, it seems more natural to consider that all the involved parties have at least one role, such that an accessed entity exports a set of interfaces to an accessing entity, within a semantically bounded domain given by the service session. The relationship can be represented by the following mapping formula.

$$\text{R\_import} \times \text{R\_export} : (\text{I\_type} \rightarrow \text{Privileges})$$

The formula can be seen from two different viewpoints. In the first viewpoint, the product of two roles, one of the accessing entity (*R_import*) and the other of the accessed entity (*R_export*) defines an operator (a function), which return access privileges to a given interface type in a semantically bounded domain.[5] In the second viewpoint, the product of roles defines a new type, in the sense of type system. In particular, the new type includes $\varnothing$, a null element, when two roles can have no interactions in the designated semantic domain.

The mapping function can be completely specified by the following 5-tuples.

(R_import, R_export, I_type, Privileges, SecSpace)

- *R_import:* the role of the accessing entity (object).
- *R_export:* the role of the accessed entity (object).
- *I_type:* the type of the interface being exported from Role_export to Role_import. It may represent a set of interfaces associated with a service type.
- *Privileges:* access privileges given to the accessing entity of *R_import* with regards to *I_type*. A wild card '*' may mean all operations in the interface are accessible.
- *SecSpace:* security space designator. A security space is a semantically bounded domain of related security interests such as service session. A wild card '*' may mean the mapping function specified by the tuple applies to all the interactions within a given service session.

For example, the VOD service SR of Figure 4-1 can be represented as follows.

*(VOD-U, VOD-CP, VOD_control, *, *)*

4.5 Role Certificate

Role certificate certifies that the holder of the certificate is given a role by the issuer of the certificate. For a structure to be called a role certificate, it has to provide the following five fields.

- *Role ID:* obviously, it has to contain what role is being given to the holder.

4. The role revocation is an operation that a certain role is revoked by the issuer (provider or owner) within a given security space-time. It is similar to CRL in X.509, as such it poses the same technical difficulty.

5. We will later call it a security space.

- *Holder ID:* it must contain the holder ID, to whom the certificate is issued to, to prevent replay of the certificate.
- *Security space ID:* to prevent replay of the certificate by the holder himself, it has to show the security space, where the role is valid. In the previous example of Figure 4-1, session ID would serve the purpose.
- *Time stamp:* for the same reason as above, the validity of a role has to be limited within a certain security space-time. Time stamp combined with time-to-live value would serve this purpose.
- *Digital signature by the issuer:* a certificate is meaningless unless digitally signed by the issuer: anybody, including the holder himself, can forge a certificate otherwise.

Access privileges may or may not be part of the certificate. Since role descriptions (5-tuples of Section 4.4) can be made public, i. e. anybody who may contact the retailer can derive the information, the receiver of the certificate (in Figure 4-1, User 1) can tell the holder's access privileges without asking them explicitly represented in the certificate.

Role ID may be a string, or an alpha numeric code, or an algebraic specification which we shall discuss in Section 6.2. In either case, both the holder and the receiver (and also the issuer) must understand the same role and its semantics, indicated by the role ID.

4.6 Implementation Example

As the role certificate may incur fairly high overhead, it should be used sparingly, only when roles are assigned, updated, or switched. It is expected that role assignment does not occur so frequently, usually only at the beginning of the service session. Once roles are assigned, the rest of the access control can be turned over to CORBA Security compliant DPE.

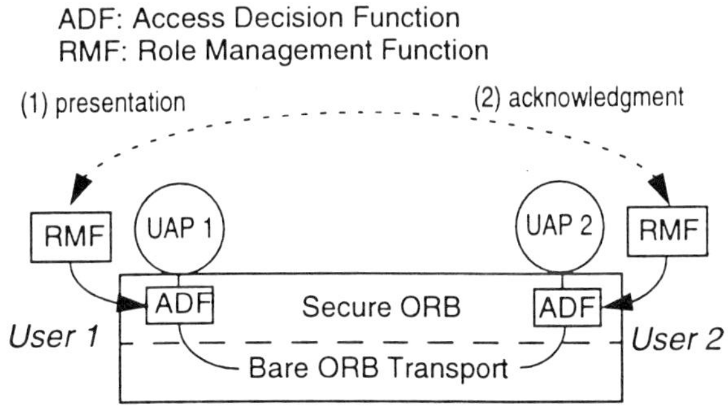

Figure 4-3. An Implementation of Role-based Access Control

Figure 4-3 illustrates an implementation of role-based access control. This is an engineering level representation of the scenario of Figure 4-1, where User 2 is the holder of the role certificate and User 1 is the receiver.

1. *Presentation:* User 2 presents its role certificate to User 1. To be more precise, the role management function (RMF) of User 1 sends its role certificate issued by the retailer to the RMF of User 2, in combination with the object reference of UAP 2. UAP 2 is a CORBA object, which will be acting as VOD user. Formally, User 2 RMF will send the following structure:

   $$\{Role\_Certificate,\ Object\_Reference(UAP\ 2)\}_{User\ 2}$$

   $\{\}_{User\ 2}$ implies that items inside the parentheses are digitally signed by User 2. Whole the structure may be encrypted, and may be sent to User 1 RMF by using ORB itself.
2. *Acknowledgment:* User 1 returns acknowledgment to User 1, in combination with interface reference which UAP 1 exposes to UAP 2. Formally, User 1 RMF returns the following structure:

   $$\{Interface\_Reference(UAP\ 1)\}_{User\ 1}$$

   $\{\}_{User\ 1}$ implies that items inside the parentheses are digitally signed by User 1. It should be noted that the interface of UAP 1 is the interface type requested by the role User 2 is assuming.

Upon completion of above exchange, each party inserts the following entry into respective access decision function (ADF) of secure ORB:

$$(Object\_Reference,\ Interface\_Reference,\ Privileges)$$

Privileges are obtained from the corresponding public role description. In short, ADF acts like a filter, which cuts off all the unauthorized accesses not listed in its entries.

5. Security Space and Related Concepts

In the previous section, a security space is defined as a semantically bounded domain of related security interests. It may contain objects representing service components, their interfaces, a series of actions and responses between components, which implement service logic. It may also include authentication of users, key management protocol exchanges, which would be seen as a part of the security space, as far as they are inseparable and necessary part of establishing and maintaining security interests of the concerned parties. The concept is not entirely new, as related, similar concepts have been accepted and widely used in recent literatures. It would be worthwhile to clarify similarities and differences between these related concepts.

5.1 Domain

Domain [11][14] can be defined as a set of objects under a certain, common rule. Its most typical usage appear as management domain, where objects follow a set of management rules known as policies. Although domain and security space are structurally similar, can often be identified with, in certain cases they clearly differ. For example, in the service session example as shown in Figure 3-2, the entire service session can be seen as a common security space, but it would be rather unusual that all the involved domains, both user and retailer domains, belong to a single security domain and a set of common policy rules.

5.2 Secure Channel

Secure channel [11] represents a secure communication link between two parties, typically between a subject and a target object. Although secure communication based on cryptographic techniques is an underlying assumption of the security space, the two concepts are not necessarily equal. For example, a security space such as a service session may require multiple secure channels or no secure channels at all, depending on its purpose and usage, as service sessions do not always be made secret.

5.3 Security Space

We start from a typical setting of service session.

1. Users are requested of their authentication upon entering the system.
2. After the completion of authentication, a secure channel is usually set up between the user and the provider.
3. When selecting a service type and entering a service session, the user requests a role, or is invited to take a role, to participate in the session.
4. A session key, or a sequence of session keys may be delivered among the session participants, using an agreed key management protocol. An ID is usually reserved for the entire service session for reference in later time.

A tuple of session ID and a sequence of session keys

$$C \equiv (SessionID, \{SessionKeys\})$$

are sufficient to give characteristics C of a security space. *SessionID* and session keys are similar in their roles in the characteristic; the difference, however, is that SessionID can be made public, but session keys are usually kept secret, and they must be cryptographically strong.

Security space is thus named due to the impression given by the following space-like properties; it is not a mechanism, rather a conceptual framework to describe security phenomena in session management.

- *Container:* it acts like a container of objects, interfaces, actions and responses, with different abstraction levels and opaqueness of contents.
- *Non-interference:* activities in two separate security spaces would not interfere with each other. Even if their interface references are accidentally revealed, actions crossing the space boundaries would be blocked, or simply have no effect. Although it is desirable, communications in a security space may or may not be made secret, depending on the nature of the application.
- *Partitioning:* a space can be partitioned, or partitioned spaces can be joined, increasing its usefulness as container.

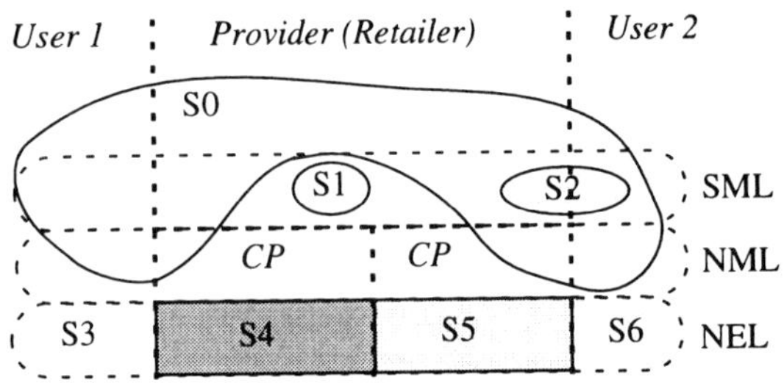

Figure 5-1. Domain, Layering and Security Space

Figure 5-1 illustrates the relationships between domain, layering and security space. In the figure, CP denotes Connectivity Provider.

- *S0:* this space contains a large collection of objects across multiple domains and layers. A service session may look like this when it is represented by a security space. The space can be made transparent, i. e. objects and interactions among objects are made plain public, if there is no visible security threats.
- *S1, S2:* a collection of objects and interfaces can be made an opaque security space, where all interactions are encrypted and made secret to the outside of the space. S2 are nested within S0, as it represents different security interests within a session. For example, S1 represents an object group [15], a set of critical resources in the Retailer domain.
- *S3, S4, S5, S6:* intersections of domains (User 1, CPs, User 2) and a layer (NEL) are made separate security spaces. Their degrees of opaqueness can be made different.

As it is seen in *S0*, security space is a loose container compared to domains and object groups, as it does not carry administrative responsibility nor life cycle/configuration relationships.

6. Role Creation from Role Class Hierarchy

6.1 Role Class Hierarchy

One of the clear advantages of the role-based control is that the functionality of SSM can be made minimal, i. e. the management of SRs and roles, and service dependent interactions are left to UAPs with prescribed roles. In the previous example of Figure 4-1, the SSM does not interfere with

User 1 using VOD_control interface of User 2 (VOD Contents Provider), though the SSM may help setting up kTN and TN connections between the two parties.

What remains is the management of roles, in particular dynamic creation of roles from federation/composition requirements. For example, User 1 in Figure 4-1 itself is a user from a federated service, whose control operation is restricted to viewing forward only, no rewinding allowed. Roles can be added together to create a new composite role. A role can be divided into sub roles, which may correspond to different versions of VOD service specification. Our proposal is to use role class hierarchy for the role management, such that the role management itself can benefit from O-O paradigm. A new role can be created by algebraic specifications on roles, which are equivalent set-theoretic operations on 5-tuples corresponding to the role.

It should be noted this role class hierarchy is different from the information model inheritance hierarchy of SSG as it was originally proposed [1]. Although the original information model can be mapped onto the proposed role class hierarchy, the fixed schema of information model hierarchy does not allow dynamic role creation.

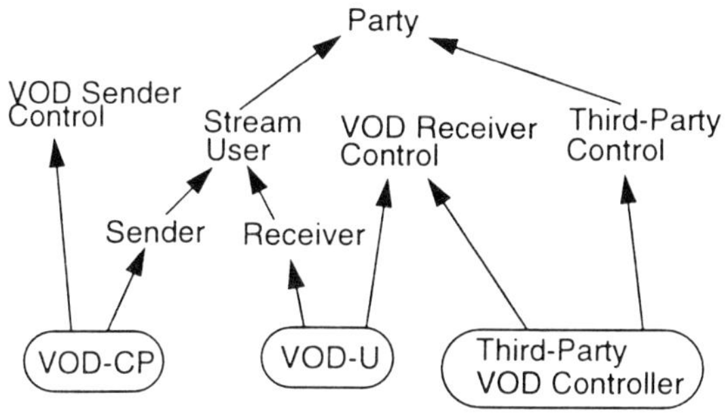

Figure 6-1. Role Creation from Role Class Hierarchy

Figure 5-1 illustrates role class hierarchy and dynamic role creation. New roles can be created with a simple algebraic equation as shown below.

$$VodCP = VodSenderControl + Party::StreamUser::Sender$$

The equation specifies that the new role (*VodCP*) will inherit access privileges from the two existing roles, which are given by role-access privilege mapping formula (Section 4.4). In other words, the algebraic specification of the new role is interpreted by the 5-tuple form, and the 5-tuple form of the new role is given by simply summing the 5-tuple forms of the two existing roles.

6.2 Algebraic Role Specification

Although the above example is a little simplistic, as privileges of two roles may differ, or two roles are mutually exclusive (in that case, the operation would result in $\varnothing$, a null role). It is possible, however, to define precise semantics of algebraic role specification from operations on the corresponding 5-tuple representation based on sensible security guidelines, e. g. least privilege principle. Following operators are currently considered useful. The formula would look like $NewRole = Role_1 op Role_2$.

- '+': set sum operator creates a role with all the access privileges of two roles being added.
- '-': set difference operator creates a role with privileges of $Role_1$, but not in $Role_2$. The situation may arise, when a tele-conference session is split into two separate sessions, and making them two separate security spaces by disallowing the participants in the second session to access the objects in the first session.
- '*': this operator creates a role, whose privileges are logical product of two roles. A privilege belong to the new role, only when it is implied by both of the two original roles.
- '|': this 'pipe' operator creates a new role, as if the two original roles are cascaded in sequence. When there is a common set between interface types exported by $Role_1$ and those imported by $Role_2$, the cascaded role would result in a non-null role. The imported interface types of the new role are a subset of those of $Role_1$, and its exported interface types are subset of those of $Role_2$. Precise use of this operator is yet to be found, but it seems to be useful when roles are to be cascaded, as they may occur in the context of delegation, service federation, and composition.
- '@': this operator creates a new role, where the new role is conditional upon the existence of $Role_1$. In other words, an entity can take up the new role only when it holds $Role_1$ first. This operator is expected to be useful, when sequential requirements are to be imposed upon roles.

A main purpose of this algebraic approach is to create a variety of roles from a fewer number of basic roles and well-defined operations on tuples. By doing so, it is hoped that interactions between roles can be statically analyzed, and possible side effects of introducing new services (and therefore roles) into the system can be made predictable.

6.3 Roles and Service Creation

The role management presented in this section is based on role class hierarchy, which is a generic template independent of individual service, session, and user instances. As it can be observed from Figure 6-1, introduction of a service (VOD) necessitates creation of roles (VOD-CP and VOD-U), which in coordination implement intended service logic within given service sessions. Therefore, it is natural to correlate role management with service creation, in such a way that their respective lifecycles (service and role) are synchronized.

On the other hand, when a user subscribes to a service, he should be entitled to the roles associated with the service. Subscription management must interact with role management, in such a way that roles are granted from ACA, when subscribed users request them.

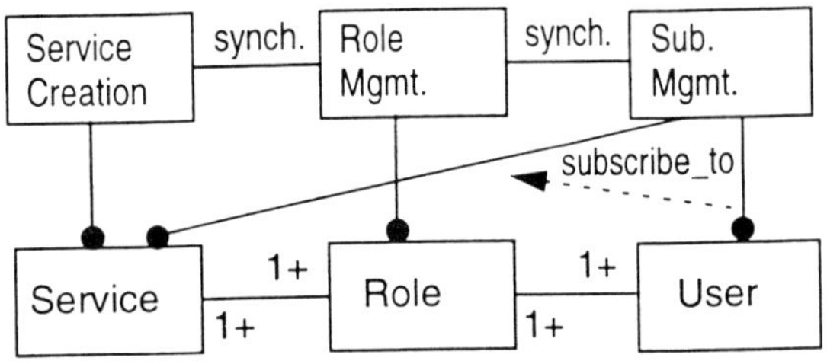

Figure 6-2. Service, Role, and User

Figure 6-2 represents the above relationships between service, role, and user. In the figure, they are represented as information objects.[6] Mappings between service, role, and user are all many-to-many, as the same role may be reused by different services, and a user may subscribe to many different services, and so on. The figure also intends to show that respective management functions, i. e. service creation, role management, and subscription management need to be synchronized with each other.

It also means to show that service and user relate to each other via role, i. e. service is not accessible by user without obtaining a role first. This may sound like a constraint, but it effectively enforces a tight security control by using role-based access control in combination.

7. Strongly-Roled Systems

Not only role-based control is useful, we can go a step further by requiring that all the entities (thus the users) must have at least one role at a time. The role must be one of the basic roles, or a composite role acceptable by the algebraic specification. Therefore, there will be no 'illegal access' to an interface, as far as roles and role-based access are concerned. Although it may sound a bit strong requirement, the conditions of network services based on TINA service session mandate exact knowledge of interface types, available operations, and sequence of interactions.

7.1 Role Mapping in Federation

Besides static security analysis of interactions among objects, advantage of strongly roled system can be made explicit when used in service federation and composition.

Figure 7-1 illustrates a possible federation scenario using role mapping. As a note, this scenario is highly simplified version of the original federation scenario,[7] to elucidate the role mapping in federation.

1. User Application (UAP) in the primary domain requests a service federation to the secondary domain. The UAP is aware of the service type provided by a federated service component (FSC) in the secondary domain and possible interface types, but the UAP can not obtain a valid role in the secondary domain on its own. We assume the UAP is already given a role, which qualifies the UAP to access a federated service type within the primary domain.
2. Federated session manager (FSM) in the primary domain forwards its request to a counterpart FSM in the secondary domain. FSM checks the UAP's role in the primary domain, and if it is valid, it maps the role to an equivalent role in the secondary domain, which qualifies for the service type offered by FSC.
3. The secondary FSM requests a certificate of the mapped role to access control agent (ACA).
4. The ACA issues a role certificate.
5. The secondary FSM returns the certificate to the primary FSM.
6. The primary FSM returns the role certificate to the requesting UAP.
7. and 8. The UAP accesses the FSC in the secondary domain, using the certificate of the mapped role in the secondary domain.

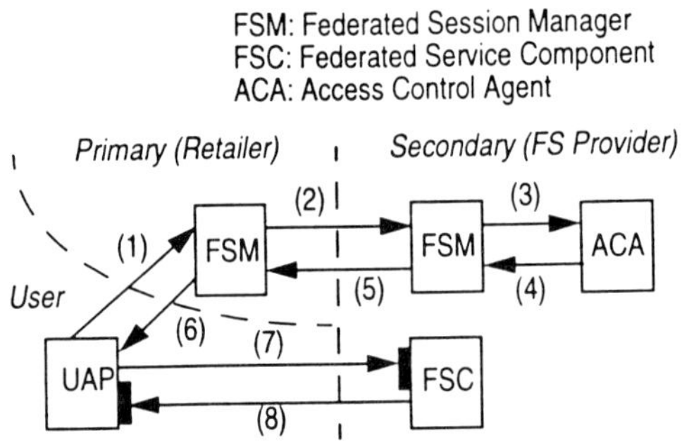

Figure 7-1. Role Mapping in Federation

7.2 Role Mapping vs. Identity Mapping

One of the reasons we think strongly roled systems are more suitable for service federation and composition is that mapping of roles is probably more favorable than mapping of user identities. When the user is a subscribed user in the primary domain, the primary provider has no reason to doubt the user's identity, when authentication of the user is successfully done.

For the secondary provider, however, the situation can be

6. This figure only intends to show relationships between concepts and entities; it is not accurate in details.

7. The original version resorts to heavy use of service session graph (SSG). As we mentioned in Section 4.2, one of the purposes of this paper is reduce the complexity of current TINA session management through a better use of security framework.

quite different. Even if authentication of the user can be done independently of the primary provider, the secondary provider may or may not have subscription information (address, credit history, social security number, company name etc.) of the user,[8] which would provide supportive facts about the identity of the user. Therefore, even if authentication of the user is successfully done, the secondary provider does not have strong reasons to trust the user's identity.

It is most likely that the secondary provider accepts service federation requests from the primary provider, because; (1) the secondary provider trusts the identity of the primary provider, with sufficient proof of its business stability backed up by some financial guarantee given by the primary provider, (2) the subscriber status of the user, in particular when a user is given a role for federated service, he/she must have subscribed to it and have been qualified by the primary provider.

Besides delegation can become potential performance bottlenecks in the system, because of the above reasons, delegation does not appear in the presented federation scenario; delegation would require mappings of identities at intermediaries between the UAP and the FSC. The secondary provider is still capable of controlling access privileges of the user from the primary domain, by controlling the role mapping.

7.3 Role Mapping Function

Suppose a role mapping function between two domains is defined as

$$D_1 \times D_2 : (Role_1 \rightarrow Role_2)$$

where D_1 and D_2 denotes primary and secondary domains, respectively. Role R_1 in D_1 is mapped into role R_2 in D_2 by a mapping function,

$$\pi_{D_1 D_2}(R_1) = R_2$$

For a composite role, role mapping will be applied recursively to the elements of the role composition, till it reaches the mapping of base types. For example,

$$\pi((A + B) \mid C) = \left(\pi(A) + \pi(B)\right) \mid \pi(C)$$

where A, B, and C represent roles in the primary domain. If the mapping of a composite role results in a null role, it implies that the service federation is not possible in the secondary domain.

Note also that the strongly roled property of the secondary domain is preserved under role mapping. Though we did not assume the primary domain to be strongly roled, but when it is, the role of FSC must be reversely mapped into the primary domain in the sequence of Figure 7-1, for the service federation to be possible.

8. Passing these information to the secondary domain without the user's consent is likely to violate privacy.

8. Service Session Management Examples

TINA service management is burdened to achieve two seeming contradictory objectives; consistent multi-domain QoS guarantee throughout a service session and a support for dynamic, flexible management services. One of its consequences are two complementary service management mechanisms built-in TINA service session architecture. Service transaction and management context are conceived [20][1] to achieve the first objective. Service session graph [1] and dynamic role creation proposed in this paper are conceived to achieve the second objective.

Although subscription contract and management context, which is an embodiment of service management agreement, are made solely between the user and the provider, certain management services for multi-party session can be practically done only through a distributed representation of management responsibility by using a service session graph and associated SRs. For example, when shared-billing is combined with online billing, effective charging rate to each session member changes when one of session member leaves the session, even when the charging rate to the entire session does not change throughout the lifetime of the session.

There are more complicated and intricate management service examples when optional services are combined with federation and composition principles. For example, a third-party may take call-control of a call set-up, but the retailer may have to provide other FCAPS management.

8.1 Third-party Call Control

Public broadband network can be more effectively utilized when it is combined with broadband intranet. Integration of internet and intranet management is strongly called for, to accommodate more flexible call management needs arising from business applications. CSM is designed to handle multi-party connections across multiple domains, and is most amiable for technology independent connection management. Although interfaces of CSM are currently not accessible from user applications, it has a potential to be an ideal tool for third-party call control. In fact, CSM belongs to Retailer domain, which is already a third-party to the connectivity providers.

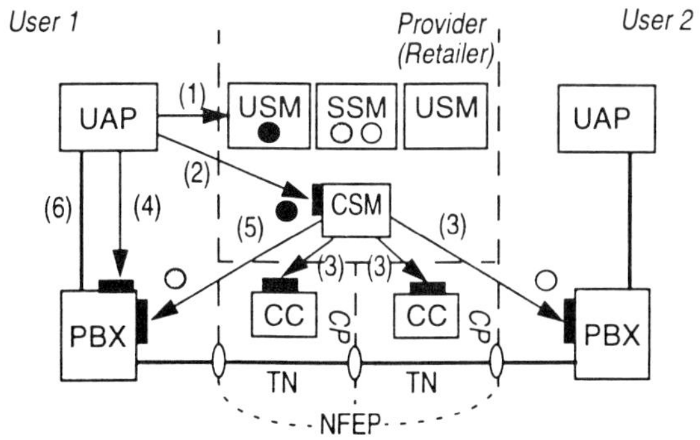

Figure 8-1. Third-party Call Control with Accounting

Figure 8-1[9] illustrates a third-party call control combined with accounting management.[10] User 1 and User 2 are two branches of a company, who wants to set up a broadband transport network between two PBXs. The company wants to manage the entire connection as if it is part of their private network. PBX in User 1 domain provides TINA compliant network management interface, such that it can receive accounting events from CSM, which summarizes accounting reports from connection coordinators (CCs) in respective CP domains. PBX in User 2 domain provides a TINA compliant call control interface, such that CSM is capable of making connection within the user domain, as if the PBX is another CC. Under all these circumstances, the user wants to set up a virtual private network, of which accounting reports are to be collected to the PBX in User 1 domain.

The roles played by the involved parties are fairly complex.

- *Role 1 (black):* User 1 controls connection management interface of CSM.
- *Role 2 (grey):* CSM (Retailer) reports accounting summaries to User 1 PBX notification interface.
- *Role 3 (white):* CSM controls connection control interface of User 2 PBX.

Suppose role-based control is available. The major task is to assign correct roles to the involved parties in the service session using USMs and SSM. The rest of the third-party control can proceed almost independently from the service session itself. When the need for VPN is ended, CSM is teared down and the service session ends. In the figure User 1, Retailer, and User 2 would belong to the same security space, as they recognize each other's roles in the sequence. Two CPs, on the other hand, do not need to be in the same security space. In fact, CCs are controlled using connection management interfaces specified at ConS reference point, where normal authentication/secure channel would be utilized.

1. User 1 requests a third-party control as sketched above. The role (black token[11]) is assigned to User 1 UAP, such that it is allowed to access call management interface offered by CSM.
2. By using the role 1 token (black), the User 1 UAP issues a connection request, such that a connection be set-up from the User 1 PBX to the User 2 UAP, whose local connection is made by TINA-compliant, User 2 domain PBX.
3. CSM issues connection requests to relevant CCs and User 2 PBX. Connection is made as if the trail extends to the end of User 2 domain, the User 2 UAP. To obtain access to TINA compliant call control interface of User 2 PBX, role 3 token (white) is used.
4. User 1 UAP controls User 1 PBX, which only has a non-TINA legacy control interface. A local connection is made, such that User 1 UAP is connected to NFEP of User 1 domain.
5. Upon completion of requested connection, the CSM starts reporting accounting summaries to the User 1 PBXs notification interface. To obtain an access to the interface, the role 2 token (grey) is used.

The entire scenario is quite flexible. As soon as the User 1 PBX provides a TINA-compliant call control interface, its role can be re-assigned, such that the entire connection (from User 1 UAP to User 2 UAP) can be managed by CSM.

8.2 Flexible Billing Management

One of the advantages of the role-based control is that roles can be added or re-negotiated, as necessity arises. This added flexibility is particularly useful for billing, as the service charge may have to be split among the parties, or charged to a third-party.

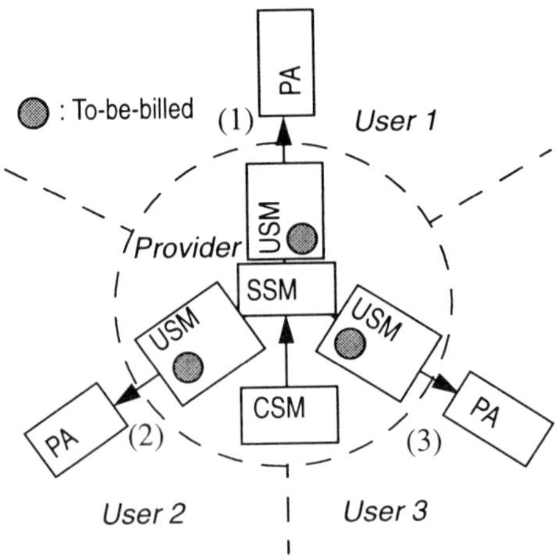

Figure 8-2. Three-Party Session with Shared Billing

Figure 8-2[12] illustrates a three-party session with shared on-line billing. In this session, billing events sent from connectivity providers (not shown in the figure) are first sent to CSM, and are then forwarded to SSM. Billing events from third-party providers such as VoD providers (not shown in the figure) are sent to SSM, to which the billing

9. In Section 4.4 we defined that privileges are given by role pairs of accessing entities and accessed entities. For the sake of the simplicity, roles of the accessed entities are not shown in the figure.

10. CP denotes Connectivity Provider.

11. In the following, we use 'certificate' and 'token' interchangeably.

12. Roles of the accessed entities are omitted.

events from CSM are added, and on-line charges are calculated at SSM. Although on-line billing is not mandatory, TINA accounting management is already designed to support on-line billing [22].

In any service session, there should be at least one party member, whose account the bill of the session should go to. In the figure, the grey token represents 'to-be-billed' role, which represents financial responsibility of the session. The billing events are to be sent to UA, where they are recorded in a billing database. When on-line billing is used, the same billing events are also sent to PA, for immediate availability of billing information in the user domains.

1. User 1 is the sole party member to be billed. The role token allows the provider to access notification interface of PA, to which on-line billing events will be sent.
2. User 2 joins the session, and agrees to share the role. By adding the role to its USM, the SSM realizes that the bill is to be split between the two, and events to be sent to two different destinations (two USMs). The billing events are eventually sent to respective PAs.
3. User 3 joins the session. He/she may or may not share the 'to-be-billed' role initially, but eventually agrees to take the role. Now the billing events are to be split into three destinations.

Roles are always negotiable, and it is observed that the behavior of SSM and USM are made 'role-driven', i. e. their behaviors are dictated by what roles they hold.

9. Conclusions

We proposed in this paper dynamic role creation from role class hierarchy. In combination with security space representation, the proposed scheme offers a dynamic, flexible and integrated security management for TINA service session. Additional benefit of this proposal is that the functionality of SSM is kept minimal, only to role management and associated SRs, thus alleviating much of the expected complexity of SSM implementation and SSG maintenance.

We also proposed strongly-roled concept for security space, where all objects in the space must have at least one role at a time. When combined with role mapping and algebraic specification techniques, the concept seems most applicable to federation and composition.

Acknowledgment

The author wishes to express his thanks to Mr. Martin Yates of BT, for helpful discussions and suggestions on security issues in TINA service architecture. He also thanks Mr. Hiroshi Kamata of OKI Electric, Dr. Tom Rutt of Lucent, Mr. Emil Lupu and Prof. Morris Sloman of Imperial College for comments to early drafts of this paper.

Reference

[1] Service Architecture Ver.5.0, TINA-C, June 1997.

[2] J. Pavon, J. Tomas, "Broadband VPN Service Prototype: Application of TINA to Service and Network Management", proc. of TINA'96, Heidelberg, pp.145-154, Sept. 1996.

[3] P. Hellemans et al., "The ALCIN Project: TINA Validation Through Prototyping", proc. of TINA'96, pp.125-134, Heidelberg, Sept. 1996.

[4] T. Hamada et al., "TINA Security Framework", unpublished manuscript, TINA-C, 1997.

[5] CORBA Security, OMG, OMG Doc. No. 95-12-1, Dec. 1995.

[6] Authentication and Privilege Attribute Security Application with related key distribution functions, ECMA, Standard ECMA-219 2nd. ed., Mar. 1996.

[7] Common Secure Interoperability, OMG, OMG Doc. No. orbos/96-06-20, July 1996.

[8] Generic Security Service, X/Open Publication Set T402, X/Open, 1994.

[9] SESAME V4 Cryptographic Application Developer's Guide, Dec. 1995.

[10] R. Sandhu and E. Coyne, "Role-Based Access Control Models", IEEE Computer, Vol. 29, No. 2, Feb. 1996, pp. 38-47.

[11] N. Yialelis, *Domain-Based Security for Distributed Object Systems*, Ph. D Thesis, Imperial College of Science, University of London, Oct. 1996.

[12] E. Lupu and M. Sloman, "Towards a Role Based Framework for Distributed Systems Management", Journ. of Network and Systems Management, Vol. 5, No. 1, Plenum Press, 1997.

[13] E. Lupu and M. Sloman, "A Policy Based Role Object Model", proc. of EDOC'97, Gold Coast, Oct. 1997.

[14] J. Moffett and M. Sloman, "Delegation of Authority", Integrated Network Management II, I. Krishnan & W. Zimmer eds., Elsevier Science Publishers B. V. (North-Holland), pp.595-606, 1991.

[15] A. Parhar, T. Handegard, "TINA Object Groups: Patterns in Chaos", proc. of TINA'96, pp.13-24, Heidelberg, Sept. 1996.

[16] Network Resource Architecture Ver.3.0, TINA-C, Feb. 1997.

[17] TINA Reference Points Ver.3.1, TINA-C, June 1996.

[18] The Ret Reference Point Ver. 1.1, TINA-C, Dec. 1996.

[19] Connectivity Service Reference Point Specification Ver. 1.0, TINA-C, Aug. 1996.

[20] T. Hamada, "TINA Accounting Management Architecture", proc. of TINA'96, pp.193-202, Heidelberg, Sept. 1996.

[21] T. Hamada, S. Hogg, J. Rajahlme, C. Licciardi, L. Kristiansen, P. F. Hansen, "Service Quality in TINA", to be presented at EDOC'97, Oct. 1997, Gold Coast, Australia.

[22] Service Component Specification: Computational Model and Dynamics, Ver. 1.0b (Draft 0.2), TINA-C, Sept. 30, 1997.

Security in the Telecommunications Information Networking Architecture – the CrySTINA Approach*

Sebastian Staamann
Uwe Wilhelm André Schiper
Swiss Federal Institute of Technology
Operating Systems Laboratory
EPFL-DI-LSE, 1015 Lausanne, Switzerland
{staa, wilhelm, schiper} @lse.epfl.ch

Levente Buttyán
Jean-Pierre Hubaux
Swiss Federal Institute of Technology
Telecommunications Laboratory
EPFL-DE-TCOM, 1015 Lausanne, Switzerland
{buttyan, hubaux} @tcom.epfl.ch

Abstract

The article presents the first results of the CrySTINA project. We analyze and structure the security problem domain in the TINA-C architecture and present our approach to provide the necessary security functionality in the form of self-contained application-independent security services and security mechanisms as part of the DPE functionality. The DPE is assumed to be basically provided by CORBA products. Therefore, we introduce the CORBA security specification and investigate if and how the identified TINA security services can be implemented using the CORBA security functionality.

1. Introduction

An essential requirement for the Telecommunication Information Networking Architecture (TINA) is security. TINA is intended to provide a comprehensive architecture for multi-service networks that shall enable multimedia communications and access to information for business and private users. In traditional communications networks that are solely dedicated to the telephony service, not much attention has been paid to authenticity, integrity, and confidentiality of the voice data carried. Efforts for security have been nearly exclusively focused on the secure and safe operation of the network itself and the protection against toll fraud. In future multi-service networks, this situation has to be changed. If commercially valuable interactions shall take place over these networks, users will require authenticity, integrity, and confidentiality for the information transmitted. The provision of security is becoming an important issue in the competition between TINA technology and other multi-service networks, first of all the Internet. In the Internet world, several approaches on various layers, all based on cryptography, are currently discussed or already in use, e.g., [3] [10] [4] [26]. In order to be competitive, TINA networks must guarantee at least the same degree of security. An important field of application is electronic commerce. The Internet does not yet have a specific architecture for electronic commerce. On the other hand, TINA with its business model and service architecture [23] together with the know-how and the customer base of the established telecommunication network operators as one of the driving forces of the TINA effort inherently possesses the appropriate infrastructure for electronic commerce. To benefit from this advantage, the overall architecture must provide the functionality to protect the transactions and to establish the legal bindings.

In order to avoid redundancy of functionality and for the sake of interoperability, security in TINA networks has to be provided in a consistent way. The security problem domain should be structured and security functionality should be provided as much as possible through general security services. A prerequisite for these services in a multi-party environment, such as a TINA network, is the existence of a security infrastructure that provides long term keys and supports the negotiation of security mechanisms and policies between different administrative domains. The means are mainly provided by cryptography. All application-independent security functionality should be available from the Distributed Processing Environment (DPE), since application and platform independent functionality is in TINA provided at this architectural level.

The provision of the security infrastructure is the subject of the CrySTINA (Cryptographically Secured TINA) project, a joint research effort of the Swiss Federal Insti-

*Research supported by the Swiss National Science Foundation as part of the Swiss Priority Programme Information and Communications Structures (SPP-ICS) under project number 5003-045364.

0-8186-8335-X/98 $10.00 © 1998 IEEE

tute of Technology Lausanne, Siemens Munich, and the Swiss Telecom. As expressed in the name, security is realized mainly by cryptographic means. CrySTINA is part of a broader research effort aiming at a Secure and Reliable Distributed Processing Environment for telecommunication networks. This article presents the results of the analysis phase, i.e. how the security problem domain must be structured in order to provide security functionality as application-independent services and how CORBA security can be used for that purpose. It is an improved and enhanced presentation of the analysis given in [20]. Similar to the approach stated in the technical report on the TINA security architecture [22], we also strive to reuse established concepts from other standardization work, such as OSI, TMN and the OMG specifications.

In the following section, we analyze and structure the security problem domain in TINA. Then we investigate how CORBA security can be used for TINA and we identify additional security services and mechanisms that must be provided by the DPE. Section 3 presents the CORBA security specification. In Section 4 we propose how CORBA security may be used in TINA networks and identify the open issues for the TINA-DPE. Finally, in Section 5 we give an outlook on our ongoing and future work.

2. The TINA security problem domain

Security concerns all parts of a TINA system; it is pervasive and cannot be addressed in isolation. To cope with this complexity, it is necessary to structure the security problem domain in an appropriate way. All services and resources may be the subject to attacks. Attacks may be the illegitimate use of components or the modification of data, state or programs. They may occur through direct access to systems, data, or services from outside or through modification of messages exchanged between interacting components. Potential attackers are outsiders, but also other stakeholders in the TINA network. Motives of attackers may be the illegitimate use of services, fraud (e.g., in online businesses), toll fraud, eavesdropping on and observation of consumers or providers, or the deliberate prevention of service provision (denial of service attack). The ultimate goal of an attack may be achieved directly or indirectly. In the latter case, an attacker may install a backdoor during a first successful attack, which enables him later on (and possibly at multiple times) the actually intended misuse. Examples for backdoors are the modification of programs or access rights.

Each stakeholder in a TINA network has his or her own administrative domain [24]. We make the assumption that the administrative domain is the trust domain of the stakeholder. This assumption is based on the fact that in the regular case the installed hardware is under the physical control of the stakeholder and the software is installed by him or herself. The trust domain may in fact consist of various nodes under the physical control of the stakeholder that are connected by physically unsecure communication links. These links can be turned into secure channels by the use of symmetric cryptography without sophisticated management of keys, so that the connected nodes form a single trust domain. Since TINA supports personal mobility for consumers [23] [1], the current administrative domain is not necessarily administered by the consumer currently using it. However, in this article we do not cover the additional trust relationship between the consumer administrative domain and the consumer. This relationship is subject of our ongoing work. In this article, we assume complete trust between the user and his or her current administrative domain. Security within the administrative domain (intradomain security) is domain specific and is achieved by local means. Within his or her domain, the stakeholder trusts in the correctness of the installed software. Towards the outside, the administrative domain must be protected against illegitimate access. For interactions with other domains (interdomain interactions), limited trust relationships must be established. The communication channels between domains cannot be assumed to be secure. Therefore protection must be achieved by cryptographic means. Security must be provided to all parts of the TINA system that are involved in interdomain interactions. The security of each part is usually dependent on the security of many other parts in the administrative domain, nevertheless, all single parts have to be protected on their own. Figure 1 shows our structuring of the security problem domain.

System Security:
System security shall ensure that systems, mainly the hardware and the operating system, are not subject to intrusions. This concerns networking resources (e.g., network switches) and computing resources. It also includes the Native Computing and Communications Environment (NCCE) (operating system and communication ports), since intrusions may not only occur over communication ports of the NCCE that are used by the DPE, but also over other ports of the NCCE. The latter point concerns mainly the administrative domains of end users (consumers) whose Customer Premises Equipment (CPE), e.g., Personal Computers (PCs) or workstations, cannot be assumed to be exclusively used as the endpoint of the TINA network.

Service Security:
Service security is mainly concerned with the preservation of the integrity of service control. Service control includes among others the verification of whether a user is allowed to use a service (subscription) and the accounting for billing purposes. Both rely on the authenticated identity of the user. This must be supported by a protocol for the authen-

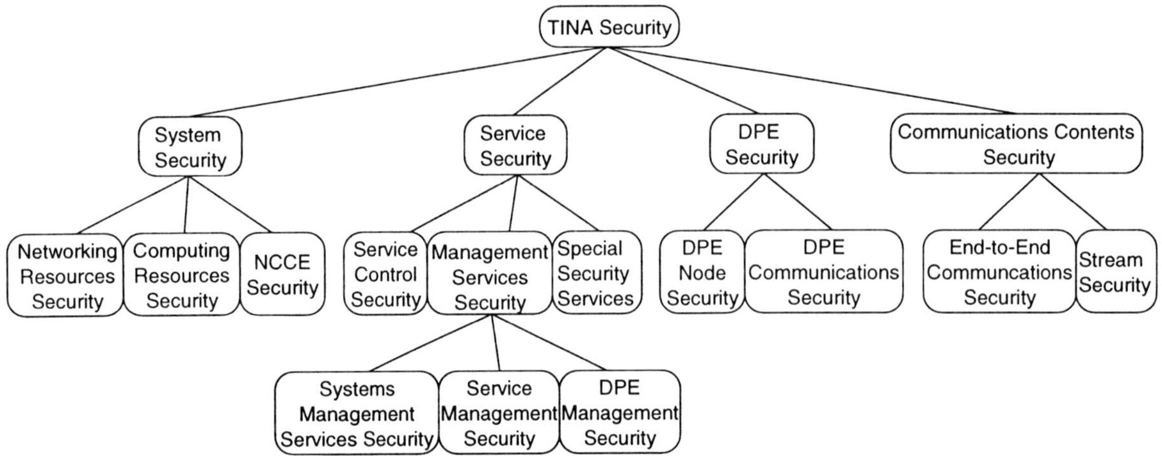

Figure 1. TINA security problem domain

tication of the user. Anonymous users of a chargeable service may be authenticated using anonymized, but chargeable (e.g. pre-paid), identities. The authentication protocol must guarantee that no secret authenticating information is revealed. This can be achieved best by mutual authentication of the user and the provider. The integrity of service control includes integrity of subscription verification and accounting. Access to the service functionality is controlled at two levels, the DPE level and the service level. At the DPE level, a coarse-grained access control based on the authenticated identities of the users involved in a session prevents attempts by others to invoke operations of the service components involved in the session. At the service level, the service logic implemented in the service component controls the access to service specific information and functionality based on the authenticated identities, context, and state information. Integrity and confidentiality of the messages exchanged between the service component's operational interfaces is achieved by the activation of the appropriate features of the DPE security services. These features must provide not only the protection of the integrity of the messages and their temporal order but also protection against interruption of the control connection itself, as we have demonstrated in [19]. Special cases of services are management services and special security services. Both require a potentially higher degree of security (e.g., stronger authentication mechanisms, longer cryptographic keys, or a physically better secured DPE node for their implementation). The special security services provide specialized security features, e.g., digital cash support, that are not present in every DPE node but are supported by dedicated providers (retailers or third party service providers) at the service level. The management services are concerned with the management of systems, services and the DPE. The security of management services is crucial, since illegal access to management functionality may be used for the implantation of backdoors. Of particular concern is the management of the DPE, which includes the management of the DPE security services.

DPE Security:
DPE security is mainly concerned with the prevention of illegal access to computational objects (CO) and CO groups as well as the protection of transmitted messages containing arguments, results, and exceptions of object invocations and notifications. DPE node security also provides the means to audit and report security relevant events on the node according to the audit specifications defined by the administrator (see also [6] [7]). DPE security includes the security of the DPE implementation and its basic services, such as the Object Services in CORBA. Since our architectural placement of security functionality allocates the general security services and mechanisms to the DPE (see Section 4), also the security of the security services themselves is part the security of the DPE.

Communications Contents Security:
Communication contents security is concerned with the authenticity, integrity, and confidentiality of the service contents information. Since all service content information is delivered in the form of streams, it deals only with streams. Streams are protected using cryptographic mechanisms, preferably stream ciphers [17] [18] or special ciphers for certain information formats, e.g., voice or video data. If the service implemented in the provider's domain does not require any modification of the stream between two users, they can have end-to-end security. Otherwise, only user-provider security can be provided. The management of the necessary keys is part of the service control.

3. CORBA security

The CORBA Security specification [13] has been released by the OMG to provide the model, architecture, as well as usage and administration interfaces for security in CORBA [12] systems. There are two levels of conformance. Some features are optional. Thus, not all secure CORBA implementations will provide the complete functionality specified.

The basic notion is the secure object invocation. For each object invocation, the request from the client object to the target object is subject to access control by the ORB security implementation. This access control may take place at the client side, the target side, or on both sides. Figure 2 shows a secure object invocation. The access is decided based on information bound to the target object, and/or information linked to the client object's request. The latter information is referred to as credentials. A credential consists of unauthenticated and authenticated attributes. Authenticated attributes are identity and privilege attributes. This general model enables a large variety of access control schemes, ranging from access control lists over capabilities to label based schemes. The scale of access control is not specified, but it can be assumed that implementors will provide access control down to the granularity of operations.

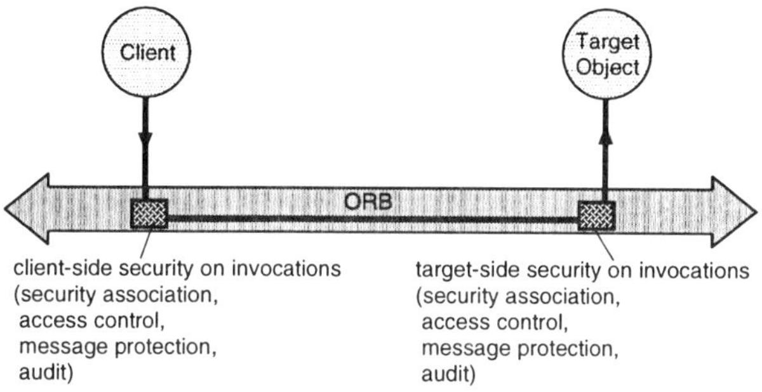

Figure 2. Secure object invocation

The client object acts on behalf of a principal. In most cases, the principal is a human user. In some cases, it may be a system entity, comparable to a system account on UNIX computers. For an object invocation, the credentials of the client object are contained in a dedicated credentials object which is referenced by the object representing the current execution context. The credentials object is created for the principal as the result of the authentication process of the principal.

Access control is only one of several concerns of a secure object invocation. A secure object invocation requires a security association between client and target object. In a security association, both parties trust the claimed identity of each other. This may require additional, mutual authentication with the creation of additional credentials, particularly if the two objects do not reside in the same ORB system. A security association will normally persist for many interactions. Depending on the security policy, integrity and/or confidentiality of requests and responses within a security association may be protected by cryptographic means. For security auditing, security relevant actions may be logged. As an option, CORBA security implementations may provide support for non-repudiation to the applications programmer.

The CORBA security document does not specify the underlying security technology itself. Instead, it defines interfaces for the use and administration of the security service(s) and interfaces to integrate security technology into ORB implementations. In a security enhanced ORB system, security is imposed at two levels, the administration level and the application level. Security policies for access control, message protection and audit are specified by the administrator. A security policy is represented by a security policy object. One security policy may be valid for one or more security domains. Each security domain contains a domain manager object which references the valid security policy object for this domain. The affiliation of an object to a security domain is determined by the administrator. The respective policies are in each case enforced on the object. At the application level, additional security measures may be enforced by the applications themselves. This may be done by the additional enforcement of administrator defined policies and/or the direct use of security features such as non-repudiation. The application enforced security measures cannot override administrator enforced policies.

There are four types of domains regarding security: security domains, security policy domains, security environment domains, and security technology domains. A security domain is the domain that is administered by one authority. A security policy domain is the scope over which a security policy is enforced. In most cases, it is identical with the security domain. A security environment domain is the domain in which the enforcement of the security policy may be achieved by local means, e.g. objects on the same machine. A security technology domain is a set of objects for which the same technology (e.g., Kerberos [11]) is used to enforce the policies.

Secure interoperation between objects depends on the membership of the objects to security technology domains, ORB technology domains, and security policy domains. Interoperability between objects in different security policy domains can only be achieved if both domains agree on a cooperation security policy for the respective interactions. This cooperation policy may be negotiated at invocation time or in advance. In the simplest case, only the security policy for the target is applied. Objects in different ORB technology domains can, technically, interact without problems using the Secure Inter-ORB protocol (SECIOP) as specified in [13], as long as the same security technology is used on both sides. (We do assume the gateway approach for inter-ORB-interoperability for the use of CORBA products in TINA. However, the following would also apply in this case.) According to the CORBA security specification, interoperability between objects in systems with different security technology requires a security technology gateway. This is obviously not a trust problem if both security technology domains are in the same security domain (one common security administration). It may cause trust problems if the boundary between security domains (with different administrations) is also a boundary between security technology domains. For instance, assume one security domain (and security technology domain using asymmetric cryptography) providing the non-repudiation service using digital signatures and another security domain (and security technology domain using only symmetric cryptography) providing non-repudiation using notary servers. In general, a security technology gateway cannot be realized without the administrators of both security domains trusting each other or a third party that runs the gateway. A less restrictive solution would be to negotiate the security technology used.

The specification for secure interoperability between ORBs extends the CORBA 2.0 standard which specifies interoperability. The information which security technology the target requires and which security mechanisms it supports is part of the interoperable object reference (IOR). The Common Secure Interoperability Specification (CSI) by the OMG [14] allows the protocols of three security technologies within the SECIOP, namely SPKM, Kerberos, and the ECMA security protocol. If the interoperability between the ORBs is based on DCE [16], the DCE security technology based on the Kerberos protocol can also be used [13]. A recent proposal [15] wants to allow to base inter-ORB security on the Secure Socket Layer (SSL) [10]. Based on the information in the IOR, a security context acceptable for both sides can be determined. The establishment of the respective security association and the protection of messages are controlled by security tokens which are added to the Inter-ORB-Protocols. Key management is not explicitly dealt with in the CORBA security specification.

4. Providing security for TINA

Security features in TINA are implemented at various levels. In our approach, the DPE offers general security services and security mechanisms[1] to the applications. In the following, we will introduce our approach for TINA security, which is based on a layered structure. Figure 3 illustrates the layering. The usage relations are as follows. DPE security services are exclusively based on the DPE security mechanisms. The implementation of these mechanisms may directly use cryptographic mechanisms or may be built on available higher level security technology, such as Kerberos [11]. The underlying security technology may use the same cryptographic mechanisms as the DPE security mechanisms or proprietary implementations. The use of cryptographic mechanisms and/or higher level security technology may be accomplished through standardized interfaces (e.g., GSS-API [2]) to facilitate the integration of existing products into the DPE.

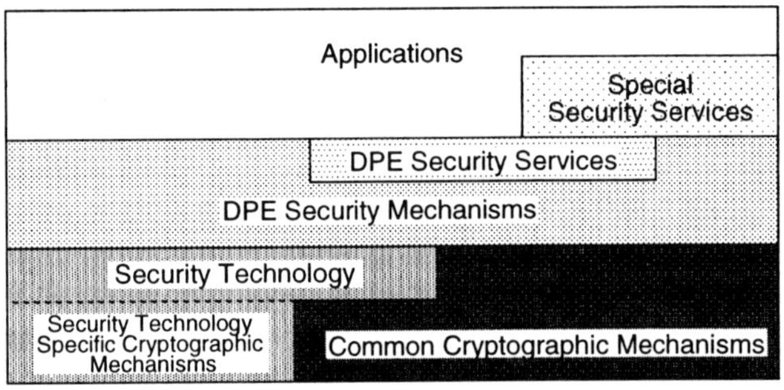

Figure 3. Layering of TINA security features

Above the DPE level, Figure 3 shows the special security services, which also rely exclusively on the DPE security mechanisms and services. They are used by TINA applications, but are application services themselves. The special security services are not implemented on each DPE node. Examples are electronic retail banking functions or notary services.

General security functionality is offered to the applications on each node as part of the DPE functionality. To ease integration in applications, as much functionality as possible should be provided as self-contained security services, i.e. the application is not concerned with how the security functionality is provided (e.g., which security mechanisms are used and on which cryptographic mechanisms or security technology they are based). The application is only required to use the DPE security mechanisms directly if the handling of the security mechanisms is service specific, e.g., if the verification of a digital signature is directly needed. The necessary DPE services and mechanisms must be provided by CORBA security and additional parts of the DPE that are addressed in the CrySTINA project. In the following, we investigate for each of the security topics introduced in Section 2 whether CORBA security functionality is sufficient and propose additional features if needed.

[1] Unlike the CORBA security specifications [13] [14], we use the term security mechanism in the sense as introduced in the OSI security architecture [5], i.e. for an abstract mechanism that can be used to provide one or more security services (e.g., a digital signature), but does usually not provide all necessary security functionality for the system.

System Security:
System security cannot be provided by CORBA security. It can only be guaranteed by a proper design, implementation and installation of the respective hardware and software as well as the protection of communication over unsecure links at the NCCE level. In operation, system security can be supported by auditing of security relevant events at the NCCE level and alarm reporting in case of serious incidents.

Service Security:
Service security relies extensively on DPE security features. In principle, the security of service control can be provided by the use of the CORBA security services in cooperation with the security relevant service logic. A crucial point is the mapping of security relevant domain types defined in the TINA-C architecture and in the CORBA architecture. Based on observations of trial implementations of the service architecture (e.g, [25]), we assume the mapping of each administrative domain in TINA onto one ORB system. This is reasonable because resulting from this mapping the interface between ORBs is a protocol interface, i.e. in the interaction with another administrative domain no potentially vendor-specific executable code from the other side is needed at the DPE level. For security reasons, we propose the following mappings of TINA domain types to security relevant domain types at the CORBA level:

- Each TINA administrative domain is mapped onto one security domain. This domain is also exactly one security policy domain. The mapping reflects that each stakeholder (i.e. administrator of a TINA administrative domain) has specific security interests and limited trust in other stakeholders and that a security domain should not contain equipment that is not under the physical control of the security authority. This mapping further implies that all interdomain interactions between service components (e.g., UAP-USM) rely on inter-ORB security and on the negotiation of a cooperation security policy.

- Each TINA administrative domain (also CORBA security domain and CORBA security policy domain) is also one security environment domain, i.e. how the security policy is enforced within the domain, is a local matter.

- Each boundary between TINA administrative domains is assumed to be also a boundary between security technology domains. This reflects that stakeholders with various kinds of CPE, varying priorities regarding security, and under possibly different national laws cannot be assumed to have the same security technology. It requires that the security technology used for interdomain interactions (including the mechanisms used) is negotiated as part of the respective cooperation security policy. This may result in the use of the same or a compatible security technology or the use of a mutually agreed security technology gateway.

Control of access to information and functionality at the service level (authorization) prevents illegitimate use of these resources within the service usage. It is based on the identities of stakeholders and authorization information regarding subscriptions and access to management functionality. The authorization process is implemented as a part of the service logic in the respective service components, e.g., the User Agent (UA) for the access service as well as the User Service Session Manager (USM) and the Service Session Manager (SSM) for the actual telecommunications service. The authorization decisions are made inside a service component after the invocation of an operation of an operational interface of the service component. These decisions rely on the authentication of the claimed identities as well as the authenticity, integrity, and (optionally) confidentiality of the messages exchanged with the other stakeholder. The three latter properties must be provided by DPE services, whereas the first, i.e. the authentication of the other stakeholder, takes place at the service level as part of the establishment of an access session between different stakeholders. CORBA security does not provide the necessary inter-ORB inter-domain authentication service. An additional authentication service must provide facilities for the mutual authentication of stakeholders. This service is used for the establishment of access sessions. Since user mobility must be supported, the authentication technology must be designed in such a way that the consumer does not need to trust the equipment in the consumer domain. This suggests the use of smart card technology. The integration of established smart card technology can guarantee that the owner (administrator) of the CPE (administrative domain) used at the moment cannot learn the secret authenticating information of the consumer.

DPE Security:
Session keys resulting from the execution of an authentication protocol at the service level may be used subsequently at the DPE (CORBA) level to prove and verify the identity of client objects. The session keys may also be used to provide authenticity, integrity, and confidentiality for the established security association using CORBA inter-ORB security. The exact security context of the security association, i.e. which security mechanisms are used, is established at association setup using the security tokens of the inter-ORB protocol. The context is derived from the cooperation security policy for the interaction between the domains, which also includes the choice of the mechanisms and the authentication servers (or public key certifiers).

Access control at the DPE level has to prevent illegitimate invocations of operations. This access control is, in contrast to the access control at the service level (authorization), not determined by the semantics of the service, but by the question whether the originating stakeholder of an attempt to invoke an operation is allowed to invoke this operation at all. The only possible results of an access decision are access permitted or access denied. Granularities are the whole interfaces or the single operation of an interface. In order to analyze how CORBA access control can be used, it is necessary to study how TINA service components are built up using CORBA. Service components are likely to be implemented as CO groups, even though the service architecture [23] does not prescribe the mapping of service components onto COs or CO groups. The service component's interfaces are in that case provided as contracts of the respective CO group. However, CORBA lacks the concept of object groups. Additionally, in contrast to TINA COs, CORBA objects have exactly one interface. We assume that each TINA CO is implemented as a set of CORBA objects and that each CO interface is implemented as a dedicated CORBA object, as proposed in [9]. The service component's interfaces (the contracts of the group) are, thus, provided as the CORBA interfaces of those CORBA objects that implement the CO interfaces that serve as contracts.

Potentially, the operations of all CORBA interfaces are accessible to everyone who has access to the kernel transport network that connects the single ORB systems. However, only few of these interfaces shall be accessible by objects acting for other stakeholders, i.e. shall be accessible for another identity than the one of the administrative domain (the owner) the object is allocated to. These interfaces are implementations of those interfaces of service components that are part of an interdomain reference point. Access to all other CORBA interfaces must be restricted to those objects that act under the same identity. This is easily achieved by ORB-local access control.

Access control for the objects that are indeed part of an interdomain reference point is more complex. However, our observation is that the functionality offered across domain boundaries is always structured as interfaces so that each instance of an operational interface is dedicated to exactly one other stakeholder, as suggested in [9]. As long as this observation holds, identity based access control can be applied at the granularity of service component interfaces, i.e. at the CORBA level to the whole object implementing the interface of the respective service component and can be realized by simply checking authenticity and integrity of the message conveying the invocation request. If the observation above does not hold, two cases with different granularities can be distinguished. If an interface of a service component is accessed by various stakeholders, but with the same rights, the unit of access control can also be the whole interface. On the other hand, if an interface is accessed by various stakeholders with different rights, the unit is the single operation of the interface. In both cases, access control should be supported by a list bound to each unit of access control that contains the identities of the stakeholders that are authorized to access the unit. This mechanism can be expected to be supported by a wide range of CORBA products.

Communications Contents Security:
In TINA, all service contents information is delivered by streams. (There is an ongoing discussion also to allow to provide contents information via operational interfaces without relying on streams. However, the following is also valid for such a delivery of contents.) Until now, CORBA does neither support streams nor stream protection. A DPE supporting streams must also support DPE security mechanisms applicable to streams. The establishment of a stream should include the establishment of a protection context for the stream. This context determines the security mechanisms and the key(s) used. The protection context is derived from the cooperation security policy of both domains. Such a policy, and possibly a session key, may already exist, e.g., if both parties are in a user provider relationship. If the parties have not authenticated each other directly (e.g., both parties are users of a common provider) and want to establish end-to-end security, they can use the authentication service mentioned above for direct mutual authentication and the negotiation of a session key.

5. Conclusion

We analyzed and structured the TINA security problem domain. It was demonstrated how the security services and mechanisms can be provided as part of the DPE functionality. Ongoing conceptual work in CrySTINA is dedicated to the identification and specification of the necessary single security services, a formal model of how administrative domains with different security policies agree on a cooperation security policy, and the use of a hierarchical public key infrastructure based on certificates as specified in [8]. The approach to TINA security presented is implemented extending the CORBA security features and using commercial CORBA products. Future work will cover the additional trust relationship between the current user of a CPE and the CPE itself that is needed for the support of personal mobility with full security.

References

[1] T.Eckardt, T.Magedanz, R.Popescu-Zeletin, M.Schulz, M.Stapf. Personal Communications Sup-

port in the TINA Service Architecture - A new TINA-C Auxiliary Project. *Proceedings TINA'96 Conference*, Heidelberg, Germany, September 1996, pp. 55-64.

[2] Internet RFC 1508. Generic Security Service - Applications Program Interface. September 1993.

[3] Internet RFCs 1825–1829, 1851, 1852. IPv4 and IPv6 Security. August/September 1995.

[4] Internet RFCs 1113, 1814, 1815. Privacy Enhancement for Internet Electronic Mail, Parts I-III. August 1989.

[5] ISO/IEC 7498-2: Information Technology - Open Systems Interconnections - Basic Reference Model - Part 2: Security Architecture (Also ITU-T Recommendation X.800).

[6] ISO/IEC 10164-7: Information Technology - Open Systems Interconnections - Systems Management: Security Alarm Reporting Function (Also ITU-T Recommendation X.736).

[7] ISO/IEC 10164-8: Information Technology - Open Systems Interconnections - Systems Management: Security Audit Trail Function (Also ITU-T Recommendation X.740).

[8] ISO/IEC 9594-8: Information Technology - Open Systems Interconnections - The Directory: Authentication Framework (Also ITU-T Recommendation X.509).

[9] B.Kitson. CORBA and TINA: The Architectural Relationships. *Proceedings TINA'95 Conference*, Melbourne, Australia, February 1995, pp. 371-386.

[10] Netscape. Secure Socket Layer. March 1996, http://home.netscape.com/eng/ssl3/

[11] C.Neuman, T.Ts'o. Kerberos: An Authentication Service for Computer Networks. *IEEE Communications Magazine*, September 1994, pp. 33-38.

[12] Object Management Group. The Common Object Request Broker, Architecture and Specification, Revision 2.0. July 1995, http://www.omg.org/corba/corbiiop.htm.

[13] Object Management Group. CORBA Security. December 1995, http://www.omg.org/library/corbserv.htm.

[14] Object Management Group. Common Secure Interoperability (CSI). March 1997, http://www.omg.org/library/schedule/Technology_Adoption.htm.

[15] Object Management Group. CORBAsecurity/SSL Interoperability. June 1997, http://www.omg.org/library/schedule/Technology_Adoption.htm.

[16] Open Software Foundation. *OSF DCE Application Guide*. Prentice-Hall, 1992.

[17] R.Rueppel. *Analysis and Design of Stream Ciphers*. Springer Verlag, 1986.

[18] B.Schneier. *Applied Cryptography*, 2nd edition. Wiley, 1996.

[19] S.Staamann. Overall Integrity of Service Control in TINA Networks. *Proceedings of Communications and Multimedia Security Conference '97*, Athens, Greece, September 1997.

[20] S.Staamann, U.Wilhelm. CORBA as the Core of the TINA-DPE: A View from the Security Perspective. *Proceedings of Object World Frankfurt 1997*, Special track *Distributed Object Computing in Telecommunications*, Frankfurt, Germany, October 1997.

[21] TINA-C Document: Authentication in TINA Access Session, Version 1.0 (Draft). Engineering Note, August 1996.

[22] TINA-C Document: Security Architecture, Version 2.0. Technical Report, March 1996.

[23] TINA-C Document: Service Architecture, Version 5.0. Baseline Document, June 1997.

[24] TINA-C Document: TINA Business Model and Reference Points, Version 4.0. Baseline Document, May 1997.

[25] H. Woo Sun, K. Eun Chul, J. Hee Kyung. Realization of TINA Service Architecture on the Internet. *Proceedings TINA'96 Conference*, Heidelberg, Germany, September 1996, pp. 235-243.

[26] P.R.Zimmermann. *PGP Source Code and Internals*. MIT Press, 1995.

Session 7

Experience Reports II

Kenichi Kitami, NTT

TINA Service Architecture: From Specification to Implementation

Patrick Hellemans (phe3@rc.bel.alcatel.be)
Hans Vanderstraeten
(straeteh@btmaa.bel.alcatel.be)
Alcatel Telecom
Francis Wellesplein 1, B-2018 Antwerp, Belgium

Juan C. Yelmo (yelmogar@dit.upm.es)
Jesús Villamor (jvillamo@dit.upm.es)
Universidad Politécnica de Madrid
Av Ramiro de Maeztu, 28040 Madrid, Spain

Patricia Lago (patricia@polito.it)
Politecnico di Torino
Corso Duca degli Abruzzi, 10129 Torino, Italy

Gianni Canal (gianni.canal@cselt.it)
CSELT
via G. Reiss Romoli 274, 10148 Torino, Italy

Abstract

The VITAL project[1] has produced an implementation of the TINA service architecture (SA), based on the SA 4.0 deliverable released by the Core Team (CT) in October 1996 [1]. In order to design a system based on these high level specifications, we were obliged to fill in all missing details required to actually come to an implemented system. This paper reports on the most important design decisions we have taken in this process. The major differences between the TINA Retailer reference point specification [2] -as defined at the time of writing this paper- and VITAL's implementation are indicated. The deployment details of the system on a distributed processing environment (DPE) platform are also detailed. Furthermore, VITAL also introduced service-specific behaviour for a number of services within the context of a teletraining application. The adopted approach for co-ordinating the service-common and service-specific parts of the service session is elaborated in the paper.

1. Introduction

The TINA consortium has released a completely revised version of its service architecture deliverable in October '96. The reader is assumed to be familiar with the concepts explained in that deliverable. The VITAL project, whose aim it is to validate TINA specifications, has used this deliverable in order to produce an implementation of it. For a description of the complete VITAL prototype, which also includes implementation of other TINA sub-architectures, including resource architecture, the reader is referred to [3]. The next section presents an overview of the VITAL service architecture system and its relation to the TINA business model [4]. The system consists of three major subsystems: access session, service session and subscription management. The major elements of the informational and computational viewpoints on these subsystems are elaborated, together with their representation at the reference point between the Consumer and Retailer business roles. In addition, the deployment details of these viewpoints on a DPE platform are also highlighted. For this purpose, a new notation is used which combines computational viewpoint with deployment aspects. This notation is explained in a separate section of the paper. A number of scenarios are included to illustrate the behaviour of the system. The conclusion section also includes the extensions to the platform that are envisaged to be realised in the near future.

2. VITAL Service Architecture Overview

The VITAL service architecture implementation is limited to components in the Consumer and Retailer domains. No service federation is supported between Retailers. Therefore, only one of the TINA reference points (RP) is supported: the Ret RP between Consumer and Retailer. The service architecture system is further decomposed into access session, service session and subscription management subsystems.

[1] VITAL (Validation of Integrated Telecommunication Architectures for the Long term) is a project partly sponsored by the ACTS telecommunications research framework of the European Union.

0-8186-8335-X/98 $10.00 © 1998 IEEE

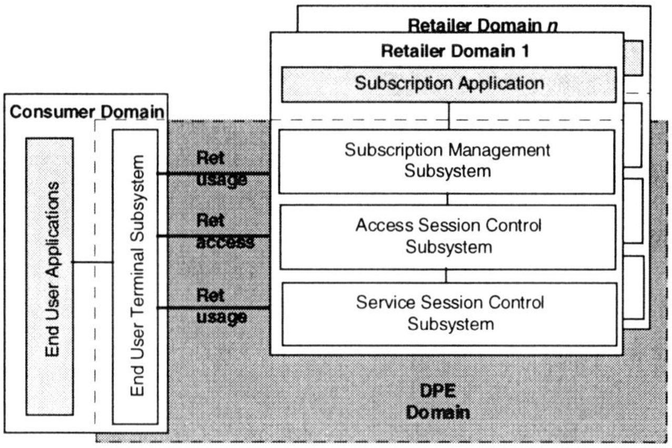

Figure 1: VITAL System Architecture Overview

The access session subsystem provides the mechanisms to establish a secure association between one Consumer and one Retailer. The VITAL implementation supports multiple simultaneous access sessions for multiple Consumers on the same domain (in the VITAL implementation, a Consumer domain consists of a single terminal.), the same Consumer on multiple domains and the same Consumer (on the same or multiple domains) with multiple Retailers. An access session needs to exist to allow a Consumer to take part within a service session.

The service session subsystem consists of both service-common and service-specific logic. VITAL distinguishes between the service offered by the Retailer, which is identified by an RServiceId, and the service used by the Consumer, which is identified by a CServiceId. In the VITAL implementation every CServiceId is mapped onto a single RServiceId. However, in principle it is possible that multiple CServiceIds are mapped onto (i.e. "make use of") the same RServiceId. Think for example of "Powerpoint Presentation" and "Word Presentation" as CServiceIds, which can be supported both by "Data Presentation" as RServiceId. In this paper, the terms service and service-specific always apply to an RServiceId, unless explicitly stated otherwise.

The subscription management subsystem assists the Retailer in offering a broad set of services and tailoring the service features to the requirements of Consumers. Consumers need to subscribe to a service before they can take part in a service session for that particular service. The participation in the service session has to reflect the agreed and customised aspects of the service subscription. Subscription features are accessible via both a Retailer application and as an ancillary on-line subscription service.

3. Deployment of Computational Objects

The TINA specifications focus on informational and computational viewpoints. The design of a TINA system also requires determination of the deployment into DPE server processes, IDL interface and implementation classes. VITAL has adopted a new notational convention for its computational specification which also reflects deployment details. First of all, we need to discriminate between service components (SC) and computational objects (CO): the latter are the smallest deployable units on the DPE, the first can themselves be decomposed into either smaller-grain SCs or COs. The VITAL notation uses a coloured background for SCs, while a white background is being used for COs. IDL interfaces are mapped onto IDL C++ classes. Attached to the IDL interfaces are grey rectangles, which reflect a certain context. Such a context within a CO is mapped onto a C++ implementation class. Within an SC the context represents a logical unit, rather than a deployment unit, and is typically mapped onto contexts in COs contained by the SC. We distinguish between static and dynamic contexts. A static context, depicted by underlining the context name, is assumed to exist when the SC or CO is created. The number of static contexts is known exactly at implementation time. A typical example is a factory context. Dynamic contexts are created and deleted dynamically during the lifetime of the SC or CO. A typical example are session related contexts. The life cycle of these contexts is typically controlled by a factory

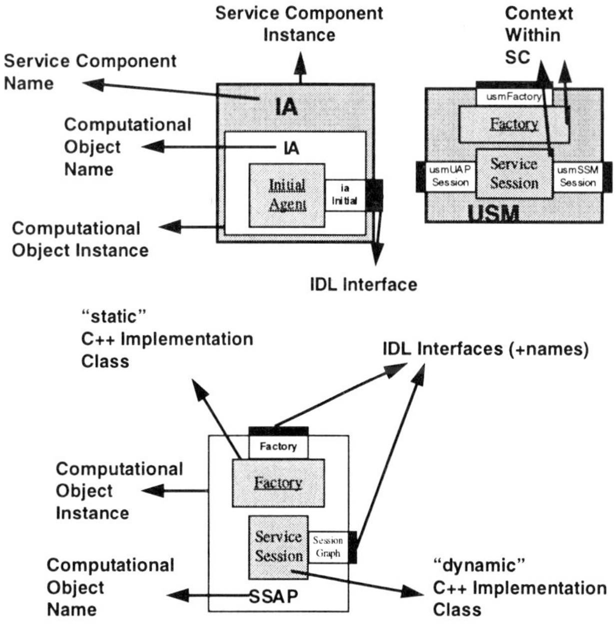

Figure 2: Computational/Deployment Notation

context. The reason why factory contexts are represented explicitly is inspired by the way DPE products work: in order to be able to dynamically create and delete interface and implementation objects inside a server process, a factory interface and implementation object is required. A factory context is not to be confused with factory-type SCs such as a service factory. The latter control the lifecycle of *other* SCs, while the first control dynamic contexts *within the same* SC. The expression "deployment of a SC instance" is used to refer to the existence of a DPE server process which executes all implementation objects belonging to that SC instance. Every implementation object is handled by a separate execution thread within a server process. Therefore, one implementation object will never block the execution of another implementation object within the same server process.

4. The Access Session Subsystem

4.1 The Computational and Deployment Model

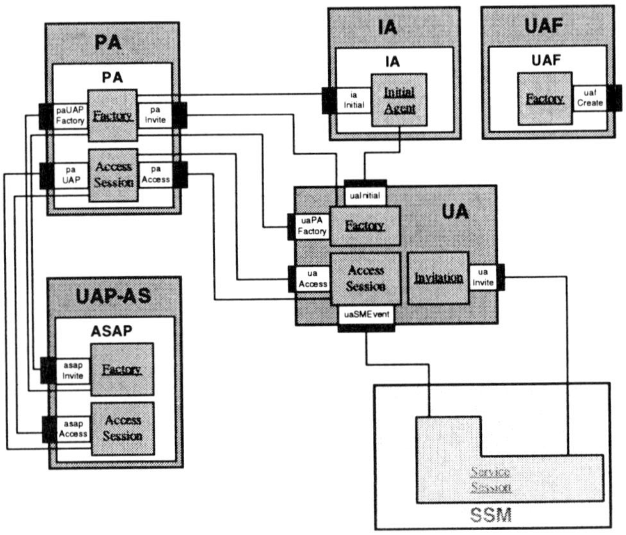

Figure 3. Access Session Subsystem - Computational Model

The *User Application - Access Session* (UAP-AS) is the part of the user application which terminates the access session in the Consumer domain.

The *Provider Agent* (PA) implements the access session contact point of the Retailer within the Consumer domain. A distinct PA SC instance is deployed in each Consumer domain *for each of the Retailers* with which the Consumer can potentially have an access session from that domain.

The *Initial Agent* (IA) implements the initial access session contact point of the Consumer within the Retailer domain to initiate the establishment of an access session with that Retailer. A single IA SC instance is deployed in each of the Retailer domains.

The *User Agent* (UA) implements the access session contact point of the Consumer within the Retailer domain for an established access session with that Retailer. . A distinct UA SC instance is deployed in each Retailer

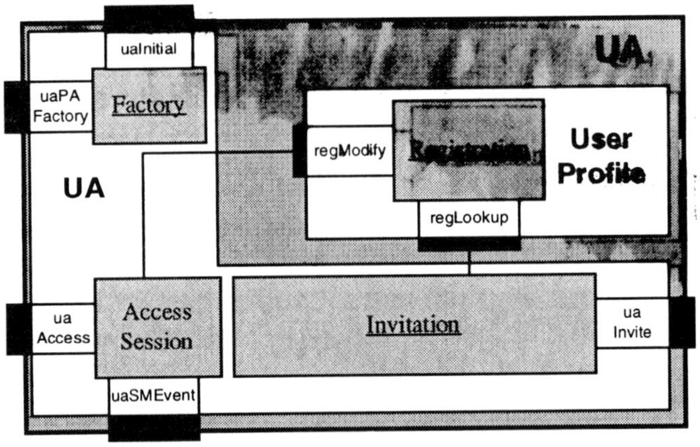

Figure 4: User Agent SC

domain *for each of the Consumers* that have a subscription with that Retailer. The UA SC is itself decomposed into a UA CO and a *User Profile* CO. The UA CO is responsible for the UA SC interaction with the PA in the Consumer domain and managing the access sessions within the Retailer domain. It also contains the *Invitation* implementation object which offers a uaInvite interface towards the Service Session Manager for handling incoming invitations to join a service session. When such an invitation is received, the Invitation object consults the *Subscription Agent* (in the subscription management subsystem) to check the service subscription. Subsequently, the invitation object consults the *UserProfile* CO to decide on the domain to which an access session is required. It consults the *Factory* object to determine whether an appropriate access session already exists. If an appropriate access session implementation object already exists, the invitation will be forwarded to that object. Otherwise, the Invitation object initiates the creation of a new access session via the Factory object. The User Profile CO implements the user registration mechanisms which provide the means to support IN CS-2 like service features, such as flexible number translation, personal mobility, screening and call redirection. The Consumer can specify on which conditions an invitation to join a service session has to be forwarded to which terminal(s). He can also set the conditions on which an invitation has to be rejected. This information is stored persistently by the User Profile CO and is consulted by the Invitation object.

The *User Agent Factory* (UAF) is the contact point for the subscription management subsystem to install/uninstall UAs for specific Consumers. The UAFactory (UAF) SC is introduced for deployment rather than architectural purposes to be able to have different server processes for the different Consumers. A single UAF SC instance is deployed in each of the retailer domains.

4.2 The VITAL Ret Access Part

The following features are provided by the VITAL retailer implementation on the Ret access RP:

- Control of the access session lifecycle
  - **create_asession_req**
    allows the Consumer to request the creation of a new access session
  - **accept_invite_a_s_session_rsp**
    allows the Consumer to asynchronously accept the creation of an access session initiated by the Retailer
  - **refuse_invite_a_s_session_rsp**
    allows the Consumer to asynchronously refuse the creation of an access session initiated by the Retailer
  - **delete_asession_req**
    allows the Consumer to request the deletion of an existing access session - note: VITAL implementation does not allow the deletion of one access session from within another access session
  - **list_asession_req**
    allows the Consumer to request for a list of access sessions between this Consumer (possibly on different domains) and this Retailer
- Control of the (user) service session lifecycle
  - **create_ssession_req**
    allows the Consumer to request the creation of a new service session
  - **join_ssession_req**
    allows the Consumer to request to join an existing service session
  - **accept_invite_ssession_rsp**
    allows the Consumer to asynchronously accept the invitation generated by the Retailer to join a service session
  - **refuse_invite_ssession_rsp**
    allows the Consumer to asynchronously refuse the invitation generated by the Retailer to join a service session
  - **leave_ssession_req**
    allows the Consumer to request the deletion of this Consumer from an existing service session (without deleting the service session itself)
  - **delete_ssession_req**
    allows the Consumer to request the deletion of an existing service session
  - **list_ssessions_req**
    allows the Consumer to request for the list of service sessions for a certain retailer service
- User Registration
  - **register_user_req**
    allows the Consumer to update the user registration data
  - **list_register_user_req**
    allows the Consumer to retrieve the user registration data

The major deviation of VITAL's implementation of the TINA Ret access part is the way invitations to join service sessions are being handled. In the TINA Ret these invitations are being delivered to the invited user's user agent, without waiting for any reply. The invited user can join the service session later at the time of his convenience. If no appropriate access session exists to deliver the service invitation to the appropriate terminal - as determined by the user registration parameters- the invitation can be stored by the user agent. In the VITAL implementation, service invitations are not just being delivered to the user, but an answer is expected within a certain time frame. If no answer is received within a given time frame, the invitation is no longer valid and the inviting user is informed of the invitation failure. This also implies that the invitation cannot be stored by the user agent, it needs to be delivered to a terminal. If no appropriate access session exists, the user agent initiates the establishment of an access session and delivers the service session invitation at the same time. The reason why we have chosen for this approach is because it is more in line of the behaviour of interactive telecom services, telephony being the most obvious ones. Users can still join existing service sessions at their own discretion, provided these service sessions have been publicly declared. This mechanism is analogous to the joinWithAnnouncement procedure of the TINA Ret.

4.3 Example Scenarios

This section presents two scenarios to illustrate the behaviour of the access session subsystem:

- Figure 5: Access Session Creation Initiated By Consumer. The IA is responsible for authenticating the user and subsequently requesting the creation of a new UA access session implementation object. The interface references offered by that new object are returned to the PA as an out parameter in the create_asession_req.
- Figure 6: Creation of a New Service Session .

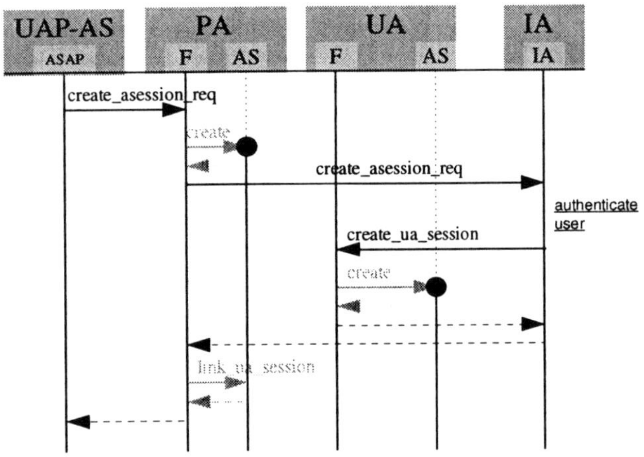

Figure 5: Access Session Creation Initiated By Consumer

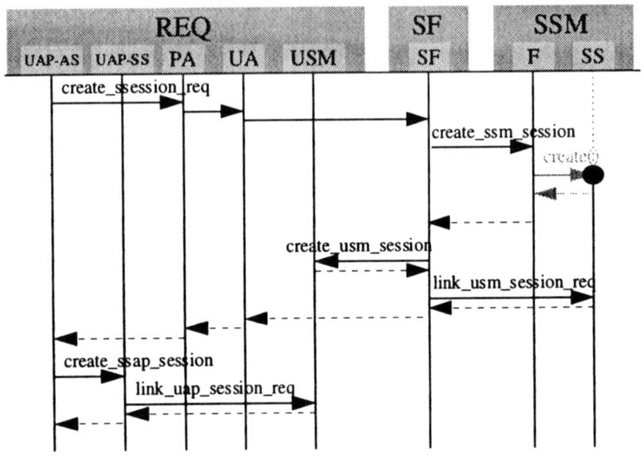

Figure 6: Creation of a New Service Session Initiated By Consumer

5. The Service Session Subsystem

5.1 The Service Session Graph

The access part of the Ret foresees the procedures to negotiate different session models when a new service session is being created. However, TINA only elaborates the session model which is based on the service session graph (SSG) information model. Figure 7 presents an overview of the VITAL implementation of the SSG (Only the classes depicted with a bold border in the SSG overview figure are actually implemented).

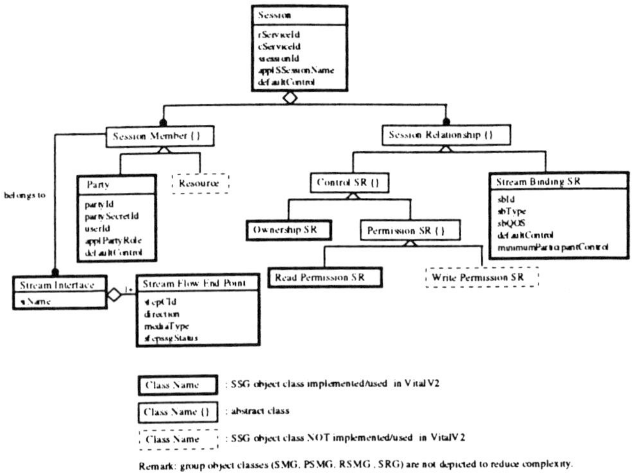

Figure 7. SSG Information Model - Overview

Every service session is uniquely identified within a Retailer domain by the sSessionId, which is stored as an attribute of the *Session* object. In addition, the user application which initiates a service session can select a name - the applSSessionName - as an informal means to identify the session. However, there isn't any guarantee that different service sessions have unique names within a Retailer domain.

A Consumer - identified by a userId - who becomes involved in a service session is represented by a *Party* object, uniquely identified within the session by a *public* partyId. In addition, another unique identifier - the partySecretId - is generated *which is only known by the Retailer and the Consumer 'behind the Party'*. The partySecretId has to be provided in every Ret usage operation invoked on a Retailer interface. Furthermore, the user application which initiates the creation a new Party object can also assign a role - the applPartyRole - to that Party.

Before elaborating the *Stream Binding Session Relationship* (SBSR), we need to introduce two supporting concepts: the *Stream Flow End Point* (SFEP) and the *Stream Interface* (SI). The SFEP represents the termination of a single information flow by a user application. It can be seen as the logical representation of a 'physical stream device I/O port' of a terminal (e.g. the outlet of a camera). SFEPs are created prior to the creation of an SBSR, never as a 'side effect' of such a creation. Their life cycle is managed by an SFEP manager which resides in the Consumer domain. An SFEP is characterised by a direction (source or sink) and a high-level media type (audio, video, ...). The SI represents a dynamic grouping of SFEPs. The user application in every Consumer domain decides on the semantics of the grouping. The SBSR expresses the connectivity requirements as required by the user applications. The

SBSR can potentially express multipoint-to-multipoint multimedia configurations. The SBSR is 'translated' into Stream Flow Connections (SFCs), which is the way of expressing connectivity in a communication session. An SBSR is characterised by a type - the sbType - and a high-level quality of service - the sbQOS. When a user application registers an SFEP for usage within an SBSR, it also specifies a name -the SFEPBindName- which enables the Retailer to determine which SFEPs can potentially by bound to the same SFC. Only SFEPs of identical mediaType which have been registered with identical SFEPBindName can be bound to each other. Every type of user application predefines the sbTypes it supports *and* the SFEPBindNames that can be used for each of its sbTypes. The SFEPBindName can be used to determine the 'topology' of the SFEP bindings, while the mediaType is used to determine the type of stream info that is sent over this 'topology'. The sbQOS consists of different elements. The sbQOS stream quality expresses a high-level qualification of the SFCs (e.g. "standard", "low", "high"). It is up to the user application to translate this stream quality into technology-specific quantifications. An SBSR can potentially be resolved into multiple SFCs, each of which SFC binds a source SFEP to one or more sink SFEPs. The sbQOS success criterion expresses whether all SFCs need to be created successfully *and* whether all SFEPs need to be attached successfully for each individual SFC.

The *Control Session Relationship* (CSR) expresses a control relationship between a Party -the controller- and a controlled SSG object. The possible controlled SSG objects are Session, Party and SBSR. The possible levels of control are *NoControl (lowest)*, *ReadPermission* and *Ownership (highest)*. This level of control expresses both the capabilities of a Party to introduce changes to the SSG (Ownership) and the visibility of SSG Objects as perceived by a Party (ReadPermission). A higher level of control always includes the capabilities implied by a lower level of control (e.g. Ownership implicitly implies ReadPermission). The Ownerships are one of the means to determine which Parties have to acknowledge the processing of certain operations on the SSG. The ReadPermissions determine which parties are automatically informed of certain SSG modifications *and* they are checked when a Consumer explicitly retrieves SSG information. A Party will never receive any information about objects for which it has no ReadPermission. Each object that can be controlled is attributed by a defaultControl setting, which is used to determine the level of control between a Party and the controlled information object *for those parties that haven't explicitly declared a CSR between themselves and the controlled object*. In addition, the SBSR is also attributed by a minimumParticipantControl setting, which expresses the minimum level of control each of the participants in the SBSR has on the SBSR. It has to be *at least* as high as its defaultControl setting and must be *at least* ReadPermission. If there exists a CSR between the controller Party and the controlled SSG object, the level of control exercised by the controller Party is expressed by that CSR control level. If there is no such CSR defined, we need to distinguish between the different types of controlled objects: the level of control on the Session is equal to its defaultControl setting; for a Party and an SBSR, the level of control is determined by the CSR between the controller Party and the Session Object *if such a CSR exists*. Otherwise, the level of control is determined by the defaultControl setting of the controlled objects.

5.2 The Computational and Deployment Model

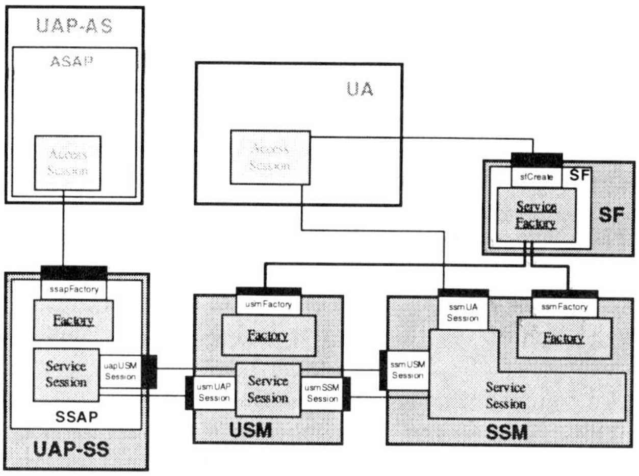

Figure 8: Service Session Subsystem - Computational Model

VITAL implements three Retailer services: Desktop Audio-Video Conference (DAVC), Shared WhiteBoard (SWB) and Slide Presentation (SP). DAVC uses the service-common behaviour offered by the service session graph, without adding additional service-specific semantics within the Retailer domain. SWB extends the service-common IDLs with extra operations to handle the management of the whiteboard content. An information model is introduced to represent the whiteboard content and a copy of its current status is maintained by the centralised SWB session manager. SP extends the service-common IDLs with extra operations to control the display of slides, but it uses stream bindings to transfer the actual slide content end-to-end between the relevant participants.

The *User Application - Service Session* (UAP-SS) is the part of the user application which terminates the service session in the Consumer domain. The UAP-SS is both CService- and RService-specific.

The *Service Session Manager* (SSM) is the RService-specific component which is responsible for the coordination of all -control- activities that take place within the context of a service session. The SSM consists

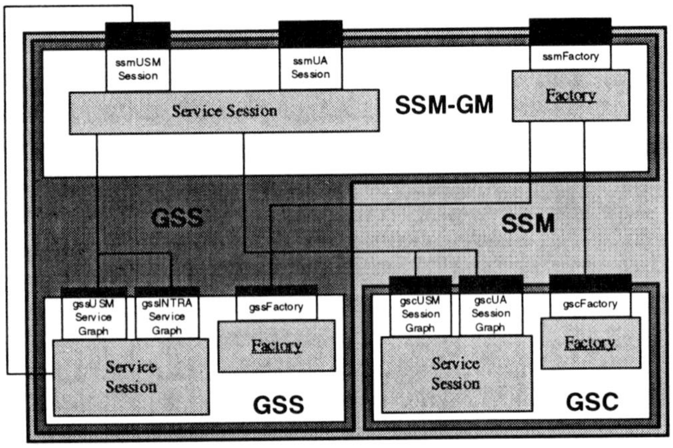

Figure 9: Service Session Manager SC

of a service-common *Global Session Control* (GSC) and a service-specific *Global Service Segment* (GSS) SC.(Various SSM and USM decomposition models have been studied by VITAL, including models taking advantage of the OO inheritance mechanism. Only the selected model for the SWB and SP service is elaborated in this paper.). The GSC implements the centralised management of the service session graph information model. One single GSC SC instance is deployed in each of the retailer domains. The GSS SC is itself decomposed into the *SSM-Group Manager* (SSM-GM) and GSS COs. The SSM-GM is the only CO within the SSM whose interfaces are offered externally to the SSM. Therefore, all invocations on the SSM are actually processed first by the SSM-GM. The SSM-GM "routes" the invocations to either the service-specific (GSS) or service-common (GSC) part of the SSM. All IDL interactions between the GSS and GSC COs are also invoked via the SSM-GM. A distinct GSS SC instance is deployed in each of the Retailer domains *for each of the services provided by that Retailer.* Every service session context within the SSM SC is mapped onto an implementation object within the SSM-GM, GSS and GSC COs.

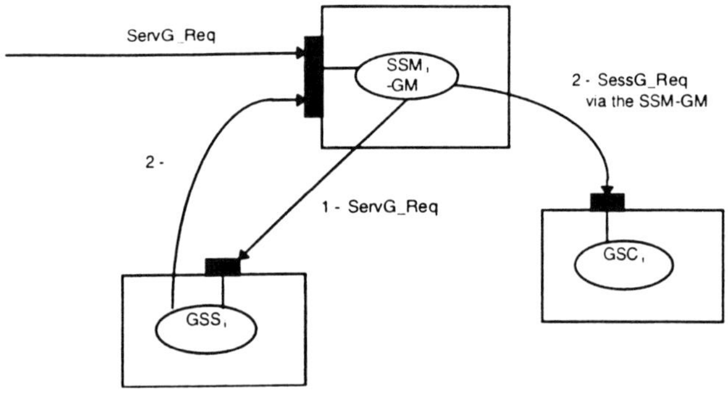

Figure 10: Service-Specific Operation Co-ordination

When a service-specific operation is invoked on the SSM (refer to Figure 10) , the SSM-GM forwards the invocation to the appropriate GSS instance. If the service-specific operation semantics requires service-common functionality, the GSS instance invokes the relevant (service) session graph related operation on the SSM interface.

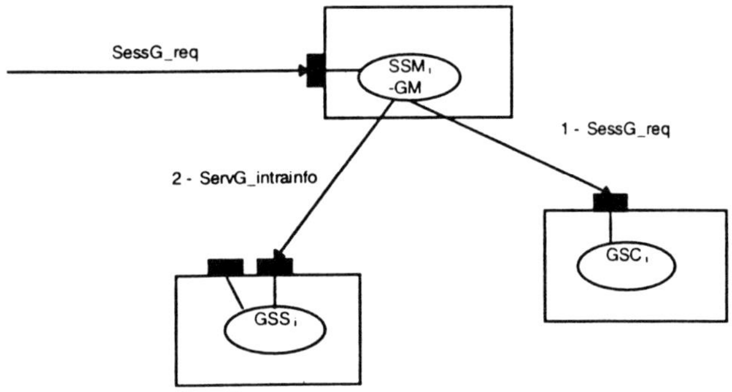

Figure 11: Service-Common Operation Co-ordination

When a (service) session graph related operation is invoked on the SSM (refer to Figure 11) , the SSM-GM forwards the invocation to the appropriate GSC instance. After completion of the service-common operation, the SSM informs the appropriate GSS instance of the modification, by invoking an operation on the service-specific intra-component information interface of the GSS. It is up to the GSS to determine whether the SSG modification is also relevant to its own logic.

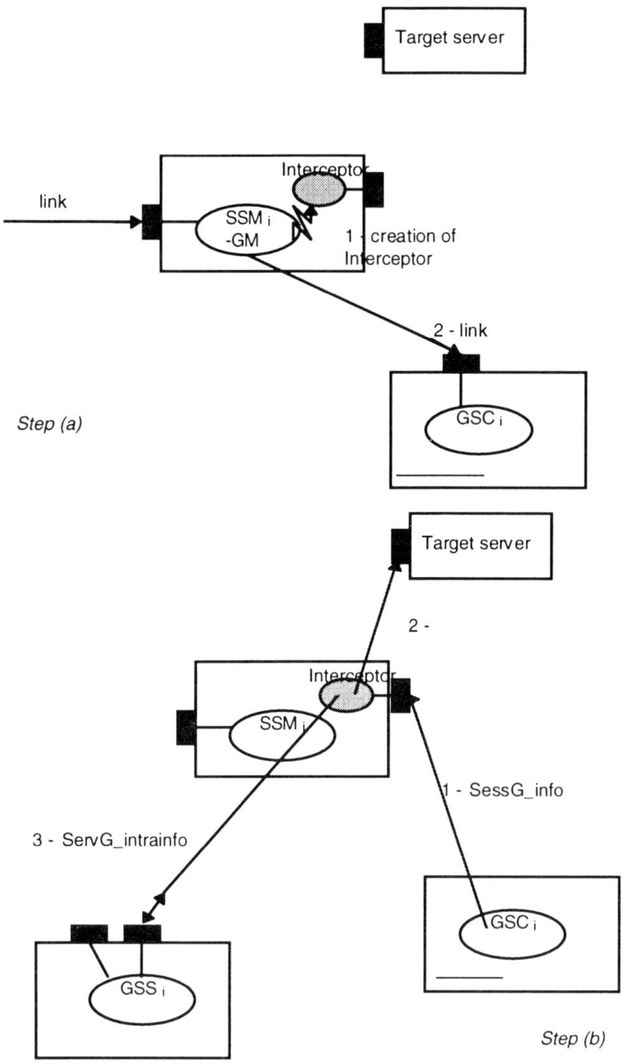

Figure 12: Interceptor for Operations Invoked by the GSC

When the interface reference of a server to be used by the GSC is communicated to the SSM (refer to Figure 12) , the SSM-GM creates an interceptor instance, and forwards the reference of the interceptor to the GSC. When the GSC thinks it invokes an external server, it is actually using the interceptor reference. As a result, the interceptor can filter the invocation, and notify the GSS of the invocation.

The *User Service session Manager* (USM) is the RService-specific component which implements the personalised management of the service session for a single party in that session. The USM is decomposed into a service-common *User Session Control* (USC) and a service-specific *User Service Segment* (USS) behaviour part, analogously to the SSM. A distinct USS SC instance is deployed in each of the retailer domains *for each couple [(R)service provided by that retailer, Consumer subscribed to that RService].* A distinct USC SC instance is deployed in each of the retailer domains *for each Consumer which has a service subscription with that retailer.*

The *Service Factory* (SF) is responsible for the creation and deletion of USM and SSM service session instances for a specific service. Every RService implements its own SF component *but the sfCreate interface has to be identical for all RService implementations.* This is due to the fact that the UA, which is the client of the sfCreate interface, must be RService-independent. A distinct SF SC instance is deployed in each of the Retailer domains *for each of the RServices provided by that Retailer.*

5.3 The VITAL Ret Usage Part

The following features are provided by the VITAL retailer implementation of the SSG on the Ret usage RP:

- multiparty features:
  - **add_party_req**
    allows the Consumer to request the addition of a new party to the service session - it results in sending a service invitation to the invited user
  - **delete_party_req**
    allows the Consumer to request the deletion of a party -not associated with the requesting Consumer- from the service session - it results in sending a session leave to the deleted user
  - **list_parties_req**
    allows the Consumer to retrieve information about the parties in the service session - only parties which are visible to the requester are returned
- streambinding features:
  - **add_iSBSR_req**
    allows the Consumer to request the creation of a new streambinding
  - **delete_iSBSR_req**
    allows the Consumer to request the deletion of a streambinding
  - **add_participants_iSBSR_req**
    allows the Consumer to request the addition of new participants - different from the requester- to the streambinding
  - **delete_participants_iSBSR_req**
    allows the Consumer to request the deletion of participants -different

from the requester- from the streambinding

join_iSBSR_req

allows the Consumer to request to join a streambinding, i.e. to add the requester as a participant

leave_iSBSR_req

allows the Consumer to request to leave a streambinding, i.e. to delete the requester as a participant

register_SFeps_iSBSR_req

allows the Consumer to request the registration of new SFEPs

withdraw_SFeps_iSBSR_req

allows the Consumer to request the withdrawal of SFEPs which have previously been registered by the requester

rebind_iSBSR_req

allows the Consumer to request the explicit triggering of the streambinding algorithm

withdraw_SIs_req

allows the Consumer to request the withdrawal of all SFEPs that have previously been registered with any of the indicated SInames to any streambinding in the service session

list_iSBSR_req

allows the Consumer to retrieve information on the streambindings which are visible to the requester

list_participants_iSBSR_req

allows the Consumer to retrieve information on the participants of a particular streambinding - only participants which are visible to the requester are returned

list_connInfo_iSBSR_req

allows the Consumer to retrieve an overview of the connectivity of a particular streambinding - for each of the SFEPs registered by the requester and attached to a SFC, the list of participants which are "connected" to it are included

list_SFepSSGStatus_iSBSR_req

allows the Consumer to retrieve an overview of the connectivity of a particular streambinding - for each of the SFEPs registered by the requester and attached to an SFC, the list of participants which are "connected" to it are included

- control session relationship features:

set_control_req

allows the Consumer to request the setting of a CSR

list_control_req

allows the Consumer to retrieve all CSR information visible to the requester

list_owners_req

allows the Consumer to retrieve the parties which have ownership on a specific object - only parties visible to the requester are returned

list_readers_req

allows the Consumer to retrieve the parties which have -at least- read permission on a specific object - only parties visible to the requester are returned

A major difference between the TINA Ret RP and the VITAL implementation is the fact that terminating an individual participation to service session is being handled only on the Ret access and not on the Ret usage. The reason behind this is twofold. First we have made the design decision that the « delete » operations need to follow the same route as the « create » operations. Since a Consumer's participation to a service session is always initiated on an access session of that Consumer, the termination of that participation is also only available on the access session. Second, the request which results in the deletion of an implementation object should never be invoked on an IDL interface attached to that object, because it causes implementation difficulties. The actual provisioning of service-specific behaviour is considered to be out of scope of TINA and is therefore completely VITAL-specific material.

5.4 Scenario

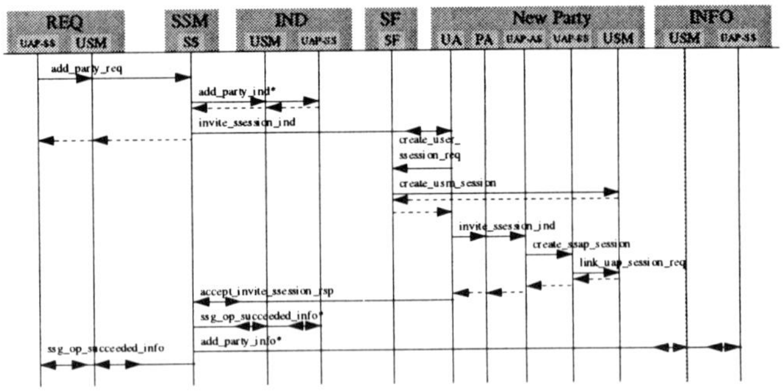

Figure 13: Add a Party to a Service Session

Figure 13 illustrates the behaviour of the service session subsystem when a Consumer requests the addition of a new party to a service session.

6. The Subscription Management Subsystem

6.1 The Computational and Deployment Model

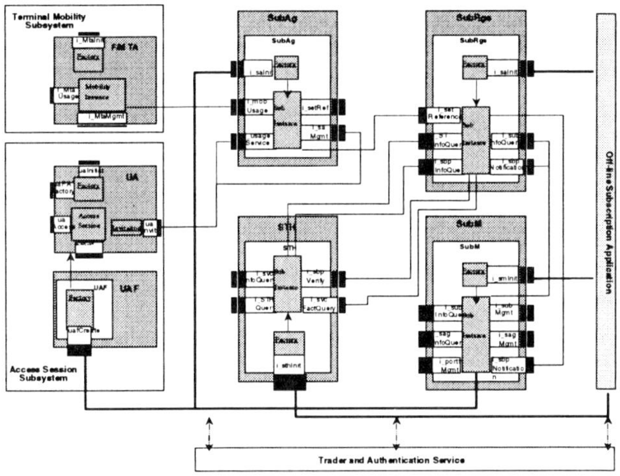

Figure 14: Subscription management computational model

The *Subscription Registrar* (SubRgs) maintains the *Subscription contract*, *Subscription* and *Service profile* objects for one single service.

The *Subscription Agent* (SubAg) keeps track of the services one particular user is subscribed to subscriber and uses this information to obtain service subscription information from a particular SubRgs. User Agents interact with this component to a list of available services and subscription information for a specific user. There are as many SubAg as User Agent instances.

The *Service Template Handler* (STH) handles one or more *Service Templates* that represent service capabilities provided by the service provider.

The *Subscriber Manager* (SubM) is responsible for managing the information of *Subscribers*, *Subscription Portfolios* and *Subscription Assignment Groups*. It is independent of any particular service.

6.2 The VITAL Ret Usage Part

In addition to a Retailer subscription application, VITAL also implements a subscription ancillary service which allows access from the Consumer domain to the subscription management information, according to the user role:

- *End-user*. Control over the user life-cycle in the customer domain, customisation of the service profile
- *Subscriber*: Control over the subscriber life-cycle in the customer domain, customisation of service attributes and subscription data, on-line subscription to new services.
- *Retailer*: On-line subscription management in the retailer domain, i.e. the same functionality as for the off-line subscription mechanism.

The subscription ancillary service capabilities are offered as *optional* operations on the Ret usage part, i.e. compliance to the TINA Ret doesn't require support of these operations.

7. Conclusion

The VITAL demonstrator consists of a huge amount of software which has been designed and implemented by a large team of engineers at geographically distinct locations. The adoption of object orientation and distributed processing as key technologies/methodologies as promoted by TINA - i.e. the ODP (Open Distributed Processing) framework for system specification/design and CORBA for component deployment - has proven to be extra beneficial in such an environment.

The TINA SA4.0 turned out to be a very useful starting point to deliver an implementation of a TINA system. Based on this material, and the experience gained in producing an early prototype of the TINA SA [5], the VITAL project has produced a detailed design and at the same time used this experience to contribute to the definition of the Retailer reference point specification. VITAL has significantly contributed to the definition of the usage part of the Ret RP spec. The remaining discrepancies between the final Ret spec and the VITAL implementations will be resolved in a next phase of the project. Furthermore, the impact of service federation and composition on the current system will be evaluated. At the time of writing this paper, we believe that this impact may be limited to the addition of new interfaces, without having to modify existing interfaces considerably. However, more study is required to be able to confirm this. Finally, it is also foreseen to introduce further service management capabilities, with an emphasis on accounting and service life cycle management.

Acknowledgements

Part of this work has been done in the VITAL (AC003) project, a research project partly funded by the European Commission within the ACTS framework.

References

[1] Service Architecture, Version 4.0, Document no. TB_RM.001_4.0_96, TINA-C, October 1996.

[2] Ret Reference Point Specifications, TINA-C, August 1997

[3] Deployment of TINA Technology on a Pan European ATM network, H. Vanderstraeten et al, ISS '97

[4] TINA Business Model and Reference Points, Version 3.1, Document no. TB_RCJ.03.0_3.1_96, TINA-C, June 1996.

[5] The ALCIN Project: TINA Validation Through Prototyping, P. Hellemans et al, TINA '96, September '96, Heidelberg

Data-Driven Implementation of TINA kernel Transport Network

Hiroaki NISHIKAWA
Institute of Information Sciences and Electronics, University of Tsukuba
Tsukuba Science City, Ibaraki 305 JAPAN
nisikawa@is.tsukuba.ac.jp

Souichi MIYATA, Shin'ichi YOSHIDA, Tsuyoshi MURAMATSU
Multimedia Processor Business Project Team, Tenri IC Development Group, Sharp Corporation
2613-1 Ichinomoto-cho, Tenri-shi, Nara 632 JAPAN
smiyata@icg.sharp.co.jp, shin@icg.sharp.co.jp, muramatu@icg.sharp.co.jp

Hiroshi ISHII, Yuji INOUE, Ken'ichi KITAMI
NTT Multimedia Networks Laboratories
3-9-11 Midori-cho, Musashino-shi Tokyo 180 JAPAN
ishii.hiroshi@na.tnl.ntt.co.jp, yuji@wins.hqs.ntt.co.jp, kitami@magnet.netlab.ntt.co.jp

Abstract

To realize the actual TINA-based telecommunications network, the performance of kernel Transport Network (kTN) such as availability, reliability, throughput and load tolerance becomes more crucial than for existing computer networks. The authors have been studying and developing a super-integrated data-driven processor to be applied to the TINA kTN nodes and network interfaces in CUE (Coordinating Users' requirements and Engineering constraints) project. Since the processor is primarily designed to be a scalable VLSI component, it is easily interconnected with each other to form a super-integrated chip and multi-chip system for achieving the performance and reliability demanded in a TINA environment. In this paper we first examine the requirements for kTN. A stream-oriented data-driven architecture is then proposed with special emphasis on effective multi-processing capability with overload tolerance. After that, we demonstrate that autonomous load balancing among super-integrated data-driven processors without adding any runtime overhead to achieve effective and reliable multi-processing is possible by utilizing the overload tolerance of the processor. Finally, this paper shows preliminary performance estimations of the super-integrated data-driven processor currently being developed to perform efficient multi-processing in protocol handling such as TCP/IP.

keywords: TINA kTN, Data-Driven Processor,pipeline processing, multi-processing

1. Introduction

Although TINA specification can be defined independently of the underlying hardware and Operating Systems (NCCEs: Native Computing and Communication Environment), the processing capabilities of the NCCEs are important when the actual network is managed and controlled by TINA applications [1]-[3]. Inter-object message passing may often be executed between physically remote nodes so that high capabilities of processing and message transfer are required to NCCEs and kTN (kernel Transport Network) connecting NCCE of each physical TINA node. Several TINA nodes in which special DPE (Distributed Processing Environment) objects such as Trading function of CORBA Object Services reside may become overloaded and requires real time processing in NCCEs. Hence, the availability, reliability, processing capability of the NCCEs and kTN are very important features of NCCEs and kTN.

Most recent computing platforms operate on Von Neumann style computers. Most problems potentially caused by a lack of processing capability can be well covered by drastic improvement in CPU power and by progress in VLSI technology. It is, however, questionable whether these improvements will be fully reflected in actual system performance. For example, an improvement in the processing power of a processor may not be linearly reflected in total system performance, especially in multi-processing capability. This

0-8186-8335-X/98 $10.00 © 1998 IEEE

is mainly due to runtime overhead such as caused by context switching. A data-driven execution scheme, especially a dynamic one, is the most natural way to achieve parallel execution including concurrency, pipelining, and multi-processing. These characteristics of data-driven processing scheme will well cover the requirements to NCCEs and kTN raised above and the applicability of a data-driven processor to TINA environment seems growing [4]-[9].

In this paper, we first examine the requirements for the NCCE and kTN and show a defect of current processing scheme by measuring performance of an actual TINA system demonstrated in Telecom '95 as an example. A stream-oriented data-driven processor with overload tolerance is then proposed to be used as a part of NCCE which interfaces the kTN. After that, we will demonstrate that autonomous load balancing among data-driven processors without adding any runtime overhead is possible due to the elastic nature of the processor. Finally, this paper shows effectiveness of the super-integrated data-driven processors to achieve efficient multi-processing in protocol handling such as TCP/IP [10]-[12].

2. Necessity of data-driven implementation in TINA environment

2.1. Requirements for implementation of TINA NCCE and kTN

TINA architecture consists of: 1) TINA application that provides versatile telecom service and controls/manages transport network, 2) DPE, Distributed Processing Environment and 3) NCCE, Native Computing and Communication Environment.

The application objects can communicate with each other via DPE without considering physical locations. The kernel Transport Network (kTN) interconnects each implementation of NCCE of physically remote node accommodating DPE and applications over it. As shown in Figure 1, the kTN transports all messages and data (called operational data) that are exchanged among applications over DPE. The messages and data will cover, for example, connection control message from TINA control node to switching node, management messages issued from TINA management node to the transport network or exchanged between TINA nodes, and service session control message between user and transport. Since the NCCEs and the kTN play very important role to process and transfer the messages to control and manage transport network, they are indispensable in achieving high reliability.

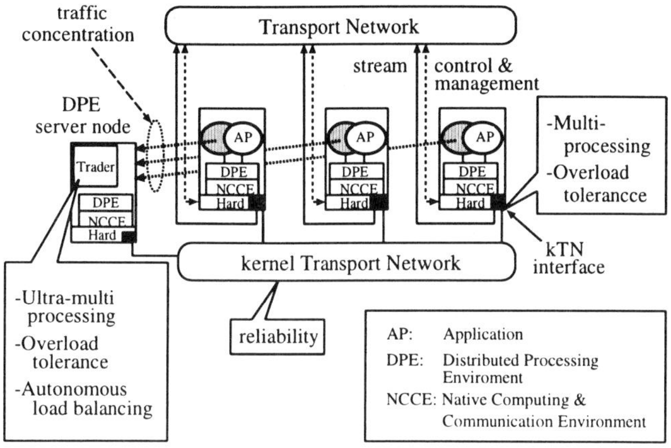

Figure 1. TINA Based Network

Specific requirements to implementation of NCCE and kTN are:

(1) High-speed signal processing capability
To process and transfer broadband and high-speed multimedia stream, effective and high-speed processing capability of multimedia signal in nodes (NCCE implementation) and high-speed transfer capability of kTN are required.

(2) Overload tolerance
Some of the nodes such as ones accommodating Object Service of DPE will suffer from traffic concentration. Hence overload tolerant processing capability for node is highly required.

(3) Effective multi-processing
The kTN interface will have to handle multiple sessions and connections simultaneously. And also DPE server nodes must handle multiple enquiries at the same time. So, multi-processing in the protocol processing in kTN interface and node processing accommodating DPE servers are required.

(4) Fault tolerance
Since the kTN transfers all the messages for network management and service control, the reliability of the kTN is very important requirement.

(5) Fault management
To maintain availability of TINA network even in case of fault occurrence, autonomous fault management functionality is required.

2.2. Experimental evaluations on kTN interface

To show ineffectiveness of the current processor, we have experimentally measured an actual TINA network performance based on current technologies. Telecommunication nodes must in nature process multiple calls, connections and messages simultaneously. In addition to call and protocol processing, management processing such as diagnosis and failure restoration must be simultaneously executed hopefully independent of normal processing. In other words, same kind of multiple contexts and multiple kinds of contexts must be handled simultaneously in the telecommunication node.

Especially, the TINA nodes accommodating TINA-DPE servers such as trader and name server will have many simultaneous contexts, because many clients will access those servers. Unfortunately in the current computing architecture, increasing the processing capability will not solve the problem but increase inefficiency. It is because as the number of contexts grows, the overhead of context switches also increases. To show the problems of multi-processing in kTN interface and NCCE implementation of DPE server nodes, we have measured several kinds of processing time of CORBA (Common Object Request Broker Architecture)-based client-server system on the ordinary workstations (Von Neumann processors), which was actually demonstrated in Telecom '95 in Geneva as TINA World Wide Demonstration system. The measured items are relation between Processing time in server node and data transfer time of kTN versus number of threads. The system configuration and measured items are shown in Figure 2.

(1) System Configuration
Two workstations accommodating CORBA(Common Object Request Broker Architecture), one of which is a client and the other is server, are interconnected through Ether network.

(2) Experiment condition
The client generates threads and each thread sets up a TCP connection. On each TCP connection, the client requests process execution to the server. At every moment when a request arrives at the server from each client thread, server generates a thread. Several kind of processing time per thread were measured 10 times by changing number of threads from 1 to 45 for the two cases: one is lighter processing (CPU time less than 1 msec for several byte parameter) and the other is heavier processing (CPU time is about 10 msec for about 100 byte parameter). Figure 3 shows the average time of 10 times measurements.

(3) Measured items
Actually, following five kinds of time was measured;

- T1: Total turn around time
- T2: TCP Connection setup time
- T3: time from start of data transfer to arrival of acknowledgment at the client side
- T4: Time from arrival of request to sending of acknowledgment at the server side
- T5: Process execution time at the server From these measured time, following time were estimated.
- Total turn around time: T1
- Marshalling (C) (Data transformation time from application data to ORB data at the Client side): T1-(T2+T3)
- TCP connection setup: T2
- Data transfer (TCP/IP data transfer time on Ether): T3-T4
- Marshalling (S) (Data transformation time from application data ORB data at the Server side): T4-T5
- Server processing (Application process execution time at server): T5

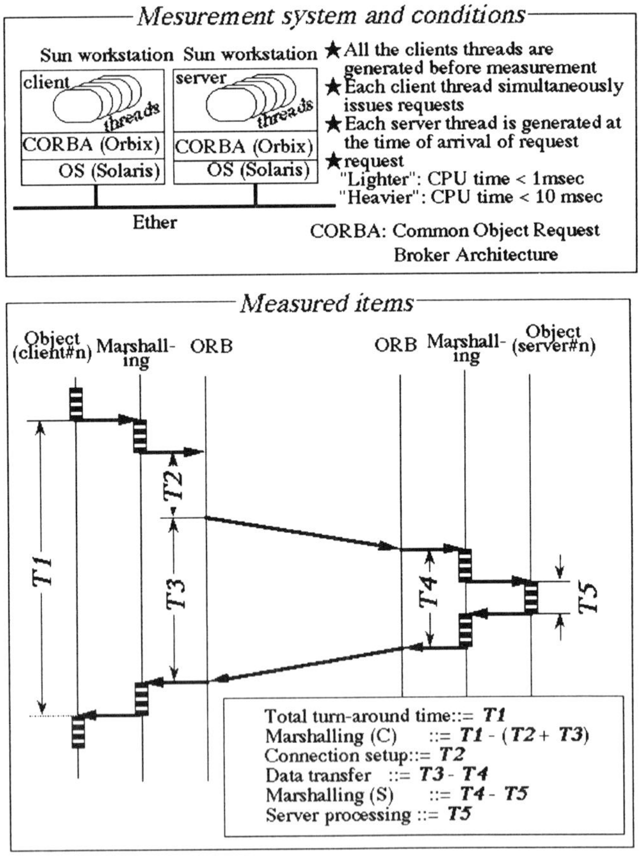

Figure 2. System Configuration and measured items in experiment of Von Neumann process execution

Figure 3 (a) and (b) show measured results for the cases lighter processing and heavier processing, respectively. Server application processing time and Marshalling (S) do not depend on the number of threads in case of lighter processing but increase in case of heavier processing. TCP connecting setup and data transfer time increase in proportional to the number of threads not dependent on either lighter or heavier processing. The reasons are considered as follows:

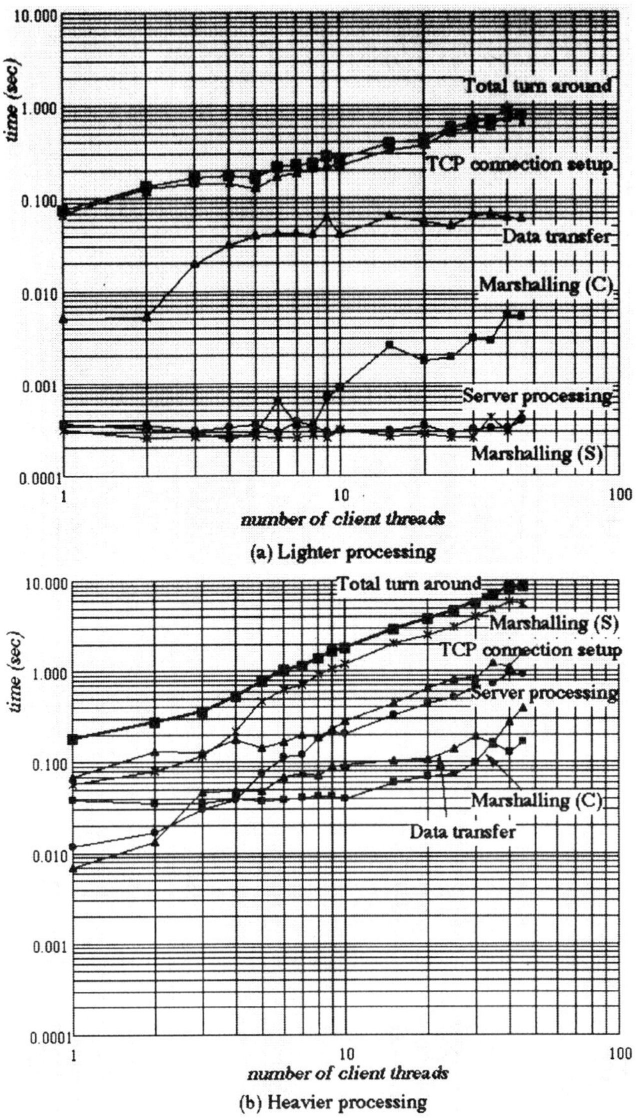

Figure 3. Experimental results of Von Neumann process execution

(1) Processing at the server
The tendency of increase of application processing time and marshalling time in heavier processing comes from following reason: in case of heavier processing, according to increase of number of threads, the probability of intervention to thread processing becomes high to achieve multiprocessing and overhead of thread exchange and process execution waiting time increases. On the other hand, in case of lighter processing, since process execution time is short and intervention probability is low, the tendency does not appear. In the TINA environment, it must be considered that processing load on the DPE becomes high and load is concentrated to the specific server. Hence, it seems very effective to apply data-driven processor to the environment because the data-driven processor has in nature no overhead for thread exchange that is seen in Von Neumann processor. Moreover, overloaded condition caused by instantaneous load increase must be taken into account in real implementation. For this situation, the stream-oriented data-driven processor based on elastic pipeline processing that is overload tolerant is highly applicable [8].
(2) Protocol Processing at the kTN interface
Figure 3 (a) and (b) show that connection setup time and transfer time increase in proportional to number of threads not dependent on the processing load (lighter or heavier processing that correspond to length of parameter, i.e. data length to be processed). And both setup and transfer time in Figure 3 (a) are almost same as those in Figure 3 (b) for the same number of threads. It is trivial that connection setup time is independent of data length. On the other hand, data transfer time is to be dependent to data length. However, since Ether network used in this experiment has several Mb/s transfer capability and the MAC packet size is 1.5 kbyte, data from the client is accommodated in a single MAC packet and the data length (several or 100 bytes) is not dominant for data propagation time. Hence, transfer time is influenced by data length but packet header processing for packet assembly and protocol processing at sending side, MAC header analysis at transit node, and packet disassembly and protocol processing at receiving side. Therefore, the reason TCP connection setup and data transfer time increase comes from that thread exchange overhead increases in multiprocessing. From this perspective, stream-oriented data-driven processor is highly applicable to kTN protocol processing.

Through above discussion, pseudo multiprocessing as extension of sequential processing in Von Neumann processor must suffer from overhead in nature. In other words, substantial problem is interference among threads to be multiprocessed. On the other hand, in data-driven processor, each process is assured to be independent from other processes. This characteristics means interference never occurs in case of not only simple multiprocessing but parallel processing of threads each of which is dependent. Figure 4 shows process execution time for each thread in case of multi-thread execution by using data-driven processor. This figure shows that any overhead caused by inter-thread interference does not occur independent of both thread execution time (load) and number of threads. Authors, based on this result, are planning to apply the stream-oriented data-driven processors into kTN interface and large scale server node processor as one of the TINA implementation activities driven by authors, which are called TTT (The TINA Trial) systems. In this case here, considering the fact that we have many existing systems, we are planning to apply the processor to applicable parts by keeping legacy interface to facilitate introduction.

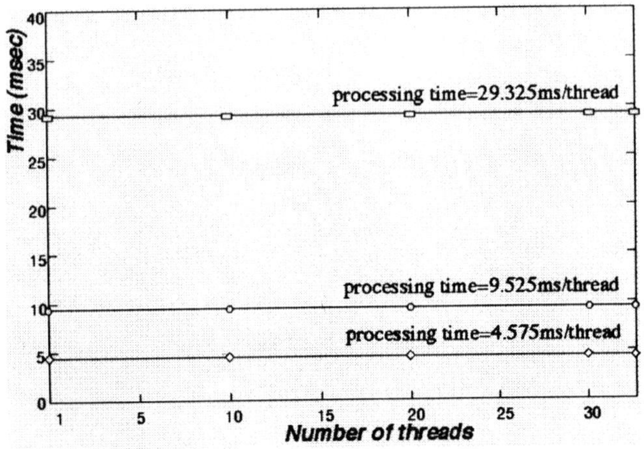

Figure 4. Processing time of stream-oriented data-driven processor

3. Super-integrated data-driven processor applicable to TINA environment

3.1. Stream-oriented data-driven architecture

(a) Requirements to process high flow-rate stream.

In order to build an information system environment based on the TINA which can fully exploit the communication capacity of multimedia services including not only voice, but also still and moving pictures, performance requirement to receive and perform necessary processing on data stream exceeding 150Mb/s becomes necessary. An introduction of pipeline processing scheme cannot be avoided even when considering the remarkable progress in VLSI technology since with the pipeline scheme, receptability of data volume can be enlarged by subdividing stages.

In order for this pipeline scheme to ideally operate, the following requirements must be satisfied simultaneously:

- Pipeline must not be starved, i.e. data flow must be sufficiently supplied to pipeline input.
- Data in the pipeline must not be flushed, and data flow volume must be maintained.
- Momentary fluctuations of the input to the pipeline must not degrade the pipeline throughput.
- Momentary fluctuations in the pipeline must not obstruct input to the pipeline.

As long as constant data flow is supplied, processing delay is negligible in the data-driven scheme which is an excellent feature that maximizes the pipelining efficiency. Also by the data-driven firing rule, processing is always in a forward direction, never requiring flushing. Furthermore by employing dynamic data-driven scheme, streams are realized in a parallel processable format as tagged token queues which can be effective in avoiding starvation.

Accordingly, the key to realization of an effective stream processing based on the dynamic data-driven scheme lies in how momentary data flow fluctuations to and in the pipeline is absorbed to maintain efficient data flow.

(b) "Flow-Thru processing" scheme.

As it is relevant from the firing rule's parallel processing which adheres to the data-driven principle, functional processes or memory accesses which involve history-sensitive processes proceeds by sending and receiving input/output tokens in conjunction with data-driven firing control mechanism. In general, data-driven execution of concurrent processes have data fluctuations due to partial ordering among processes. Therefore, in order to maintain an average data flow rate, there needs to be some buffering mechanism between mutually connected functions. Furthermore, in order to maximize the advantage of the pipeline processing scheme, this buffering mechanism must be realized so as to avoid bottlenecks in processing functions in distribution of load and function.

The authors, in order to satisfy these requirements, focused on the following issues:

- Realizing a comprehensive autonomously controlled elastic pipeline structure for the entire data path necessary for data-driven execution which can then effectively buffer fluctuations in data flow

Further, in order to locally and autonomously control these buffering effects,

- Enabling functions to the branch/ junction/ routing mechanisms so that the data can autonomously travel the appointed path

With these fundamental concepts in mind, the stream-oriented data-driven processor with "flow-thru processing" scheme has been developed.

(c) Super-integrated data-driven architecture.

Figure 5 shows the throughput for the circular elastic pipeline with respect to the pipeline occupancy. The elastic nature of our data-driven architecture is essential to flexibly form a modular multi-processor system that performs well and reliably in the TINA environment. In particular, the processors tolerated a considerable amount of instantaneous overloading due to their inherent flow-fluctuation absorption capabilities distributively located throughout the system, and were thus able to accept high-flow-rate data streams with considerable fluctuation. However, if such an overload continues, it would inevitably result in an overflow. Our data-driven processor must therefore provide a sizable amount of buffering capability for token-sequences, or streams, which contain various-length messages. Therefore, autonomously managing this overload through the use of data storage will result in the highly parallel execution essential to directly achieving a stream-oriented processing scheme. Experimental studies using the emulation/simulation facility have shown that history-sensitive data are easily prefetched. Furthermore, experiments have also shown that the average amount of active data at runtime is much smaller than the amount of stored history-sensitive data. Fortunately, in the data-driven processor, latency problems associated with memory operation can be tolerated due to the dynamic mode of the data-driven execution as long as a data-driven firing principle, such as split-phase transactions, are performed[9].

Consequently, a method must be implemented to effectively augment memory capability when a high volume of data must be handled by utilizing the above-mentioned features to achieve a well-balanced data-flow in the stream-oriented data-driven processor. Figure 6 schematically illustrates the basic execution structure of the super-integrated data-driven processor currently being studied in our project. As shown in Figure 6(b), the circular pipeline consisting of the Firing Control & Program Memory (FCP) and Functional Processing Unit (FP) is the dynamic part of the processor in which active data and instructions are processed and forwarded through the pipeline in the form of packets. The Data Storage (DS) section buffers incoming/outgoing messages from/to application objects. That is, the DS performs an elastic buffering function to manage instantaneous overflows during continuous overloading. However, since stream-oriented processing incorporating memory operations might amplify token-flow fluctuations, unique variable length queue buffers (QB1, QB2, and QB3) are used to enhance the elastic nature of the circular pipeline. That is, if the packet-flow through the downward or output line is choked by such disturbances as short-cut transfers through the bypassing paths in the lower part of the downward line, the top-most packet in the upward line continues its upward movement until it meets an empty downward stage whereupon the packet is diverted shown in Figure 6(c). Therefore, the structure functions as a variable-length pipeline, enhancing the elastic nature of the overall system.

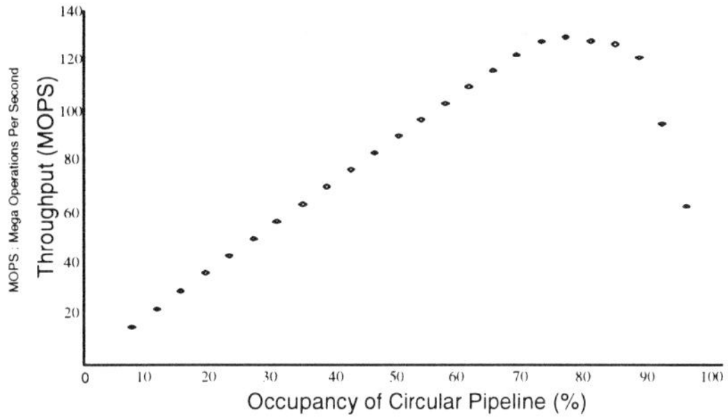

Figure 5. Throughput in Circular Pipeline

3.2. Autonomous load-balancing in multiprocessing

Figure 6(a) shows an experimental environment for verifying the super-integrated data-driven processor; it uses a data-driven processor simulator. The environment consists of five data-driven processors: one for shared memory and the others for stream-oriented processing. The processors coordinate their activities through split-phase transactions in the shared memory. Typical packet behavior in the circular pipelines are shown in Figure 7. The horizontal axis shows processing time normalized by an elastic pipeline tact. The vertical axis shows the packet count in the circular pipeline, consisting of QB1, FCP, and FP in Figure 6(b). As was easily understood, the data flow is well balanced among the four processors. The experimental result showed that the elastic nature of the QB is effectively employed to absorb the token-flow fluctuations and smoothed out uneven data flow among the processors. The number of tokens in the pipeline is higher than the minimum number of pipeline stages (32 including 2 stages in the QB). This means most pipeline stalls are completely avoided. Thus, the stream-oriented data-driven processor can achieve autonomous load balancing and overload tolerance without adding any run-time overheads, which are essential for realizing effective super-integrated data-driven processor.

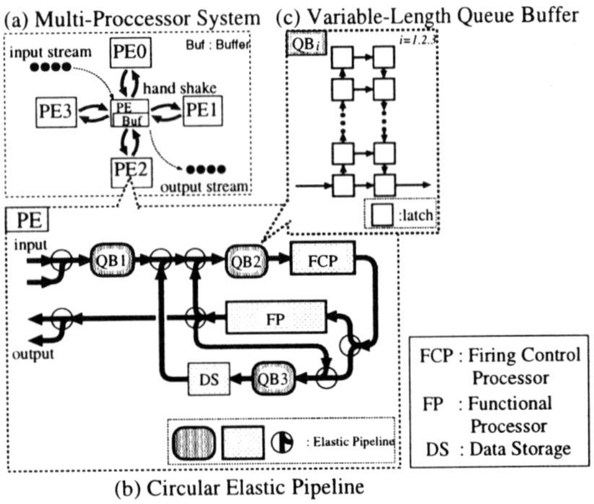

Figure 6. Stream-Oriented Data-Driven Architecture

One of the general ways to perform fault surveillance of systems is to send a test message to the system and survey the turn around time. Suppose the situation that there are two processors, one of which sends test message to the other to perform surveillance. To realize this test, the tester must multiplex testing and surveillance job and ordinary jobs, which will degrade the system throughput in case Von Neumann type processor is used. But the data-driven processor will keep ordinary throughput even when test process is executed by virtue of multiple processing capability without any overhead.

This virtue will be applicable to fault detection mechanism to realize high reliability of TINA NCCEs and kTN.

Figure 7 shows turn around time of a message, driving a sample test program which is executed in parallel with existing ordinary processes, fluctuates but keeps around minimum flow-through time.

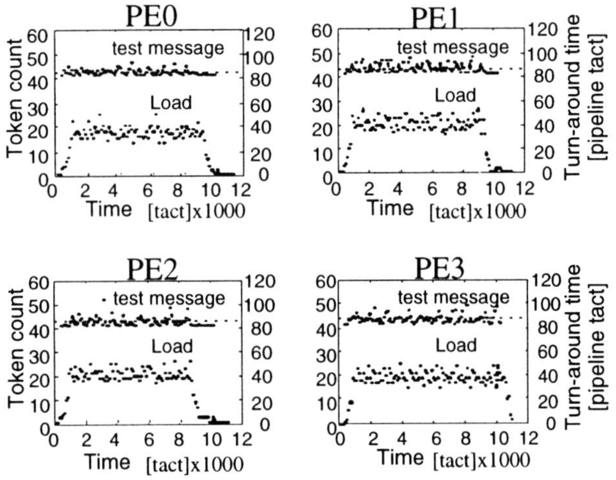

Figure 7. Autonomous Load Balancing and Turnaround Time of Test Packets

3.3. Preliminary performance estimation

As was discussed in Sec.2, efficient multi-processing in kTN interface is one of the most crucial issues for realizing TINA environment.

Currently, our project is studying super-integrated data-driven processor to implement TCP/IP over ATM using super-integrated data-driven processor system which has already been developed and demonstrated high-definition video signal processing in real-time. Prior to full implementation of data-driven TCP/IP processor, we have compared multiprocessing capabilities of Von Neumann and our data-driven processors by implementing IP packet transmission part.

To simulate Von Neumann processor, we generated sequential version of IP packet transmission part, in which only pipeline processing of packets or stream are realized using buffering memory and sequential accesses to the memory. Furthermore, the sequential version adopted iteration construct for stream processing. There are many control operations necessary to the iteration. As a result, multi-processing beyond 80Mb/s input stream rate is impossible as shown in Figure 8.

Considering the above results, we generate concurrent version of IP packet transmission part in which not only pipeline but also concurrent processing of the stream are realized without buffering memory. Furthermore, the iteration constructs are transformed into unfolded concurrent process. There are therefore no control operations, as a result, number of operations are drastically decreased. As shown in Figure 8, multi-processing without any overheads are possible upto 180Mb/s input stream rate using a single super-integrated data-driven processor.

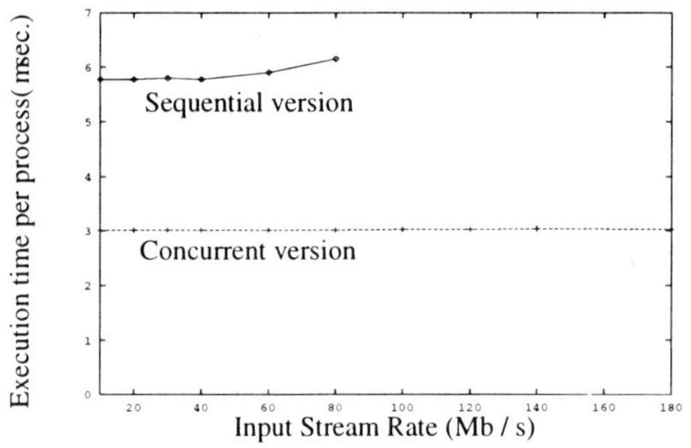

Figure 8. Preliminary performance estimation of TCP/IP

Through this study, it is shown the concurrent version taking advantage of data-driven scheme is more suitable for TCP/IP processing. Hence we start full implementation of TCP/IP as shown in Figure 9.

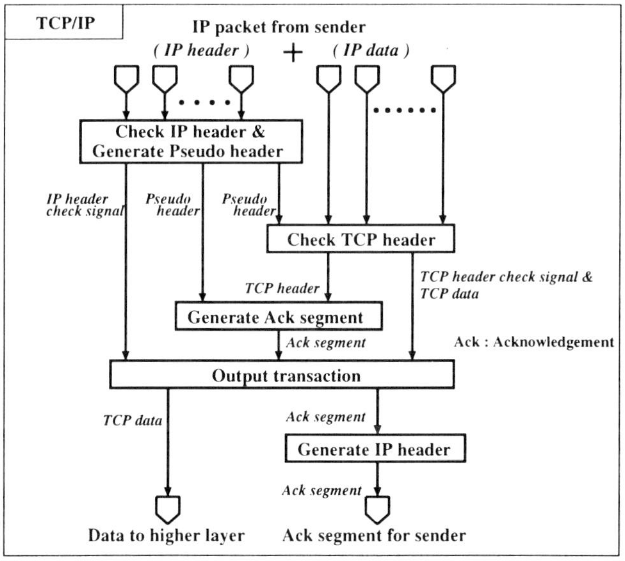

Figure 9. Data-Driven Implementation of TCP/IP

In this implementation, execution time per process keeps constant as shown in Figure 10. This demonstrates that efficient multi-processing is possible in data-driven implementation of TCP/IP.

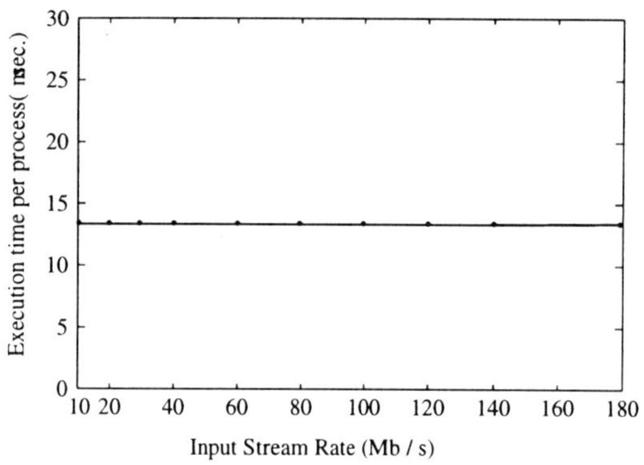

Figure 10. Execution Time in Data-Driven Implementation of TCP/IP

4. Conclusion

In this paper, we described applicability of the data-driven processor to TINA kTN. We also give a brief discussion of target performance of the super-integrated data-driven processor. Actual packet transfer rate of each data driven processor cores will achieve around 200 Mega packets per second by fabrication improvement. Eight cores will be super-integrated into a single chip. The resulting super-integrated data-driven processor can achieve 1.6 Giga packets per second transfer rate. Taking into consideration that, on average, execution of one operation in this data-driven processor equals to approximately 1.5 primitive operations, it is expected that the processor performance will lead to effective performance of 2.4 Giga Operations Per Second (2.4 GOPS). This performance is more than three times that of current super-integrated data-driven processor. The authors are hopeful that the new single chip integration will realize TCP/IP over ATM as a data-driven implementation of kTN interface.

Acknowledgments

Although it is impossible to give credit individually to all those who organized and supported the CUE project, the authors would like to express their sincere appreciation to all the colleagues in the project. The authors would also like to give thanks to R.T.Shichiku for his efficient proof-reading in preparation of this paper.

This research is partially supported by the Scientific Grant of the Japanese Ministry of Education, the Telecommunication Advancement Foundation and International Communications Foundation.

References

[1] G. Nilsson, F. Dupuy, M. Chapman, "An Overview of the Telecommunications Information Networking Architecture," TINA95, Melbourne (Feb. 1995).

[2] E. Kelly, N. Mercouroff, P. Graubmann, "TINA-C DPE Architecture and Tools", TINA95, Melbourne (Feb. 1995).

[3] W.Takita, Y.Yasutake, S.Chikara, and O.Miyagishi, " A distributed processing environment architecture for TINA world wide demonstration," IEICE Technical Report, (Nov. 1995).

[4] H. Nishikawa, H. Terada, K. Komatsu, S. Yoshida, T. Okamoto, Y. Tsuji, S. Takakura, T. Tokura, Y. Nishikawa, S. Hara, and M. Meichi, "Architecture of a one-chip data-driven processor: Q-p," Proc. of the 16th ICPP, pp.319–326 (Aug. 1987).

[5] H. Nishikawa, H. Ishii, and Y. Inoue, "A Stream-Oriented Data-Driven Processor Realizing Hyper-Distributed Systems", Proc. of the 8th IASTED Int'l Conf., pp.47–51 (Oct. 1996).

[6] H. Nishikawa, S. Miyata, S. Yoshida, T. Muramatsu, H. Ishii, H. Kobayashi and Y. Inoue, "A Data-Driven Implementation of Telecommunication Network Systems", ISADS, pp.51-58, (Apr.1997).

[7] S. Yoshida, R. T. Shichiku, Y. Matsuura, T. Muramatsu, T. Okamoto, S. Miyata, "Video Signal Processing Oriented Data-Driven Processor", Tech. Report of IEICE, ICD95-158, pp.39–46 (Nov. 1995) [In Japanese].

[8] H. Ishii, H. Nishikawa, Y. Inoue, "Data-Driven Fault Management for TINA Applications," IEICE Trans, Commun,(Jun.1997).

[9] Arvind, R. S. Nikhil, "Executing a Program on the MIT Tagged-Token Dataflow Architecture", IEEE, Trans. on Computers, Vol.39, No.3, pp.300-318 (Mar. 1990).

[10] S J. Postel, "Internet Protocol," IETF RFC 791,(Sep.1981).

[11] S J. Postel, "Internet Control Message Protocol", IETF RFC 792,(Sep.1981).

[12] S J. Postel, "Transmission Control Protocol", IETF RFC 793,(Sep.1981).

The Ret Reference Point Definition and a Prototype Implementation

Patrick Farley
BT
pads@tinac.com

Richard Westerga
KPN Research
richard@tinac.com

Abstract

This paper provides an outline of the TINA Ret Reference Point (Ret-RP) [1], a specification of the business relationships between a consumer and a retailer. It describes the functionalities of the access and usage parts defined in this reference point. Explanations are given on the concepts of session models and feature sets together with suggestions on how retailers can extend the specifications to incorporate their own capabilities, e.g. different session models or interfaces. Furthermore the paper describes details of a prototype implementation which was developed to validate, and show the applicability of the Ret-RP specification.

1. Introduction

The Ret-RP specification document has been produced in response to a Request for Refinements [3], issued by the TINA Consortium in August 96. At that time, a number of responses to the Request were produced by TINA-C partners, and an evaluation team set up to combine the responses into a definition of Ret-RP which could be used as an interoperable inter-domain reference point between a TINA consumer and retailer.

Figure 1 shows the scope of the Ret-RP with respect to a set of business roles and business relationships, defined in [4].

The **Consumer** business role uses services provided in a TINA system. Consumers use services provided by the retailer business role.

The **Retailer** business role serves stakeholders in the consumer business role, by providing them with access to services. A retailer may use other providers to support the provision of services to consumers.

The **Third party service provider** aims to support retailers or other third party providers with services. This can be regarded as "wholesale" of services, as the third party service provider does not have a contractual relationship with stakeholders in the consumer business role.

The Ret reference point is used between stakeholders in the consumer business role and the retailer business role. It is used to support the consumer's needs for access to services offered by the retailer.

The Ret-RP specification is in the form of non-formal specifications, in terms of plain text and diagrams, and of semi-formal specifications, using the OMG IDL language [5]. Its purpose is to allow systems in the consumer and retailer domains to interoperate, using a well-defined set of computational interfaces. As such, the specification is a candidate to be the basis for conformance evaluation of TINA-oriented products, for the Ret reference point.

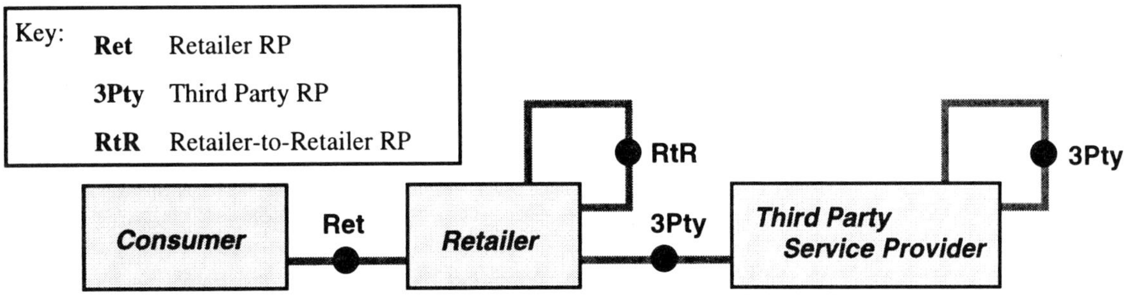

Figure 1. Scope of the Ret-RP.

0-8186-8335-X/98 $10.00 © 1998 IEEE

2. The Ret Reference Point

The Ret-RP is separated into an *Access* part and a *Usage* part. For Ret-RP, the access part describes how a consumer accesses a retailer to make use of services it provides. The usage part describes the interactions between the consumer and retailer domains during the use of a service.

The access part of Ret-RP offers the following capabilities:

- initiation of dialogue between the consumer and retailer domains,
- identification of the domains to each other (either domain can remain anonymous dependent on the interaction requested),
- establishment of a secure association between the domains, (an access session),
- set up of the default context for the control and management of usage functionality (service sessions),
- discovery of service[1] offerings,
- listings of access sessions, service sessions and subscribed services
- initiation of usage between the domains, (starting a service session),
- control and management of service sessions (e.g., stop, suspend, resume, join, notify changes, etc.).

The access part largely addresses the establishment, and use of a secure association between the domains, termed an Access Session [2].

The usage part of Ret-RP offers the following capabilities:

- control and management of service sessions (e.g., announce, stop, suspend, invite, notify changes, negotiate parties' control rights, etc.),
- control and management of stream flow bindings,
- domain management (e.g., subscriber management, service management) through management services, that are supported consistently with user services.

The usage part addresses the interactions between the domains as part of a service session [2]. A Service Session encompasses the users (parties) of a service, that are sharing the service in some common goal. (E.g. users talking to each other in a video conference are all parties involved in the same video conference service session. Other users may also be using the same video conference service, but be involved in separate video conference service sessions.)

Conformance to Ret-RP is separate for the access and usage parts. This means that a system can claim conformance to the access part of Ret-RP only, to the usage part of Ret-RP only, or to both parts of Ret-RP. The main reason for this is to allow retailers to offer 'legacy' services, through the access part, which do not conform to the usage part of Ret-RP, thereby gaining benefits of consistency of access and management for all their services.

Both access and usage parts define mandatory and optional features. The access part defines a profile of mandatory interfaces that must be supported by each domain, and optional interfaces. The usage part defines a set of profiles in terms of session models and feature sets. TINA defines a single session model comprising a number of feature sets. Each feature set provides some facet of session control, and defines interfaces to make this accessible. They can be combined to provide the specific functionality required by the service. More details on these are given in the relevant sections below.

3. Access Part of Ret-RP

The access part defines how a consumer accesses a retailer to make use of services it provides. In order to allow this, the access part defines interfaces supported by the consumer and retailer domains.

The process of accessing the retailer can be divided into 3 phases:

- **Initial Contact:** The consumer wishes to contact a retailer. He initiates contact with an interface of the retailer, and both domains identify and authenticate themselves. This phase ends with the establishment of a secure association (an access session) between the

1. These services can be primary (e.g. Video on Demand (VoD)), ancillary to the primary (e.g. configuration management for VoD) or administrative (e.g., subscriber management for VoD). See [2] for definitions.

domains. The interfaces for this phase are offered only by the retailer and are always available outside of an access session.

- **Access Session:** Once an access session is established, the domains exchange information and interface references to define the 'context' for the access session. The consumer can now use retailer facilities, such as discovering services; and starting and managing service sessions. The interface references exchanged are available for use within this access session only. After the access session ends, the interfaces cannot be used.
- **Contact outside an Access Session:** The consumer may wish to allow the retailer to contact them when they don't have an access session established. The consumer registers some of his interfaces for use outside of an access session. This allows the retailer to forward invitations to join service sessions; information about access and service sessions the consumer is involved in; and request the establishment of an access session.

The interfaces are grouped according to these phases. Figure 2 shows the interfaces defined for the access part of Ret-RP. The interfaces exported over Ret-RP are prefixed by i_Consumer, or i_Retailer, to denote the domain which supports them.

3.1 Example Scenario of Access Part of Ret-RP

This section is an example of the use of the access Ret-RP interfaces. It describes a consumer making use of retailer interfaces to establish an access session. The consumer can now use the retailer facilities, and register to receive invitations outside of an access session.

Initial contact:

1. A consumer domain contacts the retailer by gaining a reference to a **`i_RetailerInitial`** interface[2].
2. The consumer wishes to establish an access session with the retailer as a known user, so invokes the **`requestNamedAccess`**() operation on **`i_RetailerInitial`**. (The consumer can alternatively remain anonymous).

2. The mechanism by which the consumer gains this interface is not prescribed by Ret-RP. The mechanism used in our prototype is described in Section 7.

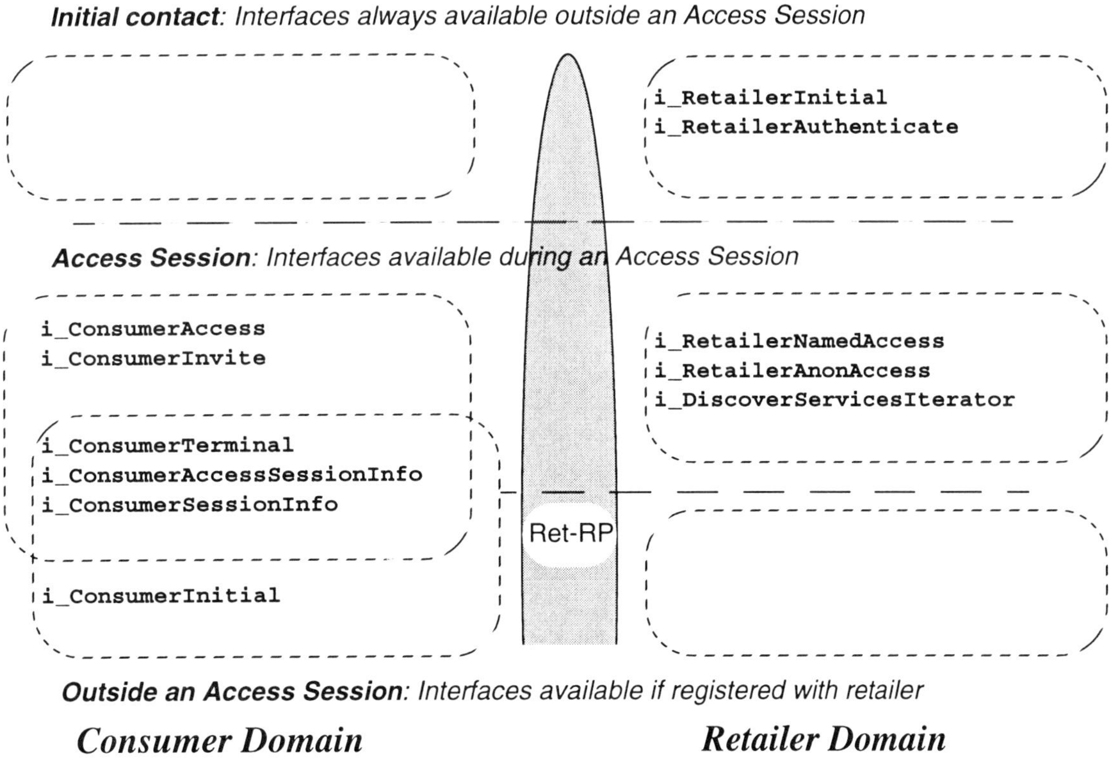

Figure 2. Interfaces in Access Part of Ret-RP.

2.a If CORBA security services have been used, then both domain's credentials and other authentication information will have been exchanged, and both consumer and retailer will have been authenticated to each other. The call to **requestNamedAccess**() returns a reference to a **i_RetailerNamedAccess** interface. (An access session has been established between the consumer and retailer domains.)

2.b If CORBA security services are not used, then the call to **requestNamedAccess**() fails, and an exception is raised. This exception contains a reference to a **i_RetailerAuthenticate** interface, which is used to authenticate the domains. Once authenticated, the consumer gains a reference to the **i_RetailerNamedAccess** interface.

Access Session:

3. At this point, an access session has been established, and the consumer domain has a reference to the **i_RetailerNamedAccess** interface.

4. The consumer domain informs the retailer domain of its interfaces and terminal configuration by calling the **setUserCtxt**() operation. The retailer is given references to all the consumer 'access' interfaces for use within this access session.

5. The consumer can now use the **i_RetailerNamedAccess** interface to:
   - discover services offered by the retailer;
   - subscribe to those services (by starting a subscription service);
   - list the access sessions and service sessions they are currently involved with;
   - start a new service session;
   - suspend, resume, join and end existing sessions[3];
   - gain references to retailer-specific interfaces;
   - register interfaces for use outside of an access session; and more...

6. The retailer can:
   - gain references to retailer-specific interfaces (using the **i_ConsumerAccess** interface);
   - invite the consumer to join a session (using the **i_ConsumerInvite** interface);
   - discover the terminal configuration (using the **i_ConsumerTerminal** interface);
   - inform the consumer of changes in their access and service sessions (using **i_UserAccessSessionInfo**, and **i_UserSessionInfo** interfaces);

Contact outside an Access Session:

7. If the consumer wishes to receive invitations when he has no access session with the retailer, he registers the **i_ConsumerInitial** interface for use outside of an access session, (using **i_RetailerNamedAccess**). Then he ends the access session, and can no longer make requests to the retailer.

8. The retailer can still invite the consumer to join a service session, using the **i_ConsumerInitial** interface.

9. If the consumer wished to join the session they've been invited to, then they would have to establish another access session, (as in step 1.).

4. Usage Part of Ret-RP

The usage part of Ret-RP describes the interactions between the consumer and retailer domains during the use of a service session.

In fact, the usage part of Ret-RP can be used for service sessions across other reference points. For this reason, the domains are considered as party and provider domains. Parties are users of the service, (human or otherwise), and providers are the providers of the service. For Ret-RP, a consumer corresponds to a party domain; a retailer to a provider domain.

Also a service session can involve multiple parties. The usage part defines interactions between a single party and a provider. However, it recognizes that there may be many instances of this reference point for a single service session, and so defines interactions that cope with actions from other parties.

3. Service sessions which conform to the usage part, can be fully controlled by the access part. Those which do not conform may, or may not, fully support these control operations. In either case, they can be started using the access part.

Services offered by a retailer will come in many different types, each offering functions which a user wants to use. The usage part of Ret-RP is not about these service-specific functions, and does not want to restrict the sort of functionality offered by services. It is about sets of generic operations which many services, and classes of services, will want to offer, in a consistent and interoperable manner.

The usage part of Ret-RP defines a mechanism called Session Models to encompass these generic operations, and allow service sessions to offer them across an interoperable reference point.

Session models define how service session components in each domain can interact in a generic manner. These session models allow components which have been designed and implemented separately to interact to support a service session. A session model defines an information model for the session, and relates how operations on interfaces affect the information model, and so the behaviour of the session. Sessions may support one or more session models to describe their behaviour.

Ret-RP defines a single session model, the TINA Session Model. This model provides a generic set of operations, applicable to single and multi-party services, for: ending and suspending the session; inviting users to join the session; managing stream bindings between parties; etc. The operations are grouped into 'self-contained' feature sets, allowing a session to support only those generic operations that are applicable.

Other session models, which are more applicable to other classes of services, can also be defined. Ret-RP allows services to support other session models. These session models may be defined by a variety of organisations, or be specific to a particular retailer. (E.g. T.120 defines a specific session model applicable to conferencing services). Other session models may, or may not, define feature sets to compartmentalize their session models.

Ret-RP can also support services which do not claim to support any session model. In this case, the service does not need to conform to the usage part of Ret-RP, but can still benefit from the facilities provided by the access part.

4.1 TINA Session Model

The TINA Session Model defines interfaces supported by the party and provider domains for use over Ret-RP and other reference points. It also defines the Session Graph [2] information model. This information model can be used to define the behaviour of operations on the interfaces.

The TINA Session Model groups the interfaces offering these operations into feature sets. Each feature set is a 'self-contained' set of interfaces, which are supported by the session to provide one or more of the generic operations, and so expose restricted parts of the information model for manipulation or examination.

A feature set defines interfaces to be supported by the party and provider domains. Each domain is required to support different types of interfaces. In general, the provider domain supports interfaces on which requests for generic operations can be invoked. The party domain supports interfaces that allow their provider domain to execute an action, or inform them of actions that will be, or have been taken. Remember, sessions often involve multiple parties, and so some actions will have been requested by another party. In general, all domains in the session support the same feature sets.

The TINA session model defines a basic feature set (BasicFS) which must be supported by all sessions which support the TINA session model. BasicFS provides sufficient functionality to control a single party session.

It allows a client application in the party's domain to:

- discover the interfaces, session models and feature sets supported by a session;
- retrieve the interfaces supported by the session, (including service-specific, and 'feature set interfaces');
- register the client's own interfaces and session models with the session;
- end and suspend the session.

The basic feature set allows domains to exchange and agree the use of additional non-standardized interfaces. This is essential to obtain service-specific interfaces and value added interfaces above or instead of TINA standard session control.

BasicFS supports a client-server paradigm, where the party domain is the client, and the provider domain is the server. All requests using BasicFS interfaces are initiated from the client applications, and are serviced by the provider domain components. This means that the simplest service that can be implemented using BasicFS is a single party service, that has interfaces on the provider domain components, with no interfaces on the party domain applications. This does not mean that services using BasicFS are restricted to single-party services, or only client-server interactions. Service

Table 1: TINA Session Model Feature Sets

Feature Set	Description	Dependant on
BasicFS	Support end and suspend session requests. Allows the party domain to discover interfaces and session models supported by the session.	Mandatory for TINASessionModel
BasicExtFS	Allows the provider domain to discover interfaces and session models supported by the party domain components.	BasicFS
MultipartyFS	Allows the session to support multiparty services. Supports requests for: - information on other parties, - ending/suspending a party in the session - inviting a user to join session - announcing the session Also supports notifications session of events e.g. another party has suspended; and the session asking the party domain to execute an action, (e.g. the party is being suspended, and the party domain components need to perform some actions before the party is suspended.)	BasicFS
MultipartyIndFS	Allows the session to indicate to the party domains, that a party has made a request, and it is going to be processed. (Usually the party will be able to give their 'opinion' on whether this action is allowed, (e.g. through VotingFS, or service specific operations).	MultipartyFS
VotingFS	Supports parties voting to determine if a request should be accepted, and executed.	MultipartyIndFS
ControlSRFS	Session relationships define the ownership and permissions of parties to perform actions on session entities, (i.e. parties, resources, streams, etc.). This feature set allows parties to change these ownerships and permissions.	MultipartyFS
ParticipantSBFS	Participant type stream binding feature set: provides capabilities to setup, modify, and delete stream bindings between members of the session. Stream bindings are described in terms of members participating in the stream binding.	BasicFS
ParticipantSBIndFS	Indications for Participant type stream binding feature set, similar to MultipartyIndFS	ParticipantSBFS

sessions may support additional feature sets, and service-specific interfaces that support multiple parties in the session, and peer-to-peer interactions.

It also defines additional feature sets which may also be supported by the session. The feature sets which a session supports is available through the access part of Ret RP, and through operations on the interface defined by the mandatory basic feature set.

4.2 TINA Session Model Feature Sets

The table provides a brief description of each of the feature sets. All feature sets are 'optional', (except BasicFS). That means that a session can support the TINA Session Model and only support the feature sets that are useful to it. For example, a single party service (e.g. Video-on-demand) would not wish to support the interfaces associated with the multiparty feature set, but may wish to support the functionality of multimedia streams from the ParticipantSB feature sets.

Feature sets generally rely on other feature sets. E.g. the multiparty feature set provide operations to support inviting new users, and ending a user's participant. But it doesn't provide operations to end the session, as these are supported by the basic feature set. This means a session that supports MultipartyFS must also support BasicFS.

The feature sets define interfaces that must be supported by the appropriate domain. The behaviour of the operations on each interface is also defined. However, the precise behaviour of an operation can be affected by the service, and other feature sets supported. For example, a request to end the session, if it returns successfully, will always mean that the session has ended. However, if the multiparty feature set is supported, it also means that the other parties are told to execute any actions required to end the session. If MultipartyIndFS is supported, then the other parties will be informed that a party has requested that the session end, and may be allowed to vote (VotingFS), refuse (through service specific interactions). Or the request may be automatically refused because the requester doesn't have permission to end the session (ControlSRFS).

5. Extending Ret-RP with Retailer specific functionality.

The Ret-RP provides a well-defined set of interfaces to enable inter-operability between the consumer and retailer domains. However, it cannot hope to define interfaces which support all the functionality which retailers would wish to offer. Retailers will wish to distinguish themselves from other retailers, by offering 'value-added services' both in the access and usage part of Ret-RP. Examples of these value-added services, include management of a user's preferences for their services, tailoring of services to the consumer domain capabilities, etc.

The Ret-RP provides a number of mechanisms that help the retailer to extend the reference point to help them provide better services to their consumers. The most obvious way to do this is through specialized retailer services. These services are accessed in the same way as all services, but may provide some retailer specific functions, e.g. for downloading user applications.

The following are access related features of Ret-RP, that can be used to extend the reference point:

Retailer-specific authentication:

The i_RetailerAuthenticate interface provides a generic mechanism for the consumer and retailer domains to agree on an authentication protocol, and transport authentication information. It also enables the use of 'unusual' hardware authentication mechanisms to be supported by the consumer terminal.

Retailer-specific interfaces:

The access part allows both the consumer and retailer to offer retailer-specific interfaces which are not defined by Ret-RP. It provides generic mechanisms to offer and register these interfaces with the other domain, and to define properties for these interfaces. This means that generic components can interoperate with all retailers, but also allow these components to use retailer-specific functionality, if they also support retailer-specific components.

Retailer properties:

Some Ret-RP defined operations allow retailer-defined properties to carried. These properties can contain extra information for the retailer, or specific instructions to be executed. They are interpreted by retailer-specific components, and so can be used as the retailer see fit. However some example uses are:

- inform the retailer of specific capabilities of the consumer domain, (e.g. if some retailer-specific applications have been downloaded);
- pass user preferences or commands for the retailer;

Service Type properties:

Retailer-defined properties can also be associated with a service type. These can be used to:

- aid in searching for services when the consumer is attempting to discover new services to subscribe to, by defining properties and quality characteristics of a service;
- pass service specific information or commands to the service session, when it is started, resumed or joined by a new party.

The following are usage related features:

Retailer-specific Session Models and Feature Sets:

A retailer can define new feature sets in addition to those of the TINA Session Model. These would be supported by a range of services, by building generic components to provide the functionality for the feature set. For example, a retailer offering many share price services could define a feature set to search for shares, which could be re-used in share viewing and share analysis services.

In fact a retailer, or standards organization, can define complete new session models to support particular classes of service.

Service-specific interfaces:

The usage part also provides a similar facility to the access part to offer service-specific interfaces which are not defined by Ret-RP. Service specific interfaces can be defined as part of a session model which is used to provide a set of generic capabilities to a class of services, or as individual interfaces offered by a particular service type. The access part allows these interface to be passed across Ret-RP, when a service session is started, resumed, or joined due to an invitation or announcement. The TINA session model provides generic mechanisms to make these interfaces available during the session.

Service session properties:

Similar to the access part, some service session operations allow service-specific properties to carried. These can be used to define types of parties and assign them to the parties in a session. The party types may give permissions or behaviours to parties of a particular type.

Additionally, all the Ret-RP interfaces can be extended through inheritance, as is normal for IDL interfaces.

6. Ret-RP specification for other Reference Points

The Ret-RP specifications detail the interfaces that are offered across the Ret reference point. However, other reference points require similar functionality to that provided across Ret-RP, and will want to make use of similar interfaces.

The Ret-RP supports interactions between consumer and retailer domains where these domains act in User and Provider roles [2] respectively. These roles are defined where one domain wishes to make use of services offered by the other domain. Many other reference points address a similar situation, such as Retailer-to-Third Party Service Provider (3Pty), Retailer-to-Connectivity Provider (ConS), Consumer-to-Broker (Bkr) and in some cases Retailer-to-Retailer (RtR). In fact these roles can be split according to the access and usage parts of a reference point.

The access part defines a set of generic access interfaces, and the Ret-RP interfaces inherit from these. The generic interfaces corresponding to the access user and access provider roles are prefixed by `i_User` or `i_Provider`. Any operations that are specific to Ret-RP are defined on the Ret-RP interfaces. Other reference points can also define their own interfaces and inherit the generic operations from the user/provider interfaces.

In the usage part, the usage party and usage provider roles are used. However Ret-RP does not define interfaces which are specific to this reference point. This is because the interactions between components of a service session are the same, whether they are across Ret-RP or another reference point. So usage interfaces are prefixed by `i_Party` or `i_Provider`.

7. Prototype Implementation of Ret-RP

As part of the definition of Ret-RP, it is important that the specified interfaces can be used to build inter-operable implementations of consumer and retailer systems. Currently a prototype is being built which has helped to ensure that the Ret-RP specifications result in realistic, buildable systems, which address at least the basic capabilities required by consumer and retailer.

7.1 The General Idea

The prototype involves an internet-based retailer, which can be accessed by any Java-enabled web browser. This means that consumers do not require any TINA or retailer-specific software to be installed before they contact the retailer. This helps demonstrate the applicability of TINA to the internet environment. Although this was not necessary to proof the validity of the specifications, it does have the advantage of showing the applicability of the specifications. Also the fact that the consumer terminal can be anything from a high-end workstation to a simple PC puts emphasis on the practical use of TINA specifications.

At the time of writing of this paper the prototype is based on the latest available Ret-RP IDL specifications. The usage part of the reference point is still under development. This means the prototype mainly implements the access part. Further developments of the prototype include implementing (part of) the usage part, implementing necessary service components according to the latest service component specifications [8] and possible integration with the existing implementation of connection management [9] in which the ConS reference point [10] is addressed.

7.2 The Configuration.

The prototype involves the consumer and retailer role, each having requirements for hardware and software. These requirements are listed below (with the actual prototype configuration between brackets). The entire configuration is depicted in Figure 3.

Requirements for Consumer Domain:

- A platform running a Java enabled web browser, connected to the internet. (PC, Windows 95, Netscape)
- A video stream player that can be launched in the browser environment. (VDOLive plug-in for Netscape[4])

Requirements for Retailer Domain:

- A platform connected to the internet as a server. (SPARC Ultra-1)
- An installation of Orbix. (Orbix MT 2.1)
- An installation of OrbixWeb. (OrbixWeb 2.0.1)
- Implementation of retailer components. (C++)
- Implementations of downloadable consumer components (Java, JDK1.0.2)
- A video stream server. (VDOLive server[5]).

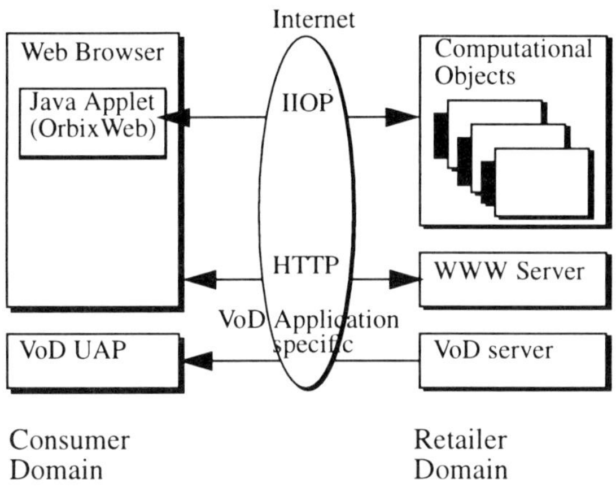

Figure 3. Prototype configuration

7.3 Initial Contact

The (internet based) access scenario for the prototype comprises three initial steps:

- **Finding the retailer**. In [4], it is suggested that a broker could be used by a consumer to find a business that acts as a retailer. Fortunately the internet makes things much simpler. It already provides excellent mechanisms to find certain content, e.g. using search engines or yellow pages (broker!). Retailers can advertise their Uniform Resource Locator URL [6], using every possible medium.
- **Contacting the retailer.** The consumer uses the URL to contact the retailer. As the consumer's browser loads the retailer homepage, a Java applet implementing the consumer interfaces (a combined Provider Agent (PA) and access session related User Application (as-UAP)) is downloaded. A parameter for the applet, contained also in the homepage, is an object reference (Interoperable Object Reference) to a retailer interface. The retailer can put in a reference to an available `i_RetailerInitial` interface instance (allowing for some load balancing). This mechanism is shown in Figure 4, upper section (1). Until now, HTTP [12] is used for requests and transport of data.
- **Identification, authentication and authorization**. The second section in Figure 4 shows the sequence of messages used in the prototype. The specifications provide a generic mechanism to perform these tasks. Obviously, this mechanism can only be used on the application level. Without some sort of CORBA security implementation no secure communication can be established. Fortunately there is a trend to use the Secure Socket Layers technology (SSL [13]) when using CORBA communication over the internet and thus providing security, transparent for the application. Examples of ORBs that use, or will use, this technology are VisiBroker for Java ORB by Visigenic and future releases of OrbixWeb. Since the used ORBs do not provide this, scenario 2.b in Section 3.1 is used in the prototype, as shown in Figure 4. The `requestNamedAccess`() operation returns an exception to indicate that authentication is required. A new `i_RetailerAuthenticate` interface instance has then been created and a reference is passed to the consumer domain. Using this reference, authentication takes place. After that, `requestNamedAccess()` is invoked once again, but now successfully. All communication so far uses IIOP [5] in order to allow non-java as-UAP and PA implementations to make use of the components in the retailer domain.

4. The plug-in can be obtained at http://www.vdo.net

5. The server can be obtained at http://www.vdo.net

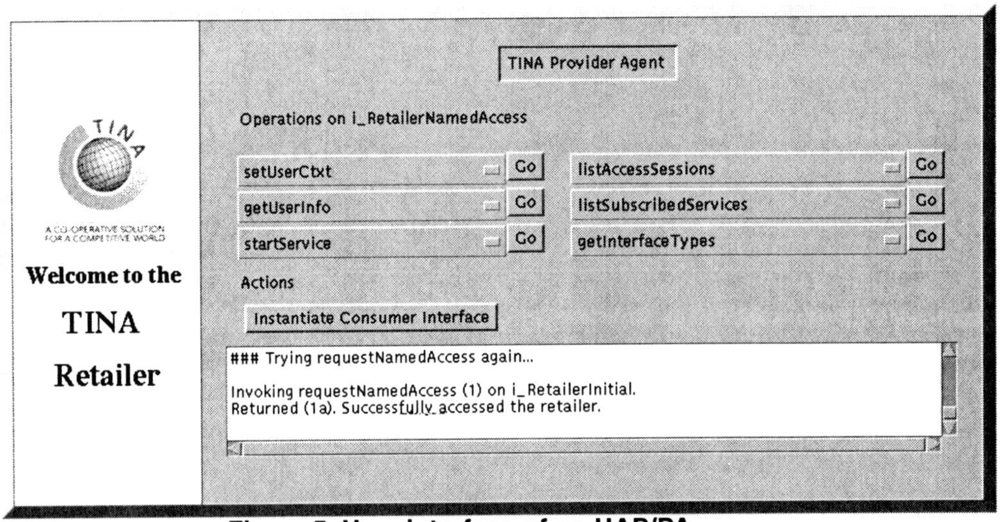

Figure 5. User interface of as-UAP/PA

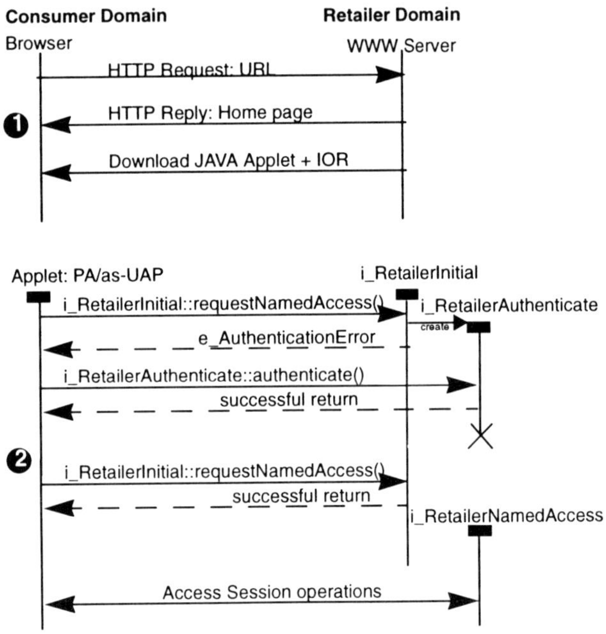

Figure 4. Typical Access Scenario

After these steps an access session with the retailer has been established and the consumer can perform the actions defined in Ret RP such as discovering new services, listing sessions and invitations and starting and managing service sessions.

7.4 Visualization.

Once the consumer has successfully accessed the retailer, he is presented with the user interface as shown in Figure 5. This interface represents the access session related User Application (as-UAP), commonly used to access TINA retailers.

This user interface is built for testing and development purposes only. It does not hide the details of execution for the end user, e.g. it visualizes the messaging between the client and servers. In Figure 5, the white area at the bottom shows information about the messages, replies and progress statements. The upper section allows the user to select whichever operation available on the **`i_RetailerNamedAccess`** interface and to execute it. Also, the as-UAP allows the consumer to instantiate the consumer interfaces, e.g. **`i_ConsumerAccess`** and **`i_ConsumerAccessSessionInfo`**. Instantiating an interface will show a separate monitor window in which incoming messages are reported.

Some of the most interesting features of the prototype are:

- The consumer can establish multiple access sessions, e.g. one access session on a terminal at home and another at work.
- The consumer can manage the access sessions. For example, from within the current access session other active access sessions can be listed and cancelled.
- The consumer can manage the subscribed services.

- From the subscribed services, the consumer can select and launch a service.

The consumer can select a simple video-on-demand service from the list of subscribed services, resulting in the activation of the consumers video stream player (a service session related User Application (ss-UAP)) which requests the retailer to send a video stream. As shown in Figure 5, the application and server use an application specific protocol over the internet. At the time of writing of this document, the access session (applet) has no control over the service session. Since legacy applications in general do not run on top of an ORB, it is hard to communicate with TINA components. An alternative approach could be used with the Netscape browser. LiveConnect, available in Navigator v3.x, extends the platform-independent Java object model to non-Java objects. This lets Java, JavaScript, and plug-ins share a common object and messaging model[6]. Since the access session is implemented in Java, plug-ins could be used for service session applications and could still be controlled from the access session. Also, the application then could communicate with the PA. The main disadvantage of this approach is the mandatory use of Netscape and the fact that the PA and application have to run in the same (netscape) environment. The same goes for the other well know browser, Microsoft Internet Explorer. Using the browser as a container for ActiveX implementations, service session applications could be implemented in similar fashion as with Netscape.

Another similar problem is the communication with components in the consumer domain outside the browser environment. The Java interpreter limits the scope of the communication between components in the browser environment to the host where the applet classes came from, in this case the retailer. Unless this can be configured (e.g. the HotJava browser[7]) other solutions have to be found.

7.5 Implementation Details

Although the interfaces are drawn individually in Figure 2, they are part of service components [8], e.g. the **`i_RetailerInitial`** and **`i_RetailerAuthenticate`** are part of the Initial Agent (IA) component and the **`i_RetailerNamedAccess`** interface is part of the User Agent (UA) component. The implementation in C++ allows one component to have several distinct interface types of which several instances can exist, e.g. for the **`i_RetailerAuthenticate`** interface, each time a consumer accesses the IA component and needs to authenticate himself, a new instance of the **`i_RetailerAuthenticate`** is created. Also, for every access session a new **`i_RetailerNamedAccess`** interface is created on the UA component[8]. The components are implemented using the guidelines described in [7]. Basically, a component exists of three classes, as shown in Figure 6.

- **The Core class** manages the interfaces, e.g. it can instantiate interfaces and remove them.
- **The Interface class** implements the methods of the interface. There can be more than one class of this type thus allowing multiple interfaces to be implemented.
- **The Container class** holds the data of the component. This class ensures that every interface can access the data of the component while still enforcing encapsulation of data.

The real inheritance structure of a component is more complex, e.g. the Interface class inherits from generated classes produced by the IDL compiler.

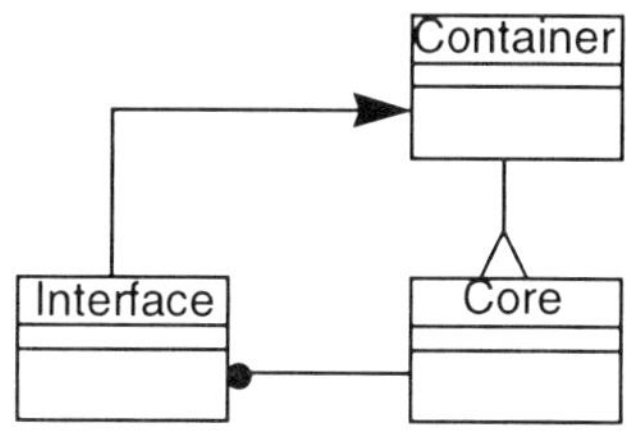

Figure 6. Class structure of a C++ component

The PA/as-UAP implementation in Java is more straightforward. The class structure is as in Figure 7.

6. http://cgi1.netscape.com/comprod/products/navigator/version_3.0/building_blocks/liveconnect/
7. http://www.javasoft.com
8. This is probably not the way a commercial implementation would handle multiple access sessions. Each instance requires (at least) one socket and thus consume a lot of resources.

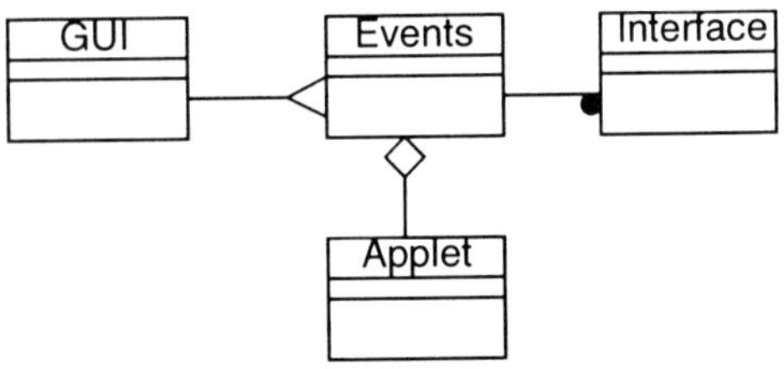

Figure 7. Class structure of the java component.

- **The GUI class** implements the as-UAP user interface. Note that the as-UAP is integrated in the PA component instead of being separate.
- **The Events class** implements the event handling of the user interface and is in fact the main class of the PA component.
- **The Applet class** is merely a wrapper to be able to run the component as an applet.
- **The Interface class** implements the methods of the interface. There are more than one interface class since the PA supports several i_Consumer interfaces.

Some other classes, which are not shown in the figure, are needed to process incoming messages. The two approaches in C++ and Java are quite similar, except for the lack of a container class in Java. The reason for the different approach in Java is based on the pointer manipulation and pure virtual declarations needed in the C++ approach (see [7]), something which is hard, if not possible, to do in Java.

7.6 Conclusion

The prototype has helped to detail the specifications and provide proof of applicability. Future extensions will add usage implementation in order to test (some of) the feature sets.

So far the implementation work has validated parts of the specifications of Ret-RP, and along with the implementation efforts of other TINA partners will help to produce a reference point which can be used to test inter-operability of TINA systems between a consumer and retailer.

8. References

[1] *The Ret Reference Point*, Version 0.8, TINA-C, Oct 1997, File:/u/tinac/97/integration/rfrs/RFR-96-01/interim/draft0.8/doc/ret.ps
Authors: P. Farley (ed.), S. Hogg, L. Kristiansen, C. Licciardi, M. Mampaey, R. Minetti, S. Pensivy, C. Smith, E. Utsunomiya, R. Westerga, M. Yates.

[2] *Service Architecture,* Version 5.0, TINA-C, June 1997; public.
Authors: C. Abarca, P. Farley, J. Forslöw, J. C. García, T. Hamada, P. F. Hansen, S. Hogg, H. Kamata, L. Kristiansen (ed.), C. A. Licciardi, H. Mulder, E. Utsunomiya, M. Yates.

[3] *Ret Reference Point,* Request for Refinements and Solutions, Version 2.0, TINA-C, August 1996; TINA-C internal.
File: /u/tinac/96/integration/rfrs/RFR-96-01/rfrs_ret.ps.
Authors: TINA-C Core Team.

[4] *TINA Business Model and Reference Points,* Version 4.0, TINA-C, May 1997; public.
File:/u/tinac/97/integration/viewable/bm_rp.ps.
Authors: M. Yates, W. Takita, R. Jansson, L. Demounem, H. Mulder.

[5] OMG, *The Common Object Request Broker Architecture and Specification*, Ver2.0, July 1995
http://www.omg.org

[6] IETF, RFC 1738 *Uniform Resource Locators*, 1994.
http://info.internet.isi.edu:80/in-notes/rfc/files/rfc1738.txt.

[7] *Guidelines for Computational Modeling in CORBA Environments*, H.Korte, R. Westerga, 1997, proc ISADS'97.

[8] *Service Component Specifications*, 1997; draft internal version 0.1
Authors: C.Abarca, P. Farley, J. C. Garcia, T. Hamada, P.F. Hansen, H. Kamata, C.A. Licciardi, E. Utsunomiya, M. Yates.

[9] *Network Resource Architecture*, 1997 version 3.0
Authors: C. Abarca, J. Forslow, T. Hamada, S. Hogg, H.B. Jeon, D.S. Kim, H.Y. Lee, N. Natarajan, F. Steegmans.

[10] *ConS reference point*, Nov. 1996, version 1.0.
Authors: TINA-C Core Team.

[11] *Hypertext Markup Language*, version 3.2
http://www.w3.org

[12] *Hypertext Transport Protocol*, version 1.0
http://www.ics.uci.edu/pub/ietf/http/rfc1945.html

[13] *Secure Sockets Layer*
www.netscape.com

Session 8

Legacy Integration

Fabrice Dupuy, France Telecom

The CAMOUFLAGE Project - Introduction of TINA into Telecommunication Legacy Systems

E. Holz, O. Kath
Humboldt University Berlin
Axel-Springer-Str. 54a, 10099 Berlin, Germany
{holz|kath}@informatik.hu-berlin.de

M. Geipl, G. Lin, V. Vogel
Deutsche Telekom AG
Am Kavalleriesand 3, 64295 Darmstadt, Germany
{geipl|vogel}@fz-telekom.de

Abstract

The CAMOUFLAGE Project is a national german research initiative led by the Deutsche Telekom. It aims at the provision of a TINA conform distributed processing environment based on existing legacy telecommunication systems. The envisaged target architecture is a switched ATM network deploying B-ISDN. Using an integrative approach a mutual stimulation between the TINA technology and the B-ISDN technology is expected. The paper presents the overall design of the CAMOUFLAGE DPE, the CMA for the B-ISDN and gives a detailed description of the B-ISDN based transport network for operational interactions.

A general mechanism for the introduction of different interoperability protocols (Open Communication Interface) to CORBA is proposed. The B-ISDN protocol GFP has been selected as the foundation for the implementation for such an interoperability protocol. It allows not only a fast inter-ORB communication, but can also - due to its different modes - be tailored to the actual needs (speed, quality, amount of data).

Within the scope of the Transport Network a Connectivity Management (CyM) basing on the TINA NRA has been developed for a B-ISDN environment. Due to the fact that B-ISDN does contain already parts of the functionality required by the NRA, CyM is a subset of NRA, extended by additional objects to control the protocol stacks and the hardware.

Different scenarios for the introduction of CyM have been specified, requiring only simple or no extensions to the existing B-ISDN networks.

1. Introduction

In 1993 the TINA-Consortium has been founded with the mission to „... provide a consistent reference architecture for open telecommunications, encompassing services, operations/management architecture, and technology management" [11]. Today after 5 years of work most parts of the TINA framework are finished. The task is now to bring the work into practice by providing platforms supporting TINA and applications utilizing TINA. Most of the projects dedicated to these areas today are concentrating on the implementation of the SERVICE ARCHITECTURE (SA) [16] and on the development of a Distributed Processing Environment based on the OMG-CORBA ([21], [22]). One among these projects is TANGRAM [29].

While the projects do conform to the SA, many aspects of the CONNECTION MANAGEMENT ARCHITECTURE (CMA) [14] have been omitted. This is due to the following facts:

- The CMA did not take into account the architecture of existing telecommunication networks and technologies.
- CORBA 2.0/IIOP has been used by most of the projects as the baseline kernel transport network, as transport network for streams proprietary solutions have been applied.

As a result, the efficient utilization of legacy communication infrastructures is limited. The new NETWORK RESOURCE ARCHITECTURE (NRA) [13] is an advancement of the CMA, going beyond its scope and taking critics into account.

After an initial feasibility study [26] the CAMOUFLAGE project has been set up between Humboldt University Berlin, Technical University Ilmenau and Deutsche Telekom AG with the goal to:

- provide a TINA DPE [15] basing on a legacy communication infrastructure (B-ISDN),
- enable the re-use of existing software components (CORBA, TANGRAM-SA),
- develop strategies for a graceful and gradual introduction of TINA technology.

2. B-ISDN

The B-ISDN protocol suite ([1], [2]) has been developed with the goal to deploy a connection-oriented signalling protocol on top of ATM-networks. The B-ISDN signalling development enlarges the advantages of ATM (scaleability and time transparency) with the concept of dynamic establishment, modification and deletion of calls and connections.

0-8186-8335-X/98 $10.00 © 1998 IEEE

Depending on the capability set, B-ISDN supports a wide range of connections (switched, permanent, semi-permanent, point-to-multipoint, point-to-point, multi-party, multi-connection) ([18], [19], [20]). In contrast to connectionless protocols as IP (Internet Protocol) B-ISDN guarantees connection characteristics for a connection from its reservation till its explicit release. Due to this guaranteed Quality-of-Service (QoS), the connection-oriented properties and the high bandwidth B-ISDN is exceptionally suited for the application in the area of multimedia.

3. CAMOUFLAGE architecture

The CAMOUFLAGE architecture reflects the integrative approach applied for the development of a B-ISDN based TINA-DPE. The whole design aims at a high degree of reuse of existing components. A CORBA-2 Object Request Broker will be used as the central building block for the DPE, and B-ISDN for the interoperability network work operational interactions and for stream interactions. The main work of the CAMOUFLAGE project concentrates on the connection of these two technologies by implementing the operational interaction transport network as a connection between CORBA islands and the implementation of a B-ISDN based CMA for the TN.

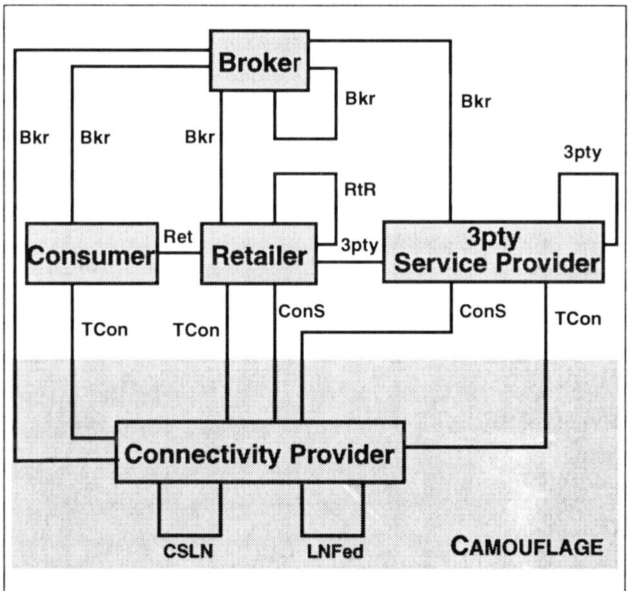

Figure 1 CAMOUFLAGE and the TINA business model

Aspects of the SA and DPE Services besides the Object Communication Support are investigated by CAMOUFLAGE at this time only marginally. For the future a combination with other projects, which address these aspects is foreseen (TANGRAM). Seen from a session oriented point of view CAMOUFLAGE can be clearly assigned to the Connectivity Session.

4. Support for operational object interaction

The principle of operational interactions in TINA is strongly related to the operation call principles defined by the OMG. Therefore the application of CORBA is favoured by CAMOUFLAGE for the operational interactions between objects within the same node and the work currently concentrates on mechanisms for the interaction between objects within different nodes. In other terms this means the provision of generic mechanisms for an inter-ORB communication. For inter-ORB communication at the invocation transport level the OMG has defined the General Inter-ORB Protocol (GIOP) [21]. At this time, only the TCP/IP based realization IIOP of the GIOP is standardized.

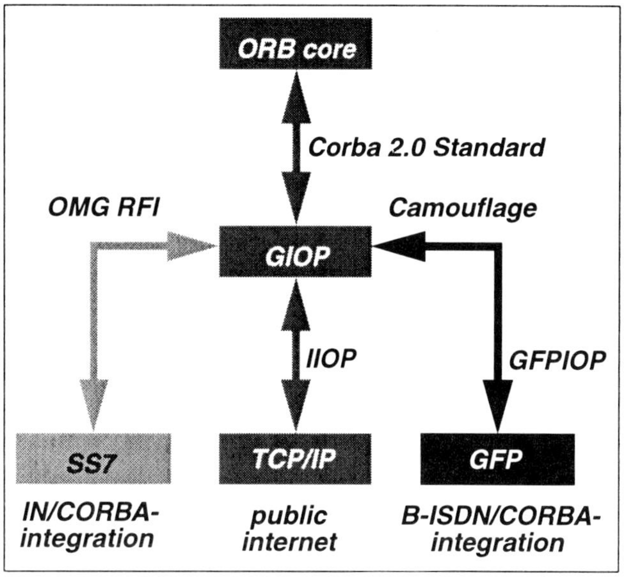

Figure 2 Inter-ORB protocols

With the OMG RFI "Issues for Intelligent Networking with CORBA" [23], which addresses CORBA/IN interworking aspects and the interconnection of CORBA islands by IN, the OMG shows one direction for the integration of CORBA and legacy telecommunication systems.

CAMOUFLAGE extends this development by widening the scope to the interconnection of CORBA islands via B-ISDN and aspects of B-ISDN/CORBA interworking. For that purpose a new specialisation - GFPIOP - of the GIOP based on the B-ISDN Generic Functional Protocol (GFP) was defined (see Figure 2).

Within an ORB domain, the default ORB transport layer based on IIOP is applied (see Figure 3). For Inter-ORB communication via B-ISDN UNI signalling protocols, the defined GFPIOP protocol is used.

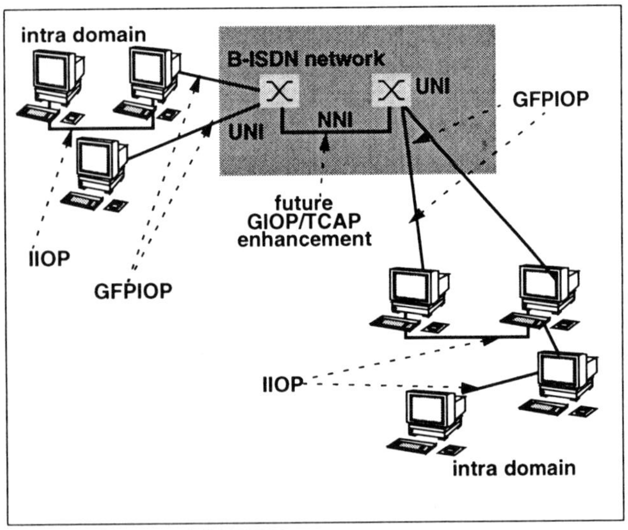

Figure 3 Usage of GFPIOP

4.1. The Open Communication Interface

A direct inclusion of alternative mappings into the ORB would require for each protocol new changes to the ORB core. To be as open as possible and to enable the use of different transport protocols by the same ORB, a well defined but small interface between the ORB and the transport protocols for GIOP messages would decouple the ORB from the underlying interoperability transport network. Such an approach will also allow the dynamic addition/removal of transport protocols.

The Open Communication Interface (OCI) is defined here with the main goal is to provide a generic approach for the interconnection of DPE nodes using existing telecommunication networks and protocols. The OCI is located between GIOP and the transport protocol for GIOP mes-

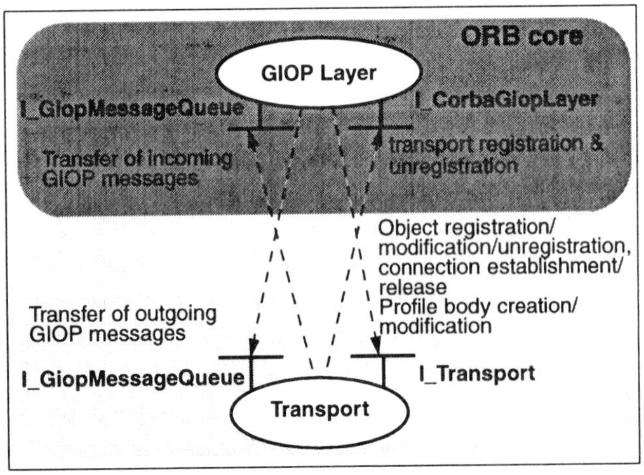

Figure 4 Interfaces and functional parts of OCI

sages. The functionalities of the OCI are supplied by entities within the CORBA runtime environment (interfaces **I_CorbaGiopLayer** and **I_GiopMessageQueue**) and within the transport service (interfaces **I_Transport** and **I_GiopMessageQueue**) respectively.

The **I_Transport** interface provides methods for the registration, unregistration and modification of objects, for the creation and modification of address information parts in object references (Profile Bodies) and for the establishment and release of connections representing a binding to a particular object.

The **I_CorbaGiopLayer** interface of the CORBA runtime environment contains operations for the registration and unregistration of a transport service. Registration in this context denotes the information of the CORBA runtime environment, that the binding of interfaces by a particular network technology is available. The **I_GiopMessageQueue** interface of the transport service stores GIOP messages to be sent to other DPE nodes and is filled up by the GIOP Layer. The **I_GiopMessageQueue** of the GIOP Layer on the other side stores incoming GIOP messages and is filled up by the transport service. The **I_GiopMessageQueue** provides additional operations to manage and store GIOP messages within a fifo data structure.

In the following sections these parts will be described in more detail, including their specifications in IDL.

a) Registration of transport services

The **registerTransport** method provided by the GIOP layer at the **I_CorbaGiopLayer** (Figure 5) allows a

```
interface I_CorbaGiopLayer
{
// exception definitions omitted
readonly attribute
OCI::I_GiopMessageQueue MessagesToReceive;
OCI::CorbaGiopLayerId id( );
boolean registerTransport
        (in  I_Transport transporter,
         in CORBA::IOP::ProfileId profileid,
         out TransportId id);
boolean unRegisterTransport
    (in TransportId id )
    raises ...;
oneway void handleMessagesToReceive ();
};
```

Figure 5 Methods for transport registration

transport service to register itself as an entity, capable to transfer GIOP messages via connections within a particular network and to provide therefore physical interface binding representations through this network.

The transport service invokes **registerTransport** with its own reference and an indication of the supported network technology (e.g. **TAG_INTERNET_IOP** for TCP/IP) as in-parameters and receives an identification, under which this transport is registered within the CORBA runtime environment.

The unregistration of a particular transport services results in the cancellation of all interface bindings provided by the requesting transport service. The **unregisterTransport** method should only be called by the transport service in case of major network problems.

b) Object Registration/Modification/Unregistration

For the creation of an application object the CORBA runtime environment provides mechanisms to register this object within the system. A local unambiguous identification (object key) will be assigned to that object and an object adapter is created in order to accept bindings to that object by clients. The object adapter uses the **registerObject** method provided by **I_Transport** interfaces of one or multiple previously registered transport services to notify, that incoming GIOP messages with an identification of this object have to be sent to the referenced GIOP layer. With the **changeObject** method the assignment of a new object key for a particular object is possible.

The **unregisterObject** method should be called if an object is deleted or to release existing bindings at the network level.

c) Profile Body Creation/Comparisons

Included in the profile body an object reference contains location information. These location information may consist of e.g. an IP address and TCP port in case of IIOP, or a phone number and an object adapter identification in case of GIOP over ISDN. To create such location information entries, the operation **createProfile** of the **I_Transport** interface is used by the CORBA runtime environment. It is called with the identification of the object (object key) for which the profile body should be created, and returns a profile body for that object.

d) Connection Establishment/Release

In order to establish a binding between a client and an object implementation, the operation **connect** of **I_Transport** is invoked by the CORBA runtime environment. The implementation of this method depends on the network technology applied. In case of TCP/IP, an TCP connection between the clients host and the TCP port of the object implementation, specified within its object reference

```
interface I_Transport
{
// exception and attribute definitions omitted
OCI::TransportId id( );
CORBA::IOP::TaggedProfile createProfile
    (in OCI::ObjectKey object)
    raises ...;
boolean compareProfiles
    (in CORBA::IOP::TaggedProfile profile1,
     in CORBA::IOP::TaggedProfile profile2);
boolean registerObject
    (in OCI::ObjectKey key,
     in OCI::I_CorbaGiopLayer orb);
boolean changeObject
    (in OCI::ObjectKey old_key ,
     in OCI::ObjectKey new_key)
    raises ...;
boolean unregisterObject (in OCI::ObjectKey key)
    raises ...;
boolean unregisterCorbaGiopLayer
    (in OCI::CorbaGiopLayerId id)
    raises ...;
boolean connect
    (in CORBA::IOP::TaggedProfile adress,
     in OCI::Duration timeout,
     out OCI::ConnectionKey conn_key)
    raises ...;
void close (in OCI::ConnectionKey conn_key)
    raises ...;
oneway void handleMessagesToSend
    (in OCI::ConnectionKey conn_key);
};
```

Figure 6 Interface specification I_Transport

is established, while e.g. a bearer independent call is requested in case of B-ISDN usage. Parameters of the **connect** operation are address information (in terms of a profile body) indicating where to reach the object, and a timeout value of type **duration** indicating the time frame within which **connect** has to succeed. The operation terminates with a value of type **boolean** and a connection identification of type **ConnectionKey**, which unambiguously references this connection. To release a particular connection, the method **close** provided by **I_Transport** has to be used.

e) GIOP Message Transfer

After the establishment of a connection, the exchange of GIOP messages between client and object implementation is possible. If a client requests the execution of an operation by a server object, a GIOP request message and message body including the object identification of the server object, the name of the requested operation, an invocation id and the operation invocation parameters have to be transmitted from the client side to the server (Figure 7).

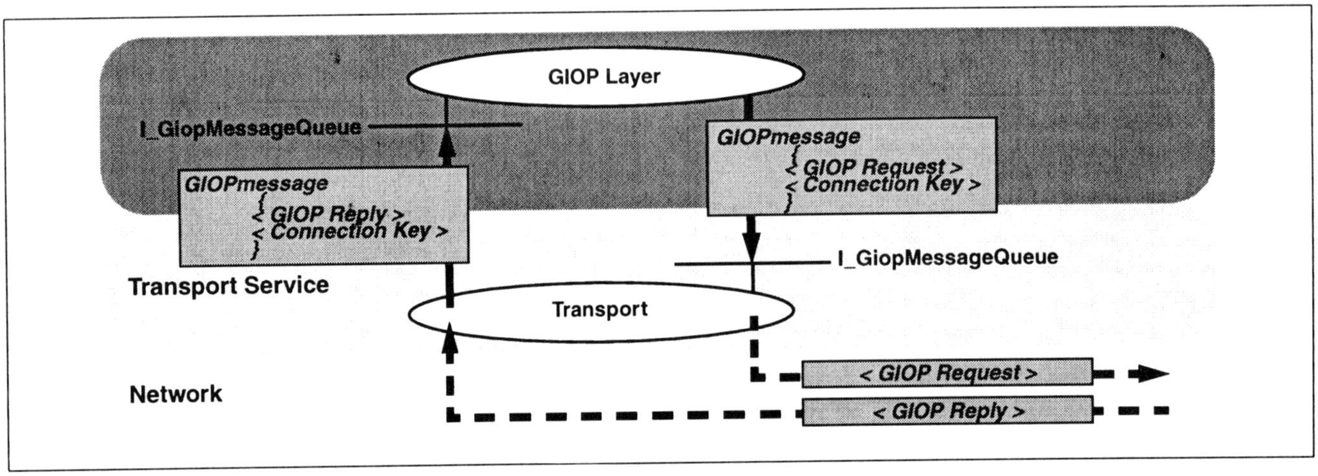

Figure 7 GIOP message flow at the invoking side

To initiate the transmission of a GIOP Request message, the CORBA runtime environment uses the method **put** (Figure 8) to append a **GIOPmessage** to the **I_GiopMessageQueue**[1]. Such a message consists of a **Request** structure and a connection identification of type **ConnectionKey**. The **Request** structure itself contains the GIOP Message Header, the GIOP Request Header, a timeout value of type **duration** and the parameters of the requested operation. Finally, the method **handleMessagesToSent** of **I_Transport** is called to start the actual transmission. Alternatively, the **put** operation may raise an event, resulting in the start of the transmission of that GIOP message by the transport service through the connection referred in the message.

At the object implementation side, the GIOP Request message is received. Because objects are registered previously through the **registerObject** method of **I_Transport**, the appropriate **I_CorbaGiopLayer** and therefore the appropriate **I_GiopMessageQueue** can be determined by the transport service. The **GIOPmessage** is put into the **I_GiopMessageQueue** using the **put** operation and subsequently evaluated by the object adapter, resulting in the invocation of the requested operation.

```
interface I_GiopMessageQueue
{
exception queueisempty {};
exception queueoverflow {};
boolean isempty ();
void setSize(in unsigned long size);
boolean reset();
boolean put( in OCI::GIOPMessage message )
    raises ...;
boolean get ( out GIOPMessage message )
    raises ...;
};
```

Figure 8 Message transfer using a I_GiopMessageQueue

If the operation is an interrogation, the its result of it is transmitted back to the client using the **put** operation of **I_GiopMessageQueue** with a **GIOPMessage** consisting of a **Reply** structure and the connection identification of the transport service. The **Reply** structure itself contains a GIOP Message Header, a GIOP Reply Header, the Reply body (containing the termination parameters). These components are transmitted back to the client via the identified connection and are queued in **I_GiopMessageQueue** of the CORBA runtime environment within the clients domain.

The same mechanism as used for GIOP Request and Reply will be applied for the remaining GIOP messages.

While the **put** operation of **I_GiopMessageQueue** is used to append a **GIOPMessage** to the message queue, the **get** operation removes the first **GIOPMessage** from it. With the **isempty** method provided by **I_GiopMessageQueue** it is possible to check, whether the queue is empty or not. The **reset** operation removes all entries from a queue and destroys them. With the **setSize** method the maximum number of possible entries can be set. This operation has been defined for optimization purposes.

4.2. The GFP inter-ORB protocol

a) The General Functional Protocol

The ITU-T Recommendation Q.2932.1 [2] defines the GENERIC FUNCTIONAL PROTOCOL (GFP) as a common mechanism for the exchange of information between B-

1. Note, that the queue for GIOP messages to be sent to the network is under control of the transport service, while the queue for GIOP messages received from the network is under control of the GIOP Layer.

ISDN-applications at the User Network Interface (UNI). The procedures defined by the GFP do support additional basic call characteristics, supplementary services and other functions independently from or in association with an existing call/bearer. Reflecting this, the GFP offers its users three different transport mechanisms:

- bearer-related,
- connectionless bearer-independent (CLBI), and
- connection-oriented, bearer-independent transport (COBI).

The CLBI and COBI transport mechanisms fit well in the operational inter-object communication scenarios of the TINA architecture. COBI requires the establishment of a signalling ATM adaptation layer (SAAL) connection and an assured transport association between a client and a server signalling application entity. Since COBI initiates a call establishment, a unique call reference number (i.e. without bearer connection) is used as a means to associate unambiguously among the related protocol messages. In contrast with COBI, the CLBI mechanism does not need a transport association between the signalling applications. Compared with COBI, the CLBI capability will be easier to implement, and is expected to be faster concerning the performance in an environment where a large number of TINA objects are involved.

At the network node interface (NNI), the function provided by the GFP at the UNI must be mapped into an equivalent function in the network signalling system. The Transaction Capabilities Application Part (TCAP) protocol of the Signalling System No.7 turns out to be very suitable for that purpose. Within the network several routing strategies can apply, e.g. the Signalling Connection Control Part (SCCP) or the Message Transfer Part (MTP) routing, which depends on the kind of addressing.

In [1] the GFP is specified as a realization of the Remote Operations Service Element (ROSE), according to the ITU-T recommendations X.219 [8] and X.880 [9]. ROSE supports interactive applications in a distributed open systems environment. In this context, the GFP provides a means to exchange ROSE Application Protocol Data Units (APDU) on behalf of signalling applications in peer entities. To carry these APDUs, the GFP uses a special kind of Information Element (IE) in the signalling protocol messages, the Facility IE. The basic principles of remote operations are the following: A remote operation is requested by an invoking entity, and the invoked entity attempts to perform the operation and reports the outcome. The supported protocol procedures for remote operations are:

- invocation,
- return result,
- return error and
- reject.

b) GIOP Mapping onto GFP

In order to use GFP as transport protocol for GIOP messages a mapping of GIOP onto GFP has been defined by CAMOUFLAGE. A precondition for objects to be reached over GFPIOP is an additional tagged profile to their IOR. This profile contains location and routing information, such as B-ISDN phone number, interface reference coded by subaddresses etc. Other information concern the binding characteristics (transport mode, QoS). A simplified IDL specification is given in Figure 9.

```
module GFPIOP
{//.....
enum GFPTransport {
    NotSpecified,
    ConnectionLessBearerIndependent,
    ConnectionOrientedBearerIndependent,
    BearerRelated};
struct BindingCharacteristics {
    GFPTransport transport;
    // some more QoS characteristics
    };
enum TypeOfNumber {
    unknownTypeOfNumber,
    internationalNumber,
    nationalNumber,
    networkSpecificNumber,
    subscriberNumber,
    abbreviatedNumber};
enum NumberingPlanIdentification {
    /* according to national and international
       standards like ITU-T E.164 */ };
struct PhoneNumber {
    TypeOfNumber numbertype;
    NumberingPlanIdentification numberingPlan;
    string numberDigits;};
struct ProfileBody {
    Version gfpiop_version;
    PhoneNumber phonenumber;
    BindingCharacteristic binding;
    unsigned short boaid;
    sequence<octet> object_key; };
    };
```

Figure 9 GFPIOP profile

Operation invocations and terminations, i.e. the appropriate GIOP messages, are mapped onto GFP Facility IEs and transmitted by GFP according to the information of the profile. Due to the fact that GFP provides different transport modes, one of these modes has to be selected for the transmission. This may be done within the profile body (binding characteristics). However, the main problem here was the development of a method for dynamic selection of one of the GFP transport mechanisms, if no preference is given.

Due to its runtime efficiency, the default transport mechanism will be connectionless bearer-independent (CLBI). In contrast to IIOP, no advance connection setup for the invocation of operations is needed. Furthermore, no instantiation of call handler state machines within the GFP protocol stack is necessary (see Figure 10).

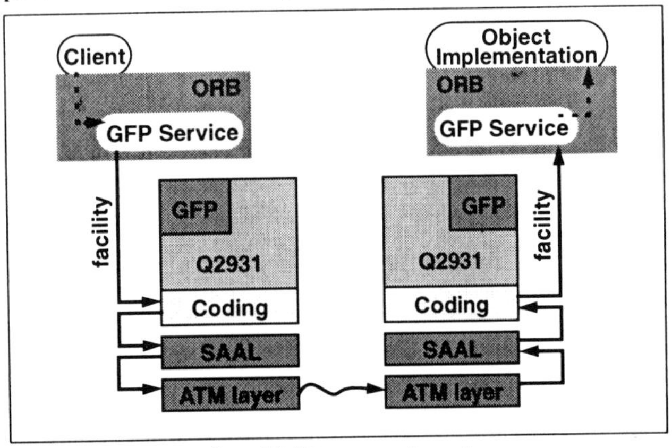

Figure 10 Control flow CLBI transport mode

The connection-oriented bearer-independent (COBI) transport mechanism is used for information exchange secured by protocol entities. This concerns e.g. explicit binding of operational interfaces and interactions between objects managing stream flow connections in the TN. For the COBI transport mode, the instantiation of call handler state machines is mandatory. The Facility IE of the initial COBI-setup protocol message contains the operation invocation request, implying at arrival time of a setup message at the called side the operation invocation. Furthermore, a GFP call will be set up. The termination of the operation and further operation invocations are transmitted by facility protocol messages within that call. On explicit request (e.g.

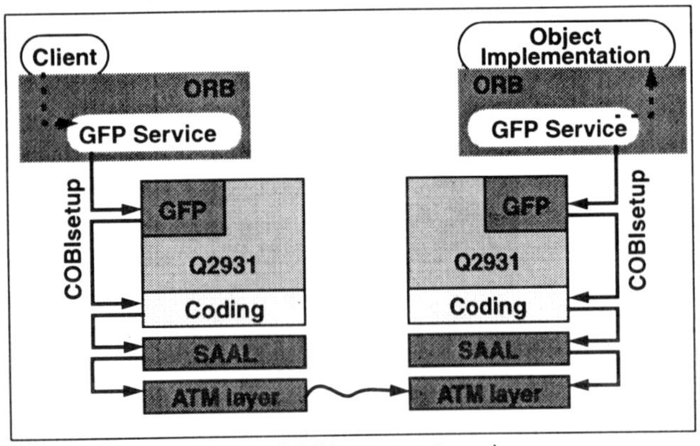

Figure 11 Control flow COBI transport mode

GIOP CloseConnection message) or at the deletion time of on of the participating objects, the interface binding and therefore the COBI call will be deleted.

In case of large operation parameters the bearer related transport mode is selected (see Figure 12), which ensures the fast and secure transmission of these parameters. An example for this case is a pure operational implementation of an ftp service: The invocation of the *get*-operation will

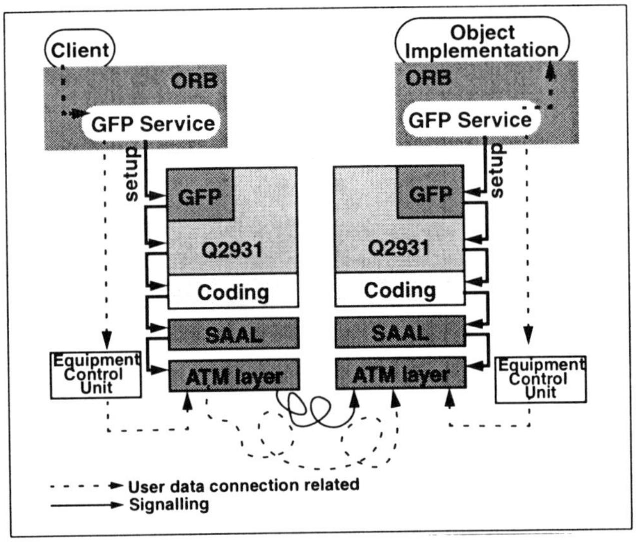

Figure 12 Control flow bearer related mode

result in an instantiation of a user connection. The operation parameters (i.e. the requested files) are submitted over this connection, while the GIOP message and request/reply headers are transported via the facility IE of the appropriate protocol signals.

c) Technology aspects

The Camouflage project has prototypically implemented the GFPIOP as well as its related protocol stack entities. As ORB OMNIBROKER was used. Its source code was enhanced with the aim to

- decide, whether IIOP or GFPIOP will be used for a concrete binding,
- interact with objects, which implement the GIOP to GFP mapping and communication between the ORB and the GFP protocol stack (GFP service).

The communication between GFP service within the OMNIBROKER ORB and the GFP protocol stack is based on SDL signals with ASN.1 data parameters.

On the bottom of the protocol stack, objects implemented within a CORBA environment are responsible to control and manage the ATM hardware. These objects communicate with the SAAL protocol entity in terms of SDL signals. The GFPIOP protocol stack entities and the kind of communication between these entities is shown in Figure 13.

Some object services, like naming and object life cycle service are utilized for boots-trapping and management of the platform.

The GFPIOP functionality is tested using some OMNIBROKER examples and applications developed within the project, like a file transfer service and a print service (using the bearer related transport mechanism). The main application of GFPIOP, especially of the COBI transport mode, is

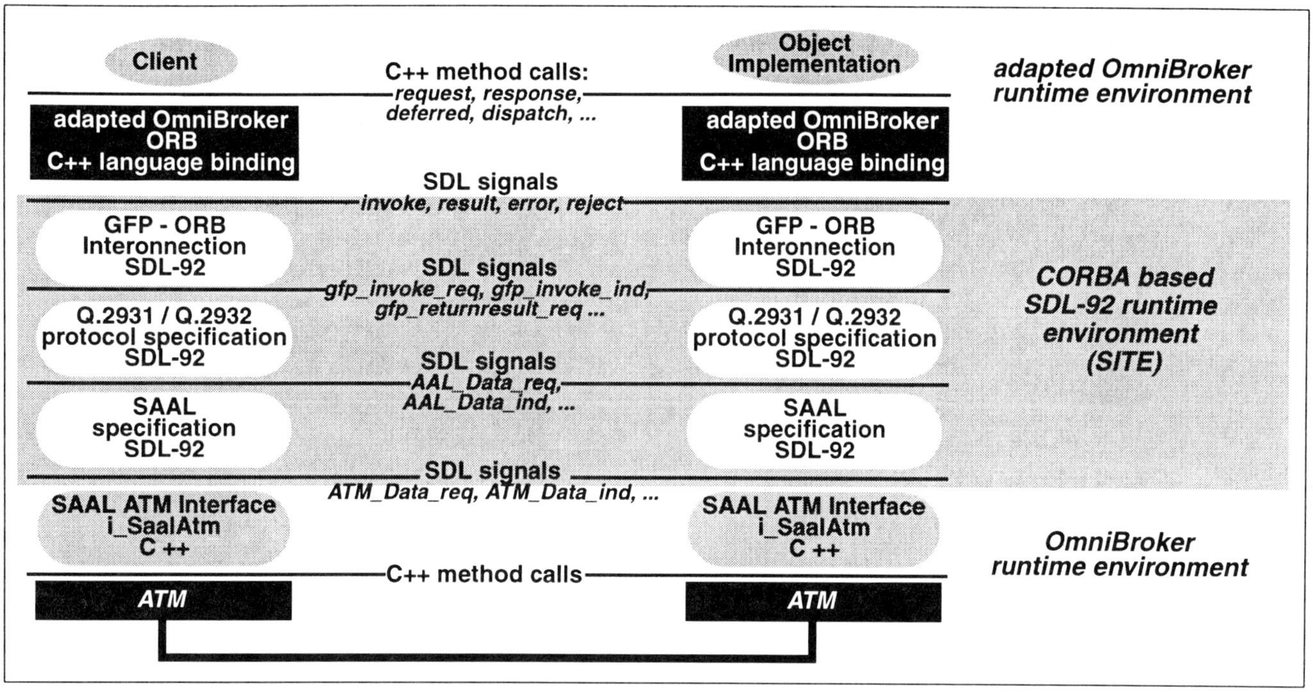

Figure 13 GFPIOP protocol stack and entities and interfaces

the communication of objects within the CAMOUFLAGE Connectivity Management (cf. section 5.), providing the management framework for stream based object interaction.

4.3. Other transport protocols

Besides the GFP currently also other telecommunication protocols are investigated by CAMOUFLAGE for their application as interoperability transport protocol. This concerns on one side protocols for N-ISDN [4] and on the other side protocols from the area of Intelligent Networks (IN, [7]). Again a similar approach as done for GFP is feasible here, basing on the Generic Procedures of Q.932 [5] (N-ISDN) resp. on TCAP [6] (IN). Moreover, also the application of the user-to-user signalling mechanism of N-ISDN seems to be a promising candidate.

5. Stream related work - CMA implementation

The TN is an abstraction of the transport network capabilities. Bindings through the TN (stream bindings) are set up and managed by objects of the communication session on request of objects of the service session. The communication session acts again as a client to objects of the connectivity session. Within CAMOUFLAGE the work does concentrate on the connectivity session.

A general architecture for the connectivity session has been defined. The starting point was the Connection Management Architecture defined by TINA-C [13], however to exploit the capabilities provided by the B-ISDN, this framework was adapted to the selected technology. A simplified computational view on the CAMOUFLAGE connectivity management (CyM) is given in Figure 14.

As it can be seen, the Connection Performer Objects (CP) are the access points to the protocol stack. As well network as user site initiated calls are supported by the CP objects. A Tandem Connection Manager as defined in the NRA is not necessary because the ability to cross administrative borders between subnetworks is part of B-ISDN.

Two additional types of objects have been introduced to manage and control the communication hard- and software:

- ECU (Equipment Control Unit) controls the ATM switch, the ATM cards of the CPE and other special hardware components,
- NEM (Network Element Manager) controls and manages the components of the protocol.

The CP objects do mark the borderline between the two technologies applied for the implementation:

- All objects of the Connectivity session are specified using IDL resp. ODL [12] and implemented in C++.
- All protocol objects are specified using SDL-92 ([10], [17]) (as SDL-blocks) and then automatically translated to C++.

Additionally the CP specifications serve as ODL-wrappers for the protocol objects, to integrate the protocol into the overall architecture.

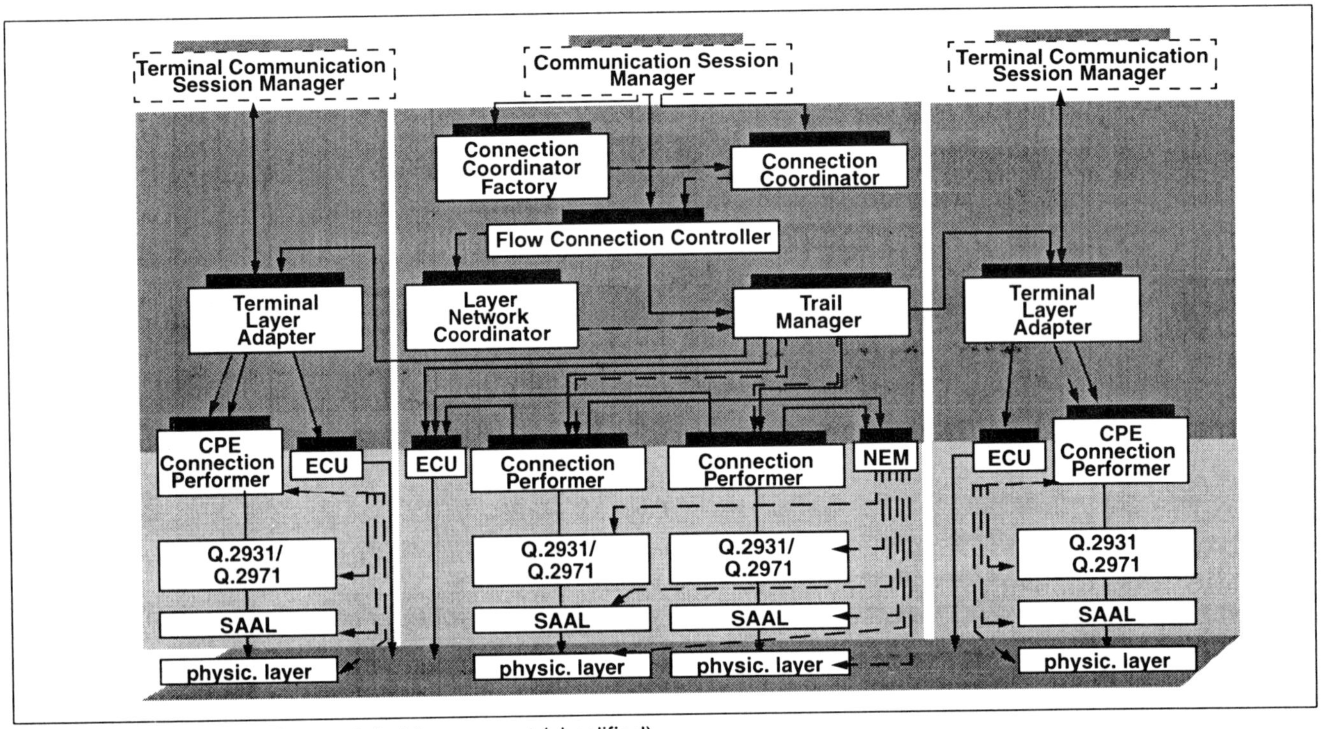

Figure 14 CAMOUFLAGE Connectivity Management (simplified)

5.1. Research Status and Availability of Products

The concept of an open communication interface is studied within the project in cooperation with Object Oriented Concepts Inc. While Object Oriented Concepts Inc. provides the ORB implementation part, the CAMOUFLAGE/ SIGTINAL project develops prototypes of transport services for the B-ISDN, N-ISDN and in future also for IN. The protocol parts have been specified in SDL and tested using the SITE SDL simulator [24]. All SDL components are automatically translated with SITE to C++, which is also the programming language for non-SDL components.

6. Conclusion

An integrative approach to introduce TINA ensures the investments of the telecommunication industry in their networks and communication software by keeping on the same time pace with the new developments in communication and information technology. The CAMOUFLAGE project does cater this need by combining the advantages of TINA with the advantages of B-ISDN to enable interworking and interoperability. A set of deployment scenarios for the gradual introduction of the CAMOUFLAGE architecture has been defined enabling the Deutsche Telekom a step-wise transition to TINA. These scenarios range from a deployment without any changes to the network up to the explicit integration of DPE and CMA functionality into the public B-ISDN network.

Currently CAMOUFLAGE is in its prototype phase, the next tasks will be the test under real-life conditions on a ATM platform and the integration of CAMOUFLAGE and TANGRAM. CAMOUFLAGE cooperates with other european projects within the EURESCOM Project P715 and participates in the TINA-DPE working group.

7. Acknowledgments

CAMOUFLAGE is the central part of the research project SIGTINAL of the Deutsche Telekom AG and is carried out in cooperation with Technical University Ilmenau. We would like to thank our colleagues at these institutions for their support in preparing this paper.We would also like to thank Mr. Anastasius Gavras (TINA-Core Team) and Mr. Mark Laukien (Object Oriented Concepts Inc.) for the helpful discussions with them.

8. References

[1] ITU-T Rec. Q.2931:B-ISDN.DSS2. User-Network-Interface (UNI). Layer 3 Specification for basic Call/Connection control, ITU-T'95

[2] ITU-T Rec.Q.2932.1: B-ISDN, Digital Subscriber Signalling System No.2 (DSS2), Generic Functional Protocol. Core Functions., Geneva, August '96

[3] Draft ITU-T Rec. Q2932.2: B-ISDN, Digital Subscriber Signalling System No.2 (DSS2). Generic Functional Protocol. Additional Constructs, Temp. Document, Miyazaki, Februar '96

[4] ITU-T Rec. Q.931: Digital Subscriber Signalling System No. 1 (DSS 1) - ISDN User-Network Interface Layer 3 Specification for Basic Call Control, ITU-T 1993

[5] ITU-T Rec. Q.932: Digital Subscriber Signalling System No. 1 (DSS 1) - Generic Procedures for the Control of ISDN Supplementary Services, ITU-T 1993

[6] ITU-T Rec. Q.771: Signalling System No. 7 - Functional Description of Transaction Capabilities

[7] ITU-T Rec. Q.1211: Introduction to Intelligent Network Capability Set 1, ITU-T 1993

[8] ITU-T Recommendation X.219. Remote Operations. Model, Notation and Service Definition. Geneva, 1988.

[9] ITU-T Recommendation X.880. Data Networks and Open System Communications. OSI Applications. Remote Operations. Geneva, 1994.

[10] ITU-T Rec. Z.100: CCITT Specification and Description Language (SDL), ITU-T '92

[11] TINA-C „Press release 13. February 1995", TINA-C, 1995, http://www.tinac.com

[12] TINA-C „Object Definition Language - Manual Version 2.3", TINA-C, 1996

[13] TINA-C „Network Resource Architecture, Version 3.0", TINA-C, 1997

[14] TINA-C „Connection Management Specifications", TINA-C, 1995

[15] TINA-C „DPE Architecture", TINA-C, 1994

[16] TINA-C „Definition of Service Archtitecture, V.5.0", TINA-C, 1997

[17] ETSI Project Team 87:„Formal specification (SDL) for B-ISDN DSS2 CS2", ETSI 1996

[18] RACE II MAGIC Deliverable 6: „Stage 2 Specification", September '93

[19] RACE II MAGIC Deliverable 10: „Protocols and Concepts of B-ISDN Signalling ", Oktober '93

[20] RACE II MAGIC Deliverable 3: „Service Description Framework and B-ISDN Service Description", Oktober '93

[21] OMG:"Common Object Request Broker: Architecture and Specification" (CORBA 2.0/IIOP), OMG Technical Document Formal/97-02-2

[22] OMG:"Common Object Request Broker: Architecture and Specification" (CORBA 2.1/IIOP), OMG 09/1997

[23] OMG:"Telecom Domain Task Force - Request for Information", OMG DTC Document telecom/96-12-02

[24] H. Böhme, M. v. Löwis, R. Schröder: "SDL Integrated Tool Environment", http://site.informatik.hu-berlin.de

[25] Object Oriented Concepts: "Omni Broker", http://www.ooc.com

[26] E. Holz, O. Kath: „Abschlußbericht Einsatz von B-ISDN-Netzen im Rahmen einer TINA-konformen Kommunikationsplattform" Project CAMOUFLAGE/S, Humboldt University Berlin, 11/96

[27] O. Kath: „TINA-C und B-ISDN-Netze", Master Thesis, Humboldt-University Berlin '96

[28] O. Kath, E. Holz, J. Fischer, M. Geipl, V. Vogel:"Provision of TINA Object Interaction through B-ISDN in ATM networks", to appear GLOBECOM-97

[29] K.-P. Eckert, M. Festini, P. Schoo, G. Schuermann:"TANGRAM: Development of Object-Oriented Frameworks for TINA-C Based Multimedia Telecommunication Applications", Proc. of ISDAS'97, Berlin

Implementation of a RPC/IDL to CMIP/GDMO Gateway

Eduardo Sudol Forbici
Cefet-Pr
Manoel Camillo Penna
Cefet-Pr

Abstract

This paper describes the design and implementation of a gateway that maps interactions between management applications developed using RPC/IDL to agents developed using CMIP/GDMO. In order to match both technologies, several mechanisms are necessary, and we show how to combine them in a coherent module, which can have most of these components generated by automated tools. Particularly, we demonstrate its use in a study for the integration of distributed management applications, developed over the OSF/DCE distribution platform, accessing management information on OSI managed objects, through the HP OpenView management platform.

1. Introduction

The development of systems using platforms derived from different models is a complex task, due to the lack of compatibility and operability that arises on the market. The use of the facilities offered by incompatible platforms demands the knowledge about both platforms and efforts in order to perform the interaction between them. The ideal solution would be changing the models prescribed by the different environments allowing full compatibility. A lot of effort has been spent in the development of the existing models and solutions for integrating legacy models are extremely desired. Indeed, the integration of platforms not designed to operate together is a possible solution, but their differences must be complementarily exploited by adequate integration mechanisms.

This work presents the design and the implementation of a gateway that performs at run-time, the data mapping between OSF/DCE [1] and the HP OpenView management platform [2]. The proposed approach looks for an agreement between RPC/IDL technology to be used in the development of distributed applications, and CMIP/GDMO technology, which is the basis for building OSI agents.

2. Motivations

The demand for new sophisticated services requires a network infrastructure into which services can be introduced easily and quickly. TINA-C is dealing with those issues by an open architecture for telecommunications services. This architecture is based on distributed computing and object orientation, and supported by a distributed environment (DPE) [3]. Possible candidates for building DPE are OMG compliant object request brokers (ORBs), and OSF distributed computing environment (DCE). TINA architecture also includes DPE services which can be supported by services included on those platforms (e.g, OMG COS [4]).

On the other hand, to address the problems that arise in management of telecommunications networks, the ITU introduced the TMN concept. In order to accommodate the management capability of multiple technologies, the TMN architecture should support general and flexible interfaces. ITU selected OSI system management technology as the basis for TMN interfaces, defining the

0-8186-8335-X/98 $10.00 © 1998 IEEE

Q3 interface for the communication between building blocks. Q3 interface relies on OSI CMIP protocol and standardized object models for supporting management interactions.

The ideal solution for building management applications over distributed environments like DPE, would be that interfaces and protocols of TMN and TINA were the same, but in many cases this will not be the adopted solution. Telecommunication equipments with Q3 interfaces are now a reality and will be part of the market for several years. Moreover, management applications, referred as Operation Systems (OS) on TMN standards, are the more complex building blocks of TMN architecture. They are inherently distributed, and would be more efficiently developed over distributed environments like DPE. OSF/DCE is a robust industrial implementation of a distributed computing environment, and is a candidate for implementing DPE kernel. In this scenario, the integration of Q3 network elements into TINA environment would require the development of mapping mechanisms between Q3 and the protocols and information model of the distributed environment supporting DPE. Consequently, a gateway enabling the mapping between RPC/IDL and CMIP/GDMO environments arises as an important tool.

Independently of the platforms chosen to build the gateway, this work presents concepts and experiments which can be used to build different gateways. Integration of CORBA and OSI technologies is an important issue too, the first representing the main trend for implementing information systems based on distributed objects. This approach is also being considered at TMN Laboratory at CEFET-PR, and the related results will be presented in future papers.

3 Run-time Data Mapping Module

The development of distributed management applications based on a distributed computing platform like OSF/DCE, which is capable of accessing OSI managed objects over a management platform like HP OpenView, demands the construction of a gateway between both environments. In this scenario, the distributed management application are not directly connected to OSI agents, allowing that people involved in their construction, not to be concerned with programming details of OSI agents and vice-versa. This approach keeps the efficiency of the DCE distribution mechanisms added to the advantages of OSI standardization.

The elaboration of this interface demands the mapping of data types and managed object operations to IDL type definitions and remote procedures declarations, and also the conversion of exchanged data at run-time. The full mapping consists first in translating from ASN.1 [6] definitions to DCE/IDL type definitions, and second in translating operations defined according the guidelines for the definition of managed objects GDMO [7] to definitions of remote procedures. In this paper we don't describe how those definitions are mapped to each other, but we concentrate in describing the run-time mechanism for data exchanging. We have also developed a set of mapping rules and the corresponding algorithms, which are being used to implement the tools for automating the implementation of gateways [8].

The conversion mechanisms must be implemented in the run-time gateway for the involved managed objects, including support to the management functions like negotiation, filtering and scoping. Moreover, the gateway must provide access to attributes and the execution of the actions defined in GDMO templates.

4. Gateway Components

As the gateway is the mechanism which provide the interoperability between distinct computational environments, it must run in distinct environments simultaneously. The first step on defining its architecture is to decompose it in modules with the necessary functionality to perform the tasks in each environment.

A DCE application is composed at least by one client and one server. The client asks for services that are executed by the server. On the other hand, a HP OpenView application is composed by at least one agent and one manager. The last one targets management operations and requests for information to the first. According to this view, a distributed management application based on DCE environment would be composed by a collection of client and server modules, where one of them would be the gateway. The requests coming from client modules arrive to the gateway and are delivered to the corresponding agent on the OSI environment. Conversely, notifications coming from OSI agents arrive to the gateway, and are delivered to the corresponding DCE modules.

An OSI agent would see the gateway as an OSI manager, and the DCE management applications would see the managed objects as a DCE server. The gateway must have the necessary functionality allowing it to act in both environment, making good use of the advantages offered by them.

A management application that is willing to interact with managed objects should call a remote procedures in the gateway server. It just sees an IDL interface in the gateway, that corresponds to the GDMO specification of the managed object. It must know the managed object class and its Relative Distinguishing Name (RDN). The Gateway should only know the GDMO interface of managed objects existing in the agent.

As mentioned before, in a first level, the gateway is decomposed in two main parts, one running in the DCE environment and the other running in HP Open View environment. They communicate with each other by means of a transference channel (see figure1). The gateway part running in the DCE environment is itself decomposed into three distinct modules: the first one, responsible by the server initialization, is named **server**, the second one, responsible by the implementation of the remote procedures is referred as **implementation of remote procedures** (IRP), and the third one, responsible for handling notifications, is named **notification manager** (NM). The gateway part running in the HP Open View environment has a single module, which acts as a OSI manager, and is referred as **manager**.

A transference mechanism is necessary to perform the communication between the modules on different environments, and it is named **transference channel**. There are two kinds of transference channels, one for handling operations and the other for handling notifications. Figure 1 illustrates the components of Gateway.

Much of the functionality of the server and manager will exist in every gateway and can be easily generated by automated tools. This characteristic makes easier the implementations of gateways by defining algorithms and reusable code that can be grouped into library of routines.

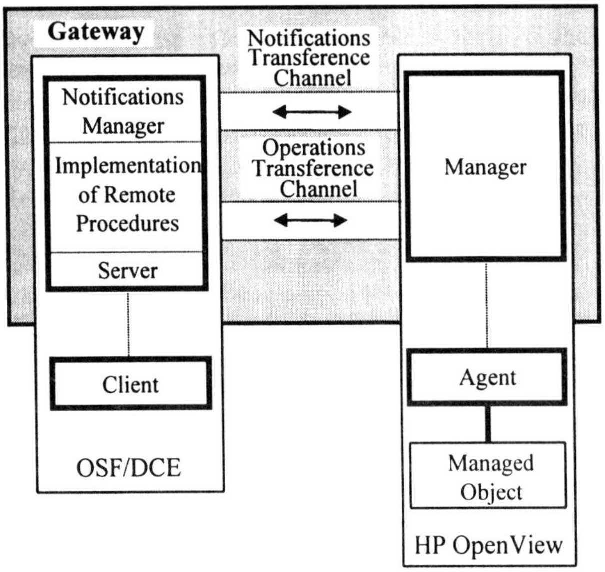

Figure 1 - Architecture of gateway

4.1. Server Module

The server module contains the initialization code executed before the processing of any remote procedure call. This code includes invocations to the RPC runtime, to prepare the server to receive the remote procedure calls. The first step consists in verifying if the transference channel is correctly activated. In case of failure, the server will not execute the initialization procedure and will abort. Otherwise, and after the transference channel verification, the server records its interface in the RPC Runtime and selects one or more protocols that will be used by the RPC Runtime in the network environment. Then it requests a binding handle to the RPC Runtime, and registers it in the endpoint mapper. Finally, it exports the binding information to the Cell Directory Service and waits for calls.

As multiple gateways can exist in the DCE environment, we selected the explicit binding method. Using this method makes easier to the management application to find the corresponding gateway at execution time. Moreover, this method makes easier to the gateway to obtain the identification a management application host, when handling notifications.

4.2. Implementation of Remote Procedures Module

The IRP module implements the management operations to be performed on managed objects. When an operation is called, the corresponding remote procedure is invoked in the IRP, which then uses a transference channel to relay the invocation to the manager. The manager will then prepare a CMIP/GDMO invocation and target it to the appropriate agent.

Typically the IRP module can be multi-threaded, each thread dynamically started by the stub code to attend an incoming RPC. On the other hand, the manager is not multi-threaded, and we were faced to three alternatives for implementing the IRP:

In the first alternative (figure 2), the implementation is mono-threaded and there is just one transference channel. The consequence of this choice is a lower performance on attending client requests and a more complex solution for handling notifications (see section 3.5). Despite of this, we made this choice for our prototype implementation, as discussed in section 5. In the second alternative (figure 3), the implementation is multi-thread and there is just one data transference channel. Because the need of sharing of the channel by multiple threads, its access must be controlled by a synchronization mechanism like semaphores or condition variables.

In the last alternative (figure 4), the implementation is multi-thread, but there are multiple transference channels and managers. A manager, transference channel pair is used by a RPC thread in the IRP. In this case the gateway will become more efficient in attending the requests, but increasing resources utilization.

Alternative 1:

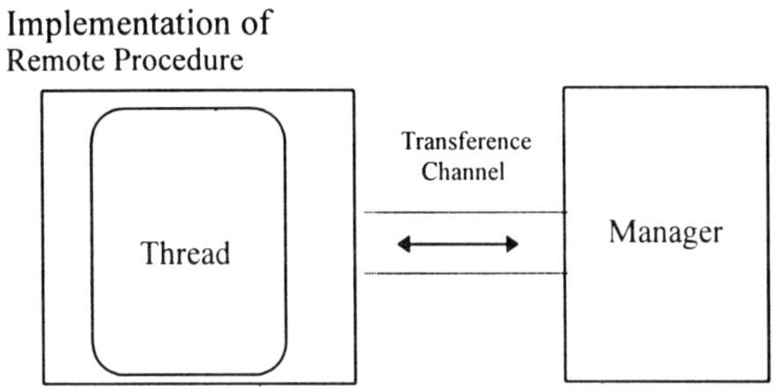

Figure 2 - Mono-thread Implementation with one manager and transference channel

Alternative 2:

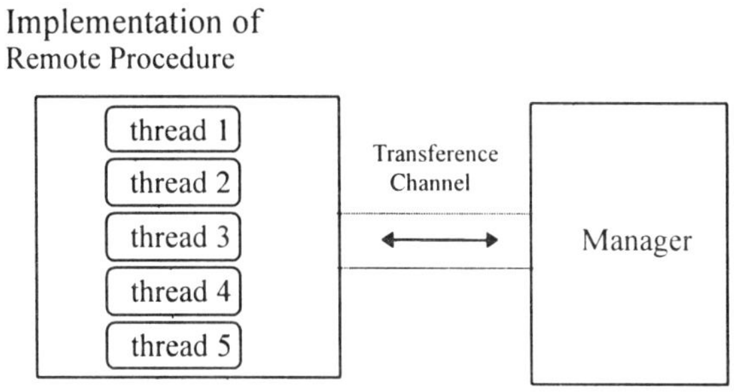

Figure 3 - Multi-thread Implementation with one manager and transference channel

Alternative 3:

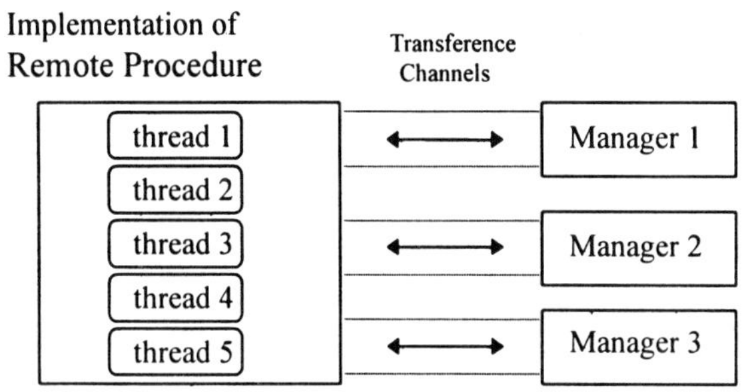

Figure 4 - Multi-thread Implementation with many managers and transfence channels

4.3. Manager Module

The manager gets invocations from a transference channel, maps them to XOM/XMP requests, and targets them to the appropiate agent. It also gets notifications coming from agents and relay them to the NM. The first task of the manager after its initialization is to check the situation of the transference channels. If the channels are working, the manager will perform the typical tasks of an HP OpenView manager: initialize the XOM workspace, negotiate the interface characteristics (CMIP) and open a management session with the communications infrastructure (CI). After these steps, it will be ready to perform the management tasks, that is, handling requests coming from management applications and notifications coming from the agents. If the transference channels are not activated the manager will abort without performing any management or initialization task.

4.4. Transference Channel

The transference channel is the link between the IRP and NM modules in the DCE environment, and the manager module in the OSI environment. Operations and notifications are transmitted from one environment to another through the channel. The channel can provide communication between processes placed in the same host or in different hosts, and can be implemented by several different mechanisms, for example, FIFOs, pipes, and sockets [9].

Before being transmitted, operations and notifications must be serialized and converted to a common format, and the inverse process must be done at the receiver side. For that, the sender and the receiver must agree on common rules for marshalling and unmarshalling the data. The BER (Basic Encoding Rules)

Computing) Sun Microsystems are standards for encoding, and we selected the last one for implementing our prototype. A common message format for specifying which operation or notification is being transmitted should also be defined [8].

4.5. Notifications Manager Module

A notification is an unsolicited message sent by the agent to the manager. The notification is collected by the manager and transmitted to the NM through the notification transference channel. Two mechanisms, similar to that suggested in the event services from CORBA [12] are considered for handling the notifications in NM.

In the first one, called "Pull Model", NM makes available a remote procedure where management applications seek the notification information when it is necessary. In this model, NM keeps the notifications stored until all management applications that are interested on them have pulled them.

In the second model, called "Push Model", NM sends the notification to the management application as soon as they became available. In this model the management application must provide a remote procedure that is called by the NM when notification arrives.

In both models (pull and push) NM must provide a specific remote procedure to allow management applications to register the intention for receiving specific notifications. In the Push model, NM will need to find the interested management applications, hence the first parameter of the registration procedure is a binding handler. Having this parameter the NM has the necessary information to find them.

5. Prototype

The ideas presented in this paper were validated through the implementation of a prototype. All modules were implemented in C running over HP-UX operating system. An initialization process starts the server, NM and manager modules, creates a pipe between the server and the manager, and another between the NM and the manager. The corresponding file descriptors are passed as arguments to them. The IRP was implemented using the first alternative discussed in section 3.2, and the NM was implemented according to the pull model. The whole code contains approximately 6.000 lines of C code, and behaves with acceptable performance.

6. Conclusion and future work

This paper presented the architecture and the mechanisms used for building a gateway for integrating DCE management applications with OSI based agents. The proposed components were implemented to validate the architecture, based however on the simplest choice. The overall performance was acceptable and we believe that the proposed architecture is general to be applied to different technologies. The same concepts are being applied to build a gateway between CORBA based management applications and OSI based agents.

Bibliography

[1] John Shirley, Wei Hu, David Mazid, Guide to Writing DCE Applications, O'Reilly & Associates, Inc., 1994.

[2] Hewlett Packard, HP OpenView Distributed Management Developer's Guide, U.S.A, 1994.

[3] Nilsson, G.; Dupuy, F.; Chapman, M.; The TINA Consortium: "An Overview of the Telecommunications Information Networking Architecture", proceedings of TINA'95, February 1995.

[4] Object Management Group, Object Management Architecture Guide, John Wiley & Sons, New York, September, 1992.

[5] Object Management Group, The Common Object Request Broker: Architecture and Specification, OMG Document Number, Version 2, July 1995.

[6] ITU-T Recommendation X.208, Specification of Abstract Syntax Notation One (ASN.1), Fascile VIII.4, 1988.

[7] ITU-T Recommendation X.722, Guidelines for the Definition of Managed Objects (GDMO), September 1991.

[8] Eduardo Sudol Forbici, Construção de um Módulo de Mapeamento entre RPC/IDL e CMIP/GDMO, Dissertação de Mestrado, CPGEI-CEFET/PR, maio de 1997.

[9] W. Richard Stevens, Unix Network Programming, Prentice-Hall, 1990.

[10] ITU-T Recommendation X.209, Specification of Basic Encoding Rules (BER) for Abstract Notation One (ASN.1), Fascile VIII.4, 1988

[11] Sun Microsystems, RFC1014 - XDR: External Data Representation Standard, Internet Network Working Group, 1988.

[12] Object Management Group, CORBA Services, OMG Document No. 91.12.1, Framinghan, Mass., 1995.

A staged approach for TMN to TINA migration

Daniel Ranc, Alcatel Alsthom Corporate Research Centre, France
George Pavlou, University College London, UK
David Griffin, University College London, UK
Alex Galis, University College London, UK

Abstract

The TINA architecture is often perceived as a service layer specialized framework; however a detailed examination of the architectural context and ongoing work in the field demonstrates that it is not so, and that the TINA architecture can be envisioned as a Network Management framework as well if care is taken about some TMN specificities. Having acquired that point, the paper proposes a conceptual architecture for interworking TMN and TINA based systems. It then describes a component architecture for a migrating OSI MIB that demonstrates concretely the feasibility of building a TMN-able agent running in a TINA-based environment.

1. Architectural Context

This section is devoted to an examination of background material regarding CORBA and TINA architectures, specifically from a Network Management architecture and requirement point of view.

CORBA

Overview

The OMG (Object Management Group) aims at designing an architectural framework supporting detailed interface specifications that will drive the industry towards inter-operable, reusable, portable distributed software components based on standard object-oriented interfaces. This resulted in the definition of an abstract reference model called the *Object Management Architecture* (OMA), and the specification of a concrete platform called the *Common Object Request Broker Architecture* (CORBA).

The Common Object Request Broker Architecture has been designed by the Object Management Group (OMG) with the goal to provide a truly open object distribution architecture; thus moving one step further both:

- regarding proprietary architectures such as OLE;
- regarding non-object oriented environments such as DCE.

The CORBA architecture (in a snapshot) is composed of a *software bus*, the Object Request Broker (ORB), on which sit CORBA *clients* (referencing distant objects) and CORBA *servers* (holding objects). The clients and servers share one unique definition of their interface, which is specified in a particular language: the *Interface Definition Language* (IDL). IDL is independent from implementation languages, thus the developer has a choice of *IDL bindings* (to C++, to SmallTalk, to ADA.) from which to select.

In addition, the CORBA architecture offers a number of services which are, from bottom to top:

- core object services which are generic: naming, persistence, events etc. which can be thought of as augmenting the functionality of the ORB;
- Common Facilities which define application-level frameworks: Information Management, Task Management etc.
- Business Objects, which are at application level and perform business-specific functions.

The TINA-C consortium integrated much of the OMG concepts into its telecom architecture. For instance its specification language, ODL, is by essence an extension to IDL. Furthermore, although TINA does not specify the communication media between software entities, existing implementations of the Distributed Processing Environment all are based on CORBA in various flavours.

CORBA represents a certain semantic shift from the state-of-the-art middle-ware approach. Two reasons appear:

0-8186-8335-X/98 $10.00 © 1998 IEEE

- The manipulation of the objects changes:
    - Classic middle-ware (XOM/XMP is a good example) provides to the programmer an API allowing to manipulate distant objects, however these objects are not in the user's name-space; instead, they are in the framework's name-space.
    - CORBA, on the other hand, brings these objects within the scope of the implementation language directly as generated code (through the IDL compiler). Operations defined in IDL are translated into target language operations. This way, the distant object is manipulated exactly as if it were local: a transparent access is possible.
- The place of the protocol changes. Whereas in known middle-ware the protocol is part of the API, in CORBA the protocol is largely hidden to the user.

These characteristics give a high level of abstraction to distributed programming. However, it has still to be studied if this abstraction does not introduce performance drawbacks; scalability is another topic to be studied as well.

The interface language definition: IDL

The way in which CORBA envisions the distribution of applications through its architecture is largely different from the OSI approach. Within the CORBA architecture, interoperability is based on the use of a software bus, the ORB (Object Request Broker), instead of stacks as in the OSI paradigm.

Definition of interfaces of software components (clients and servers in the CORBA paradigm) are made using the IDL specification language.

IDL allows to describe operations signatures. It is self-contained and does not need further specs for value transportation (the ORB insures value interoperability and marshalling).

Compared to GDMO, which conceptually lies in a similar sphere in the OSI paradigm, IDL differs from GDMO in following ways:

- GDMO has been specifically designed for TMN purposes, whereas IDL intends to be generic. This explains the constructs and entities that are aimed at the specific requirements of TMN that are found in GDMO and which are, naturally, absent from IDL. An example of such features is the Managed Object Class, which is conceptually on a different level than the class notion of IDL.
- GDMO bears features such as the containment concept (`NAME BINDING` template), Attribute Value Change specifier, packages, etc. that are not found in IDL.
- Some of the features of GDMO are tightly coupled to the TMN architecture. These features are naturally not found in IDL. An example is the containment relation, which is linked to the concept of MIB. Such features do not exist in IDL (although they can, of course, be provided by external services e.g. a relationship service as proposed by the OMG).
- GDMO, due to a historical semantic evolution, shifted slightly from a strict interface definition language, towards a system definition language: the GDMO compilers are indeed directly used to build agents and managers out of the GDMO interface specification, as if GDMO did a conceptual intrusion *within* the Operations Systems instead of staying at their interface. IDL is not, and will not, although the ODL language, an extension of IDL used by TINA-C, can be considered as going in a similar direction.
- GDMO has been extended with a powerful modelling tool, the Generic Relationship Model (GRM) which is a formalism intending to model any kind of relationship between GDMO entities, along with consistency modelling templates. IDL will clearly never need such a model.

The grammar of IDL is largely inspired from C++, however it introduces some particular constructs dedicated to interface specification; on the other hand it naturally does not feature any procedural elements. An IDL specification looks somewhat like a C++ header file: declaration of a class, its attributes, methods.

The mapping between GDMO and IDL that is chosen within the study is based on the standardisation work done by XoJIDM in its static translation proposal. The XoJIDM mapping can be thought of as a way to allow information transfer between the two types of models - however it does not intend to map the spirit behind both formalisms.

Naming

In CORBA the address space unit is the server. A server could handle one or several MOs, depending on the implementation strategy and scalability. OSI naming, on the other hand, makes use of the universality of the X.500 scheme to denote each and every entity throughout the system.

In a CORBA environment, naming and addressing are different issues. The *naming* consists mainly of the host name and port number that allows to reach the CORBA server while the *addressing* concerns the object reference which doesn't depend on the implementation at all.

And since an object reference depends on the name of the server, the IP address and the port, it is not location transparent and the object can not be moved in another location transparently. The role of a Naming Service consists of keeping the association between logical names and object references.

In OSI, the naming tree is responsible for retrieving the MO and, through scoping and filtering it allows multiple object selection. CORBA is optimised for single object selection. To provide this functionality the CORBA Naming Service [COSS] is used and the root object of the OSI naming tree should be registered in the CORBA Naming Service.

In a nutshell:

Compared to previous technologies or designs, CORBA realises three shifts:

- *from proprietary to open, inter-operating*
- *from an API-based approach to an object-oriented approach*
- *from explicit remote objects to implicit remote objects.*

Joint interdomain management (JIDM)

XoJIDM produces two documents:

1- The Specification Translation [JIDMs] relates to the static/compile-time environment. It defines a mapping of ASN.1 to IDL types, using exclusively 'typedef'. Complex constant values cannot be represented in IDL, they are defined as operations returning the constant value. The generated IDL types ignore some ASN.1 specifications: defaults values, tagged types, constrained types and subtypes appear only in IDL comments. Those features do not need to appear in the IDL, which is concerned only by the way the data is coded and transferred, not the constraints applied on their use. The same document presents a translation from GDMO to IDL. A managed object class maps to an IDL interface with the same name, plus two additional interfaces which support multiple replies and notifications. We will extensively use this work in our architecture, for it carries out a mandatory step for the migration from OSI to CORBA TMN.

2- The Interaction Translation [JIDMi] relates to the dynamic/run-time environment. It describes how the CMISE services can be performed by CORBA entities. However, this JIDM work is still in a draft status at the time of delivery of this article.

TINA architecture from a TMN point of view

The TINA architecture as seen from a TMN perspective, delivers a number of powerful features to network management:

- thorough object orientation

The OSI framework, while defining a genuine object model for the information modelling itself, is not object oriented in its component architecture nor in its handling of inter-component communication. In contrast, the TINA framework is object oriented in all of its aspects, including the computational architecture and the way in which the DPE handles communication (since it is based on CORBA features in this context).

- platform

The DPE realizes a qualitative step beyond existing management platforms of the ISO/ITU-T world. Its CORBA-based distribution infrastructure hides physical implementations, technologies and DCCE details (e.g. Kernel Transport Network).

- NRIM information model

The NRIM information model brings a technology-independant view of the network that can be mapped onto existing ISO information models. The corresponding management level in ISO terms is the Network Management Layer (NML).

- Business Model

TINA proposes a sophisticated Business Model filling the relative weakness of OSI's specifications regarding the Service Layer. This model is largely in line (although not equal to) with the NMF definitions at this level. The Business Model based applications and components represent as well a streamlined opportunity to integrate the latter with existing OSI based systems in a hierarchical applicative layout (see below).

2. Migration perspectives and proposed approach

2.1 Possible perspectives of TMN-TINA co-existence

When examining the different scenarios of TMN systems migration toward the TINA architecture, different co-existence options come to mind. These perspectives, or migration paths, differ regarding the relative topology of OSI and TINA systems one to another on one hand, and regarding the way in which information is really exchanged between both worlds on the other.

As will be perceptible through the following descriptions, the applicability of one particular option depends on the applicative context, and on the degree of coupling between OSI and TINA/CORBA entities.

Applicative vs. peer-to-peer

The exchange of information between OSI/TMN-based components and TINA-based components may involve two different semantics :

- In the *applicative* relation, a TMN-based component offers a specific, application-oriented access to a TINA component. This topology makes interoperation possible through a particular (ad hoc) interface for a specific application.

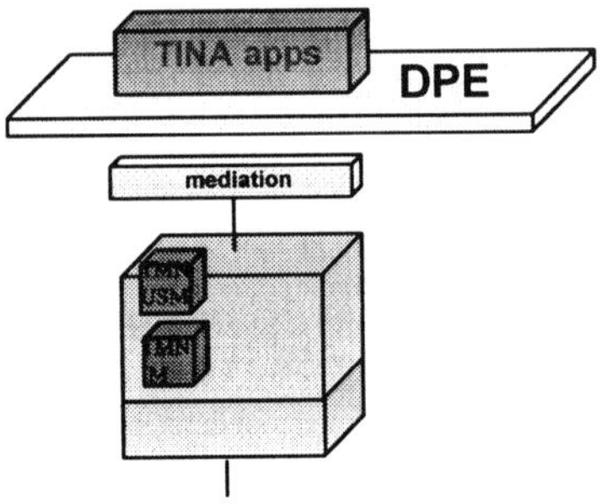

applicative relation.

The applicative relationship between TINA and OSI components represents a realistic migration scenario from TMN to TINA, because it inherently preserves existing OSI installations and allows to provide TINA service layer applications to the customer at the cost of a mediation device (CMIP-CORBA gateway).

- In the *peer-to-peer* relation, the TMN-based components and the TINA-based ones behave exactly in the same manner, with respect to the DPE. In this case, all applications have access to the managed objects disregarding whether they reside in TMN agents or in TINA components.

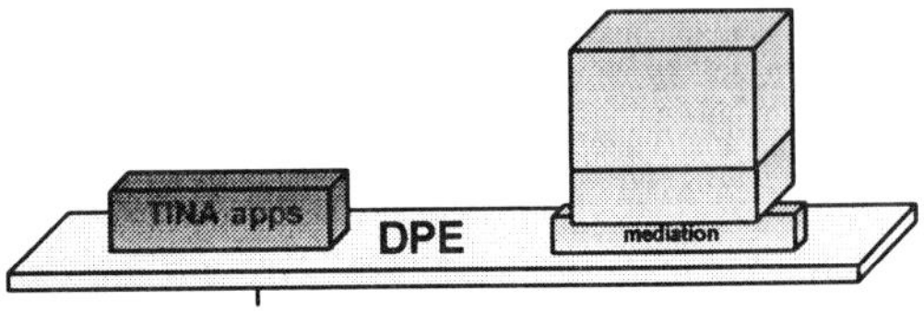

peer-to-peer relation.

The peer-to-peer relation projects mixed TMN systems bearing both ISO components and TINA components on the same level of responsibility regarding ressource management. The DPE insures communication between all components. A mediation couples the DPE with the OSI components.

loose vs. tight coupling

Depending on the tightness of the relation between TMN and TINA components, two layouts can be distinguished :

- In the *loose coupling* layout, the two types of components reside in distinct protocol domains, like « pure » entities (they communicate each in their own protocol). The interoperation requires therefore an additional component, the mediation. Loose coupling enables out-of-the-box or existing applications to communicate easily without any modification. This allows to envision CORBA interoperability in the context of deployed TMN software at a minimum redesign cost.

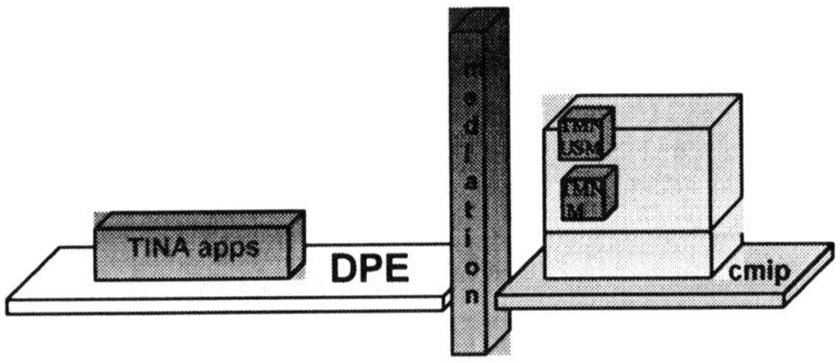

loose coupling layout.

The loose coupling layout attempts to keep either interoperating system completely unaffected by the respective other protocol, concentrating all the translation mechanisms into the mediation component. Loose coupling enables to reuse existing OSI components directly.

At the opposite, in the *tight coupling* layout some bi-domain entities are able to communicate both through an ORB and through CMIP. In a sense, these entities encapsulate their own mediation. In this situation, a CORBA manager is able to communicate directly with a CMIP agent. Tight coupling, when combined with the ability to recompile/relink existing application code, makes existing applications access transparently new CORBA entities. Indeed, as this layout proposes really an extension of the communication classes of the TMN framework towards CORBA, tight coupling brings the advantage of reuseability of existing TMN software.

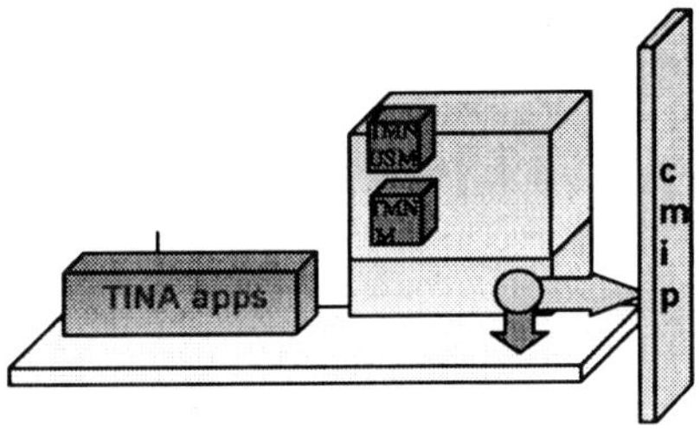

tight coupling layout.

Tight coupling introduces a new kind of components not any more bound to one distribution technology. More precisely, the ability to communicate with respects to these protocols is a property of the object instances themselves, instead of a centralized scheme where some dedicated service would provide communication capabilities. Implementations make use of the inheritance and polymorphism properties of OO languages in order to confer to the objects the new behavior in a transparent manner.

2.2 A migrating OSI MIB

In OSI Systems Management (OSI-SM), managed elements or management applications that assume an agent role provide management interfaces consisting of the formal specification of management information and of an access service/protocol that is mapped onto a well defined protocol stack. While the management information specification provides the MIB *schema*, object discovery and multiple object access facilities allow applications in manager roles to dynamically discover existing object instances. Operations to objects are always addressed through the supporting agent, which provides query facilities in a database-like fashion.

The CORBA operational paradigm is different to that of OSI and Internet management, as it originates from the distributed system world. CORBA objects are specified and accessed separately, in contrast to the managed object cluster administered by an agent. CORBA objects are most commonly addressed by *type* and not by *name*. This is due to the nature of distributed systems where, typically, instances of the same type offer exactly the same service e.g. printer servers, statistical calculation servers, etc. Of course this does not mean that there are no support mechanisms to distinguish between instances of the same type (name servers, traders). It means though that the whole framework is optimised towards a "*single object access, address by type*" style of operation, in contrast to the manager-agent model which is optimised for "*multiple object access, address by name*" style of operation.

If CORBA is used as the underlying access and distribution mechanism for a TMN-based management system, managed objects could be mapped one-by-one onto CORBA objects, accessed by client objects in managing roles. The key difference is that, in CORBA, clusters of managed objects logically bound together are not seen collectively through an agent. As such, an important issue is to provide object discovery and selection facilities similar to OSI scoping and filtering. Such facilities are very important in management environments where many instances of the same object type typically exist, and names are not always known in advance e.g. objects representing current connections.

The use of the ODP / OMG CORBA model for network and service management presents a number of difficulties to overcome due to the dynamic nature of management information and the number of managed objects typically present in managed elements. On the other hand, the OSI-SM model scales much better and has already been used for managing large telecommunications infrastructures e.g. SDH/SONET, ATM, etc. Based on this observation, an ideal framework would combine the expressive power of OSI-SM and the programmability, portability and distribution aspects of ODP / OMG CORBA. In order to specify such a framework it is important to be able to map management information specifications in GDMO to computational interfaces in IDL. The information modelling aspects of the two frameworks are similar and the X/Open Joint Inter-Domain Management task force (XoJIDM) has defined rules for this mapping [JIDM].

In OSI Management, managed objects can be accessed collectively through the CMIS/P scoping and filtering facilities. These may be used for discovery services and they minimise the management traffic incurred on the managed network. In addition, the same operation may be performed on many managed objects which is not only an engineering-level optimisation but also allows a higher level of abstraction to be provided to managing functions. Discovery facilities may be provided through naming servers and traders in CORBA but the efficiency of such mechanisms, with potentially thousands of transient managed objects in network elements, needs to be evaluated. In addition, the CMIS/P operational paradigm with potentially multiple operations expressed through a single request is lost, unless similar facilities are provided over CORBA.

Our proposal is that the operational framework of OSI management is retained over CORBA through Management Brokers (MBs). A logically bound cluster of managed objects similar to an OSI/TMN agent application is administered by a management broker providing multiple object access facilities similar to CMIS. Of course, managed objects may be also accessed directly in the standard CORBA fashion. Event management is provided by event discriminators and logs through filtering, in order to overcome the relevant CORBA limitations. Finally, the rest of the OSI Systems Management Functions [SMF] are maintained as generic CORBA objects that may be instantiated within a cluster. This approach essentially maintains the OSI operational model over CORBA but replaces the access (i.e. CMIS/P) and distribution (OSI directory) mechanisms. As such, it retains the OSI management expressive power, event model and generic management facilities while it benefits from the distribution, portability and easy programmability of CORBA. The approach is depicted in the figure below.

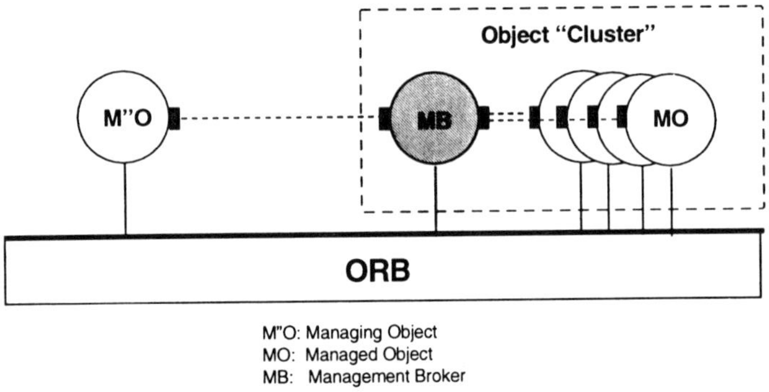

The OSI-SM Operational Model Over CORBA

A Management Broker does not only have a CMIS-like interface but also interfaces to be "informed" about new objects it needs to administer and to receive notifications from the managed objects it administers. In addition, every MO inherits from a *ManagedObject* interface through which the MB notifies it that it administers it. A MB and the objects it administers may be physically distributed but they logically form a "cluster" which can be collectively accessed. A managed object may belong to more than one MB domain. In the case of managed network elements, at least one MB needs to be physically located together with the local managed objects so that it provides optimised access and event dissemination facilities with minimal management traffic. In general, it is not necessary to provide separate IDL interfaces for every managed object as this may not be technologically feasible with the current state of CORBA implementations. In this case, the managed objects and the broker may interact through a local mechanism e.g. internal C++/Smalltalk interfaces if they share a common address space (an engineering "capsule" in ODP terms).

3. Long-term perspective

Rather than using dedicated OSI-SM-like Management Brokers, it is possible to provide similar clustering, searching, notification, etc. facilities through the use of CORBA services. A combination of the services provided by name servers, traders and event/notification servers (with some extensions) could reproduce the necessary features of OSI-SM. One problem with the use of CORBA is that federation is a key aspect in order to achieve scaleable management systems. It would be necessary to have dedicated name servers, traders and event/notification servers for every logical cluster of managed objects, e.g. in every managed element, in order to reduce traffic and increase real-time response. These "low-level" servers will be unified by "higher-level" servers in a hierarchical fashion but federation issues have not yet been worked out and are not simple. In addition, even with such facilities in place, the generated management traffic will be at least twice that of OSI management. With CORBA, matching object references will be returned to the client object and the operations will be performed on an object-by-object basis. In OSI management, the multiple object access request will be sent in one packet while the results will be returned in linked replies, one for each object accessed.

An example of the use of hierarchical traders to provide global filtering features is given next.

A trader [X9xx] supports sophisticated queries, matching sought properties of the target object(s). Objects can export their interfaces to the trader together with a list of *attributes* and a list of *properties*. Clients may request the object references of a particular type that match assertions on attributes and properties. The difference between the latter is that attributes may change dynamically while properties are fixed during the lifetime of an object instance. As such, the trader needs to evaluate assertions on attributes by retrieving them from all the instances of the type associated with the query. The function of the trader is very similar to filtering in OSI management. A key difference is that only interfaces of a particular type can be found through the trader. An additional difference is that filtering is tightly coupled with OSI managed objects through the supporting agent while the ODP/OMG trader is a separate server.

Traders can be federated in order to be able to cope with big object spaces and different administrative domains, as demonstrated in the following figure.

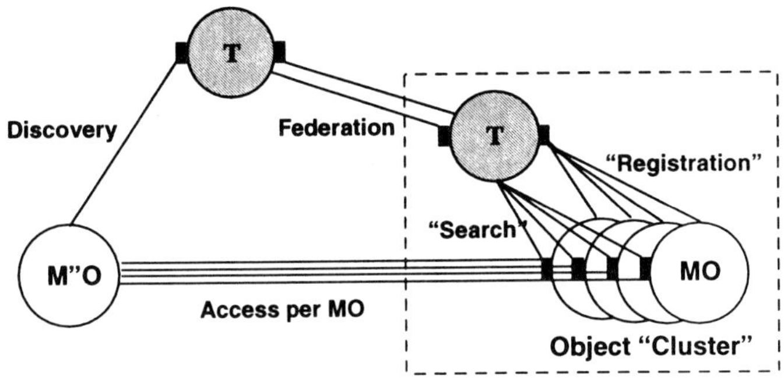

4. Conclusions and remaining issues

The perspective of concretely using TINA as the architecture for deployed TMN systems raises, from the perspective of TMN requirements, a number of issues related to the specificities of TMN applications. Indeed, TMN systems require a degree of availability and scalability seldom encountered in IT.

The scalability issue hides, in turn, a number of architectural questions. Today, deployed systems are able to handle on the magnitude of 1^E6 managed objects while providing a high level of resiliency. The core property enabling this functionality is, fundamentally, the *structure* of the objects through the naming tree, naming scheme (inherited from X.500) and containment relationship.

While being well aware that these notions are partially taken over by the TINA/CORBA architecture (e.g. XoJIDM naming emulates largely X.500 naming, but without the notion of naming tree as such), present support components such as the trader, which take in charge the retrieval of managed objects according to search conditions, are still not validated regarding full scale applications. Other research directions include a methodology to actually group managed objects into scalable *clusters* or *sets* which reproduce to some extent the configuration of an OSI agent within the TINA paradigm, while hiding the vast majority of object references to the DPE.

The migration of the management systems presented retains the operational model of OSI-SM/TMN over a CORBA-based Distributed Processing Environment. This approach is based on the results of the XoJIDM work for mapping GDMO specifications to IDL but it focusing on a native CORBA-based Open Distributed Management Architecture. This approach provides a smooth migration path from current TMN-based management systems to target systems operating over CORBA-based DPEs. While there is plenty of ongoing research regarding OSI-SM/TMN and CORBA migration and interworking, the approach presented in this paper retains the relevant advantages of TMN for network management while it is both compliant and complementary to the JIDM approach.

References

[X701] ITU-T Rec. X.701, *Information Technology - Open Systems Interconnection - Systems Management Overview*, 1992.

[SNMP] J.Case, M.Fedor, M.Schoffstall, J.Davin, *A Simple Network Management Protocol (SNMP)*, RFC 1157, 1990.

[X901] ITU-T Rec. X.901, *Information Technology - Open Distributed Processing - Basic Reference Model of Open Distributed Processing - Part 1: Overview*, 1993

[CORBA] OMG, *The Common Object Request Broker Architecture and Specification (CORBA)*, 1991.

[M3010] ITU-T Rec. M.3010, *Principles for a Telecommunications Management Network (TMN)*, SG IV,1996.

[TINA] *An Overview of the Telecommunications Information Networking Architecture (TINA)*, TINA'95 Conference, Melbourne, Australia, 1995.

[X720] ITU-T Rec. X.720, *Information Technology - Open Systems Interconnection - Structure of Management Information -Management Information Model (MIM)*, 1991.

[X722] ITU-T Rec. X.722, *Information Technology - Open Systems Interconnection - Structure of Management Information: Guidelines for the Definition of Managed Objects (GDMO)*, 1992.

[IDL] OMG, *Specification of the Interface Definition Language (IDL)*.

[COSS] OMG, *Common Object Services Specification (COSS) - Event, Life-Cycle, Name, etc.*, 1994.

[X710/11]ITU-T Rec. X.710/711, *Information Technology - Open Systems Interconnection - Common Management Information Service Definition and Protocol Specification (CMIS/P) Version 2*, 1991.

[IIOP] OMG, *CORBA Internet Inter-Operability Protocol.*

[X9xx] ITU-T Rec. X.9xx, *Information Technology - Open Distributed Processing - Trader.*

[SMF] ITU-T Rec. X.730-750, *Information Technology - Open Systems Interconnection - Systems Management Functions.*

[JIDM] X/Open / NMF, *Joint Inter-Domain Management (JIDM) Specifications - SNMP SMI to CORBA IDL, ASN.1/GDMO to CORBA IDL and IDL to GDMO/ASN.1 translations*, 1994.

Session 9

DPE Integration

Gerd Schuermann, GMD FOKUS

An extensible and modular Real-Time TINA DPE Kernel

Frédéric Dang Tran, Bruno Dumant, Jean-Bernard Stefani
France Télécom CNET
DTL/ASR, 38-40 rue du Général Leclerc, 92794 Issy Moulineaux Cedex 9, FRANCE
frederic.dangtran@cnet.francetelecom.fr

Denis Métral-Charvet, Stéphane Carrez, Christian Jacquemot
Chorus Systemes
6 Avenue Gustave Eiffel 78182 Saint-Quentin-En-Yvelines CEDEX FRANCE
denis@chorus.fr

Abstract

This article describes a minimal and modular TINA-compliant DPE Kernel architecture which allows the construction and configuration of DPE instances tailored to the specific needs of their environment and which provides the necessary handles for the execution of real-time applications with end-to-end QoS constraints. The proposed architecture features a flexible binding framework allowing the introduction of arbitrary forms of binding beyond the implicit binding model assumed by CORBA. In particular, the framework described here allows the explicit establishment and control of QoS-constrained bindings such as multi-party stream bindings between multimedia stream interfaces. The implementation of this DPE architecture within Chorus COOL-ORB is covered in detail.

1. Introduction

Telecommunication systems involve the deployment of complex distributed applications over a wide range of computing and networking equipment from resource-constrained devices such as portable phones or interactive TV set-top boxes, up to ATM switches, PBXs and mainframes. These systems can be qualified as open real-time systems to the extent that the set of applications together with their real-time constraints is a priori not known in advance. In terms of communications, these applications typically resort to various kinds of protocols: signaling protocols, protocols adapted for wireless communications, real-time transport protocol for continuous multimedia flows etc. All these technical requirements make it highly unlikely that a one-size-fits-all strategy is the right approach for defining a Telecom DPE architecture. Flexibility is needed to adapt and tailor the platform to the specific needs of its environment.

With the advent of interoperability protocols, CORBA2.0 ([1]) is poised to become the de-facto cross-platform, multi-language middleware for distributed object-oriented applications. However the CORBA2.0 architecture, as is, does not completely meet the requirements of real-time telecommunication applications. A large part of the CORBA2.0 architecture remains largely monolithic and disallows adapting or extending the ORB internal machinery. As a result, it is not clear how to provide adaptations required to support multimedia and real-time applications, to scale down ORBs to fit resource-constrained devices or to support multiple communications protocols. The work pursued in the context of the ACTS ReTINA project [6] is to specify and implement a flexible DPE architecture whose machinery is exposed to the programmer so that it can be populated with various binding[1] and communication mechanisms as needed. The reader is invited to refer to [15] for a thorough description of the generic binding model underpinning the proposed architecture and to [8] for a full description of the DPE architecture.

The approach adopted is to consider CORBA2.0 as a particular "personality" (i.e. a set of APIs and language mappings) and to decouple the specifics of the CORBA2.0 API from the personality-independent kernel interface. In this regard the DPE framework proposed in this paper can readily be used to built non-CORBA personalities such as Java-RMI-like platforms. This work has been inspired by the Subcontract mechanism in the Spring distributed system ([9]) and the SOR system ([10]). Both systems provide

[1] *Binding* should be understood as both the process of interconnecting different objects and as the end-result of this process, i.e. the "binding object".

0-8186-8335-X/98 $10.00 © 1998 IEEE

a flexible framework for client-server communications. The framework proposed here generalizes these approaches to arbitrary types of interactions and in particular multi-party multimedia communications.

This paper is structured as follows: section 2 gives an overview of the proposed DPE architecture and contrasts it with CORBA. Section 3 describes the core abstractions of the DPE framework. Section 4 describes the binding framework which provides a set of reuseable components for building communication objects. The main resource abstractions of the DPE kernel are covered in section 5. Section 6 describes the current implementation of this DPE framework within COOL ORB including a multimedia stream binding library which illustrates how the proposed framework can be used to instantiate complex forms of bindings. Finally, section 7 concludes.

2. Overview

The main design principle which has driven the proposed architecture is to separate basic interface management responsabilities from communication aspects. The resulting minimal DPE architecture is summarized in figure 1.

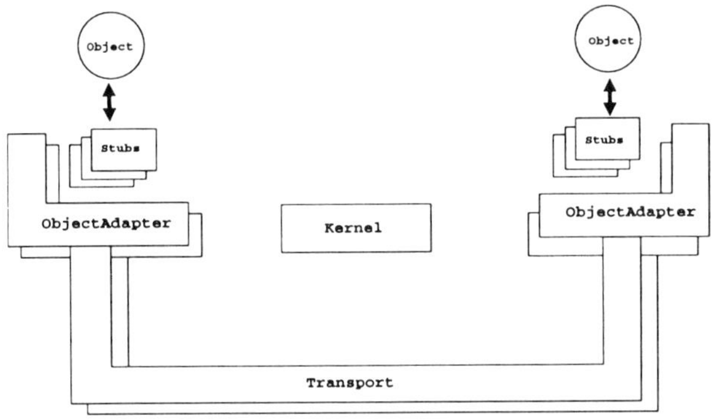

Figure 1. An open DPE architecture

The DPE base system proper is reduced to a *Kernel* which acts as a backplane for *object adapters* and transport protocols. The kernel also embeds a minimal name server used for synthesizing interface references for bootstrap purposes.

The CORBA notion of *object adapter* is extended to cover not only server-side binding mechanisms but the full end-to-end computational binding between interacting objects[2]. Object adapters encapsulate communication resources including specific messaging and transport protocols as well as specific families of stubs. The establishment of a binding between a given set of interfaces is realized by invoking an operation on an object adapter. Both implicit (as assumed by CORBA where object adapters are not visible at the client side) and explicit forms of binding are possible. Examples of object adapters include:

- A simple client-server object adapter encapsulating the IIOP protocol
- A client-server object adapter supporting atomic transactions
- A Multimedia stream adapter allowing the set-up of multiparty connections between a set of media producer and consumer objects.

Exposing the DPE binding machinery in such a way makes it possible to introduce new communication mechanisms and invocation semantics encapsulated in object adapters written by application developers and possibly discovered and installed at run-time.

The next section will describe the interface of an object adapter and its related abstractions.

Object adapters provide a first level of "replaceability". The proposed DPE framework includes a binding framework which makes it possible to reuse binding components across object adapters. This will be covered in section 4.

3. Core Framework

The foundation of the proposed DPE framework is made up of the abstractions summarized in figure 2 (together with a possible implementation of a CORBA personality).

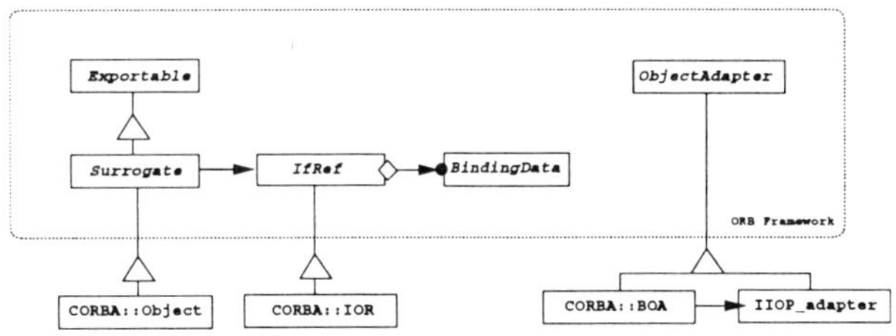

Figure 2. DPE framework basic types

The *Exportable* abstraction[3] is the central abstraction of the framework. The type *Exportable* is the supertype of all types of interfaces that can be exported outside their address space. A *Surrogate* is a representative for a (possibly) remote interface. Note that *Surrogates* are not necessarily stubs. Each object adapter is free to implement surrogates directly as stubs (as in typical CORBA platforms) or to defer the instanciation of stubs at an object-adapter-specific

[2]As shown in figure 1, these objects are not necessarily clients and servers; for multimedia applications, they could be stream endpoints.

[3]The terminology used for naming the different framework abstractions adheres to the vocabulary of the ODP Reference Model [13].

time (e.g. as a result of the first invocation on the surrogate or as a result of an explicit binding request). The latter option enables a late binding at the stub level and can be interesting when the establishment of a binding is a costly operation (e.g. multimedia stubs embedding media compression functions).

As a representative of a remote interface, a surrogate must provide a way to unambiguously designate that interface. For that purpose, each *Surrogate* is associated with an *IfRef* which abstracts on-the-wire representation of interface references. Each *IfRef* is associated with a key which unambiguously identifies the interface within its address space. An *IfRef* can be populated with *BindingData* which provide object-adapter-specific information necessary to create a binding towards the referenced interface (a communication endpoint address,...). CORBA IORs are a specific implementation of on-the-wire interface references.

The two key functions of an object adapter are *export* and *bind*:

```
interface ObjectAdapter {
  Surrogate export(
       in Exportable it,
       in Type type);
  Exportable bind(
       in sequence<Surrogate> targets,
       in QoS qos);
       ....
};
```

An object adapter can export an interface, i.e. register a local interface so that it can be transmitted outside its address space. Actually, the interface submitted to the object adapter is not necessarily a local interface but can be a surrogate representing a remote interface. The latter possibility makes possible a recursive binding process. A *Surrogate* can be the representative of a remote interface which is itself a surrogate for a remote interface and so on. Such a chaining and referencing scheme greatly simplifies the construction of bridges since there is no need to construct application-level proxies.

The object adapter offers a *bind* operation that is used to explicitly create bindings. It takes as parameters a set of targets represented by surrogates and a `qos` parameter representing the expected quality of service[4] to be guaranteed, and returns an interface that represents the binding and that will be used to interact with the targets. In the client-server case, there is only one target, and the returned interface is a surrogate of the target, usually implemented as a client stub. This operation uses the ifrefs of the targets to establish a communication channel to the specified targets.

[4] the `qos` parameter is currently a placeholder. Specifications of quality of service constraints is beyond the scope of this work.

4. Binding Framework

The binding framework is the part of the overall DPE framework which provides a set of cooperating abstractions for the construction of object adapters and their associated communications resources. This framework needs to fulfill the following requirements:

- support for both RPC-like protocols (e.g. IIOP) and stream protocols
- protocol independence of stubs
- arbitrary and configurable degree of resource sharing between bindings (e.g network connections)

A binding object is made up of a stub layer responsible for (un)marshalling invocation data[5] and a transport layer responsible for transferring data across address spaces. The current approach attempts to maintain a complete separation between stubs and the underlying protocols, so that protocol-independent stubs could be reused across object adapters. However the framework does allow an object adapter to provide its own family of specialized stubs.

Each object adapter is associated with:

- A registry of stub factories
- a *Marshaller* which provides the basic abstraction for (un)marshalling rules
- a protocol stack (see section 4.1 below)

A given object adapter can reuse functionality of an existing objet adapter either by derivation or through delegation.

4.1. Communication framework

The communication framework of the ReTINA DPE is greatly inspired by the x-kernel framework [16].

A binding consists of a chain of *session* objects. Session objects are created by *protocols* on the request of object adapters.

- protocols are abstractions of protocol machines at a given site; they manage the establishment and release of sessions; the protocol abstraction directly represents protocols like TCP, IP or GIOP. Protocols have higher and lower interfaces. Methods of the higher interface of a particular protocol may be invoked by protocols placed over it in the protocol stack, or by an object adapter. Conversely, methods of the lower interface may be invoked by protocols placed below it in the protocol stack;

[5] Here the term "stub" refers both to client-side and server-side stubs (skeleton in CORBA speak)

- sessions are logical communication channels that obey the particular communication protocol that dynamically created them.

These abstractions are very convenient to build arbitrary protocol stacks and protocol independent stubs. Stubs here simply constitute the local ends of bindings.

The aim of this framework is to let arbitrary protocol graphs be maintained, with arbitrary sharing of resources. In the following, we will use the situation depicted on figure 3 as an example of a particular protocol graph; TCP, UDP, RTP, IP, GIOP correspond to the well known protocols; TORB represents an arbitrary invocation protocol, and Stream a specific protocol for manipulating video and audio streams; Magic is a very simple protocol whose purpose will be explained later.

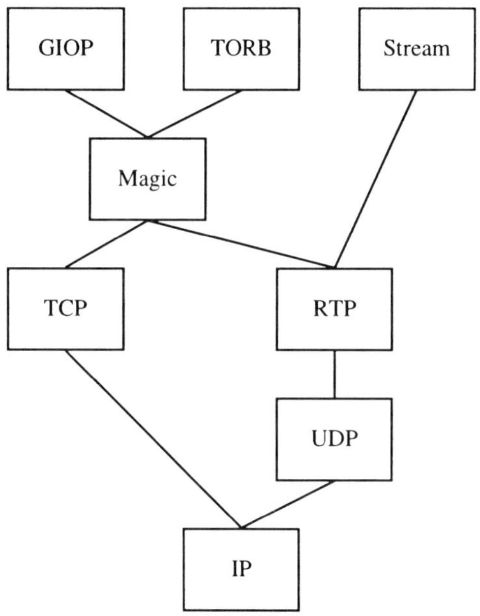

Figure 3. Protocol graphs

To the `export` and `bind` operations on object adapters correspond `export` and `bind` operations on protocols that are carried by the higher protocol interfaces:

```
interface Protocol_High {
  EndPoint export(
      in Protocol_Low hlp,
      in sequence<byte> pid,
      in sequence <Protocol_High> stack);
  Session_High bind(
      in Protocol_Low client,
      in sequence<byte> pid,
      in sequence<EndPoint> participants);
  Session_High bind(
      in sequence<EndPoint> participants);
}
```

`export` is used by a protocol to inform a lower level protocol that it is willing to accept a connection using the specified protocol stack. It is also invoked by the object adapter on the higher interface of the protocol at the top of the protocol stack to initiate the export process. `hlp` represents the lower interface of the invoking protocol, while `protocol_stack` represents the stack of protocols to which the invoked protocol will in turn export itself. The `pid` argument is an identifier that the lower level protocol may use to designate the higher level protocol.

In our example, the GIOP protocol would call the export method on the higher interface of the Magic protocol specifying the four bytes 'GIOP' as a protocol id, and the stack would consist of the two protocols TCP and IP. This call dynamically creates, if necessary, the resulting protocol stack, illustrated by the path GIOP $\rightarrow$ Magic $\rightarrow$ TCP $\rightarrow$ IP in figure 3. In the same example, the only purpose of the "magic" protocol is to add a header (consisting of the protocol id of the exporting protocol) to the messages enabling the coexistence of several protocols on top of the same TCP connection.

The `export` operation returns an `EndPoint` object: an `EndPoint` object represents a participant to a binding at a given level. The `EndPoint` type is refined by each protocol so that protocol specific information (host name, port, object key...) can be added to it. This information can typically be encoded as binding data in an ifref.

The `bind` operation is used by a protocol to establish a communication channel with a set of participants described by `EndPoint` objects. This operation comes in two flavors, depending on whether the client is willing to use this channel only to send messages, or also to receive some. The result of the bind operation is a session object, or more precisely a chain of session objects corresponding to the different protocols used that let a client send messages in the communication channel using a particular protocol stack.

Figure 4 summarizes the binding creation process in the case of the creation of an IIOP binding. The binding process is initiated by an object adapter that takes as input an ifref, e.g. provided by a stub . The object adapter decodes the IIOP binding data from the ifref, creates an endpoint representing the target object, and passes it as a parameter in the bind invocation on the GIOP protocol, that will in turn call the bind operation of the next protocol in the stack, i.e. Magic. The process continues until the IP protocol is reached. The IP protocol returns an IP session, that will be used by the TCP protocol to "extend" a TCP session, and so on until the GIOP protocol returns a GIOP session encapsulating all the lower sessions.

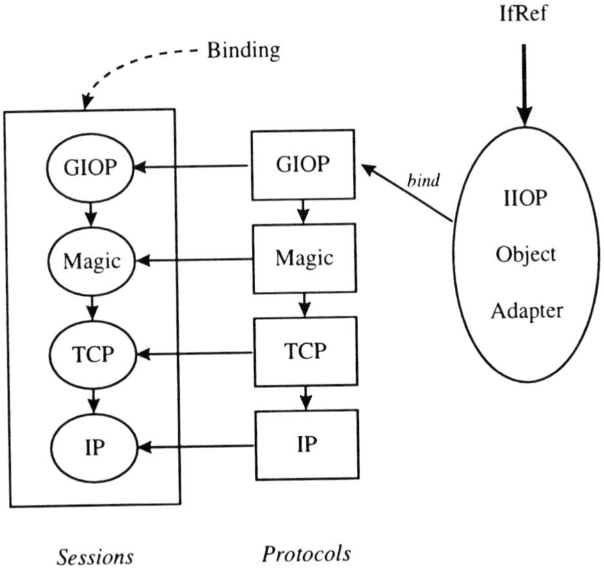

Figure 4. Creating a binding

5. Kernel resource abstractions

Resource abstractions encapsulate operating system resources and are defined in order to provide applications with fine-grained control over all computing and networking resources. This is an absolute prerequisite for the construction of applications with end-to-end QoS constraints. The main design principle is to separate mechanisms from policies. In this regard, the present real-time DPE kernel does not mandate any particular policy such as rate monotonic scheduling policy or priority inheritance. Instead it defines a basic mechanism layer on top of which particular policy modules can be plugged in. The main kernel resource abstractions are:

- *Capsules* which represent an address space and a unit of allocation for *threads*
- *Activities* which represent a unit of sequential execution; each *activity* can be configured with policy-specific scheduling parameters
- synchronization abstractions (*mutex, semaphore, entry*). These resources can notify particular policy modules which control how threads are queued and how scheduling parameters are propagated.
- *entries* which allow to control the dispatching of invocation and the level of concurrency on a per-interface basis.

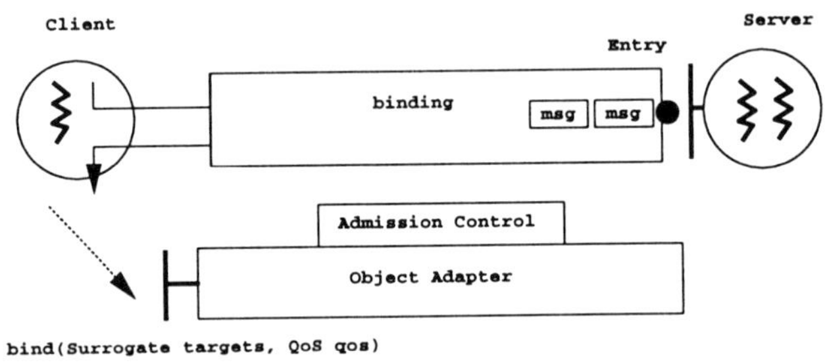

Figure 5. A QoS-constrained explicit binding

Figure 5 illustrates how these abstractions are utilized in the case of a simple client-server QoS-constrained binding. A client explicitly establishes a binding toward a given target interface by invoking a *bind* operation on a particular object adapter (e.g. an adapter encapsulating ATM networking protocols). This adapter attempts to create the binding. This phase includes possible interaction with specific admission control modules (e.g. at the ATM network level and for both CPUs) to ensure that the end-to-end QoS requirement can be met. The present resource framework allows to select specific resource management policies all the way including (i) how threads are scheduled on the client and server capsule, (ii) how server threads are queued behind the server *entry* and (iii) how messages scheduling parameters are propagated to the server thread. As an illustration of this kernel resource framework, the reader is invited to refer to [12] for a description of an admission control scheme for end-to-end QoS-constrained bindings.

6. The CHORUS/COOL ORB DPE Kernel

COOL ORB has been used to develop the ReTINA DPE kernel, in the context of the ACTS ReTINA project. The current COOL ORB implementation does not yet cover the full set of the ReTINA DPE Kernel specifications but it integrates progressively the different features described in the previous sections. One of the demonstrations of this integration is the Multimedia Object Adapter introduced in section 6.2.

6.1. COOL ORB Overview

COOL ORB provides a CORBA2.0 personality on top of a partial incarnation of the ReTINA DPE Kernel as depicted in figure 6.

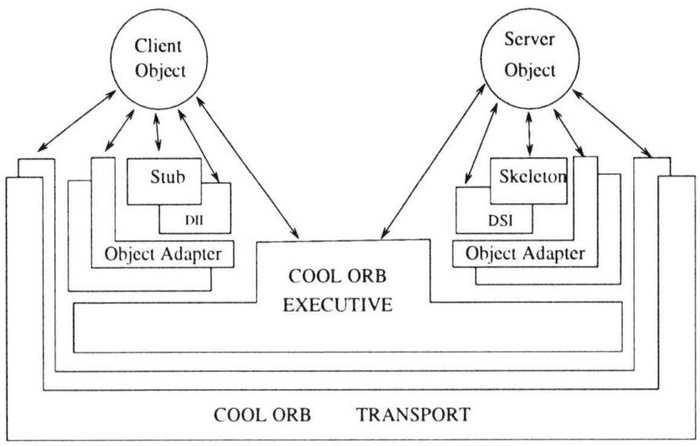

Figure 6. COOL ORB Architecture

The COOL ORB Executive component corresponds to the DPE Kernel proper (see section 2). It provides a set of platform independent abstractions allowing applications to control the usage of execution resources: Following the ReTINA resource abstraction specifications introduced in section 5, the notions of *capsules*, *activities* as well as the synchronisation tools such as *local or distributed mutexes, semaphores or multiple-reader-single-writer-locks* and *Pools of activities* are supported by COOL ORB . Portable interfaces are exported for all these abstractions.

In conformance with the kernel architecture covered in section 2, the COOL ORB Executive component can be configured with several object adapters and provides an interface to select and parameterize an object adapter at bind time. The multimedia stream adapter described in section 6.2 is an example of an implementation of an application-level object adapter.

The COOL ORB Transport Component conforms to most of the specifications described in section 4. This component includes the following features:

- *Multi-protocol support* allows the co-existence of multiple RPC-like protocols (currently COOL protocol or IIOP are supported) and the manipulation of different transport protocol stacks such as TCP/IP, OSI or CHORUS/IPC. In addition, COOL features a *Transport and connection management* service which provides a programming interface to access and modify the transport characteristics (such as buffer sizes) and allow the explicit control of the established connections by providing open/close operations on the connections.

- *Variable sized object references* are a partial implementation of the *IfRef* abstraction covered in section 3. The CORBA Object interface is extended to allow an application to associate arbitrary binding data with CORBA object references as illustrated in figure 2. Binding data are defined as components identified by tags and containing data marshaled using *Marshaler* objects.
This concept is typically exploited to build multimedia stream interface references as explained in section 6.2.

- *Marshaler* objects (see section 4) encaspulate a given marshaling format and allow reliable marshaling and un-marshaling of the arbitrary data stored in a component a variable sized object reference.

- *Proxies* enable to modify or extend the behavior of an object invocation by deriving application oriented stubs from the default ones.

- *Interceptors* are application objects inserted at runtime at the different steps of the processing of an object invocation. Interceptors provide means for applications to modify the normal semantics and may for instance be used to determine quality of service parameters, implement security policies, gather instrumentation data, etc.
Different kinds of interceptors can co-exist:
  - *Invocation protocol interceptors* called before and after the marshaling of arguments,
  - *Object adapter interceptors* called prior to object localization and after an invocation,
  - *Transport protocol interceptors* called when requests are sent or received, when connections are closed or opened, etc.

To conclude this overview of the COOL ORB based ReTINA DPE Kernel implementation, it should be noticed that the internal structure of COOL ORB follows an object oriented design approach where features are isolated in independent components. These components interact through well-defined internal interfaces, which allow advanced users to replace or complement existing ORB components with their own or this allows to add new components such as new object adapters.

6.2. A multimedia stream communication framework

This section describes a multimedia stream communication framework which is a specialization of the proposed ORB framework in the case of the transmission of continuous media flows. This framework allows the construction of multiparty stream binding objects involving the transmission of continuous flows with the ability to plug transport protocols as well as media coders. The objective of this framework is to devise a "multimedia-aware" ORB, i.e. an ORB with native support for "programmable stream interfaces". Currently stream interfaces are specified using standard CORBA IDL. Such a framework could be used as the foundation for a full-fledged multimedia binding service for controlling and managing multimedia streams which is beyond the scope of this work.

The key component of this library is a Multimedia Stream Object Adapter (*MMStreamOA*) which is a specialization of the Object Adapter abstraction described in section 3. Figure 7 shows how this adapter fits in alongside CORBA components. Standard operational client-server interactions go through the CORBA personality components (i.e. stubs and skeleton, BOA, IIOP protocol stack), whereas stream interaction is mediated by the *MMStreamOA*.

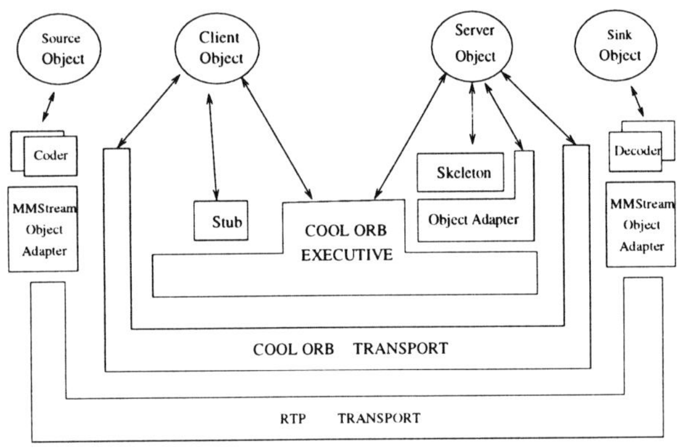

Figure 7. A Multimedia ORB

The *MMStreamOA* is associated with (i) an extensible registry of stubs which act as software video/audio codecs and (ii) a set of transport protocols suited for the transmission of continuous media flows. The current implementation is restricted to video flows and features software codecs for various types of video formats (JPEG, H261, etc.) as well as an implementation of the RTP protocol on top of multicast IP. How these protocols fit into the communication framework is illustrated in figure 3.

The *MMStreamOA* is specialized for a limited set of predefined IDL interface types through which media frames are received or transmitted :

```
interface VideoFlowSink {
      void consume(VideoFrame frame);
};

interface VideoFlowSource {
      void produce(VideoFrame frame);
};
```

The current `MMStreamOA` enforces an explicit binding model for stream interfaces. Upon unmarshalling, say a `VideoFlowSink` interface, an unbound surrogate is instantiated. This surrogate will raise an exception if invoked directly. Thus, the cost in terms of resources (network connection, codec buffers, protocol code in case of dynamic linking, etc.) is only incurred for applications actually using the stream interfaces for transmitting media flows, not for third-party objects.

The establishment of a stream binding is achieved by invoking a specific operation on the stream adapter. For instance a video source object will invoke the following operation on the stream adapter:

```
VideoFlowSink bind_source(
          VideoFlowSink sink, QoS qos);
```

Various binding scenarios are possible where the binding initiator can be a source object, a sink object or a third-party object. A source object does not necessarily know whether a target sink interface represents a unique interface or several sink interfaces in the case of a one-to-many stream binding.

The stream adapter is associated with a specific stream interface reference format. Stream interface references generated by the *MMStreamOA* are fitted with the following multimedia-specific binding data (in the case of CORBA IORs, this is mapped to a private protocol profile) containing protocol information as well as the media formats supported by the referenced object.

```
// CORBA IDL

struct MMStreamBindingData {
 ProtocolId proto_id;
 string network_address;
 Role   role;
 sequence<media_t> media_formats;
};
```

This allows a lightweight binding set-up since all the information needed to connect to the remote stream interface is cached in the interface reference.

Figure 8 depicts an established video bridge between two video connections conveying a video flow in JPEG and H261 format respectively. This application is an illustration

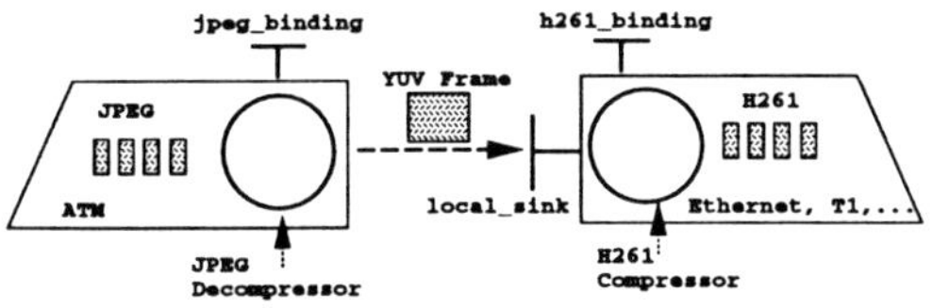

Figure 8. A video bridge

of the bridging mechanisms covered in section 3. The two video binding objects are directly connected together without the need to create an application-level proxy. The H261 compressor acting as a client stub behaves as an implementation object as far as the JPEG video binding is concerned.

7. Conclusion

This article has presented a high-level framework for distributed object-oriented systems from which it is possible to derive highly flexible and configurable TINA DPE platforms whose machinery is made up of replaceable components and which allows arbitrary forms of binding beyond the client-server model of CORBA.

The main tenets of this DPE framework have been implemented in the proof-of-concept TORBoyau platform [11] (written in C++) which features a minimal CORBA API along with a multimedia streaming API. In the context of the ReTINA project, a Java implementation of this framework is under completion. The dynamic linking and late binding capabilities of Java combined with the flexible binding framework of the DPE contribute to achive a very high degree of integration between system and language-level binding. For the specific case of CORBA2.0, opening up its architecture may result in portability interfaces beyond what the CORBA specifications currently cover so that ORB sub-components (e.g. stubs, protocol stacks) could be reused across ORBs from different vendors.

The COOL ORB product [5] from Chorus Systemes [4] provides today the level of flexibility described here as illustrated by the multimedia stream library covered in section 6. This library has been used in the context of a distributed information service demonstration.

The modular real-time ORB architecture described in this paper is currenlty being used as input for answering RPFs issued by the OMG Real-Time Special Interest Group: the Minimum ORB RFP ([2]) and the Real-Time RFP ([3]).

References

[1] The Common Object Request Broker: Architecture and Specifications, CORBAV2.0. Object Management Group, July 1995.

[2] Minimal ORB RFP, OMG Document: orbos/97-06-14, July 1997.

[3] Real-Time CORBA 1.0 RFP, OMG Document: orbos/97-09-01, September 1997.

[4] http://www.chorus.com

[5] CHORUS COOL-ORB r4 product description, CS/TR-96-207

[6] http://www.chorus.com/Documentation/retina.html

[7] ReTINA, "DPE Kernel Requirements Specifications", deliverable D1.02, June 1996.

[8] ReTINA, "DPE Kernel Architecture Specifications V2", deliverable D1.05, June 1997.

[9] G. Hamilton, M. Powell, J. Mitchell: "Subcontract: a flexible base for distributed programming", Proceedings of the 14th Symposium on Operating Systems Principles, Asheville NC, December 1993.

[10] M. Shapiro: "A binding protocol for distributed shared objects", 14th International Conference on Distributed Computer Systems (ICDCS), Poznan, Poland, June 1994.

[11] F. Dang Tran, V. Perebaskine: "TORBoyau: architecture and implementation", Research Report CNET/PAA/TSA/TLR/4587, CNET, Issy, France, December 1995.

[12] L. Leboucher, J.B. Stefani: "Admission control for end-to-end distributed bindings", Proceedings of 2nd International COST 237 Workshop on Teleservices and Multimedia Communications, Copenhagen, Denmark, November 1995 - Lecture Notes in Computer Science 1052, Springer Verlag 1995.

[13] ITU-T | ISO/IEC Recommendation X.903 | International Standard 10746-3: "ODP Reference Model: Architecture", January 1995.

[14] F. Dang Tran, Denis Metral-Charvet: "A Multimedia Communication and Binding Library for the ReTINA DPE", ReTINA Technical Report RT/TR-97-7 , March 1997.

[15] F. Dang Tran, V. Perebaskine, J.B. Stefani, B. Crawford, A. Kramer, D. Otway: "Binding and Streams: the ReTINA approach", TINA'96 conference proceeding, September 1996.

[16] L.L. Peterson, N. Hutchinson, S. O'Malley and M. Abbott. RPC in hte x-kernel: Evaluating new design techniques. Proc. of the 12th ACM Symp. on Operating Systemes Principles, pages 91-101, Litchfield Park, AZ, USA, November 1989.

A CORBA Framework for Multimedia Streams

Damian McGrath and Martin Chapman
IONA Technologies, Plc.
8-10 Lower Pembroke St.,Dublin, Ireland
Tel. +353 - 1 - 6625255; Fax +353 - 1 - 6625244
Email: {dmcgrath,mchapman}@iona.com

Abstract

This paper describes a CORBA-based framework for the control and management of Multimedia Streams. The concepts of streams and flows are introduced along with the abstractions used for multimedia devices. Examples of usage for point-to-point and point-to-multipoint streams are given. The OMG specification describes two profiles for implementers of the framework, a light profile and a full profile. The differences between these profiles are described. The Simple Flow Protocol (SFP) for transporting IDL- typed and untyped flows is briefly outlined. Finally, it is shown that the full profile of the OMG streams specification can be used to implement a TINA CSM interface with only minimal changes to TINA-C IDL specifications.

1. Introduction

This paper describes the OMG's recently adopted specification for A/V Streams [1,2]. It will present an overview of the main components of the architecture, and show how the architecture can be used to implement a TINA CSM. This approach will allow TINA implementers to use off-the-shelf components based on the OMG standard for building their systems.

The general approach taken is illustrated in Figure 1. IDL interfaces are used for controlling and managing the streams, for example, setting them up, tearing them down, stopping and starting them. The streams themselves are carried on separate transport connections and so can be considered 'outside' the ORB. This is somewhat analogous to the split between signalling and voice channels in telecommunications. The approach has much in common with previous work done by the IMA [3] and by TINA in the CSM [4] specification but differs somewhat to traditional ODP interpretations where stream 'binding' is an intrinsic part of a flexible binding architecture. This paper aims to demonstrate the advantages of the OMG approach, and show how they can be leveraged by the TINA implementer.

2. Architectural Overview

The specification defines a set of interfaces that implement a distributed multimedia streaming framework. The principal components of the framework are:

- Virtual Multimedia Devices and Multimedia device -

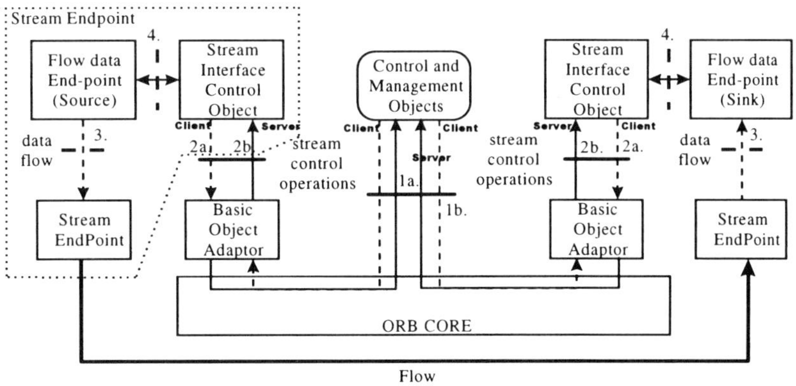

Figure 1 - The outline streaming architecture, as proposed in the OMG's RFP

0-8186-8335-X/98 $10.00 © 1998 IEEE

represented by the VDev and MMDevice interfaces respectively

- Streams - represented by the StreamCtrl interface
- Stream endpoints - represented by the StreamEndPoint interfaces
- Flows and flow endpoints - represented by FlowConnection and FlowEndPoint interfaces respectively.

A *stream* represents continuous media transfer between two or more *virtual multimedia devices*. A *stream endpoint* terminates a stream within a multimedia device. A simple stream containing two flows is illustrated in Figure 2. A stream can contain multiple *flows*. An operation on a stream (for example, *stop* or *start*) can be applied to all flows within the stream or to a subset of the flows. A *multimedia device* abstracts one or more flow endpoint sources and/or sinks. A source/sink could be terminated either in hardware (for example, a camera) or in software (for example, a graphical user interface). A multimedia device can support more than one stream simultaneously. It acts as a factory which creates a VDev and a StreamEndPoint for each new stream to be terminated. The transport connections for a group of flows, as abstracted by a StreamEndPoint, are 'bound' to a group of multimedia devices, as abstracted by the VDev object. The relationship between the StreamEndPoint and the VDev can be considered similar to the TINA concept of a Nodal Connection Graph (NCG).

The StreamCtrl object allows the application programmer to stop, start or otherwise control the stream as a whole or individual flows within the stream. Not all flows supported by an endpoint need participate in a stream connection. Flow endpoints which do not participate in a stream are said to be 'dangling'. A practical example of where this might be used is a video server which provides flows for Video, English language audio, French subtitles or French dubbed audio. The client may choose to connect, for example, Video, English audio and French subtitle flows while leaving the French dubbed audio flow unconnected.

3. Stream Setup

The StreamCtrl is used by the application programmer to set up and manage a point-to-point or multipoint stream. Typically the application programmer would do something like the following to establish a point-to-point stream:

```
StreamCtrl_var aSC;
...
aSC->bind_devs(aMMDevice,
       bMMDevice,
       someQoS,
       aflowList);
```

The MMDevices act as factories which create a VDev & a StreamEndPoint object to represent the endpoint of a particular stream. The VDev and StreamEndPoint abstract the device specific and transport specific aspects of a stream termination respectively. Each VDev may contain a number of objects representing multimedia devices which originate or terminate flows. Similarly, each StreamEndPoint contains a number of FlowEndPoint objects which represent the transport endpoint for the flow. After the VDevs and StreamEndPoints are created, the StreamEndPoint asks both VDevs to talk to their peers to establish device compatibility for each of the flows to be connected from the list aflowList. This is done by calling the operations:

```
aVdev->set_peer(bVdev);
bVdev->set_peer(aVdev);
```

It is the responsibility of each VDev to ensure that the flows it originates can be decoded by its peer VDev. For example, suppose aflowList contains a list of four flows <"video1" "video2" "audio1" "audio2">. The flows "video1" and "audio1" originate in aVdev and terminate in bVdev. The "video1" flow is in MJPEG format therefore the aVdev object must inform its peer

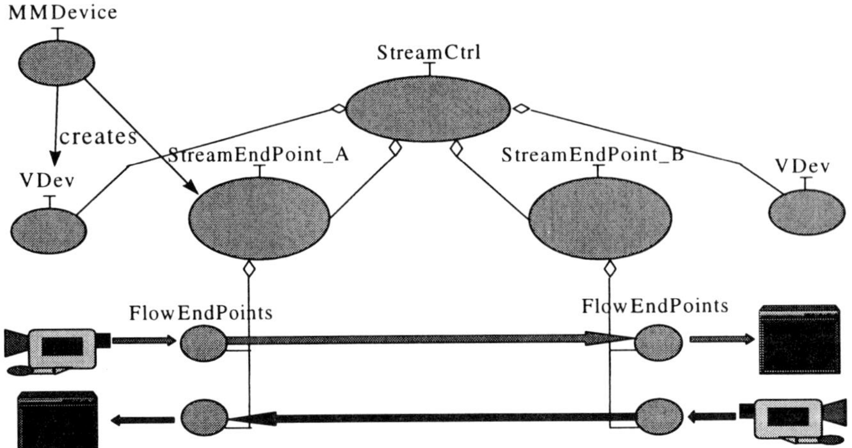

Figure 2 - Main objects in a typical stream scenario

that it should decode the video in this format. This is done by making a call:

```
bVDev->set_format("video1",
      "MIME:/video/x-mjpeg")
```

Note, the use of the MIME content-type description provides a standardised way for communicating formatting/encoding information. There is also an operation `set_dev_params()` on the `VDev` interface which provides a way of setting specific format parameters. For example the call to `set_format()` could be followed by a call

```
bVDev->set_dev_params("video1",
      <{"video_framerate" 25}
      {"video_colourDepth 8"}>)
```

This tells the video device that the framerate at which information is being transmitted is 25 fps and the colour depth is 8 bits. A number of common parameters for audio and video devices are included in OMG standard. The standard also provides for an extensible system similar to MIME content types where new values can be registered as they are needed. The device compatibility checking mechanisms are one of the principal advantages of this system for establishment of streams. In particular, it allows hardware manufacturers to supply objects which support the IDL for VDev or FlowEndPoint. These devices can then negotiate compatibility with their peers.

Once compatibility is established, the transport connections can be set up by the `StreamCtrl` calling `connect()` on one or both of the `StreamEndPoints` in both of the `StreamEndPoints` (in a point-to-point scenario). The `connect()` call works by calling `request_connection()` on its peer `StreamEndPoint`. This call informs the peer of which flows it wishes to establish a transport connection for and what QoS is to be used for the connection. The call returns listening transport addresses which can be used to connect the appropriate flows. Having done this the `bind_devs()` call returns to the application programmer who can then start and stop the flows by calling

`aSC->start(flowNameList)` and

`aSC->stop(flowNameList)`

respectively.

The A/V Streams specification provides two 'profiles' which may be implemented. The 'light' profile is optimised for efficiency with minimal number of invocations needed to establish a multi-flow stream. The 'full' profile contains all the functionality of the light profile but also contains finer grained interfaces for more flexible stream establishment. The light and full profiles are full compatible with each other. In full profile implementations, `FlowEndPoints` support IDL interfaces whereas in the light profile they do not. The `FlowConnection` interface is similar to a `StreamCtrl` but can only be used to setup and control individual flows. The transport connection is set up when the `FlowConnection` object calls `go_to_listen()` on the sink `FlowEndPoint`. This returns a listening transport address and the `FlowConnection` then calls `connect_to_peer()` on the source of the flow with this address as a parameter. The `FlowEndPoint` IDL also supports the operations `set_peer()`, `set_format()` and `set_dev_params()`. It can therefore be used to perform device format compatibility checks. In full profile implementations it is relatively straightforward to override the `bind_devs()` operation so that special algorithms can be used to match `FlowEndPoints` to one another. Following the matching process, the `StreamCtrl` which supports the full profile can create a `FlowConnection` for each flow established. Alternatively, `FlowConnections` can be created individually and the aggregated under a `StreamCtrl` by calling `set_flow_connection()` on it. In section 8 of this paper it is shown that the full profile of the A/V Streams specification is particularly well suited to implementing TINA CSMs.

4. Multicast Support

The A/V Streams specification has full support for multipoint streams. An interface called `MCastConfigIf` is used to distribute device configuration information to sink stream/flow endpoints. Both ATM-style multipoint connections and UDP-style multicasting are supported. A specialisation of the `MMDevice` interface can be used to abstract a conferencing bridge. This is one way to implement an n-to-n stream. Another way is to use UDP-style multicasting. The flow endpoints can be configured to receive all packets on a multicast address and then must use source identifiers contained in the packets to demultiplex data from the different senders.

5. Properties and QoS Issues

The OMG specification for A/V Streams makes extensive use of the OMG Property Service specification. All of the major interfaces inherit from the `PropertyService::PropertySet` interface. This allows the state of the objects supporting these interfaces to be queried. The A/V Streams specification defines sets of mandatory properties which must be supported by each interface. For example, the `MMDevice` interface supports properties which name all of its flows, states their directionality (source or sink) and what media formats those flows support. A `MMDevice` could therefore be used, for example, to 'advertise' all the flow types available on a particular terminal. The `VDev` interface supports a similar set of properties which can be used by the `set_peer()` operation to establish the compatibility of a peer `VDev`. The `VDev` interface also supports a property "flowname_related_MediaCtrl" which returns a Media Controller object for each flow. A Media Controller can be used to provide device specific control for operations such as stop, fast forward or rewind. There is a standard media controller interface, similar to the one defined in the IMA MSS, provided as part of the A/V Streams specification. Alternatively, the application programmer could devise her own for a particular task. For example:

```
interface myRewinder{
  void rewind();
  void stop_rewind();
};
```

She would then need to provide a specialised `StreamCtrl`, for example

```
interface myStreamCtrl
      : StreamCtrl{
  void rewind(
       in string flowName);
  void stop_rewind(
       in string flowName);
};
```

The implementation of the `rewind()` method in `myStreamCtrl` will simply retrieve the MediaController associated with the flow by querying "flowname_related_mediaCtrl". This returns an object reference which can be narrowed to a `myRewinder` control.

QoS is also represented using property lists. It is envisaged that QoS will normally be represented to the application programmer in terms which she can understand, like "video_framerate" and "audio_numChannels". These are translated by the MediaStreams runtime to network level QoS parameters. One of the central difficulties in using properties to represent QoS parameters is that CORBA Any types are very difficult to deal with when it is not known what kind of values may be expected. For this reason, the specification, stipulates a set of standard network level QoS parameters. These cover values like network token rate, jitter and bandwidth. Values may be specified with an associated range, for example, Bandwidth and Bandwidth_min where the former is the preferred bandwidth and the latter is the minimum acceptable bandwidth. The types and units of each of the standard QoS parameters is specified. The specification also includes recommended higher-level QoS parameters for video and audio.

6. Multiprotocol support

The A/V Streams specification is designed to use potentially any transport protocol, for example TCP, UDP and ATM-AAL5. The specification also includes a protocol, SFP (Simple Flow Protocol), which is specially designed to meet the needs of the Media Streaming system. It provides the ability to transport information which assists in the timely delivery of data. The information transported in the protocol includes timestamping for allowing inter-flow synchronisation and reconstruction of timing relationships at the source. It also includes sequence numbering so that the correct ordering of packets can be reconstructed at the destination. The SFP is capable of performing simple credit-based flow control so that a point-to-point source does not overwhelm a slower sink. Importantly, the SFP can also be used to transport IDL-typed data as well as 'untyped' data such as raw MPEG. There is a special mapping which can be used to map many of the SFP protocol fields directly down to their RTP equivalents so SFP may be carried most efficiently over RTP.

7. Typesafe stream interfaces

IDL has not yet been extended for specifying streams and flows, the A/V Streams specification however provides a set of rules for deriving typesafe specialisations of the main interfaces in anticipation of such extensions. For example, if a stream of type `videophone` were specified in this extended syntax it would lead to the generation of an interface `videophone_StreamCtrl` which is derived from `StreamCtrl`. This will contain operations like `bind_videophone_devs()` which can only be used to bind `videophone` multimedia devices. Similarly, the `videophone` interface inherits from `MMDevice`, `Vvideophone` which inherits from `VDev` and `videophone_A` and `videophone_B` inherit from `StreamEndPoint_A` and `StreamEndPoint_B` respectively.

8. Implementing TINA CSM using OMG A/V Streams

Using a relatively simple modification to the current TINA NRA 3.0 IDL [4] it should be possible to implement a TINA-C Communication Session Manager (CSM) object using OMG streams. This carries the advantage that TINA implementers can use standards-based shrink-wrapped components.

The CSM abstracts a Logical Connection Graph which consists of multiple Stream Flow Connections (SFCs).

Each SFC can be either unipoint or multipoint. An SFC will therefore have one source Stream Flow End Point (SFEP) and N sink SFEPS. A flow can be either activated or deactivated.

The SFC concept in TINA maps to the `FlowConnection` concept in OMG streams. The SFEP concept maps directly to the `FlowEndPoint` concept in OMG streams while the concept of activation/deactivation of an SFC maps to that of stopping and starting a `FlowConnection` in OMG. With this in mind it is relatively simple to change the NRA IDL to accommodate a CSM implementation either by traditional TINA means or using OMG streams.

Replace the following TINA IDL

```
struct t_SFepRef{
       t_IntRef     tcsmRef;
       t_Sfep_Cid   SFepId;
};
```

with:

```
struct t_SFep{
       t_IntRef     tcsmRef;
       t_Sfep_Cid   SFepId;
};

union t_SFepRef switch(long) {
       case 1 : FlowEndPoint
       case 2 : t_SFep;
       default : t_SFep;
};
```

The main caveat to usage is that a CSM cannot support SFCs with a mixture of SFEPs and `FlowEndPoints`. A given SFC must be implemented either using a TINA SFC or an `FlowConnection`.

With this knowledge a TINA Service programmer who needs to use the OMG/CSM interface might typically do the following for setting up a connection. Use a trader to find an `FDev` on the appropriate terminal with the appropriate flow type. Use an OMG `FDevs` (the flow level equivalent of an `MMDevice`) to create the desired `FlowEndPoints`. Make a call to the CSM to set up the connections (OMG device negotiation will occur automatically, ensuring that the flow endpoints support compatible encoding/decoding). Activate the connections via the CSM, starting the flows. Release the SFCs. Destroy the `FlowEndPoints` by calling `destroy()` on the appropriate `FDevs`.

9. Summary

The OMG A/V Streams provides a flexible and open way of implementing multimedia service delivery systems. It includes extensive mechanisms for determining compatibility between multimedia devices. The specification makes extensive use of the OMG property service and defines mandatory properties and standard sets of network level QoS parameters. It makes provision for extending the set of allowed properties to aid compatibility checking, multimedia device configuration and QoS specification. At least one implementation of the standard will be available shortly. Because its general applicability, the OMG streaming standard has provoked interest not only in the telecoms community but also in areas such as biomedicine, entertainment and security. Widespread use in multiple domains will further consolidate the advantages of economies of scale and reliability which the TINA implementer will get by using a 'shrink wrapped' product which conforms to an industry standard.

References

[1] *Control and Management of A/V Streams*, telecom/97-05-07

[2] *Errata on Control and Management of A/V Streams, telecom/97-06-04*

[3] *IMA Recommended Practice, Multimedia Systems Services Part 1: Functional Specification*, September 1994.

[4] *Network Resource Architecture, Version 3.0,* TINA-C, February 1997

[5] Douglas C. Schmidt, *Measuring the Performance of Object-Oriented Components for High-speed Network Programming*, www.wustl.edu/~schmidt

[6] *RTP: A Transport for Realtime Applications*, RFC1889, March 1996.

[7] *OMG RFP5 Submission, Trading Object Service,* orbos/96-05-06, Version 1.0.0, May 10, 1996.

[8] J. Postel, *Multipurpose Internet Mail Extensions (MIME) Part Two: Media Types,* RFC 2046, November 1996.

[9] C. Partridge, *A Proposed Flow Specification,* RFC 1363, September 1992.

A Notification Service for TINA

Michel Ruffin, Alban Couturier, Olivier Potonniée, Marcel van der Meulen

Alcatel Corporate Research Center†

Sabine Habert

O2 Technology‡

Abstract

This paper presents Alcatel's Notification Service which has been designed in the context of TINA. It has been partly developed in the framework of the TINA auxiliary ACTS projects ReTINA (AC048) and VITAL (AC003). A first prototype has been issued by mid 1996. Based on the users feedback, a second version has been designed and will be delivered by end 1997.

This paper presents the rationales for such a service and its main features: generic and typed notification, push and pull communication models, filtering, subscription, federation, fault-tolerance, and quality of service. For each feature, a comparison is done with the four actual OMG proposals for a Notification Service and the rationale for our choices in the light of the TINA use of the service is stressed.

1. Introduction

The Notification Service is a DPE service based on the CORBA technology allowing objects to emit asynchronous information - called notifications - without being aware of consumer objects. Similarly, it enables an object to receive notifications without having to interact with supplier objects. The service acts as a broker between suppliers and consumers.

Managing notifications is a standard issue for telecommunication applications. A CORBA-based generic Notification Service should fulfil multiple needs in the context of TINA. For instance, the service can be used to propagate resource configuration updates, accounting information or alarms for reacting to failures.

OMG has specified a service to transmit events between objects. This service, the « Event Service », provides some of the functionality required to dispatch notifications. However, it misses essential properties that make it unusable as it is in the telecommunication domain: filtering and quality of service. Filtering events should drastically reduce the event traffic which can be very high in large scale systems. The Notification Service must also allow to associate data management to events. This data will be known and used by the service to manipulate events in a specialised way. For instance, a priority associated to each event will allow the Notification Service to order events.

This paper presents a global design of a Notification Service for TINA, based on Alcatel's experience in designing and implementing such a service. It compares this design to the current four proposals answering the [OMG RFP 97] for a Notification Service: [DTSC/Nortel, GMD, TID/HP, NEC].

Alcatel's Notification Service has been developed in the framework of the TINA auxiliary ACTS projects [ReTINA] and [VITAL]. The requirements on the service are coming from the TINA architecture designers of these projects but also from Alcatel's experience in introducing CORBA into TMN. A first prototype has been issued by mid 1996. Based on the users feedback, a second version has been designed and should be delivered by end '97.

By prototyping a Notification Service, Alcatel's goal is not to develop a product, but to understand the requirements on such a service for telecommunication uses and more specifically for TINA uses in order to encourage the development of the resulting features in future products of DPE service providers.

Section 2 presents the requirements for a TINA Notification Service. Section 3 describes the basic principles of a Notification Service: the architecture of the service, the definition of a notification, the push and pull communication model and the filtering mechanism. Then the different features of a Notification Service are presented: subscription, federation, fault-tolerance and quality of service. Finally, section 5 concludes and presents the future work.

The work presented in this paper has been partly funded by the European Union ACTS projects ReTINA (AC048) and VITAL (AC003).

† Contact : Michel Ruffin, Alcatel Alsthom Recherche, Route de Nozay, 91460 Marcoussis, France. Tel.: +33 (0)1 69 63 13 57, Fax: +33 (0)1 69 63 17 89, E-mail: Michel.Ruffin@aar.alcatel-alsthom.fr

‡ O2 Technology, 7 rue du parc de Clagny, F-78035 Versailles Cedex, France. Tel.: +33 (0)1 30 84 77 46, Fax: +33 (0)1 30 84 77 90, E-mail: sabine@o2tech.fr

0-8186-8335-X/98 $10.00 © 1998 IEEE

2. Requirements

2.1 Needs for a Notification Service

In the Telecommunications Management Network [TMN] systems notifications (called events in the TMN context) are used to propagate data to unknown sets of components. In this context a notification or an event can be of one of two kinds :

- *Error information* used to inform that an element is close to, or has entered into an error state. The receivers of such information can then act to correct the situation.
- *Informative information*. For instance, the availability of a new service in the system, or the expiration of a deadline. This kind of data can be used in any application domain.

The needs for a Notification Service have been identified in several TINA applications:

- In Alcatel's work of introducing CORBA into TMN, which can be considered as a first step towards an industrial implementation of TINA. The Notification Service should provide the functions of traditional TMN's Event Forwarding Discriminators to propagate fault signals. Alcatel's Notification Service prototype will be used through a gateway between a CORBA TMN manager and a CMIP based agent.
- The VITAL project expects the availability of a service providing the same functionality as TMN's Event Forwarding Discriminators in the domain of fault information propagation for use in the context of a full implementation of the TINA architecture.
- In ReTINA , the Notification Service is planned to be used to disseminate accounting information. The TINA resource configuration management could also use this service to propagate the resource state modifications. Finally, the O2 database uses the notification service to propagate events, like the logging of a user or the deletion of an object, to objects managing or controlling the database.

2.2 Requirements for a TINA Notification Service

For telecommunications and other application domains, the following requirements on the notification service have been identified:

- **Scalability** is mandatory in telecommunications systems, which are large and highly distributed. Consequently, the notification service should be able to manage important numbers of suppliers, consumers, and events. In order to avoid bottlenecks, the service should not be monolithic.
- **Performances** must be kept high, despite of distribution. The communication path between suppliers and consumers must be minimal, and the number of events transmitted on the network must be reduced to the subset of pertinent ones.
- **Fault Tolerance** is essential for a service to maintain the system in a safe state. This property can be observed at different levels of the service. First, the service must be able to resist crash effects, and thus maintain a persistent copy of its state. Second, it must resist to erroneous interactions with suppliers and consumers. For example, it must check validity of filters or notification occurrences in terms of notifications types. Third, it must also ensure the continuity of the service during occurrences of events bursts.
- **Flexibility** allows users to adapt the service to their needs. This is important for a service that can be used in multiple application domains.
- **High level interfaces** simplify the developers' task by reducing the learning curve and minimising the code on the client side. Reducing the number of steps to perform an operation reduces the risk of bad manipulations.

Experience in distributed object oriented systems led to identify a set of solutions that fulfil these requirements:

- **Federation** increases the *performance* and *scalability* of the system by balancing the load of one component over multiple components.
- **Filtering** reduces the network load, increasing *performance* by allowing consumers to precisely specify the events they are interested in. Filtering is applied to the data associated to notifications. Only notifications matching filters are sent to consumers.
- **Quality of Service** parameters ensure *flexibility* by providing an interface for setting the service to the most suitable behaviour. In particular, event queue management and reliability should be customisable according to notifications nature. Quality of service should be customisable for kinds of notifications or specific instances.
- **Typed and generic interfaces**, as defined in the OMG Event Service, should both be available to communicate the event data. The choice of the kind of interface is part of the *flexibility* of the service. Typed interfaces are *high level interfaces*, whereas generic interfaces have better *performance*.

3. Basic components

This section presents the basic components of the Notification Service: the service architecture, the concept of notification, the pull communication model and the filtering mechanism.

3.1 Architecture

Alcatel's Notification Service architecture decomposes the Notification Service in two parts: a notification manager and several notification channels (see Figure 1). Since the notification mechanism has been defined for [TMN] in [X733, X734, X720], normalised concepts have been reused in order to match the telecommunication needs. The TMN Event Forward Discriminator (EFD) which constructs event reports and manages the notification propagation (e.g. filtering and routing) can easily be mapped on the notification channel concept. A notification manager is introduced because general purpose Life Cycle Service functions are not powerful enough to meet the requirements of telecommunications as defined in TMN. This manager is in charge of the global administration of the service (instanciating and removing channels, managing subscription to the service, remembering subscriber's configuration, notification type description, checking types, establishing federation between channels, ...). Subscribers register to the service for supplying or consuming a *kind* of notification (this is defined in Section 4.1). There is one notification channel for each *kind* of notification.

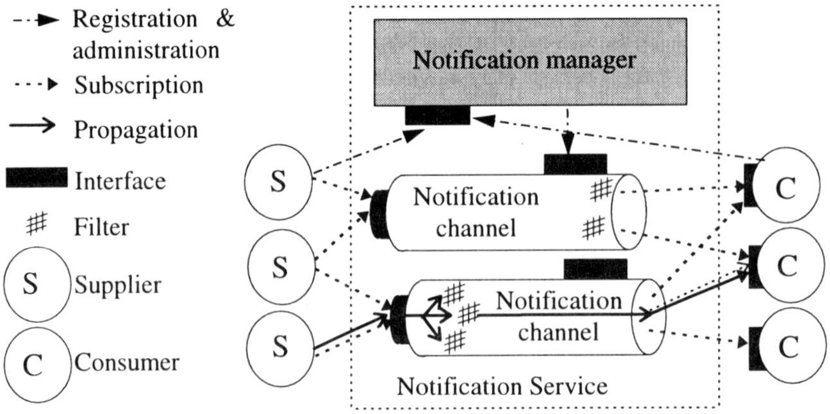

Figure 1: Notification Service architecture

Through the subscription interface, a consumer can dynamically provide a filter to the manager. Filters are used in order to reduce the amount of notifications that a consumer receives. Notification channels provide an interface to the manager for their administration. Among other information, filters are passed to the channels through this interface.

When a supplier or a consumer subscribes for a *kind* of notification and no corresponding notification channel exists, a channel is created by the manager. When subscribing, a supplier gets back in return the reference of a notification channel. Suppliers can then provide notifications to the channel by invoking its notification interface. For each of its consumers, the notification channel uses the consumer's filter to determine whether the notification has to be propagated to the consumer. Propagation is done by invoking the consumer notification interface.

In the example of Figure 1, a supplier invokes the notification channel to propagate a notification. The channel filters the notification according to the filters of each of its consumers. The notification passes through only one of the three filters and is propagated by invoking the corresponding consumer.

The OMG proposals that want to stay compliant with the Event Service specifications ([DSTC/Nortel], [TID/HP] and [NEC]) do not specify the manager part of the architecture and the administration interface of the notification channel is directly accessed by suppliers and consumers. Channel creation and destruction, and storage and management of global service information are not directly addressed. [DSTC/Nortel] mentions the use of an event type repository to store the type description of events. [NEC] defines an event network object for managing the federation of channels. Finally [GMD] does not clearly define the channel concept, the propagation of notifications is handled by a notification server, and some internal combination of computational objects assumes the role of the channels.

The manager allows to discharge users from the burden of managing channels (instanciation and destruction) and it offers high level operations on channels such as automatic federation for load-balancing. The split of the service in two kinds of entities (manager and channels) allows to discharge the channels as much as possible from administration work. The main channel task - notification filtering and routing - can thus benefit of more resources. The interface and functions of the channel administration interface are limited to the minimum since a part of the work is done by the manager. This basic choice is directly driven by scalability and performance requirements.

3.2 Notification

A notification is passed by invoking an operation on the notification interface. The notification interface is the same for a supplier invoking the service side (i.e. the channel) as for the service invoking a consumer. Interface types can be either generic or typed, corresponding to generic or typed notifications.

Both notification propagation models can be seen as multicasting asynchronously on some set of unknown objects after filtering. From the user point of view, typed notifications allow to propagate notifications as a remote invocation on the consumers with the guarantee of type checking. Generic notifications are lower-level and allow to propagate notifications like packets of data. The clients have to pack the data on the supplier side and unpack it on the consumer side. Type checking is not guaranteed and consequently this method is more error prone. The drawback of typed notifications is the complexity of a user-transparent implementation and the fact that the use of typed notifications in this implementation might be time consuming compared to the use of generic notifications.

3.2.1 Generic notifications

A generic notification consist of a user-defined set of data and a header. The generic notification interface supports a single operation (let's call it `push ()`) having as a parameter the generic notification. In Alcatel's implementation the data part of the generic notification is encapsulated by a CORBA « any » type.[1] The header may contain information such as quality of service requested, supplier identification, timestamps, notification family information (alarm, error, warning, ...).

Figure 2 presents the propagation of a generic notification. The supplier invokes the « `push ()` » operation of the channel with a generic notification (gn) as parameter. After notification filtering (assuming that the notification passes the filters) the channel invokes the same operation on each consumer.

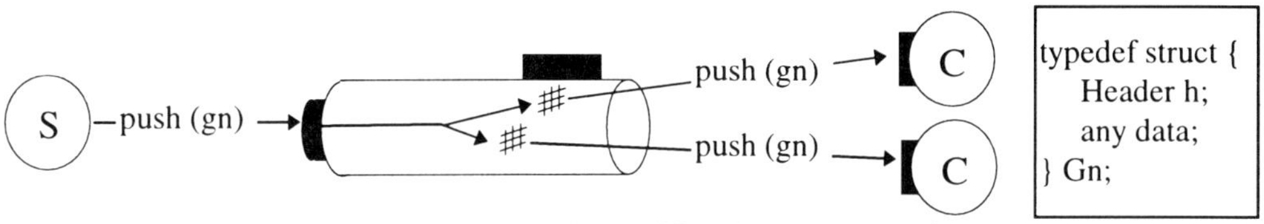

Figure 2: Generic notification propagation

3.2.2 Typed notifications

A typed notification is an operation with an arbitrary number of user-defined typed parameters. The type of the notification is the signature of the operation (operation name and type description of its parameters). A notification interface can contain several typed notifications.

Typed notifications are conceptually more object-oriented than generic notifications, since the propagation of the notification consists in the propagation of an **arbitrary user-defined** operation (see Figure 3).

Depending on the use of notifications, both kinds of notification are useful: a low-level mechanism with good performance or a high-level one less error prone.

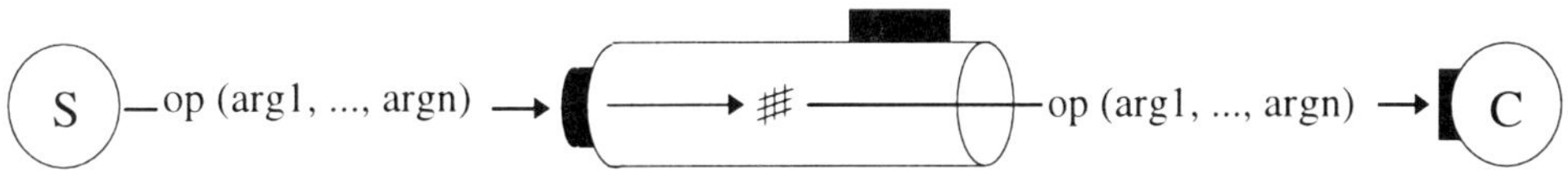

Figure 3: Typed notification propagation

In Alcatel's Notification Service, generic or typed notifications are **arbitrary** and **user-defined**. This proves to be very convenient for the user because from this point of view the channel interface is defined by the user. In order to get better performances for typed notifications we will consider predefined typed notifications later on (Section 3.2.3).

[1] The CORBA « any » type allows to invoke dynamically operations taking as parameter some data of unknown type. The « any » is composed of two fields containing the type description and the data value. Complex data can be built by for instance passing in an « any » a sequence of « any ».

3.2.3 Implementing typed notifications

Arbitrary client-defined typed notifications constitute a powerful mechanism. In Alcatel's prototype this is implemented by using the Dynamic Invocation Interface (D.I.I.) and the Dynamic Skeleton Interface (D.S.I.).

prototype. Since these predefined notifications will be known at compile-time, the use of time consuming mechanisms such as D.I.I. and D.S.I. will not be needed for these notifications.

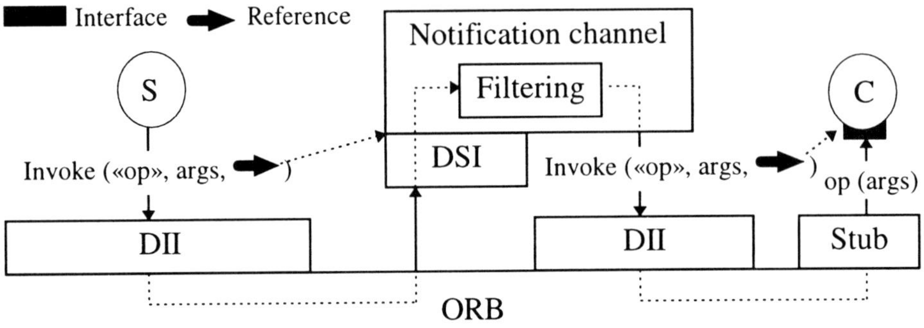

Figure 4: Implementation of typed notification

Figure 4 presents the propagation of a typed notification in our implementation. The supplier wishes to invoke the client-defined operation of name « op » with a list of arguments (noted args) on the notification channel. Notification channels are compiled once and present an interface independent of the client operations. This implies the use of the D.S.I. in order to receive invocations of operations which do not belong to the channel interface. The D.S.I. allows to receive this unknown operation and to transform it to provide to the channel the stringified name of the operations and a sequence of « any » containing the operation parameters.

In the Notification Service V1, the D.S.I. was not used. Instead, the client had to provide a « converter » object for each kind of typed notifications. This « converter » object had to receive notifications from the supplier, transform the invocations into a string and call a notification channel operation to convey the string to the channel. In addition to the burden of writing this object the user also had to provide a factory for the converter object. So the use of the D.S.I. largely simplifies the programming work for the service client.

The use of D.I.I. is needed between the supplier and the channel (see Figure 4) in our current implementation, because the supplier has a reference on a channel object that does not support the method for a specific typed notification in its interface. Between the channel and the consumer D.I.I. is needed because the channel is not linked with the stub for the interface of the consumer.

The main drawback of the use of D.I.I. and D.S.I. is the cost in time of such invocations. For a telecommunication application (and thus for TINA), the few typed notifications which have been defined for TMN can be expected to be suitable for fault-management. Since these kinds of notification can be known in advance, a set of predefined will be added in future versions of Alcatel's

3.2.4 Notification concept in OMG proposals

[DSTC/Nortel] defines generic notifications as arbitrary sequences of name/value pairs where the names are strings. Some predefined names can be « understood » by the service, allowing to form a notification header. The notification type description is identified by a particular name. Typed notifications are not specified in the proposal.

[GMD] defines four types of generic notifications corresponding to object creation, object deletion, attribute value change and method invocation. The object creation and method invocation types both contain an arbitrary sequence of name/value pairs. Compared to other proposals, there are four kinds of generic notifications with strongly typed headers. Typed notifications are not specified in the proposal but the method invocation type allows to describe an invocation and can be considered as a basic typed notification.

[TID/HP] defines generic notifications as a block of data of any type, arbitrary user-defined and without a header. Typed notifications are not specified.

[NEC] defines arbitrary generic and typed notifications similar to the generic and typed events defined for the Event Service. In addition, structured events are defined for increased filtering performance. These events are composed of a header part, a sequence of name/value pairs and an opaque body. Filtering is limited to the header and to the sequence section. Anticipating the OMG standard on passing objects by value, an event object encapsulating a structured event is defined.

To our knowledge Alcatel's prototype is the only implementation providing typed notifications that can be arbitrary defined by the user. From the users of our V1 prototype, we learned that whenever possible (when time constraints are not too strong), users prefer to use typed notifications. Generic notifications are an alternative when high performances are needed.

In Alcatel's Notification Service, the type descriptions of notifications must be provided to the service manager before subscribers and consumers can subscribe to them. This allows to avoid the problem foreseen in [NEC] concerning low performance for the filtering of user-defined types, because when providing a description, a name can be associated to notification operations (in the case of typed notification) and data fields. Names can be used for identifying the filtering fields of notifications.

Finally the definition of predefined TMN-like typed notifications allows to overcome the problem of performances for typed notifications, since in this case the use of D.I.I. and D.S.I. is not necessary.

3.3 Pull communication model

Until now only the *push* communication model has been considered: the notification is moving from object to object at the initiative of the supplier objects. Suppliers invoke the channel when a notification occurs and the channel invokes the consumers on a notification reception.

With the *pull* communication model the notification is moving between objects at the initiative of objects needing notifications: consumers invoke the channel to get notifications or the channel invokes the suppliers to get notifications. The two models can be combined as requested, suppliers and consumers choosing their communication model with the channel at registration time.

The pull communication model implies the need for an additional notification interface. This interface should be supported by the channel when consumers use this model or by a supplier if the supplier wants the channel to pull its notifications. This interface is nearly symmetric to that of the push model. For instance for generic notifications, the consumer needing a notification will invoke the `pull ()` operation of the channel. It will be blocked until a notification will be available. An additional non-blocking operation (`try_pull ()`) is added in order to return even when no notification is available.

In Alcatel's implementation the pull communication model is limited to generic notifications. This is due to a need to assess the usefulness of this mechanism in telecommunication applications before extending it to typed notifications. In the context of VITAL, ReTINA and our experience of introducing CORBA in TMN no use has yet been found of the pull model, but all the OMG proposals provide it.

One of the possible advantage of this model is the fact that a consumer does not have to be an object.

For the sake of clarity, the pull communication model will not be considered in the remainder of this paper except when a feature description requires this.

3.4 Filtering

The filtering mechanism used by the Alcatel notification service is directly derived from the EFD filtering mechanism described in [X720]. The EFD filtering description language defined in ASN.1 has been translated in IDL. Writing a filter consists in assembling IDL structures which can be send easily and dynamically through invocations. A filter is a recursive structure organised as a tree of logical expressions, whose leafs are filter items. A filter item is a union storing a constraint (equality, substring, superString, greaterOrEqual, lessOrEqual,...) and a couple of attribute names or values. A filter is a union which switches on a filter type (TRUE, AND, OR NOT, filterItem) and stores a sequence of filters (in the case of AND and OR), a filter (for NOT) or a filter item (for the filter type FilterItem). An extension to the TMN norms filter has been made providing the possibility to compare two attributes instead of an attribute and a value.

The Alcatel Notification Service does not need to parse strings describing filters, and can manipulate and forward them without string translation. Moreover, a filter can easily be incremented by applying a logical AND or a logical OR between the existing filter and a given filter. Although this filter notation comes from the TMN norms, it is very close to an IDL notation of the OMG trader constraint language. All the OMG proposal are based on this grammar.

The place of the filter is essential in a notification service, because the performance depends on the moment and the place where the notification is processed (and possibly deleted) by the filter. Placing filters into the channel is essential. This allows to optimise the combination of filtering and routing to decrease the network traffic. Only filtered notifications will be multicasted to consumer objects.

In a federated notification service, where multiple notification channels can filter the same notification successively (see Section 6), the filtering must be performed as close as possible to the supplier. It could even be in the supplier's address space itself to avoid any network traffic for unwanted events. However, this approach implies the presence of a notification channel, i.e. the encapsulation of the filtering engine, in the supplier's application, like the EFD in TMN. Through an optional advanced interface, the supplier could offer the notification manager the means to create a notification channel in the supplier's address space. This implies that the supplier application has to be developed with its own notification channel and filtering machine. This channel must implement every feature of the classical channel. The manager could manage this channel normally, and update the filter in order to match it to the real demands of the final consumers.

To avoid the emission of events at their source when there is no consumer awaiting them, the manager could set the suppliers in a mute mode through the advanced supplier interface, until a new consumer registers. This functionality is not offered in the current implementation of our service, but it is planned to be added to future releases. Nevertheless these mechanisms must only be optional features in order to preserve a light, classical and easy use of the notification service.

These filtering features allow to process the notifications as close as possible to the supplier and to send notifications only when a consumer needs them.

4. Advanced features

This section describes the Notification Service advanced features which allow the service to be integrated in telecommunication applications.

4.1 Subscription

Several points have to be considered for designing a powerful subscription interface:

- **Description of notifications.** Families of notifications can be named. Their type can be described to the system. Some tools can be provided to organise and manage the description information.
- **Description of subscribers.** Suppliers and consumers can be identified logically to recognise them in case of failures or to associate a context to them (For instance a filter can be associated to a consumer).
- **Propagation of subscription information.** When suppliers register to the service, the description of the notification that they intend to send can be made available to other components of the system. When consumers register, their filter can be forwarded to the different emitters of notifications in the system.
- **Providing filters**: dynamically or statically, by replacing the whole filter or by incrementing (see Section on Filtering).
- The possibility to **suspend** and **resume** the reception/emission of notifications for a while.

4.1.1 Notification descriptions and subjects

In Alcatel's Notification Service, notifications are organised by *subjects*. A subject is an **arbitrary, client-defined** name for notifications which are sharing the same type or a **set** of types. A subject is the minimum unit for subscription or un-subscription.

Subjects must be declared to the service before being used. When declaring the subject, the list of associated notification types should be described. In the case of a generic notification, the type description corresponds to one or several data types. Whereas in the case of typed notifications, the type description is the interface signature corresponding to the set of operations associated with the subject. The signature is composed of the list of signatures of the operations; this signature of an operation is composed of the operation name and the type description of its arguments. For a typed notification subject, the type description will be used to check at registration time whether suppliers propose to invoke the correct service interface and whether consumers provide the correct interface to the service. For generic and typed notifications, the type description is used in the filtering to identify the data field (generic notifications) or the operation name and the arguments of operations (typed notifications).[2]

A subject is an abstraction allowing to organise notifications logically. There is at least one channel initiated for each subject for which a subscriber has registered. Several channels can be instanciated for load balancing reasons (see Section 4.4.4). Quality of service can be associated to the unit of subscription (i.e a subject) because a channel cannot handle more than one subject.

Subjects can be organised hierarchically. For instance, « failures/hardware/disk » could be a subject in which « disk » is a sub-subject of « hardware ». Subjects are declared by the user and form a hierarchy organised as a forest.

Once subjects are registered by the service, suppliers and consumers can subscribe to several subjects in a single operation and it is possible to register a set of consumers or suppliers for a subject in one operation. When registering, a type description should be provided to the service and type checking is done between the provided description and the type description which has been registered with the subject.

Initially when this design has been done, it was planned that when registering for a subject an object was registering for all its sub-subjects. This is complex to implement. A possible implementation would consist in having a channel for each sub-subject and using federation for reproducing the hierarchy of subjects, a sub-subject channel being a producer of its higher level subject channel. This is an interesting feature for providing a

[2] In the current Alcatel implementation, sets of types are not yet implemented. Only one data type can be associated to a subject for generic notifications, and only one operation description can be associated to a subject for typed notifications.

better API to service subscription and it will be considered for future versions of the service.

A notion similar to subject does not appear in OMG proposals or is limited to type description. Generally quality of service is associated to channels irrespectively of what is transiting through the channel.

The notion of subject provides a clean and powerful API to the service and is possible because the service architecture offers a centralised manager able to cope with general management information.

4.1.2 Subscriber descriptions

In the push (respectively pull) communication model consumer (respectively supplier) objects should provide an object reference for allowing the channel to make invocations.

References are not guaranteed to stay unchanged when for example a fault occurs. Consequently a logical identifier is needed to identify the subscriber object independently of the object location and the object instance.

For instance after a client-side failure a new subscriber object can be instanciated which will continue the work of the previous instance.

Identification is also needed for security reasons. It should be possible to restrict notification delivery to a list of consumers. It should also be possible to restrict the acceptance of notifications depending on which supplier emitted the notification.

Identification can also allow the service manager to register configurations. For example a consumer could register a filter description that could be used for each instance of this consumer. Subscription dentification is also used for the suspend/resume mode (see Section 4.1.4).

Alcatel's Notification Service defines subscription identifiers as arbitrary user-defined strings which are used for all of these purposes except security which is not implemented.

In [NEC], [DSTC/Nortel] and in [TID/HP] a subscriber is only identified by its object reference. In [GMD] unique names are used to identify suppliers and consumers.

4.1.3 Subscription information propagation

When a subscriber registers for a kind of notification, the information corresponding to this subscription can be propagated through the channel.

When a supplier registers, a description of its subscription can be sent to the consumers. This is called « **offering** » in [DSTC/Nortel]. This « offering » concept implies that a consumer has to provide the necessary interface to support this concept to the service.

When a consumer registers, its filter is passed to the channel. This filter could be moved as near as possible to the supplier location in order to reduce the network traffic by filtering at the source. In fact to be efficient a filter should be built at each level of the transmission chain to gather the requirement of all the consumers concerned by this level. [DSTC/Nortel] calls this « **quenching** ». At the supplier side this requires an interface accepting filters and the possibility to interpret filters. The possibility of updating filters dynamically and incrementally is necessary.

Alcatel's service provides a limited implementation of quenching in the case of channel federation (see Section 4.2). Offers propagation could be replaced by the manager interface allowing to get information on subjects. This limited implementation avoids an additional interface for the suppliers or consumers.

[TID/HP] and [GMD] do not include subscription information propagation. [DTSC/Nortel] supports both « quenching » and « offer » concepts. [NEC] only mentions the possibility to propagate filters towards the event source, but does not further elaborate on how this should be done.

4.1.4 Suspend/ resume mode

To simplify the service use, a suspend mode is desirable. The suspend mode allows a consumer to stop receiving notifications for a while. When entering the suspend mode, all the notifications queued or transiting by the channel are lost for the consumer. When the consumer resumes the reception of notifications, it does not need to subscribe to the subjects or provide filters again.

Alcatel's prototype, [DSTC/Nortel], [TID/HP] and [NEC] are all providing these modes.

4.1.5 Conclusion on subscription

Hierarchical subjects, subscriber identifiers, incremental filtering, and suspend/resume mode were not present in Alcatel's V1 prototype. The features in this list corresponding to a better API have been requested by the users for the V2.

Subscriber identifiers have been added to support fault-tolerance, quality of service and security.

The concept of subjects results in a simple registration interface. This combined with the fact that the notification channel interface is defined by the user implies an easy use of the service.

4.2 Federation

In case of a lot of suppliers and/or consumers or notifications, a notification channel might get congested. A solution for this problem is load balancing, the division of the load over multiple channels. Federation is a means to perform this load balancing. Federation consists in connecting two channels, with one channel acting as a

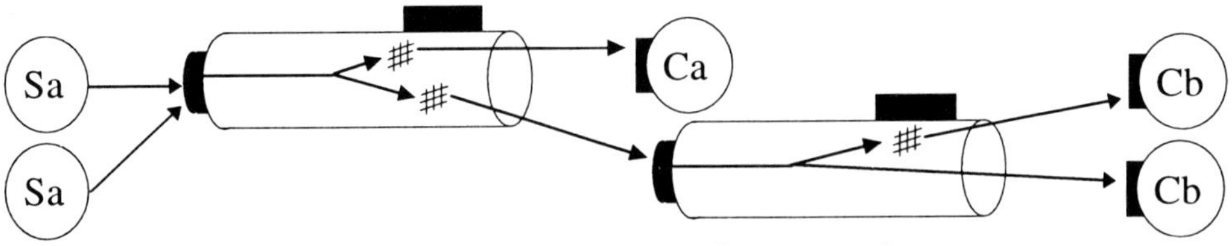

Figure 5: Channel federation example

supplier for the other channel which in turn is acting as a consumer of the first channel. Federated channels can share the load for better performance and scalability, while maintaining the initial distribution.

Figure 5 presents an example of channel federation. In Alcatel's prototype, two ways to federate channels through the notification server interface have been defined: the « like a supplier » federation, where a channel is asked to register like a supplier to another channel, and the « like a consumer » federation where a channel is asked to register like a consumer channel. The difference consists in filtering. For the first kind of federation, the channel only forwards notifications. For the second kind of federation a filter must be specified, that will be applied to all notifications passing from one channel to the other. This second kind of federation allows to reduce the notification load of a channel.

If several channels share a common part in their filter, this one can be factorised and passed on to the supplier channel. Such a filter factoring policy allows to reduce the load of the second channel when there is a large amount of notifications or suppliers. This mechanism corresponds to the [DSTC/Nortel] quenching mechanism.

Federation also allows to define topological domains: a set of servers and sites which share a common view of notifications. Federation can be intra- or inter- domain. An example of inter-domain federation is given in Figure 5. Two domains are defined: A (Sa, Ca) and B (Sa, Cb). The first channel belongs to the notification service A while the second belongs to B. The federation allows a notification emitted in the A domain to be propagated to the B domain.

Topological domains for notifications are a direct match for TINA network administrative domains.

Intra-domain federation generally aims at factorising filters or decreasing the server and site load. For example all high priority notifications corresponding to a subject could be propagated to a separate channel, whose subject could be « emergency ».

The Alcatel prototype provides federation at subject level through the notification manager interface. Subject federation is more « logical » than channel federation since multiple channels can exist per subject for load balancing purpose.

OMG proposals federate at channel level. Only Alcatel's service and the NEC proposal handle federation globally with the help of a manager. The use of a manager allows to federate subjects easily without having to know notification channel instances or their references.

In the current implementation, subject federation is done at the initiative of a service administrator. Further study needs to be done to federate channels automatically since the policy for federation depends on the service use.

Federation is essential for scalability and performance in telecommunication applications to reduce the network traffic. The fine-tuning of federation at subject-level instead of service-level allows to customise the service to a clients needs.

4.3 Fault-tolerance

In TMN, notifications are used to propagate information about failures, so the Notification Service is a building block for reliability in the system. A similar use of the Notification Service by TINA can be expected. A mechanism used for guaranteeing reliability should provide at least the same level of reliability as the expected system reliability. Consequently, the Notification Service should provide different levels of reliability according to the needs of its users.

The different levels of reliability are defined by the kind of fault or malfunctioning behaviours which are considered: it can be the overload of a consumer or a channel[3] which is not able to accept the notification in time, a communication failure between the channel and its subscribers, a site fault (the memory is lost) that can either results in a channel or a subscriber crash or a disk failure resulting in the loss of data stored on a disk.

To solve communication failures, overload of a consumer and consumer crash, notification logging can be used to discharge the channel and to re-inject the notifications later on. An additional mechanism on the supplier side is necessary to resist channel crash effects to change the old channel reference by the reference of the new channel. With the pull communication model, this mechanism will also be necessary on the consumer side.

[3] The case of channel overload is not treated here since in the case of telecommunication some specialised solution can be studied and will be developed in section 4.4.4.

To solve a channel crash, additional information should be logged to remember the channel state: the list of consumers in the push model, the list of suppliers in the pull model, and the list of filters. Some other information can be kept by the service, e.g. the description of the notification types in the case of Alcatel's Notification Service and the federation configuration. This information should be logged to resist site failure effects (manager crash in Alcatel's case). Finally logged data should be replicated to resist disk failures. The number and place of replica defines the degree of reliability and availability.

Logging allows to save the necessary data for recovery, but a mechanism should be added to do the recovery. Besides ad-hoc solutions for simple recovery situations, transactions can be considered to automate complex recovery. These solutions can be a part of the quality of service modes.

The reliability of the system depends on that of the logging mechanism. The use of a customisable logging service [Ruffin 92] can provide an adapted quality of service: Log replication provides high reliability, log distribution determines availability, and high level logging techniques provide low overhead.

The Alcatel Notification Service provides three levels of fault tolerance: the logging of notifications, the logging of channel states (list of subscribers, filters and notification logging) and manager state information logging. Control operations are provided to select these features and fine-tune them. Notification logging or channel state logging can be switched on at channel level or at service level (for each channel managed by the service). This fine-tuning allows a quality of service customised to the user needs. The quality of service can be further fine-tuned by using an appropriate transactional model. The use of transaction and logging services is considered for future versions of the service.

[TID/HP] and [DSTC/Nortel] do not define features for fault-tolerance. [GMD] only specifies the logging of notifications. [NEC] defines logging of notifications, logging of the channel state, logging of the global service configuration state and the use of the CORBA Transaction Service.

Fault-tolerance of the service is important for fault-tolerant applications and thus for telecommunications. Being able to choose the level of fault-tolerance (e.g. at channel-level, at service-level) allows to fulfil different applications needs.

4.4 Quality of Service

Quality of Service allows to select among the mechanisms offered by a service the most adapted to the needs of an application. This is particularly important in the telecommunication domain where multiple uses of a service request fine-tuned customisation.

Not all quality of service possibilities are detailed in this paper, only the most common ones are presented.

4.4.1 Ageing

Like for messages in traditional communication protocols, a limit of duration can be assigned to notifications while transiting between suppliers and consumers. A notification which has not been transmitted to consumers after this duration is discarded.

In case of federation, this mechanism (called ageing) can be used in order to avoid cycles. If some channels have been federated in a cycle, a notification can loop indefinitely. With this mechanism the notification will be eventually discarded.

Ageing prevents the buffers to get overloaded, but it can discard relevant notifications. This problem can be remedied by combining this feature with other QoS features.

4.4.2 Priority

A priority can be assigned to notifications. A quality of service mechanism allows highest priority notifications to be filtered and disseminated before lower priority notifications in the same channel. This feature enables consumers to receive the most relevant notifications as soon as possible.

This priority policy is finer than the basic mechanism where a priority level is assigned to each channel.

4.4.3 Delivery guarantee

Several situations can lead to the loss of notifications: Automatic deletion of notifications can occur when ageing and priority are combined during local overload; lower priority notifications are waiting for higher priority notifications to be disseminated and risk to be discarded if the channel does not filter them quick enough. A consumer can miss some notifications due to a communication failure or to a consumer transient failure. Some notifications can be lost when a channel managing a queue of notification fails. For all this reasons different delivery guarantees can be provided to the client depending on the kind of problem or failure it wants to resist to.

These different modes of quality of service impliy the use of the fault-tolerance mechanism detailed in the previous section.

The Alcatel prototype provides only one delivery mode called « at least once delivery » that guarantees the delivery of any notification to each consumer except in the case of channel failure. Higher level quality of service for notification delivery is planed to be studied for further versions of the service.

Ageing, priority and delivery guarantee are basic quality of service features. They are proposed by [GMD, TID/HP, NEC]. [DTS/Nortel] proposes different quality of services mechanisms: selection of the buffer limit for queuing notifications, selection of channel pull mode frequency and selection of a policy for discarding notifications when the channel is overloaded. This last mechanism is particularly important in the domain of telecommunications. As a request from our users, we are currently studying the possibilities which are presented in the following section.

4.4.4 Policy to resist event overload

Notifications can be used to propagate error states. These notifications must be treated in order to avoid a failure of the system. If the error is not processed quickly enough, it might lead to a cascade of errors, leading to a burst of events and an overload of the system. The Notification Service must be able to resist to such situations. Two directions can be investigated to prevent an overload of the system: preventive and curative.

The preventive direction consists in defining a mechanism minimising the amount of situations where a burst occurs. For example, the use of priorities allows to specify that error events must be processed before informative events. This will result in a smaller response time to errors. The filtering is also a mechanism that reduces the load of the system: only the event accepted by managers are transmitted.

But there are some cases in which the flow of events increases and the Notification Service has no control over it. In order to resist to such situations, the service must have a mechanism to reduce the event propagation. A simple policy is to discard events with the lowest priorities. This can be done automatically when the notification lifetime is set: the higher priority notification are forwarded, meanwhile the lower priority ones get old enough in the queue to die. Another possibility is to block suppliers until the load reduces.

But more complex policies can be chosen, that both keep all events, and allow suppliers to continue their work. For example, the overloaded Notification Service could send an exception to suppliers trying to send new notifications, after having registered them in a list of unsatisfied clients. Then, when the load is getting down, the service would advertise its availability to its unsatisfied clients. These suppliers would consequently resend the notifications, if they are still meaningful.

Another strategy can be the use of *shadow channels*. Classically, a subject in a notification domain is associated to one channel. But when the notification manager detects an overwhelming traffic, the critical subject could have multiple channels in the network thanks to a notification factory dedicated to create shadow channels. In the push model, a shadow channel will receive the same configuration as the classical channel concerning the consumers. Its reference will be returned in an exception « new channel created, retry on this one » by the normal channel to the suppliers it can not cope with. When the load diminishes, the notification manager warns suppliers to return to the old channel reference. This can be done with the same exception mechanism. For the pull model, supplier must be replaced by consumer and vice versa. The load of the traffic will not decrease on the network, but the load will be shared by the normal channel, and its shadow counterpart(s).

In the same way, the COS trader could be coupled with the notification service to register subjects, and associated channels, in order to offer an alternative to the management access of the notification service.

It appeared during the Alcatel's Notification Service V2 achievement that service advanced functions could be eased if the notification suppliers and consumers could offer an interface allowing some management, for instance for implementing the shadow mechanism. The supplier/consumer could be warned of the notification service state and could even receive instructions. Nevertheless, these additional interface must stay optional, in order to allows flexibility for notification service users.

5. Conclusion and future work

A Notification Service provides asynchronous multicast between CORBA objects. Compared to an Event Service, the notification service's major additions are the support for filtering and some quality of service modes. Filtering allows to decrease the traffic which is an essential point for telecommunication applications and consequently for TINA services. Performance, quality of service and fault-tolerance are major considerations for a real-time system and thus for telecommunication software.

5.1 Features for a TINA Notification Service

We believe that the following service characteristics have to be considered for the design of a TINA notification Service:

Generic and typed notification. Generic and typed notifications provide the service user with a trade-off between a low-level mechanism with good performances and a higher level mechanism less error prone. In the second case, the cost in performance can be reduced by the use of predefined typed notifications.

Filtering. Filtering should be at the channel level for performance and scalability purposes. It should be dynamic and incremental in order to provide the quenching mechanism.

Subscription. A high level API for subscription is not essential but it is important for an easy use of the service. The concept of subjects organised hierarchically is promising and needs to be developed and assessed.

Federation. Federation at subject level is essential for performance and scalability, and can support the TINA network administrative domain concept.

Fault-tolerance. Fault-tolerance is ubiquitous in industrial telecommunication applications. A customisable fault-tolerant Notification Service should be a building block for higher level fault-tolerance.

Quality of service. Quality of service is a basic feature in real-time systems and thus in telecommunication software. We still need to investigate which quality of service features are useful for a notification service dedicated to telecommunication applications.

Notification manager. A notification manager which handles channel instanciation and deletion, notification description registration, subscription, federation, high level fault-tolerance and recovery provides to the user a set of automatic high level functions, simplifying the use of the service and centralising the service state. The second advantage of the notification manager stems from its ability to perform a part of the work, discharging channel objects of a part of the administration task. Thus, channel tasks can be more concentrated on their main task of communication. This is a very valuable feature in the context of communication.

5.2 Future work

New work begins around Alcatel's Notification Service in the context of Vital and ReTINA ACTS projects. In addition to the completion of the actual features by for instance integrating the Notification Service with a logging service more advanced studies should be done.

As a service for reliability, the reaction of the service in case of critical situations such as the increasing of notification flow beyond the possibilities of the service need to be evaluated in the context of real telecom applications in order to refine the adequate solutions according to application's needs.

A second research direction for the Notification Service is its interfacing with transaction services and models. According to the integration with particular transaction models such as the open nested model or the queuing model different profiles of the Notification Services can be provided to the users according to different quality of service requirement for fault-tolerance.

The next step for telecommunication application is a real-time notification service integrating time constraints for real-time needs in communication.

6. References

[NEC] NEC, BEA, Expersoft, Fujitsu. IBM. ICL, IONA, Oracle, TIBCO, Visigenic: « Notification Service Submission », OMG TC Document telecom/97-06-01, June 3rd, 1997.

[DSTC/Nortel] DSTC, Nortel Technology: « Notification Service ». OMG document telecom/97-05-02, June 2nd, 1997.

[GMD] GMD Fokus : « OMG Notification Service » OMG TC document 97.x.x June 3rd, 1997.

[OMG 94] The Object Management Group: « Common Object Services Specification ». V I 1.0. Framingham, MA, 1994.

[OMG RFP] OMG document: telecom/97-01-03.

[ReTINA] http://www.chorus.com/Research/retina.html

[Ruffin 92] Michel Ruffin: « KitLog: a Generic Logging Service ». Proceedings of the 11th Symposium on Reliable Distributed Systems, Houston, TX (USA), pages 139-146. October 5-7, 1992.

[TID/HP] TID, HP: « Joint submission to Notification Service RFP » Edition 2.0, June 1997.

[TMN] ITU-T, M3010, Principles for a Telecommunications management network

[VITAL] http://www.mari.co.uk/vital.

[X720] CCITT X720: « Structure of management information-Management information model ».

[X721] CCITT X72: « Structure of management information-Definition of management information ».

[X733] CCITT X73: « System management -Alarm reporting function ».

[X734] CCITT X734: « System management - Event report management function ».

Session 10

Panel: What is Next for TINA?

Cesare Mossotto, CSELT and TINA Consortium

Session 11

Quality of Service

Joe Sventek, Hewlett-Packard

TINA Connectivity Services: Quality and Quantity through QoS Based Routing and DPE Performance.

Huw Oliver
Hewlett-Packard Laboratories, Filton Road, Stoke Gifford, Bristol BS12 6QZ, UK.
heo@hplb.hpl.hp.com (Currently Visiting Senior Research Fellow at Lancaster University)

Abstract

TINA's Connection Management Architecture (CMA) is characterized by a hierarchy of Connection Performers (CPs) acting over subnetworks and, at the low level, open interfaces to switch resources. This paper describes the QoS based routing approach adopted in ReTINA with its dual advantages of subnetwork centralized calculation together with truly distributed processing gains. Those distributed link-by-link performance figures are presented for the low-level switch interface.

1. Introduction

The work described in this paper was undertaken as part of the ACTS Project ReTINA (AC048).

The goal of ReTINA is to develop and demonstrate an industrial strength open distributed processing environment (DPE). This DPE will support distributed real-time multimedia applications over emerging broadband networks.

The programme of work comprises many aspects including end-system control, object models for streams, advanced DPE services and software engineering tools for service construction. Our focus is on Network Control and the feasibility of achieving this through the use of distributed object systems. In terms of technology, our part of the project will deliver a CORBA-compliant distributed object-based DPE with real-time and quality of service management capabilities.

This paper describes the derivation and use of Quality of Service (QoS) estimates for routing and higher-order admission control. We believe that the architecture exhibits the dual advantages of logically centralized calculation together with truly distributed processing gains.

2. Background

2.1 Traditional Routing

Traditional routing is only concerned with connectivity and typically uses shortest path algorithms for path computation. In order to support QoS based routing, however, a more complex model is needed where the network is characterized with multiple metrics such as available bandwidth, delay, jitter and loss probability. The basic problem is to find an optimal path with respect to these metrics.

Broadly, we can categorize routing algorithms by the location of the decision (is it distributed at each node; centralized at some central node; at the source node; at some subset of nodes), by the strategy (is it static or is it adaptive to the current load), by the decision time (datagram vs. virtual circuits), by the source of the network information and finally by the performance criteria.

Two common examples that exemplify different advantages and disadvantages are hop-by-hop routing and source routing. In source routing, a path is computed on-demand at the source and carried in the packet header. Packets are switched according to the path in the packet. Since the computation is done centrally, source routing is very flexible - algorithms of choice can be used. The disadvantage is that full routing information for the network is required at the source. In hop-by-hop routing packets are forwarded "hop by hop" at each node. Each node has a routing table used to work out the next hop. This table is updated with new routing information periodically. One advantage is that hop-by-hop routing can use fully distributed algorithms.

0-8186-8335-X/98 $10.00 © 1998 IEEE

2.2 Traditional Signaling

One of the goals of the TINA architecture is to improve on the current approaches to Connectivity Services in Broadband ISDN/ATM.

Existing work on Connectivity Services in Broadband ISDN/ATM standards bodies has concentrated on traditional signaling techniques. These come from the telephony world. For example, the ATM standard for signaling (SVC) is now Q.2931 which is based on the narrowband ISDN (digital telephony) Q.931 standard.

The limitations with this model can be summarized as

1. They are geared to running a single connection within a single call. We need complex topologies.
2. They assume calls last for minutes and that set-up latencies are correspondingly non-critical.
3. The implementations of UNI signaling have too large a footprint. All code needed to run the protocol has to be built into the terminal. The switch also needs to run the entire protocol leading to a huge amount of code running on the switch. The code is also all mixed up with session level logic about what the end parties can do.
4. More than the code, the specification of the ATM-Forum UNI 3.0 requires 143 text book pages to define the connection management signaling. This is because it must define the delivery semantics and encoding at the bit level. An equivalent CORBA based interface for slightly more functionality requires 3 pages, is mainly self documenting and automatically defines the delivery semantics and encoding.

3. Goals

We want our Connection Management to be able to respond to sender requests for special handling of real-time services. We want to support applications that require some degree of consistent throughput, delay, jitter or error rate.

One of the ways we do this is to maintain an estimate of the current state of the network and to use this

- to improve our route choosing algorithm and
- to provide a basis for a higher level call admission control scheme.

The value of this information will increase correspondingly with the increase in heterogeneity that we are seeing in today's networks. As different routes display disparate capacities and varying utilization, so it will become more important to base our routing decisions on more than the topological information alone.

4. TINA Connection Management Architecture

The large scale of today's networks and the interconnected nature of separately administered networks means that it quickly becomes impossible to maintain accurate knowledge about all node and link states throughout the collection of networks. The TINA solution to this is to progressively aggregate the network state information into a hierarchy of subnetworks.

TINA's Connection Management Architecture (CMA) is characterized by a hierarchy of Connection Performers (CPs) acting over these subnetworks, until, at the lowest level, we reach open interfaces to switch resources. This hierarchy imposes a parent-child relationship between subnetworks. A child subnetwork is represented in the parent as a single node, with only its access points shown. The further up the hierarchy we go the more the information has been abstracted. One consequence of this is that there is necessarily a loss of information as we move up the hierarchy.

The ATM Forum PNNI standard [1] follows the same approach. Here the hierarchical aggregation is very similar but routing is source based and set-up initiation is signalled along the chosen route. The state changes in the links have to be periodically collected centrally. Link Status Update algorithms are introduced for addressing this. The CMA has one distinct advantage over this scheme when it comes to maintaining this hierarchically aggregated network state information. The requests for connection and disconnection will pass through the appropriate controlling object associated with the subnetwork to which the request was made. The decision whether to accept or deny the request is also made centrally within that subnetwork. Thus, the controlling object has direct access to the changes taking place in its subnetwork.

Another advantage of separating the control and payload channels can be seen in times of network congestion. In schemes such as RSVP [2], if the control messages are in-band they are likely to be lost at times of severe congestion - just when they are most needed. Suggested solutions include assigning separate virtual channels or the use of priorities for control messages.

In summary, we believe the TINA CMA does offer advantages for maintaining QoS status information but, as with other approaches, we are still necessarily operating with inaccurate information.

5. Quality of Service Management

We now present our approach to the use of maintenance of that QoS status information in more depth before returning to the low-level switch issues.

5.1 Application Level

The QoS Manager has the task of assuring the quality of a connection in accordance with the requirements of the session customer on whose behalf the connection is being made. ReTINA has adopted a contractual view of QoS. Each application presents a description of the traffic it will offer and what QoS guarantees it requires. In order to establish a communication flow, source and sink application have to harmonize their requirements. This results in a QoS contract between the applications which determines the required QoS as flow parameters (basically throughput, end-to-end delay, and reliability).

The requirements will be stated in application-specific or service-specific terms rather than in terms of low level network parameters. We introduce a layered mapping from these terms into a generic set of parameters for use by the QoS Manager. These will subsequently be mapped into technology specific parameters (see Figure 1).

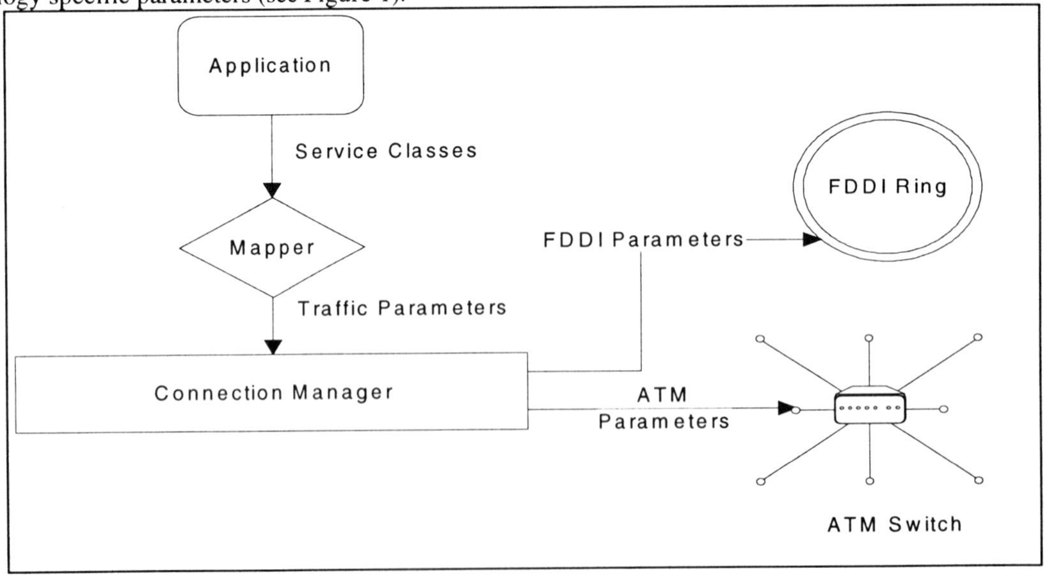

Figure 1:QoS Mapping

A second pragmatic requirement we imposed upon ourselves was motivated by the belief that the majority of multimedia traffic in the near future will be generated by Winsock 2.0 [3] applications. We have kept our QoS parameter set compliant with the Winsock 2.0 model and have implemented a Layer Service Provider with the ReTINA Connection Management. Further description will be forthcoming but is outside the scope of this paper.

5.2 Hierarchical QoS-Based Routing

A TINA network is modelled as a number of Layered Networks. A Layer Network is made up of (partitioned into) Subnetworks interconnected via links. Each Subnetwork, in turn, can be partitioned into smaller Subnetworks. At the lowest partitioning level Subnetworks model individual network elements (e.g. ATM switches). This forms the network hierarchy.

Routing under the Connection Manager is performed by a technique known as hierarchical routing. The process of routing involves making decisions at each subnetwork over which subordinate subnetworks to route upon, based on whether it can contribute to the goal of achieving the route. This is a divide and conquer solution, repeated until we get down to the subnetworks representing hardware switches.

Quality of Service is introduced into the routing model by choosing to use a subnetwork based not only on whether it contributes to our route but also whether it can provide the desired QoS characteristics.
The diagram below shows how the subnetworks divide into further networks.

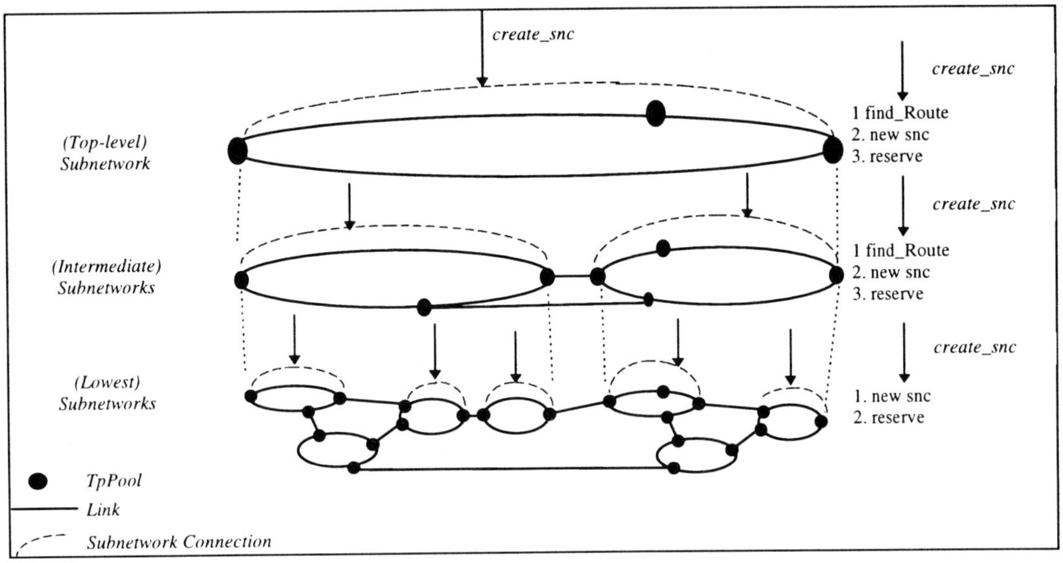

Figure 2: Hierarchical Routing

Our approach is optimistic. We assume that the subordinate layers will be able to establish the desired route. We use an open transaction model where the actions to reserve resources are visible outside the transaction. If not, and it is quite possible that they will not because our QoS estimates are, after all, estimates, then we provide compensating actions to undo the effects of the transaction.

5.3 Functional Components

At each subnetwork level routing decisions are made by a QoS Agent It consists of three components: a Router, a QoS Evaluator, and a Broker.

5.3.1 Router Component

The Router has the job of selecting possible routes from point to point through its given subnetwork. In the case of point to multipoint connections, each branch of the connection is established using a point to point connection.

5.3.2 QoS Evaluator

The QoS Evaluator does two things:

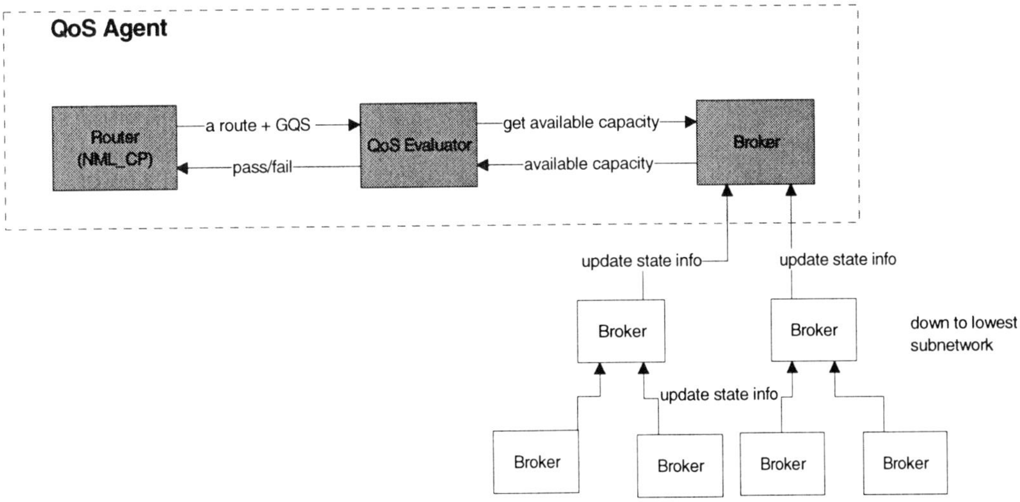

Figure 3: QoS Agent Components

1. It receives a proposed route through the subnetwork, and a QoS requirement (in terms of a Generic QoS Specification) from the Router component. It then decides whether the given route meets or fails the QoS requirement. This decision is made by querying the broker from the subnetwork about the QoS capacities involved in the route and by evaluating the effect of admitting the route.
2. It derives collapsed QoS estimates for a complete subnetwork as described in section 5.5.

Note that when a proposed route for a connection is being evaluated there are two things to consider. First, will the route be acceptable for the new connection. Secondly, will the newly accepted traffic break any of the existing QoS traffic contracts. To calculate the latter from scratch would be infeasible so we use a parameter called the "available bandwidth": the idea is that the QoS contracts will be maintained as long as the equivalent bandwidth of any new traffic does not exceed the available bandwidth.

5.3.3 Broker Component

The Broker is responsible for maintaining QoS capacity information of an associated resource (either a Subnetwork or Link). Initially we have chosen to store the available bandwidth, delay and error rate but this is easily extendible. The broker hierarchy follows the network hierarchy and a broker is also responsible for initiating if necessary the update of its parent broker.

The Router, QoS Evaluator and Broker run in the same process for efficiency but the Broker runs in its own execution thread. This allows QoS estimate updates to occur as a low priority in the background but newly arrived connection requests have immediate access to the latest status figures.

5.4 Subnetwork Status Updates

The Broker at each subnetwork holds the QoS attributes for the nodes and links for that subnetwork. The QoS attribute values can be updated either by a sub-broker deciding that its status has changed sufficiently to "push" its new estimates up a level, or, by this broker deciding to "pull" or request estimates from the sub-broker. This happens, for example, at initialization.

It is necessary to get new estimates from the child broker because the topology of that subnetwork is unknown to this broker. That is the purpose of the hierarchical abstraction. One immediate question is how frequently these Subnetwork Status Updates (SSUs) occur. We have some constraints:

1. SSUs must not propagate all the way up the hierarchy to the root. That would cause the root to become a bottleneck.
2. SSUs must happen sufficiently often to keep the subnetwork capacity estimate accurate enough to get "good" routing decisions.
3. SSUs must not happen so frequently that they overburden the processing load.

Finding a solution to requirement 1 was not difficult. There are two reasons why updates do not propagate to the root and form a performance bottleneck. First, the network capacity may be unaffected by the admission of a new connection (the new traffic may have been routed on a non-critical path) or second, the change in capacity may be too small or too transient to warrant an update to the parent broker. Our initial scheme for balancing requirements 2 and 3 is given in Appendix A. We intend to experiment with the parameter space in future work.

The QoS estimates from the inferior brokers are cached in the parent broker since these QoS values are accessed frequently during route evaluation or QoS estimation. We use object proxies to make the caching transparent.

5.5 QoS Estimate Derivation

As described above, the need for a subnetwork to re-derive the estimate of its QoS capacity can be triggered either by a request from the parent or by the QoS Agent itself. In either case the activity is the same. The entire QoS-attributed topology has to be collapsed into the set of boundary nodes with QoS capacity estimates between those nodes.

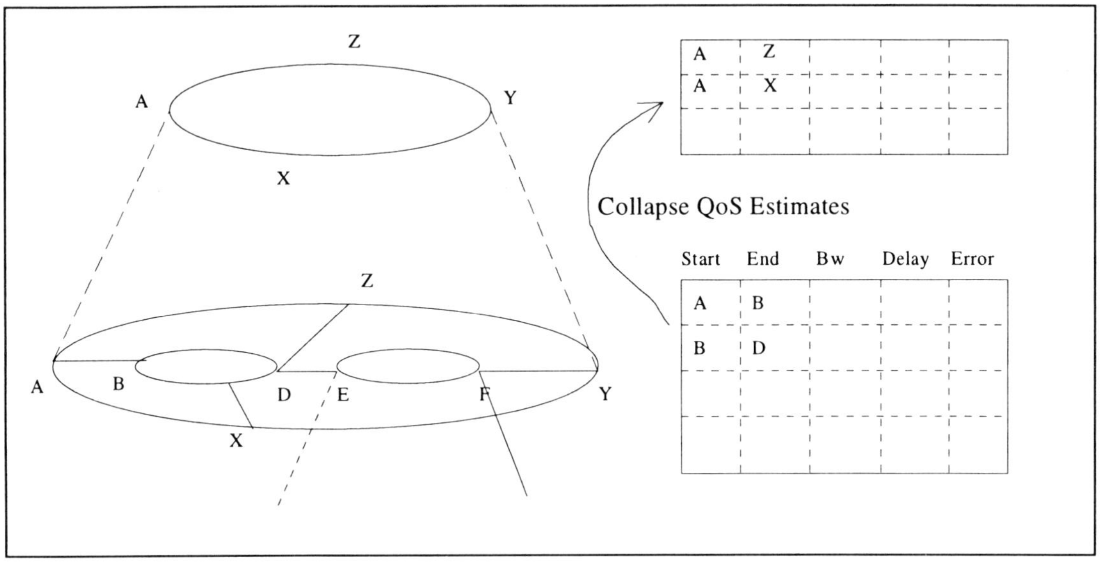

Figure 4: Deriving Subnetwork QoS Estimates

Initially we searched for heuristics or estimation algorithms that, while not exact, would quickly provide a rough estimate of QoS capacities across a subnetwork. We failed to find anything suitable and resorted to a complete analysis of the subnetwork using the different QoS parameters in turn as cost metrics for Dijkstra's shortest path algorithm. We were pleasantly surprised by the efficiency of this (not least compared to any remote object binding costs!). We are comfortable with these calculation costs, particularly remembering that they occur in a background execution thread.

Once we have a path (either from the Dijkstra algorithm or from the Router component) we can form an end-to-end value for the available bandwidth, delay and error rates. The formulae for these are given in appendix B.

6. At the Physical Level

Once the connection request reaches the leaves of the hierarchy, it is processed by EML-CPs. These are objects that can talk to the physical device (i.e. switch) directly. These separate switch set-up processes can proceed independently and in parallel.

Ideally we would like switches to export a control interface together with an advertisement of their current resources. In practice, today's switches do not do this and industry fora exist [5] to encourage the development of such open interfaces in future.

The first issue, the lack of a control interface, can be tackled in two ways. Either we can use the management interface (SNMP) to set up connections, or, in some cases, we can manipulate the switch connectivity matrix directly. We discuss each of these approaches in the rest of the paper.

The absence of advertised switch resources was tackled by building simple mathematical models of the switch. The delay and error rate models are given in Appendix C.

6.1 Switch Connectivity Matrix Manipulation

To overcome the above problem of no switch control interface, we built a Java Virtual Machine on the switch which hosted a program that directly manipulated the connectivity matrix. We were impressed by the use of the Java Virtual Machine on the different controllers to achieve platform portability and also the use of the JVM on the switch to build the server software. The use of 'Java everywhere' has allowed us to drop useful pieces of software onto the VM such as a Corba ORB. Without Java it would, for example, have been a significant piece of platform dependent work to embed an ORB on the switch.

We have experimented with different platforms for the controller (HP-UX and NT), using the same IIOP interface to the switch. Figure 3 shows two examples of the software pieces. The central block represents the switch while the right and left blocks show two alternative controllers.

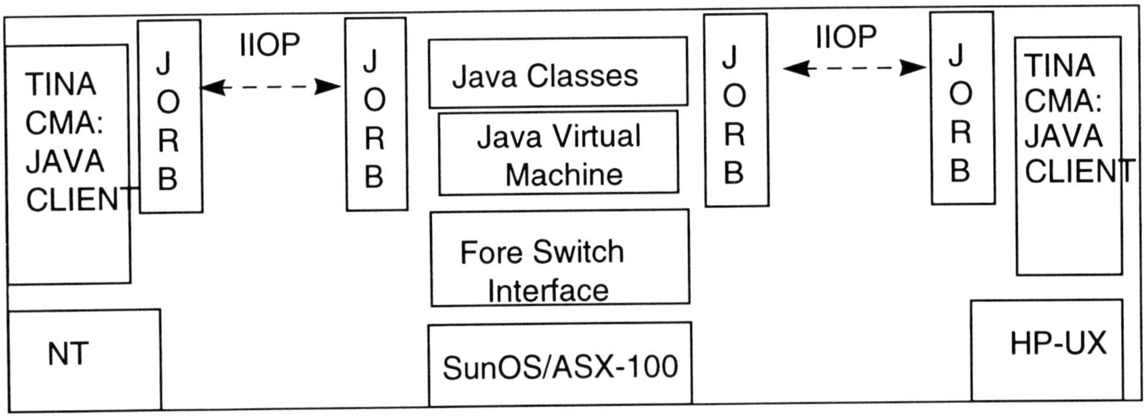

Figure 5: Java allows easy porting of the client software

This section describes a series of experiments and performance measurements carried out to assess the achievable performance of connection creation and deletion on an ATM switch. One of the comparisons we make in this section is between the use of SNMP (Simple Network Management Protocol) as a switch control interface and a more direct control interface developed at HP Labs. The overall conclusion of these measurements is that SNMP performance can indeed be bettered by an order of magnitude by means of the alternative switching interface.

However, the performance measurements described in this section also provide input to other discussions, as listed below.

1. Performance of SNMP versus alternative link by link setup mechanisms.
2. Performance of different Java implementations.
3. Performance of C versus Java in I/O bound applications.
4. Performance of CORBA versus raw sockets.
5. Deployment of TINA CM computational objects.

Communication was done by means of two distinct mechanisms:
1. A CORBA/IIOP interface.
2. A raw interface based on TCP/IP sockets.

The connection server can be configured at start-up to provide either or both of the two interfaces, allowing clients to choose the means of communication. During measurements, the server was configured to only provide one of the two possible network interfaces at a time, although provision of multiple interfaces is unlikely to affect server performance.

Measurements on pure CORBA solutions suggested that CORBA is too heavy-weight for communication with the physical switch. This realisation prompted implementation of a raw socket interface for comparison. However, to be plugged directly into the TINA framework, the server is required to provide a CORBA interface. Therefore, a hybrid solution was tested, using a CORBA proxy running on a fast workstation, and forwarding requests to the switch by means of the raw socket interface.

Two different ORBs were used, both written entirely in Java, and both providing a full CORBA 2.0 implementation:

1. JORB (Java ORB) from HP, in alpha 3 release.
2. OrbixWeb2.0 from Iona, providing the Java equivalent of Orbix 2.1.

The following table lists performance measurements of Java based solutions.

	Network Interface	Client	Server	RPC/sec
1	JORB	torino/JDK 1.01	asx100/JDK 1.01	14
2	Socket	torino/JDK 1.01	asx100/JDK 1.01	106
3	JORB	torino/JDK 1.01	brandt-s-1/JDK 1.02	107
4	Socket	torino/JDK 1.01	brandt-s-1/JDK 1.02	597
5	JORB	brandt-s-1/Symantec	asx100/JDK 1.01	16
6	Socket	brandt-s-1/Symantec	asx100/JDK 1.01	121
7	JORB	brandt-s-1/Symantec	vecheo/Symantec	401
8	Socket	brandt-s-1/Symantec	vecheo/Symantec	1330
9	JORB	brandt-s-1/Symantec	brandt-s-1/Symantec	418
10	Socket	brandt-s-1/Symantec	brandt-s-1/Symantec	1818
11	JORB	brandt-s-1/JDK 1.02	brandt-s-1/JDK 1.02	159
12	Socket	brandt-s-1/JDK 1.02	brandt-s-1/JDK 1.02	998
13	JORB	brandt-s-1/Symantec	asx100/Kaffe	10
14	Socket	brandt-s-1/Symantec	asx100/Kaffe	105
15	OrbixWeb2.0	brandt-s-1/Symantec	brandt-s-1/Symantec	600

As shown in row 1, JORB communication with the asx100 switch allows around 14 RPCs per second. This number is clearly limited by three main factors:

1. The performance of the switch itself.
2. The performance of the Java virtual machine.
3. The performance of the JORB implementation.

The impact of switch performance can be seen by comparing rows 1 and 3, respectively rows 2 and 4 (This assumes that JDK 1.01 and JDK 1.02 exhibit similar performance. This has not been tested). The performance gain from running the server on brandt-s-1, as opposed to asx100, is 7.6 using JORB, and 5.6 using sockets. Hence, performance of the switch is a bottleneck. That performance gain is larger in the JORB case must be attributed to the larger overhead of parameter marshalling, resulting in more time spent in the Java virtual machine.

The impact of virtual machine performance can be seen by comparing rows 9 and 11. Both rows reflect JORB performance with the client and server running on the same machine, in this case a 200MHz Pentium Pro PC. In row 9, the Symantec virtual machine is used, which includes a just-in-time (JIT) compiler. In row 11, the JDK 1.02 virtual machine is used. The performance gain from the JIT compiler is a factor 2.6 with JORB communication. Similarly, comparing rows 10 and 12 we get a factor 1.8 with raw socket communication. The difference in speed-up must again be attributed to the different times spent in the Java virtual machine, with the socket solution being the most I/O bound of the two solutions. In any case, similar speed-ups can in principle be achieved on the switch by applying a better Java virtual machine, although one was not available at the time these tests were made.

The impact of using CORBA is apparent from comparing any odd-numbered row to its following even-numbered row. For example, the difference between rows 1 and 2 is a speed-up of 7.6 from replacing JORB communication by socket communication. However, this overhead also depends on the CORBA implementation. For example, as can be seen by comparing rows 9 and 15, OrbixWeb performs roughly 50% better than JORB, and would therefore allow around 20 RPCs per second against the switch.

6.1.1 A Hybrid Solution

Given the above measurements, it might be worthwhile to try to combine the best of the two worlds: the speed of plain sockets with the plugability of CORBA/IIOP. To that end, we tried to insert a proxy object between the client and the server: by means of JORB, the client makes calls to a proxy running on a fast PC on top of a fast Java implementation. The proxy then forwards the request to the real server, using the raw socket interface. Measurements are shown below:

	Client	Proxy	Server	RPC/sec
15	brandt-s-1/Symantec	brandt-s-1/Symantec	asx100/JDK 1.01	97

As can be seen, this approach does in fact combine the best of the two solutions. Combined with a fast JIT based Java implementation on the switch, around 200 connection creations per seconds should be comfortably within reach.

6.1.2 A Server Implementation in C

Finally, we implemented the server in C. The performance of the server written in C was measured in two different configurations: with the client connecting directly to the server using sockets, and with the client connecting via JORB to a proxy server, which then connects to the server using sockets. The C server does not provide a CORBA interface. Performance is shown below:

	Client	Proxy	Server	RPC/sec
16	brandt-s-1/Symantec	brandt-s-1/Symantec	asx100/C	251
17	brandt-s-1/Symantec	Direct (no proxy)	asx100/C	600

Comparing rows 15 and 16, we see that rewriting the connection server in C provides a speed-up factor of 2.6 in a proxy based solution. With the client calling the server directly by means of raw sockets, the C server provides a speed-up factor of 5, as can be seen by comparing rows 6 and 17. It should be noted that this speed-up would shrink to between 2 and 3 if a faster virtual machine could be used on the switch. However, this relatively small speed-up reflects the I/O bound nature of the server. Hence, if the computational complexity of the server was larger, a larger speed-up could be expected.

6.2 SNMP

Using a Smalltalk client and communicating with the switch by means of SNMP, around 50 connection creations or deletions per second was achieved. Assuming that Smalltalk exhibits performance similar to Java (this is ignoring the fact that many Smalltalk systems include a compiler), this number is comparable to row 17, providing around 600 connection creations per second. Hence, using a connection server instead of SNMP, an order of magnitude better performance can be achieved.

Although we have only performed measurements on the ASX-100, others quote SNMP performance around two connection creations or deletions on other switches. This seems to suggest that there are large differences between the performance of SNMP on different switches, and that SNMP implementations can therefore in general be improved with respect to performance. We would expect to see optimised implementations of SNMP on switches where the interface is being used for switch control.

Another optimisation would be an attempt to overcome the SNMP feature that multiple get/set attribute operations have to be made in multiple calls to the interface. One obvious way to address this would be to have an object on the switch that could handle multiple attribute operations as a bundle and that could then make the multiple SNMP calls locally. Another would be to use SNMPv2 which supports multiple get/set operations.

To be fair to SNMP, we should not forget that it is significantly more general than our protocol, applies to most or all switches for private networks (switches for public networks may support the CMIP protocol), and was furthermore not conceived as an alternative to connection setup by means of signaling. Nevertheless, although it can be bettered, SNMP performance on the ASX-100 is probably sufficient for many purposes.

7. Conclusions

The fundamental characteristic of the system is the separation of the control and data traffic planes and the use of distributed objects to implement the former. This separation gives us access to all control messages passing through the system. This makes the maintenance of QoS estimates feasible. We still have a lot of engineering to do to explore the parameter space and the scalability issues.

Basically, the system works. We were able to switch multiple video and audio sources between multiple sinks using the set-up in Figure 6. We were able to improve route determination based on QoS estimates and it forms an effective high-level admission control.

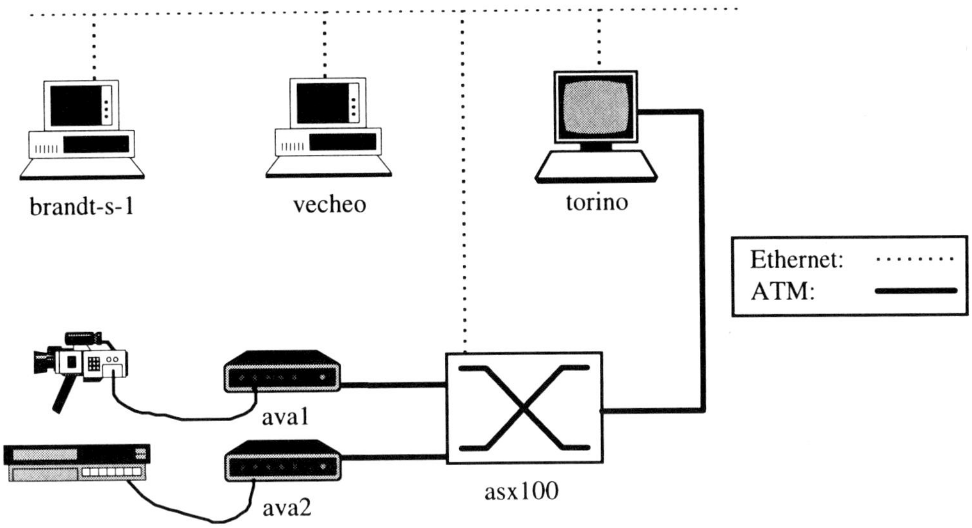

Figure 6: Sources, Sinks and Connections

At the switch level our measurements demonstrate that SNMP performance is indeed inferior to the connection server used in this experiment. On the test equipment used, SNMP achieves around 50 connection creations per second, which should be compared to the around 600 per second achieved by the connection server implemented in C.

However, only a few switches actually allow the implementation of a connection server by means of an open switching interface such as used within HP Labs. Our ability to put this interface onto the ASX-100 is an exception.

The inferior performance of the switch hardware as compared to a modern PC or UNIX workstation suggests that it would be unwise to run a full EML-CP on the switch itself, although this is indeed possible. A more viable approach is to run the EML-CP on a high-performance PC or workstation, only communicating with the switch itself when necessary to actually create or delete physical connections. This separation of concerns also makes sense from a software engineering point of view, since it allows different switches to be controlled by the same (powerful) controller hosting the corresponding EML-CP object implementations.

8. Acknowledgements

The Java and C performance figures were measure by Soren Brandt of HP Labs. The performance of connection creation and deletion by means of SNMP was measured by Matthieu Goutet of HP Labs. The ReTINA CMA and QoS Manager design was a joint effort of the ReTINA partners.

9. References

[1] R. Cherukuri, D.Dykeman (eds.), and M.Goguen (chair). PNNI Draft Specification. ATM Forum 94-0471, November 1995.

[2] R.Braden (ed.), L.Zhang, S.Berson, S.Herzog and S.Jamin. Resource reSerVation Protocol (RSVP) version 1, functional specification (draft-ietf-rsvp-spec-13.ps). INTERNET-DRAFT, Internet Engineering Task Force - RSVP WG, July 1996.

[3] "Windows Sockets 2 Application Programming Interface" Revision 2.2.1 2nd May 1997. http://developer.intel.com/ial/winsock2/specs.htm - Intel Winsock Spec WWW page.

[4] D.Comer and R.Yavatkar, Proc. 2nd IEE Symposium on Parallel and Distributed Processing (1990).

[5] http://comet.ctr.columbia.edu/opensig

Appendix A: a Subnetwork Status Update scheme

For each subnetwork define Δ_c, Δ_d, Φ_c and Φ_d as percentages of the subnetwork's capacity and delay respectively (we could also add jitter to this without loss of generality). Define an interval T for each subnetwork.

We send an SSU to the parent broker whenever

1. there is a change in capacity > Δ_c and the current bandwidth reservation level > Φ_c or the delay change > Δ_d and the current delay > Φ_d or
2. there is a change in capacity > Δ_c and the current bandwidth reservation level < Φ_c but was > Φ_c at the last SSU or the delay change is > Δ_d and the current delay < Φ_d but was > Φ_d at the last SSU or

3. there was a connection request rejection and the bandwidth reservation or delay has increased since the last SSU or
4. there has been no SSU for T seconds

Appendix B: end to end QoS values

Using t for the available bandwidth (or throughput), d for the delay and e for the error rate, we can calculate:

$$t_{total} = \min\{t_i \mid (1 <= i <= n)\}$$

$$d_{total} = \sum_{i=1}^{n} d_i$$

$$e_{total} = 1 - \prod_{i=1}^{n} (1 - e_i)$$

for the end to end figures. Note that the formulae are associative so we can calculate the values as we go along (min and sum are obvious; the proof for e is a simple exercise).

Appendix C: simple models of delay and error rates

Delay

Each switch has a constant processing capacity which expresses the number of bits switched per second. The processing capacity for switch s is denoted by c_s bits per second. Consequently, the delay factor for one bit, associated with switch s is $\Delta_s = 1/c_s$. A flow f is characterised by its requested bandwidth, regarded as the mean bit arrival λ_f

The delay incurred (by a bit b) at switch s is given by $D_b = Q_s + \Delta_s$ with Q_s the delay factor incurred by bits waiting in the queue in front of the switch. If L is the number of those bits we get $Q_s = L\Delta_s$
and hence $D_b = L\Delta_s + \Delta_s$.
With the help of Little's formula $L=\lambda D$ we get $D_f = \Delta_s (\lambda_f D_f + 1)$ and thus $D_f = 1/C_s - \lambda_f$ seconds per bit.
For a set of flows $\{f_1, f_2, \ldots f_n\}$, this generalises (proof omitted) to a total delay

$$D_{F;s} = \frac{1}{c_s - \Sigma\lambda_f}$$

Error Rates

This is trickier as we have to know something about the internals of the switch, i.e. we have to model a finite system (we would have no errors with infinite buffers).

The general formula for the probability of a full queue, i.e. when cells get dropped, is

$$P_K = \rho^K(1-\rho)/(1-\rho^{K+1}) \text{ for an M/M/1/K queue}$$

where ρ is λ/C, the "load" or "traffic intensity".

Engineering for Quality of Service

Chris Loftus (*), Edel Sherratt (*), Panagiotis Demestichas (**)

(*) Department of Computer Science, University of Wales Aberystwyth,
Penglais, Aberystwyth Ceredigion, Wales, UK SY23 3DB
(**) Intrasoft S.A., 2 Adrianiou Str., 115 25 Athens, GREECE

Abstract

Important factors that will determine the success of the players in the competitive communications market of the future are efficient service creation and deployment, and Quality of Service (QoS). In this paper we present an approach service creation and QoS handling that is followed by the ACTS project (AC227) SCREEN. In particular we identify two general problem classes that should be tackled by the service developer, in the context of QoS handling. The first problem class is concerned with the derivation of the adequate service logic that will capture and negotiate the required QoS levels, during service execution. The second problem class is concerned with the translation of QoS requirements into performance requirements of service components, and subsequently, with the deployment of the service components on the DPE. In the sequel, we specify promising solutions related to the first problem class, that are based on the definition of component contract offers and negotiation.

1. Introduction

It is envisaged that the competitive communications market of the future will offer a wide variety of sophisticated services. To a significant extent the success of the players in the market will depend on their adaptability to user demands. Another determinant factor will be the provision of advanced service features in the most cost efficient manner, and the introduction of services by minimally impacting the existing communication infrastructures. In those respects, compliance to the Telecommunications Information Networking Architecture (TINA) is the first important measure [1]. The second is efficient creation of value added end user services targeted to CORBA and TINA compliant Distributed Processing Environments (DPEs). The latter forms the primary focus of the European ACTS project (AC227) SCREEN.

Recognising that Quality of Service (QoS) will arguably constitute the most important single determinant of the commercial viability of any future advanced communications service, SCREEN has developed component oriented engineering practices that address QoS features during every phase of service creation. In this paper we present the SCREEN approach to service creation and QoS handling. A method for identifying QoS requirements, based on a general service classification scheme, is presented. The classification scheme guides the process of formulating the QoS requirements. Having modelled QoS requirements, we are lead to two general problem classes that should be tackled by the service developer, in the context of QoS handling. The first problem class is concerned with QoS representation, and negotiation. The aim of this problem class is the derivation of the adequate service logic that will comprise the pertinent functionality (which will be used during service execution). The second problem class is concerned with the translation of QoS requirements into performance requirements of service components, and subsequently, with the deployment of the service components on the DPE.

In this paper we provide promising solutions to the problem of QoS representation, translation and negotiation, and to the problem of translating QoS requirements to service component performance requirements. The solution is based on the definition of component contract offers in terms of QoS parameters indicated by the ITU model [2]. Furthermore, we present a recent approach to meeting non-functional requirements through negotiation of component contracts [3]. This provides a promising new approach to ensuring service quality by design.

The rest of this document is organised as follows. In section 2, we present the service creation process and we highlight the existence of the QoS handling problem. In section 3, we state in general terms two classes of problem related to the overall problem of QoS handling. In section 4, we present an approach to QoS handling that is based on component contracts. Finally, in section 5, concluding remarks are drawn and issues for further study are identified.

2. Service creation process

SCREEN has defined, and implements by means of existing commercial or freeware tools, a service creation process that consists of five phases (see Figure 1). As a general comment, it is envisaged that these phases constitute a single iteration of a larger process model (e.g., a spiral or incremental development model). Cross phase activities, such as quality assurance, project management, costing, configuration management and traceability are also assumed, though not illustrated in the diagram. The

0-8186-8335-X/98 $10.00 © 1998 IEEE

phases illustrated are:

Requirements gathering and analysis during which the developer assembles and documents the requirements on the service in co-operation with the user;

Service analysis during which the developer decomposes the service and identifies service components and implementation constraints;

Service design during which the developer defines the interface and behaviour of the service in accordance with implementation constraints;

Validation and simulation during which the developer verifies that the defined behaviour is in accordance with the requirements;

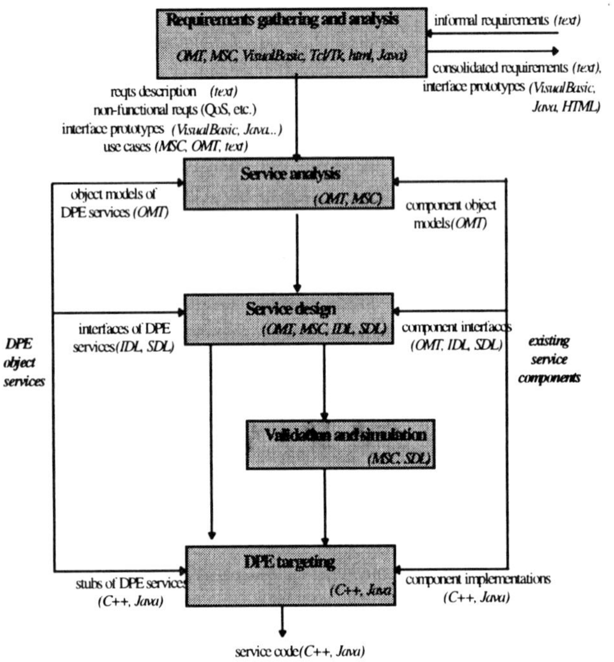

Figure 1: The SCREEN Service Creation Process

DPE targeting during which an implementation is generated from the design.

Two important notions underlying each phase of the service creation process are service composition and QoS handling. Regarding service components, the developer is provided with tools for identifying, searching, and consequently for, reusing service components.

QoS requirements are identified during the requirements gathering and analysis phase. An approach to this problem is described in section 3. In the subsequent phases, the identified requirements may be taken into account in both of the following two manners. First, the service logic should be capable of expressing and negotiating the desired QoS levels with the underlying communication facilities. This functionality will be used during service execution (i.e., at run-time). Second, some QoS aspects should be translated to performance requirements posed on the service components, and to performance constraints that should be met by the arrangement of the service components on the DPE. This problem class is addressed by the service developer during the design phases of the service.

2.1. Service creation targeted to TINA environments

SCREEN aims at the development of services that adhere to the TINA standards. In this respect, incorporated in the service creation environment that is developed in SCREEN are the following features. First, models (software libraries) representing service independent features of a service. Second, facilities for developing TINA compliant service dependent components. The aim is to enable the service developer to create the service dependent components (e.g., SSM, USM), to integrate some necessary service independent features in the service logic, and to test the service dependent components in conjunction with the service independent ones (e.g., UA).

3. Approach to QoS handling

QoS requirements may be specified using the general framework for the derivation of QoS parameters provided by ITU-T ([2]). According to this model QoS parameters refer to the typical functions performed within a service, namely, *access*, *information transfer*, and *release*. This set of functions may be extended to include other issues, e.g., *reconfiguration* functions such as new party invitation, or addition of a new information stream. Each phase is characterised by parameter sets, which may fall within the *speed*, *accuracy*, and *dependability* categories.

The wide variety of complex services that may be offered along with the diverse requirements that these services impose necessitate a service classification scheme. This scheme is required in order to obtain the QoS requirements that should be met, during the information transfer phase of the various services. In this paper we make the first step in this direction by providing the simple scheme of sub-section 3.1. This scheme is by no means exhaustive, however, it enables us in sub-section 3.2, to characterise (in a rough manner) service classes in terms of QoS. Having modelled QoS requirements, we are lead to two general problem classes that should be tackled by the service developer, in the context of QoS handling. The pertinent problem definitions are provided in section 3.3.

3.1. Service classification

A simple classification scheme is obtained by considering some of the sets of service features that exist in a service, e.g., the teleservice type namely conversation, conference, retrieve, distribution, messaging, or the set of information types that are communicated, i.e., audio, video, data, picture, text, graphics. Each service feature type is associated with specific QoS characterisation as we will see in the next-subsection.

The set of teleservices include the following types.

First, conversation or conference, which correspond to the connection-based bi-directional communication, of at least two parties. Second, retrieval which corresponds to the connection-based, point-to-point or multipoint-to-point, unidirectional communication by means of which a user may retrieve different kinds of information from particular information sources (e.g., databases). The direction of the communication is from the sources to the users. Third, messaging, which corresponds to the connectionless, point-to-point or point-to-multipoint, unidirectional communication that enables the forwarding of information units in a non real time manner (e.g., e-mail). Fourth, distribution, where information is provided by a central source (e.g., television or audio broadcast). Each teleservice may involve the transfer of one or more information types, namely, audio, video, data, text, graphics, and pictures.

3.2. QoS characterisation

In order to obtain QoS requirements that should be met in the information transfer phase of a service we may exploit the classification scheme of the previous sub-section. In other words, QoS requirements are imposed by the specific service feature included in a service. In this perspective, the conversation and conference service feature require low delay and delay variation. The retrieval service feature requires low delay, and usually, error and loss free communication. Similar remarks may be made for the distribution service feature. The messaging service feature do not require delay or delay variation constraints. However, it requires error and loss free communication. Another service feature that imposes QoS requirements is the nature of the information that is communicated. Such examples are, the categorization of information types as rate-oriented (audio and video) or unit-oriented (data, text, graphics, pictures), or some inherent characteristics of the information type (e.g., integrity is more essential for data transfer than for audio transfer).

The QoS requirements of the other phases of a service (i.e., access, release, reconfiguration, etc.) may be obtained in a more general manner (i.e., these phases may be characterised at least, qualitatively, independently of a classification scheme). Classical measures for characterizing QoS during these phases are the delay with which the respective tasks (that should be performed in this phase) are accomplished, or the ratio (probability) at which these phases succeed in successfully accomplishing their task. Since the functionality of these phases is obtained through the cooperation of a set of service components, the QoS aspects identified may be readily translated to performance requirements imposed on service components

3.3. Problem definitions

In this sub-section we identify two general problem classes that are related to the overall problem of QoS handling. Specifically, the first problem class is related to the service execution whereas the second one is related to the service design.

The aim of the first problem class is to attribute the service logic with the necessary capabilities to handle QoS during service execution. The service logic should be capable of quantifying the required QoS levels. Subsequently, the required QoS levels should be translated to performance requirements that are to be requested by, and possibly negotiated with, the underlying network. A factor that could be taken into account is the cost at which the network will provide the respective performance. In summary, the aim is to design (as far as possible) service independent components, that are capable of addressing the following generally stated problem, during service execution. Given a set of QoS requirements find (after negotiating with the underlying network) the minimum cost performance requirements that the network should guarantee so that the QoS constraints are satisfied.

The second problem class has a first objective the translation of QoS requirements into performance requirements imposed on the service components, and as a second objective to map the service components onto the targeting environment. *Execution time* and *space* [4] are two (but not the only) features that may be used as a basis for assessing component performance. The rationale is that a component may be regarded as a piece of software, performing a certain task. Both dimensions are functions of the input size. Execution time is the interval that lapses until task completion, and is often equivalently expressed as the *complexity* of the algorithm that performs the task. Space is the memory required for the task completion (during code execution). A method for performing the translation of QoS requirements onto service component performance requirements is provided in the next section.

A list of quality criteria for mapping the service components onto the DPE may be the following. First, communication cost, which implies that components which interact extensively should be placed as close together as possible, and that components that need communicate in order to accomplish a certain task should also be placed as close together as possible. Second, load balancing, which implies that some components should be distributed to particular network nodes, in order to avoid bottlenecks. Bottlenecks may deteriorate service quality and increase communication cost. Load balancing might be achieved by introducing as many local nodes as possible and by properly allocating users to service nodes while keeping the total equipment cost within acceptable range. The experimentation with other quality criteria is an issue left for further study.

3.4. QoS handling in TINA services

SCREEN aims at proposing concrete solutions to the problems that were generally stated in the previous subsections. Subsequently, SCREEN will experiment with various alternatives regarding the mapping of the QoS handling functionality to specific (TINA compliant) service components. Results from the experiments will be reported in a future study.

4. QoS handling through component contracts

An important feature of the SCREEN practices is that services are assembled from service components. The SCREEN practices require that the effect of individual components on the overall quality of the services from which they are built should be assessed. This is done by defining contract offers made by individual components, by expressing these contract offers in terms of the NFR-Framework [3,5,6], and by using the NFR-Framework to reason about the synergies and conflicts implied by the contracts offered by different components.

4.1. Component contracts

The QoS parameters associated with a component constitute a contractual offer indicating the minimum QoS that can be provided for a given DPE profile. The contract comprises a list of parameters with associated QoS thresholds. Its purpose is to assist selection of components from a Component Repository. A component is considered for inclusion within a service provided its QoS contract does not preclude QoS requirements for the overall service.

The content of the contract varies between implementation or design components.

Contracts associated with implementation components (e.g. C++, Java) define the worst case QoS that the component offers for given DPE profiles; for example, the impact of 1 service request/second versus 1000 requests/second, or the impact of connecting communicating sub-components using a 64kbps link versus using a 10mbps link.

- Contracts associated with design components (e.g. SDL) define either the QoS achieved by past implementations of the design, or else provide QoS estimates based on simulation. However, the design might be implemented in a way which does not fulfil its QoS potential. This means that new implementations must be checked as part of an overall quality assurance procedure to ensure fulfilment of QoS requirements.

4.2. Representing contracts using the NFR-Framework

NFR-F helps system designers capture non-functional requirements (such as modifiability and performance) and map these to possible design solutions. NFR-F supports reasoning about the relative merits of one design solution over another with respect to their impact on particular QoS parameters and also about conflicts and synergies amongst QoS parameters themselves (e.g. the conflict between space and time performance).

To represent component contracts, QoS parameters are treated as goals in NFR-F, and are expressed as follows:

NFR-type[parameter; importance-indicator]

For example, the highVoiceQuality QoS parameter of a service, where this QoS was considered critical, could be expressed as:

highVoiceQuality[service; 1.0]

The importance indicator is a value between 0 (no importance) and 1.0 (highest importance). Normally [3] a qualitative indicator is used such as "critical" or "very critical". However, a numeric range is more expressive and is more helpful when evaluating the degree to which overall service QoS has been met. QoS goals are represented using goal refinement graphs, like that illustrated in Figure 2.

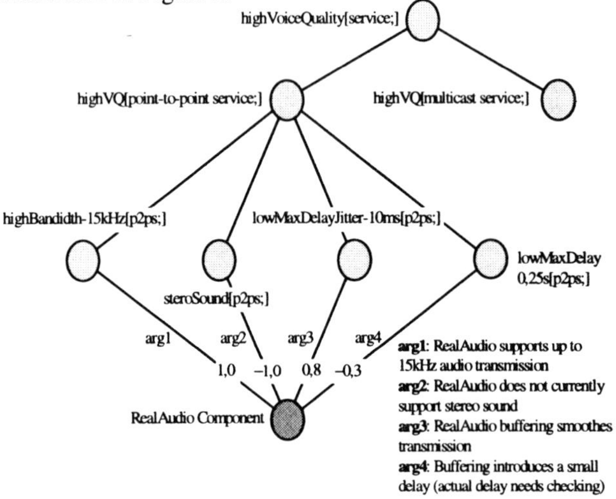

Figure 2. NFR-F graph for QoS impact of the RealAudio component

Refinement can stated be in terms of the QoS type or in terms of the parameter. For example, highQV [p2ps] is refined to stereoSound [p2ps] and highQualityVoice [service] is refined to highQV [point-to-point-service].

The NFR-F graph shown in Figure 2 represents a QoS contract offer. This contract is created by the publisher of the RealAudio component, who estimates the QoS characteristics and other NFR properties of the component and then constructs an NFR-F graph with appropriate satisificing links using a public domain tool which can be obtained from [7]. This NFR-F graph is stored as part of the RealAudio component. The publisher may associate several graphs with a component; for example, to reflect different QoS behaviours expected for different DPE profiles.

As well as the overall QoS goals, the graph represents a solution space of satisficing goals. In [5] the term "satisficing" (rather than "satisfying") is used to indicate that a solution goal might only partially satisfy an NFR (e.g. it only weakly supports it).These are connected by AND and OR links. The AND operator indicates that several solution sub-goals are to be combined. The OR operator indicates choice between alternative solution sub-goals. In Figure 2, the RealAudio component is the only satisficing goal present, and is linked to the QoS goals that it affects. These links indicate that the solution either partially supports or partially militates against the QoS goals.

The degree of satisficing is indicated by labelling the satisficing links with a numeric weighting between -1.0 and 1.0. Normally, [6] a notation using ++, +, - or – is used to represent the strength of support. However, we believe that the numeric range is more expressive.A negative value indicates that the component does not support the QoS goal. A positive value indicates support for the QoS goal. Extremes of support or conflict are indicated by the values 1.0 and -1. In Figure 2, the RealAudio component does not support stereo sound and so a value -1.0 is placed on the link. However, it can support 15kHz audio transmission and so 1.0 is used.

One kind of NFR-F goal not illustrated in the above example is the argumentation goal. Argumentation goals provide evidence and rationale for other NFR-F goals and links. Argumentation goals may be annotated with informal textual comments supported by references to case studies, engineering publications and other experience reports. Alternatively, formal statements may be used to define applicability of design-solutions or to state the semantics of a QoS goal. An example would be to formally express the following statement: For manual checking (the satisficing goal) of the accuracy of travel expenses (the NFR goal), then accuracy and integrity of these expense claims can only be assured where the checker is not the issuer of claim. Formal or informal annotations can also be used to configure satisficing link weights based on context of use (e.g. a highly unreliable Internet transport with a high packet retransmission rate might cause voice transmission delay which requires the value -0.3 to be increased to -0.5 for that specific DPE context).

NFR-F also has the concept of method which is used to capture architectural design knowledge and experience in a reusable form. There are various types: refinement methods (e.g. the refinement of performance into time performance and space performance); satisficing methods (common ways of solving NFRs); and argumentation methods (reusable pieces of rationale, e.g. "the provision of programmable APIs using script languages such as Tcl has been shown to make programs modifiable [8]"). SCREEN is considering whether methods should form components in their own right. Libraries of such methods are being developed based on industrial case studies including those presented in[3,9].

Finally, correlation rules and links capture knowledge and experience concerning tradeoffs between design alternatives or between QoS goals. For example, a strong negative correlation link could be placed between the QoS goals: stereoSound and highBandwidth-15kHz.

4.3. Component composition, QoS and NFR-F

This section describes how QoS can be found for a component aggregation. For example, suppose transcripts of all voice communication are required., illustrates use of NFR-F to reason about QoS when composition of the RealAudio component with a speech to text translator is being considered.

Several steps are needed to discover the impact of component composition on overall QoS.

1. For the each component in the service, if the component is being reused an appropriate NFR-F graph is extracted. For a new component, a new graph must be created.
2. Where a graph is reused, the engineer must:

- Examine the graph to see if all QoS requirements are dealt with. It may be that certain QoS requirements are not applicable for the given component. However, some may be missing due to an oversight by the component publisher. Also, the engineer should check the argumentation goals associated with the QoS goals to ensure that their semantics are consistent with his/her QoS requirement semantics. For example, the component publisher and re-user may have different ideas of what constitutes good QoS for a parameter; this may happen due to changes in expectations over time, and has implications on component maintenance. If necessary, update the NFR-F graph.
- Calibrate the NFR-F graph with importance indicators, illustrates this. Here the service engineer considers that stereo sound is not important and assigns the value 0.0, whereas maintaining a delay jitter variability of less than 10ms is assigned a high priority, as is keeping the overall transmission delay to less than 25ms. These values are used as amplifiers when assessing the overall QoS fulfilment of the component or components.

3. Calculate a maximum QoS value for the root goal in the NFR-F tree.

- For a single graph (where only one component is being dealt with), this involves assuming that all values on the satisficing links have a maximum value (1.0). Each of these values is propagated up the graph and multiplied by the relevant importance indicator factor. In the QoS part of the graph these values are summed where QoS refinement links merge. The summed value is then normalized to fall within the range –1.0 and 1.0. This is to prevent one subgraph from dominating other subgraphs. Without this, the value propagated to *highVQ* in Figure 3 would be 2.5 which then gets further propagated up the graph dominating the 1.0 propagated up the *auditability* subgraph. With normalization, the maximum value becomes 1.0 and the normalization factor used to achieve this scaling is 0.4 (1.0/2.5). This factor is used in Step 4.

- Where several graphs are merged, duplicate QoS goals should be factored out leaving a single QoS refinement graph for each QoS type. Satisficing links are then made from different components to the same QoS goal. Conflict resolution may be necessary. In these cases, one or other of the QoS goals must be selected and the weightings on satisficing links re-calculated. Merging QoS goals results in the merging of their associated satisficing links using the AND operator (see Figure 3). During value propagation only the AND operator results in the propagation of the lowest value (1.0 in this case since we are propagating maximum values).

4. Calculate actual QoS satisficing values for the merged graphs. This time, the values on the satisficing links are propagated. This results in *highVQ* being assigned the satisficing value of –1.3, which is then normalized using the normalizing factor 0.4 giving a value –0.52. The AND operator results in the minimum value being propagated.
5. Find the degree of satisficing for a given QoS goal by calculating the percentage difference between the normalized maximum (always 1.0) and the normalized actual satisficing value (–0.52). So the satisficing value is –52% for *highVQ*.
6. Overall QoS satisficing can be found by introducing an anonymous QoS goal at the highest level which connects to each of the other QoS refinement graphs.

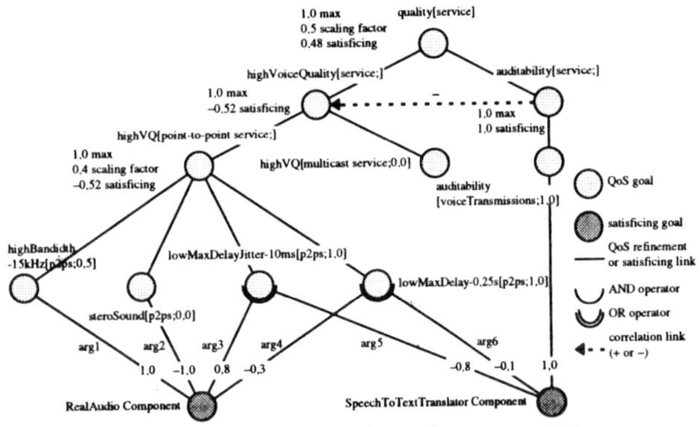

Figure 3. Aggregation of two candidate components.

It can be seen that before the *SpeechToTextTranslation* component is added, that overall QoS satisficing is 40%. Introducing the new component increases this to 48%. Although there is an increase overall, the *speechToTextTranslator* still has a strong negative influence on the NFRs associated with the *RealAudio* component. This negative influence is represented by a correlation link between the two top-level NFR goals. If there had been no satisficing links between the *speechToTextTranslator* and the NFR goals associated with the *RealAudio* component then the overall level of satisficing would have been 70%. The developer using the graph to reason about design choices may decide to use a different solution for recording voice transmissions, such as storing on file. However, this, in turn, may be found using the NFR-F, to conflict with other NFRs such as disk space conservation.

5. Conclusions and directions

In this paper we presented two problem classes that should be addressed during service creation, in the context of, QoS handling. The first problem is concerned with QoS identification and negotiation. The aim is to design the appropriate service logic that comprises the pertinent functionality (which will be used during service execution). The second problem is concerned with the translation of QoS requirements to service component performance requirements, and to performance constraints that should be satisfied by the arrangement of the service components on the DPE. Furthermore, we provide solutions to the problem of QoS representation, translation and negotiation, and to the problem of translating QoS requirements to service component performance requirements. The solution is based on the definition of component contract offers in terms of QoS.

Regarding, the approach presented in the previous section, issues for further study are the following. First, the comparison with related techniques, e.g., [10]. Second, the assessment of the benefits of the approach: reasonable stability despite novelty; tool availability, compatibility with the ITU framework for deriving QoS. Third, the investigation of the limitations of the approach. Such an example is the lack of support for reasoning about the direct impact of the functioning of one component on another. Hence, in Figure 3 if the voice stream is passed through the STT component before reaching the RealAudio component, the latter may be able to compensate for the increased delay jitter caused by STT. Consequently, the weighting on the satisficing link should be made less negative or even positive. Fourth, to pursue directions for improvement. For example, the NFR analysis could be combined with the functional analysis, or with simulation.

6. Acknowledgements

We would like to acknowledge the support provided by our colleagues in SCREEN throughout this work.

7. References

[1] TINA-C Technical Report: "Service Architecture", Version 4.1. January 17, 1996.

[2] ITU-T Recommendation, "*Quality of Service and Dependability Vocabulary*", Geneva, 1993

[3] L. Chung and B.A. Nixon, "Dealing with non-functional requirements: three experimental studies of a process oriented approach", In Proc. *17th ICSE*, pp. 25-37, 1995

[4] T. Comer, C. Leisersin, R. Rivest, "*Introduction to algorithms*", The MIT Press, Cambridge, Massachusetts, London, England, 1991

[5] J. Mylopoulos, L. Chung and B. Nixon, "Representing and using non-functional requirements: a process-oriented approach", *IEEE TSE*, Vol. 18, June 1992

[6] L. Chung, B.A. Nixon and E. Yu, "Using non-functional requirements to systematically support change", In Proc. *Requirements Engineering '95*, 1995

[7] NFR Assistant public domain tool: ftp.cs.toronto.edu/pub/eric/home

[8] R.J. Gautier, C.W. Loftus, E.M. Sherratt and L. Thomas, Tool integration: experiences and directions, in Proc. *17th ICSE*, pp. 315-326, 1995

[9] R. Darimont and A. Van Lamsweerde, "Formal refinement patterns for goal-driven requirements elaboration", in Proc. *SIGSOFT'96*, pp 179-190, 1996

[10] J. Vanwelkenhuysen, Quality requirements analysis in customer-centered software development, in Proc. *2nd IEEE International Conference on Requirements Engineering*, IEEE Computer Society, 1996

Quality of Service Negotiation in TINA

Jarno Rajahalme[1], Telma Mota[2], Frank Steegmans[3], Per F. Hansen[4], Fernanda Fonseca[5]

[1]Jarno.Rajahalme@research.nokia.com
Nokia Research Center
3 Burlington Woods Dr., Suite 260
Burlington, MA 01803, USA

[2]telma@cet.pt
CET
R. Eng. José F. P. Basto
3810 Aveiro, Portugal

[3]fste@rc.bel.alcatel.be
Alcatel Telecom
Francis Wellesplein 1
B-2018 Antwerp, Belgium

[4]pflyh@tdk.dk
Tele Danmark R&D, UTTS,
Telegade 2
DK-2630 Taastrup, Denmark

[5]Fernanda.P.Fonseca@telecom.pt
Portugal Telecom
R. Entrecampos 28
1700 Lisboa, Portugal

Abstract

A solution based on terminal capability negotiation is proposed for the previously undefined QoS negotiation mechanism in the TINA architecture.

Capability negotiation provides means to select suitable media codecs as well as session protocols and network termination points for each terminal involved in a particular communication session.

A hierarchy of QoS contexts is proposed to provide grouping of related QoS parameters. Terminal based QoS parameter mapping between the contexts isolates technology specific details to be recognized only where the technology itself is supported.

The main result of this work is that by integrating the QoS mapping with the multimedia protocol profile negotiation and actual set-up of the communication, there is no need for generalized transport level QoS descriptions within the network architecture.

The focus of this work is on the Network Resource Architecture, but Service Architecture issues are also briefly discussed in the context of an example scenario.

1 Introduction

A number of standardization bodies are currently dealing with the complex area of Quality of Service (QoS) for communication systems. Current multimedia communication recommendations (e.g. the ITU-T H.32x series [4, 6]) are specific to a particular communication technology, but they are not considering QoS implications for architectures combining a number of communication technologies.

These considerations are very important for the Telecommunications Information Networking Architecture (TINA) since it targets to be a framework for integrating any service and transport technology in a service and technology independent manner. Consequently it has to deal with QoS for all kind of services communicating via a wide range of technologies (e.g. coding, transport network). A well-defined QoS framework is therefore crucial for the success of TINA.

There are two obvious approaches towards QoS in technology independent communication architectures. Firstly, finding a set of generic network level QoS attributes applicable to all technologies, but this is unfeasible. Secondly, collecting all attributes from all available technologies, but this yields an impractical large set that can never be complete, since new technologies constantly emerge.

It appears logical to have a layered QoS framework, since most communication system architectures are layered as well. A layered QoS framework requires QoS mapping between any two adjacent layers [7, 8].

The mapping of QoS parameters from one layer to another relates closely to the negotiation of the protocol profile for multimedia communication. The selection of e.g. media encoding will have a major impact on the required transport level QoS. To enable technology independent negotiation, this paper proposes:

- A way of expressing terminal capabilities based on a given QoS layering;
- A mapping mechanism driven by the capabilities;
- A negotiation process in the TINA context.

0-8186-8335-X/98 $10.00 © 1998 IEEE

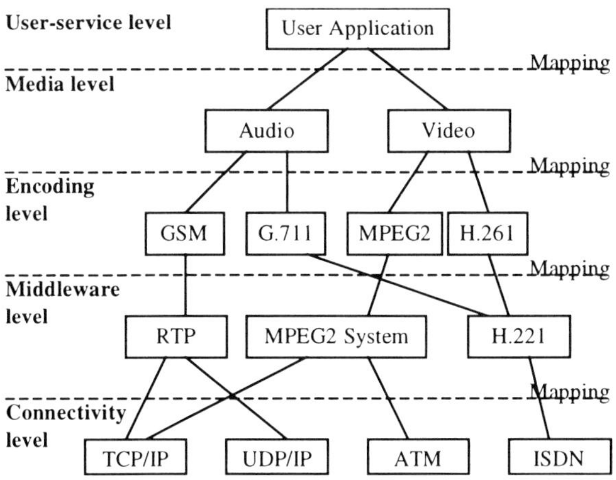

Figure 1: Hierarchy of QoS contexts (examples)

Furthermore, it also gives a brief overview of the current level of support in some parts of the architecture.

2 Quality of Service in the current TINA specifications

Some QoS support is given at various levels in the current TINA specifications and the awareness of the problem within TINA is quite evident throughout the documentation. However, the current deliverables do not have a consistent solution for dealing with QoS.

The current Service Architecture [13] gives only little help on how to treat QoS at the service level. The Service Architecture states that the QoS must be chosen before the set-up of stream flows, that management services should take QoS into account, and that service composition and federation has an impact on the end-to-end QoS of a particular service.

So far, the Network Resource Architecture [12] assumes that the ATM QoS parameters form an appropriate generic set for most purposes. Although this results in a quite excessive set of parameters, it is not complete yet. Furthermore, the semantics of these parameters are not always unambiguous. For instance, it is not straightforward how a desired audio quality has to be specified. A user probably has an idea of what "Compact Disc", "AM radio", or "telephone audio" quality means, but the mapping from these to ATM parameters is not obvious.

3 Levels of quality of service

In this paper Quality of Service is *the overall quality, as the user of a service perceives it*. Furthermore, a QoS context is *an interrelated set of parameters affecting the Quality of Service*. These parameters always depend on the QoS context in which they are defined.

Even though there might be a generic relationship between any two QoS contexts, finding such a relationship is practically impossible. Therefore a mapping functionality between two specific QoS contexts is required to represent this relationship.

Relationships between QoS contexts can be structured as a *layered hierarchy* that spans from user level to network level, see Figure 1.

Figure 1 shows the levels of QoS contexts with example contexts, and dependencies between them. Mappings need to be defined on a case by case basis for each dependency.

The **user-service level** is the highest level of QoS contexts. It includes both service dependent and independent QoS parameters. These parameters are mapped to media level parameters. Some example parameters are given in Table 1.

The **media level** contains QoS contexts for each supported media type. At this level a standard set of QoS parameters exist for each context. Mapping from user service level parameters to media level parameters is both service and medium dependent. This level serves as a common interface between user-service level and encoding level QoS parameters. This level can be bypassed if service is specific to a certain encoding.

The **encoding level** contains QoS contexts for each supported media-encoding format. All QoS parameters are context specific. For each context, a mapping exists from the media type parameters to context specific

Table 1: Examples of QoS parameters for QoS contexts at different levels

Level / Context	*Parameters*	*Values*
User service	Overall quality	Low, medium, high
	Pricing	Service dependent
	Restrictions	Preferred lower level parameters
Media		
Audio	Audio quality	CD, FM Radio, PSTN, Cell-phone
Video	Video quality	Broadcast, VHS
Encoding		
H.261	BitRate	100..1.92Mbit/s
H.261	Resolution	QCIF, CIF
Middleware		
RTP	Jitter buffer size	N milliseconds
Connectivity		
ATM	VC range	24..1024
TCP	Window size	N bytes

parameters. Each media-encoding format supports only one specific media type. Service independency is achieved at this level, as all service dependent QoS parameters have been mapped into encoding level parameters.

The **middleware level** contains QoS contexts for protocols between the connectivity level protocols and the media codecs (encoding level). These protocols can be session protocols (e.g. RTP), multiplexers (e.g. MPEG System Stream, H.221), or encryption protocols (e.g. SSL). These protocols are typically dependent on both upper and lower level protocols. For example, MPEG System Stream can only be used with MPEG audio and video codecs, H.221 requires ISDN, and RTP is typically implemented on top of UDP or TCP (see Figure 1). In some cases, no protocol in this level will be used.

The **connectivity level** contains contexts for possible transports for the multimedia streams. The QoS parameters at this level are entirely network technology specific. These parameters will be interpreted only by other components of the architecture specific to the particular technology.

4 Terminal capabilities

Heterogeneous terminals have to find a common mode of operation to be able to communicate. To achieve this, the involved terminals must share the knowledge of their capabilities with others. This requires a generic representation of the terminal capabilities. The negotiated common mode of operation is called the *protocol profile*.

The levels of Quality of Service introduced in the previous sections are used for grouping the terminal capabilities.

4.1 Media encoding capabilities

The media encoding capabilities describe what kind of media encoding the terminal is capable of processing. The encoding capabilities may have complex interdependencies among themselves, as well as towards lower level capabilities (i.e. middleware protocol and connectivity level capabilities). These dependencies may be caused by processing limitations, interconnection restrictions (in case of hardware codecs) or configured restrictions. For example, a terminal may have the limitation that an MPEG audio codec can be used only within an MPEG system stream, while another terminal might be flexible enough to allow usage of the same codec with other multiplexers as well.

ITU-T H.245 [4] provides a data structure powerful enough to express the interdependencies between media codecs, but does not address the dependencies towards middleware and connectivity level capabilities. This is because even though the recommendation addresses multiple network technologies (e.g. PSTN (H.324), and the Internet (H.323)), it is applied to one specific conferencing technology at a time. Within such a single conferencing technology, dependencies towards lower level protocols are implicitly defined. For integrated multimedia communication architectures the data structures must be extended to allow representation of explicit dependencies towards the other levels.

4.2 Middleware Protocol Capabilities

Multimedia middleware protocols are used on top of connectivity level protocols to further refine the service to be more suitable for end-to-end stream transport.

The required multimedia middleware functionality is highly dependent on the functionality provided by the used connectivity level protocols and both functional and QoS requirements of the media encoders and the application. Possible functionalities include:

1. Encryption (e.g. SSL)
2. Multiplexing (e.g. MPEG System Stream)
3. Sequencing, end-to-end flow control, jitter smoothing (e.g. RTP over UDP/IP)

Current multimedia communication systems (e.g. H.320, H.323) implicitly define middleware level protocols. In an integrated system these protocols must be negotiated, and stacking must be allowed (e.g. encryption on top of RTP stream).

4.3 Connectivity level capabilities

Since integrated multimedia systems can use a variety of transport network technologies, the transport technology used for a stream must also be negotiated.

Connectivity level capabilities express the potential network connectivity the terminal has. For each technology, at least one capability structure is presented. This structure includes the operational parameter options for network terminations using that technology. For example, an ATM connectivity capability might contain a range of VC numbers the terminal can assign, the available bandwidth of the access link, etc.

4.4 Capability IDL definition

IDL data structure for a capability is defined as follows:

```
struct t_Capability {
  t_CapabilityId     type;
  t_CapabilityKey    key;
  t_CapabilityKey    instanceId;
  t_CapabilityDir    dir;
  t_AttribList       attributes;
  t_SimultaneousDependencies
                     dependencies;
};
```

type identifies the kind of the capability in question. It is an ASN.1 OBJECT IDENTIFIER from the name space shown in Figure 2. The figure shows the hierarchy of capability types. An actual capability has a type name derived from this hierarchy. For example, for G.723.1 audio codec, the `type` would be {tinac encoding(1) audio(2) g7231(2)} assuming that the sub-tree (2) was assigned to the G.723.1 codec. `type` is encoded as a string of decimal numbers separated by dots. Also, the tinac-prefix can be left away, leaving the heading dot in place. For the G.723.1 the `type` would then be "`.1.2.`**`2`**".

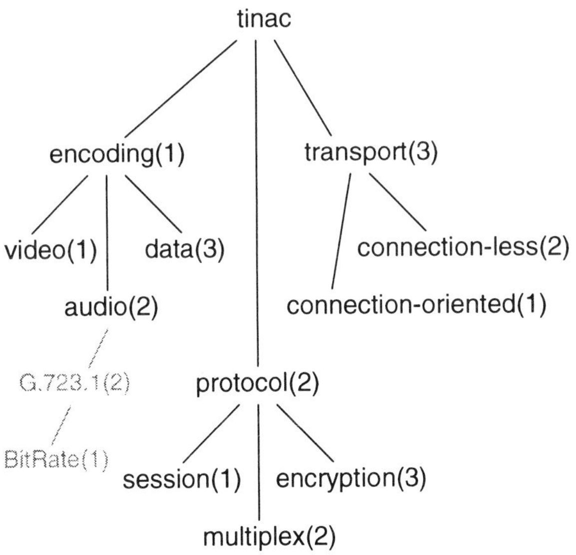

Figure 2: Capability type name space

key is terminal unique key for the capability. The capabilities of the terminal might include multiple capabilities of the same type, but have different dependencies, for example. This field enables unique referencing of each terminal capability.

instanceId is assigned for a capability when the resources required for the use of the capability in the terminal are reserved. This allows multiple simultaneous reservations of a capability with the same `key`.

dir is the directionality of the capability. This can be one of following:

1. `receiveCapability`,
2. `transmitCapability`,
3. `receiveAndTransmitCapability`.

For example, a video on demand terminal might have a MPEG2 video decoding capability, but no encoding capability, and thus would designate its MPEG2 video capability as `receiveCapability`. Instantiated capabilities are always resolved to either receive or transmit directionality.

attributes contain the context specific capability boundaries, including these for QoS. When terminal local resources are reserved, and an `instaceId` is assigned, this list will contain the actual parameters of the protocol represented by the capability.

The attributes are also identified with ASN.1 OBJECT IDENTIFIERs. The attribute names for each capability are assigned under the capability type's sub-tree. For example, the BitRate parameter of G.723.1 could be named as "`.1.2.2.`**`1`**". The value of an attribute is encoded as the IDL `any`. The actual data type (e.g. `string`, `long`, `struct`, etc.) is specific to the particular attribute. For this example it could be an `enum` enumerating the two possible bit rates.

dependencies specify which other capabilities are required in order to select this capability. The required capabilities are contained in a list of simultaneous requirements, each containing alternative capabilities. For example, if a terminal can transmit G.711 formatted audio using RTP over IP, but could transmit it directly over ISDN, then the capabilities would have the dependencies shown in Figure 3:

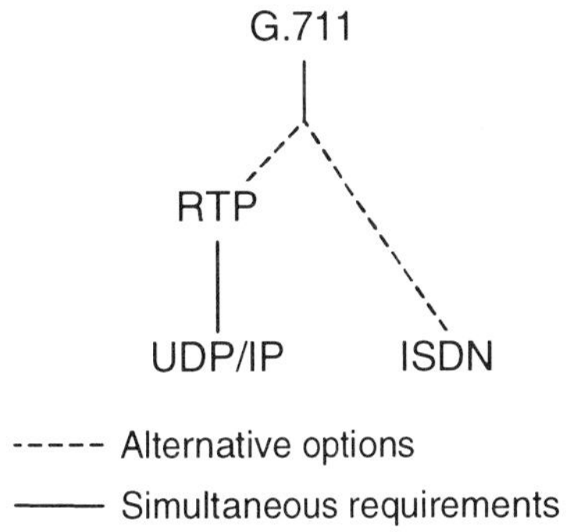

Figure 3: Example dependencies

4.5 Terminal capability set

Terminal capability set is defined as follows:

```
struct t_CapabilitySet {
  sequence<t_CapabilityDescriptor>
            capabilityDescriptors;
  sequence<t_Capability>
            capabilityTable;
};
```

`capabilityTable` is a list of all capability structures in this capability set.

`capabilityDescriptors` are simultaneous/ alternative lists containing references to the capability table. This structure expresses different allowed combinations of capabilities.

For example, capability descriptor {{H.261}, {H.261, H.263}, {G.711, G.723.1}} would mean that the terminal can operate two video streams at the same time, one using H.261, and the other using either H.261 or H.263. At the same time either a G.711 or a G.723.1 formatted audio stream could be handled.

Typically only encoding level capabilities would be referenced in this structure. Lower levels would then be referenced by the dependencies of the encoding level capabilities.

5 Terminal capability driven QoS mapping

Figure 4 shows a simplified part of the TINA computational model that is used as the framework for operational descriptions in this section.

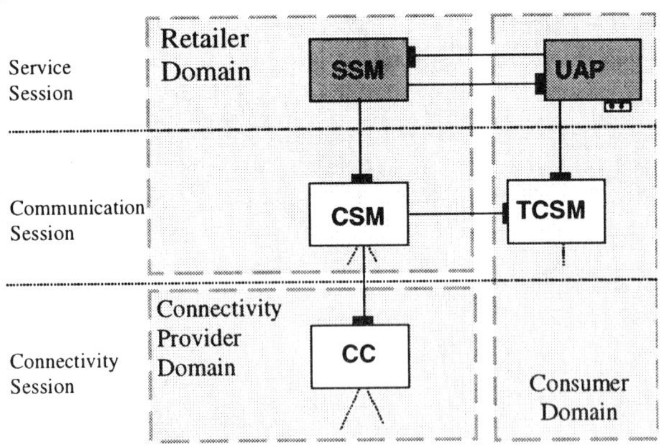

Key:
CC: Connection Coordinator
CSM: Communication Session Manager
SSM: Service Session Manager
TCSM: Terminal Communication Session Manager
UAP: User Application

Figure 4: Simplified part of the TINA computational model

5.1 QoS mapping in terminals

The proposed QoS negotiation mechanism is based on the mapping done by the terminal between adjacent QoS contexts. When different media codecs, network interfaces etc. are installed on a terminal, the terminal will also be configured with the necessary mapping information. This enables the network portions of the architecture to be independent of the different coding protocols and especially their QoS implications. Connectivity level details are encapsulated into the technology dependent portions of the network resource architecture.

The first mapping that needs to be done is typically from the user-service level to the media level. This mapping is service dependent, but must be unique within the service itself to allow service level QoS negotiation to converge. Because of this the SSM should do this mapping. However, some services might prefer UAP doing this.

The terminals provide all the other mappings, i.e. from media to encoding level, from encoding to middleware level, and from middleware to connectivity level. To do the mapping, the terminal performs the following procedure internally for each stream flow end point:

1. Get information of all encoding capabilities supporting the media type of the stream.
2. Keep the capabilities that can support the required media level QoS.
3. Ask for parameter ranges satisfying the requirements. Ask also for requirements towards the dependencies of the capability.
4. Check that at least one of each alternative dependency is available and fulfils the requirements of the capability.
5. Repeat from 3 for each non-transport dependency (recursively)
6. For each transport capability, ask for the transport parameter ranges satisfying the higher level requirements.

Mappings performed are by no means unique, but and different terminals might give different results. The mapping from the media level parameters is expected to be standardized for each encoding level QoS context. Lower level mappings should be performed as accurately as possible to not jeopardize the interoperability between terminals. For example, if bandwidth requirement of a codec is overestimated, it might be pruned from the capability set, even if no other alternatives are available. Terminals can use methods, such as described in [7], to achieve sufficient mapping accuracy. The capability set that remains contains all the terminal capabilities that can satisfy the requirements set for each stream. The capability sets of each terminal are then used in the negotiation mechanism to find a common mode of operation.

5.2 The QoS negotiation mechanism

QoS negotiation starts within the stream binding process between the SSM and the UAPs. The initiating terminal or the SSM gives the initial request in terms of the service dependent QoS parameters. These are mapped to the media level for each stream required by the service. The actual negotiation is done in terms of the media level QoS parameters. Each terminal uses the QoS mapping process outlined above to check whether it can support the requested media level QoS or not.

The negotiation results into common set of media level QoS parameters for the SFEPs, which are then passed by the SSM to the CSM as communication session parameters.

SSM passes the list of stream flow end points (SFEPs) and the associated QoS requirements to the CSM for communication session setup. From thereon the negotiation continues between the Terminal Communication Session Managers (TCSM) and Communication Session Manager (CSM) components (Figure 4). The CSM contacts each TCSM, asking for the terminal capabilities for given SFEPs. The returned capabilities are constrained by the media level QoS requirements, as described in the previous section.

When the CSM has the capability information from each terminal, it compares the capabilities and searches for a common protocol profile for each stream flow connection (SFC). If a match can not be found, the CSM has following options:

1. Bring in a transcoding unit to convert from a media format to another
2. Use a multi-point bridge to connect point-to-multi-point streams
3. Request that the connectivity provider bridges from one technology to another (e.g. ISDN over ATM)

The CSM needs to choose connectivity provider(s) to support network flow connections for all stream flow connections. This selection is based on the transport capabilities advertised by the terminals, including their QoS requirements. The transport capability choices of each SFEP are then passed to the chosen connectivity provider. The connectivity provider will then make the ultimate selection of the transport to be used. The exact means for this and the above are outside the scope of this work.

Note that in addition to the terminals and the SSM the CSM can also have other sources of information. It might be configured to prefer some connectivity provider over others, or it might have to ask the connectivity providers (represented by Connection Coordinators), whether they can provide requested connectivity or not. These and other considerations are outside of the scope of this work.

6 Example

The QoS negotiation process explained in this paper becomes clearer through the simple example presented in this section. A 3-party audio conference service has been chosen in order to show not just the QoS mechanisms but how it can be included in a realistic TINA scenario.

Figure 5 illustrates the chosen example scenario. Three participants are involved, represented respectively by Terminals A, B and C, and their related service session audio conference parts of the applications (ss-UAPA, ss-UAPB and ss-UAPC). The terminal of participant A is connected to an ISDN network, terminal B to an ATM infrastructure and C to an IP network (over 28.8k PSTN connection). Two different connectivity providers are involved in the provisioning of connectivity services, one owns an ISDN infrastructure while the other the remaining technologies considered, i.e. ATM and an IP. In order to keep the simplicity of the example, the terminals will use the following encoding capabilities: participants A and B will use a G.711Alaw64k codec while the third participant (C) uses a low bit rate G.723.1.

G.711 is the normal PCM (Pulse Code Modulation) codec used for encoding of audio in one 64kbit/s ISDN B channel. G.723.1 is a low bit-rate audio codec (supports two bit rates, 5.3 and 6.3 kbit/s). G.711 is a very low complexity codec, while G.723.1 is quite complex (about 15 MIPS). Both of them deliver "telephone quality" audio.

The interactions between the involved components are shown in Figure 5[1]. As a starting point, it has been assumed that a service session was already established between the participants. The scenario proceeds with user A requesting the ss-UAP to set-up an audio conference, involving all three participants. Then, the following service specific scenario takes place:

1. The ss-UAPa queries the TCSMa for SFEPs with media type 'audio' that support an 'Audio quality' range. TCSMa returns {Cell-phone, PSTN}, meaning

[1] For simplicity the USMs are not shown in the picture, in principle all interactions between ss-UAPs and SSM should pass through USMs, but they do not add any functionality for the QoS negotiation.

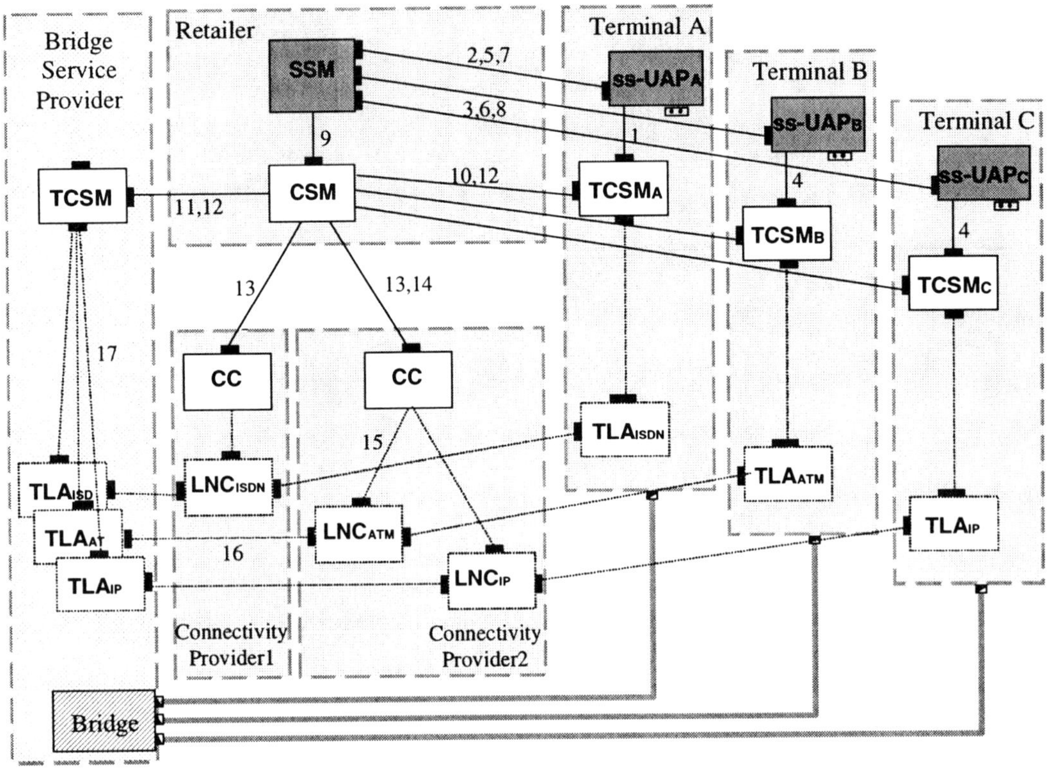

Figure 5: Example scenario

that the terminal A can support audio qualities ranging from Cell-phone to PSTN.

2. The ss-UAPa passes the SFEP information to the SSM, including the acceptable 'Audio quality' range, and requests it to provide available QoS parameters for the entire stream binding.
3. The SSM queries each of the other ss-UAPs for the corresponding information (SFEPs and QoS).
4. Corresponds to 1. Terminals B and C return SFEP information with 'Audio quality' ranges {PSTN, PSTN}, {Cell-phone, PSTN}, respectively.

As the result of negotiation the SSM calculates the range that is supported by all parties ({Cell-phone, PSTN} for A and C, {PSTN, PSTN} for B).

The SSM contains the service specific mapping between user-service level and media level QoS parameters that can be supported by the service type. In the service in this example the mapping is as follows:

'maximum'	→	CD
'high'	→	FM Radio
'medium'	→	PSTN
'low'	→	Cell-phone

This kind of mapping required especially if the service makes use of multiple media types at the same time.

The SSM uses this mapping to conclude the supported user-service level parameter ranges from the result of the above negotiation. In this way the users will only get QoS parameters presented that are actually supported. In this example this yields only 'medium-low' for terminal A and C and 'medium' for terminal B.

At this point a number of scenarios are possible, depending on the service policy for user-service level QoS negotiation. Each user could be allowed to decide for a particular part of the stream binding, all users could vote, or a particular user could suggest a QoS and the other could be allowed to accept or reject the choice. In the sequel we assume the latter.

The choices of user-service level parameters are returned to ss-UAPa, for final user-service level QoS selection. The corresponding media level parameters may be passed as well, e.g. for "expert" users who might be interested in knowing the corresponding details, or who wish to monitor the actual QoS delivered. In this example, the final set of parameters is selected by the user A. A range could be selected as well, allowing the SSM to apply some intelligence in the final choice or facilitating the voting process.

5. The SSM is informed about the choice.

6. Other users are presented with the choice for voting and the result is returned to the ss-UAPa. If it is rejected, steps 5 and 6 might be repeated.
7. The ss-UAPa finally sends the actual stream binding request to the SSM.
8. The SSM informs the other ss-UAPs about the final choice.
9. The SSM sends a set-up communication session request to the CSM. The list of stream flow connections (SFCs) and their respective stream flow end points (SFEPs) are given to the CSM together with the connection topology, the directionality of the SFEPs and the chosen media level QoS parameters, including the media type ('audio'). In this example, a multipoint-to-multipoint bi-directional connectivity is required (see Figure 6).

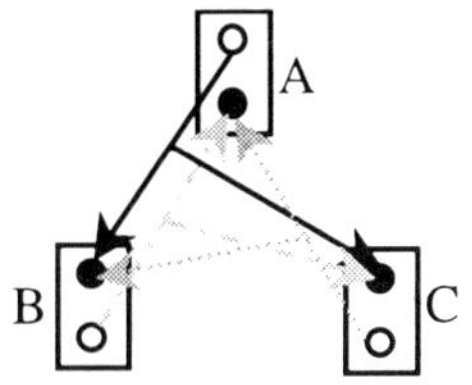

Figure 6: Stream binding topology

Since some sink SFEP are being used for terminating more than one SFC, the CSM will try to locate a special resource (i.e. audio-bridge) to 'mix' the 'audio' streams and provide only one stream to each sink SFEP.

10. Next the CSM queries each TCSM for a capability set related to the terminal's SFEPs, and the appropriate media level QoS. In this scenario the terminal A returns the following capability set:

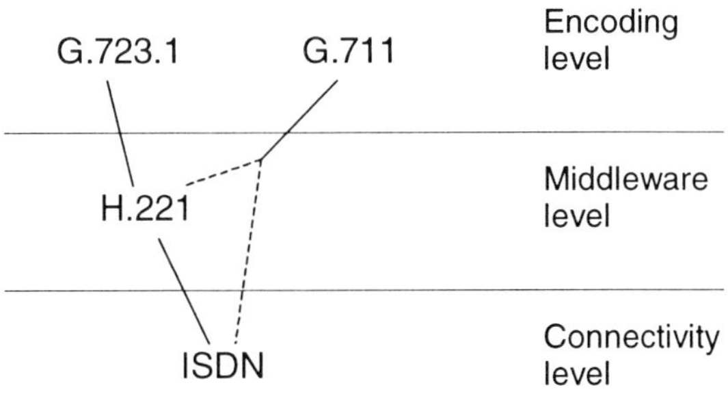

Figure 7: Audio capabilities of the terminal A

In the capability set in Figure 7, TCSMa has included all encoding capabilities supporting the required QoS for which it has the required transport resources available. In this example, terminal C only has a low bit-rate PSTN connection, and terminal B has only a higher bit-rate G.711 audio encoder.

The CSM figures out that there is no match among all the codec capabilities assigned to the different SFEPs. The CSM will use this information together with the SFC topology information (multi-point to multi-point) to find a suitable bridge.

11. The CSM will request the bridge administration to provide a multi-point bridging functionality among the SFEPs it already had. It matches the capability sets of the corresponding SFEPs and returns matching SFEPs (and capability sets) for each of the given SFEPs. The CSM can use the SFEP specific media level QoS ranges passed by the SSM to allow wider flexibility when negotiating with the bridge.

The CSM will do matching between terminal and bridge SFEPs just as it would do between two terminal SFEPs.

12. Finally the CSM will select capabilities on all the terminals and the bridge, passing the selected capability information (codec and parameters plus session protocols) for each of the SFEPs.

    Each TCSM will then fill in parameters for each possible connectivity level capability, and returns these to the CSM for network connection establishment.

At this point all the codecs have been configured. Now, the CSM will analyze the required network capabilities for each of the terminals. From this it will compose a set of NFEP pools that need to be interconnected. This is actually the SFC to NFC mapping process. In our example, the pools for the bridge will contain NFEPs pointing to respective ATM, ISDN and IP network termination points to be connected to each of the terminal. Each terminal will only have one pool with an NFEP pointing to the specific network termination it is connected to.

13. The CSM will contact both connectivity providers based on this information to set-up the three NFCs from the bridge to each of the terminals. Note that this can be more than two connectivity provider depending on the complexity of the CSM and the supported layer networks of the connectivity provider.

The rest of this scenario will demonstrate the set-up of one of the three NFCs. The NFC supported by an ATM trail will be used for the scenario.

14. The connectivity provider's CC receives the request for setting up a point-to-point bi-directional NFC between two NFEP pools. Since both pools represent network termination points of the same layer network (ATM), it doesn't need to find an inter-working unit.

15. Consequently, the CC requests the ATM LNC to set-up a ATM layer network binding to support the NFC. The CC may interpret more information (e.g. required QoS in technology specific terms) of the NFEP pools to select the most appropriate LNC domain if necessary.

    The LNC will unpack the NFEP pools to retrieve the ATM specific information. This includes the necessary addressing information for finding the responsible ATM TLA for each NFEP pool (this example considers only one domain) and possibly the termination point address (e.g. E.164). It also may include traffic requirements such as average, minimum and maximum bandwidth as well as maximum delay and maximum delay variance. If not included this information may be retrieved via the TLA. However, in some cases the CC will need this information to select the appropriate LNC domain or inter-working unit.
16. The LNC will contact the TLAs related to the NFEP pools to negotiate the channel and possibly the ATM QoS parameters.
17. The TLAs will inform the TCSMs about the channel selection, so that the network termination points can be associated with appropriate middleware protocols and/or media codecs. Finally, the LNC will set-up the necessary connections within its layer network, based on the layer network specific QoS requirements.

7 Summary

The TINA architecture aims to be generic and flexible enough to support any type of service over any type of network. In order to achieve this objective this paper proposes a generic mechanism to deal with QoS which is, by nature, service and technology dependent.

The main focus of the paper has been on the communication session level. It has been described how QoS can be handled by terminals, starting from the service level down to the level of the actual transport technologies, how QoS negotiation should be done, and how this negotiation affects the connectivity control when different technologies are involved.

When proposing a QoS framework the QoS requirements were categorized by placing them in specific contexts in a layered hierarchical model. The specified capability levels provide the necessary information to be used by the TINA architecture to perform a negotiation process for the participating terminals, and to make a choice among the available connectivity providers and technologies.

References

[1] ITU-T Recommendation G.711, *Pulse Code Modulation (PCM) of Voice Frequencies*, ITU-T, 1988.

[2] ITU-T Recommendation G.723.1, *Dual Rate Speech Coder for Multimedia Communication Transmitting at 5.3 & 6.3 kbit/s*, ITU-T, 1996.

[3] ITU-T Recommendation H.221, *Frame Structure for a 64 to 1920 kbit/s Channel in Audiovisual Teleservices*, ITU-T, 1993.

[4] ITU-T Recommendation H.245 – Version 2, *Control Protocol for Multimedia Communication*, ITU-T, March 1997.

[5] ITU-T Recommendation H.261, *Video Codec for Audiovisual Services at px64 kbit/s*, ITU-T, 1993.

[6] ITU-T Recommendation H.320, *Narrow-band ISDN Visual Telephone Systems and Terminal Equipment*, ITU-T, 1993.

[7] Knoche, H. and De Meer, H., "Quantitative QoS-Mapping: A Unifying Approach", *Proc. IWQOS'97*, New York, USA, May 1997, pp. 347-358.

[8] Nishio, N. and Tokuda, H., "Simplified Method for Session Coordination Using Multi-level QOS Specification and Translation", *Proc. IWQOS'97*, New York, USA, May 1997, pp. 335-346.

[9] *Ret Reference Point Specification*, Version 0.7, TINA-C, September 1997.

[10] Smith, A. and Grace, A., "A QoS Configuration System for Distributed Applications", *Proc. IWQOS'97*, New York, USA, May 1997, pp. 259-262.

[11] Stallings, W., *Data and Computer Communications*, 5th edition. Upper Saddle River, NJ, USA, Prentice-Hall, 1997.

[12] *TINA Network Resource Architecture*, version 3.0, TINA document no. TB_v3.0_97_02_10, TINA-C, Feb. 1997.

[13] *TINA Service Architecture*, Version 5.0, TINA-C, June 1997.

Session 12

Architectural Evolution

Max Sevcik, Siemens AG

Service session control for multimedia services - informational and computational views

Lill Kristiansen
Telenor R&D, P.O.Box 83, N-2007 Kjeller, NORWAY
lill.kristiansen@fou.telenor.no

Abstract

The TINA computational view defines operational interfaces and stream interfaces belonging to computational objects. Regarding the stream interfaces the TINA service architecture (optionally) offers explicit control of the establishment and lifecycle of stream bindings.

This paper will explain the informational and computational concepts and objects defined in TINA that relates to explicit stream bindings. In the information view this is the different objects defined in TINA service session graph. In the computational view TINA offers explicit stream binding establishment on different levels of granularity. The feature sets for participant oriented and stream flow oriented control are described.

Such stream binding control operations are useful in multi-media multi-party services. A simplified tele-education example is given to illustrate the use of the concepts, the example also shows how tailored service logic can easily be achieved.

1 Outline of the paper

TINA bases it's modelling to a large extend on the ODP separation in different viewpoints. The TINA Service Architecture [5] have chapters for business modeling (corresponding to ODP enterprise viewpoint), and chapters for informational and computational modeling. In addition TINA will produce documents for the engineering view (which is not specific to the service architecture).

This paper will explain the informational and computational concepts and objects defined in TINA that relates to multi-media multi-party services using explicit stream bindings.

1.1 Requests for two types of 'lightweight session control'

There has long been a request for 'lightweight session control' among the TINA member companies. This request arose partly because people saw that some services will not need streams at all. (Simple ticket-booking service is one example where streams are not needed and operational interfaces will be sufficient).

The other reason for requesting 'lightweight session control', was the fact that in TINA Service Architecture Delta 1995 [1] the *Service Session Graph* (SSG) was defined in the informational view, and computational interfaces for modifying these information objects was defined in the computational view, but it required a lot of operations going back and forth just to set up a point-to-point connection between two users. The operations were 'small', like 'create SI' and then 'create SFEP' and then again 'SFEP belongs to SI'.

There were also some complaints that the separation in generic session control (USC, GSC etc.) modifying the generic SSG information model) was separated from service specific control (USS and GSS). The argument was that it would be more 'object oriented' to have only one level of service session control, which could easier be tailored to handle specific service behavior. Due to the generic session control a lot of explicit control and ownership session relationships had to be established, again resulting in quite a lot of operations on the computational level between the generic part (USC) of the User Service Session Managers (USMs) and the generic part (GSC) of the Service Session Manager (SSM).

1.2 TINA's solution

Thus based on the comments received in the review of Service Architecture Delta95 [1] TINA started to have a look at a more flexible approach to service session control.

- In Service Architecture Ver.4.0 [2] the separation of service generic / service specific session control was removed from the main body of the document, leaving one level of service session control. And USM, SSM etc. became the main computational objects supporting this.
- In a face to face meeting in Red Bank February '97 to discuss the production of a unified document from the different responses to the RfR/S on the Ret reference point (between consumer and retailer), the concept of

'feature sets' emerged. This concept is used in the Ret reference point document [4] and was also adopted by Service Architecture Ver.5.0.

The feature sets was in particular designed to help out easy, lightweight computations for 'simple' services like the ticked-booking service mentioned above or single party services etc. But it also focused quite bit on simplifying computations even in cases where streams were used.

TINA has now separated out (many different) feature sets. The following are some of them:

- handling of multi parties (e.g., invitation of a new party)
- set-up of control session relationships (to enable multiple users having joint control of resources)
- streams

There are many issues related to feature sets etc. covered in [4] and[5] that could deserve a paper, but to limit the scope of this paper it will be focused on multi-party multi-media services, and in particular on the stream binding concept in TINA. However, brief mentioning of some other feature sets will occur later in the paper, and the full description may be found in [5] and in detailed specifications like [4] and the forthcoming Service Component Specification.

1.3 Focus on the streams

This paper will focus on informational and computational aspects in cases where streams are used. It will give an overview of the entities involved in both views, and then illustrate how the concepts are used by giving an example.

2 The information model for service session control

2.1 Overall view of the information model for Service Session Graph (SSG)

The service session control uses the SSG information model as described by means of the OMT diagram in Figure 1. (The dashed classes are abstract and are specialized further in[5] but details are not shown here.)

In addition to the concepts shown in Figure 1 TINA has defined grouping concepts. One may define Party Session Member Group (PartySMG) (e.g., useful if several parties shall own a resource together), or general Session Member Group (SMG) useful if one wants to group a set of more heterogeneous entities. There might also be cases where a Stream Binding Session Relationship Group (SBSRG) may be defined. SBSRs may also be grouped together with CtrSRs into a Session Relationship group (SRG).

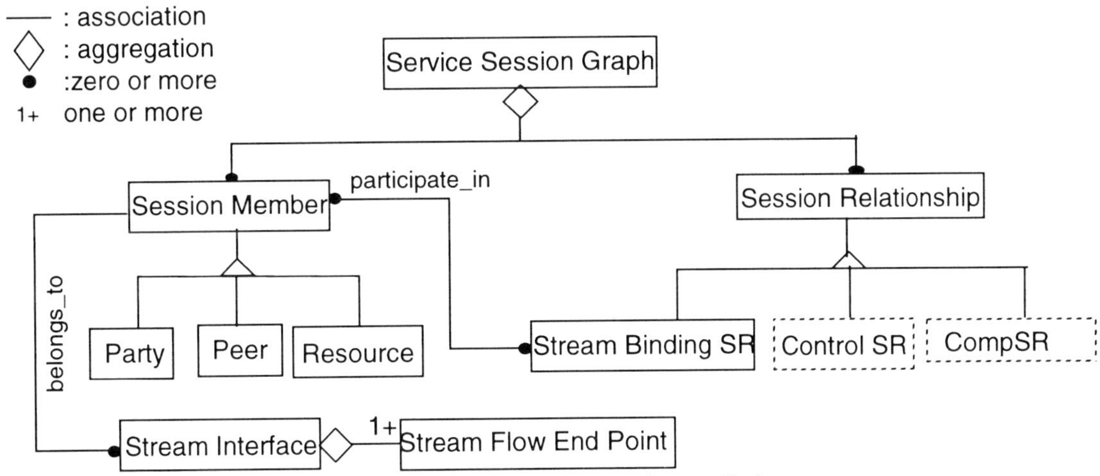

Figure 1 Service session graph overall view

Peer (Session Member) and CompS R will not be handled in this paper.
Control SR is subclassed into ownership and permission control session relationships, though these consepts are useful in relationships to streams, there use will not be shown in detail in this paper.

2.2 Stream Binding Related Information Objects

In order for a session member to express its ability to communicate with other session members by means of streams (as opposed to using operational interfaces), the session member has stream interfaces(SIs) belonging to it, as shown in Figure 1.

Some of these stream interfaces may actually be used in stream bindings, so there is a difference between 'stream interfaces *belongs_to* session member' and 'session member and stream interface are *associated*'. In the latter case the stream interface has a 'bind' relationship to a stream binding, while the 'belongs_to' relation may be established *before* any actual stream bindings are requested. Thus 'belongs_to' is useful in cases where several applications are running on the same terminal, so that there is a need for one user application (UAP) to get hold of the SI, (maybe in competition with other application running on the same terminal). The responsibility of establishing the belongs_to relation may typically be given to the TCSM, which is common for all the applications on this terminal.

The stream binding session relationship class is representing the binding between parties or their corresponding stream interfaces and stream flow endpoints as described in the following:

Stream Flow Endpoint (SFEP): represents the termination of a single flow by an application. In other words it can be seen as the logical representation of a physical stream device I/O port of a terminal (e.g. the outlet of a camera). Further details can be found in TINA Network resource Architecture [3].

Stream Interface (SI): represents a dynamic grouping of SFEPs. The session member will associate as many stream interfaces as required, the cardinality is as shown in Figure 1. It's worth noting that the session member - stream interface relationships (i.e. associated and belongs_to) are inherited by the SM's specialized classes: party, peer and resource. The stream interface class used here is defined in [3].

Stream binding session relationship(SBSR)[1]**:** can be described by a number of different information models. TINA supports three[2] stream binding models, they all rely on the notion of stream flow connection. A stream binding is representing a multi-party-to-multi-party binding, or a multi-point-to-multi-point binding. In both cases the binding maps onto a number of stream flow connections (SFCs), each SFC is representing a point-to-multi-point (or possibly a point-to-point) connection between SFEPs. The stream binding information model is related to setup, modification, deletion etc. of the binding (i.e it is used to describe the stream binding we want to set up). Thus this is TINAs explicit stream binding mechanism.

Stream flow connection (SFC): it describes point-to-point or point-to-multi-point connections between SFEPs.

Figure 2 describes stream flow connections and the relations between Session Member, SIs, SFEPs and SFC.

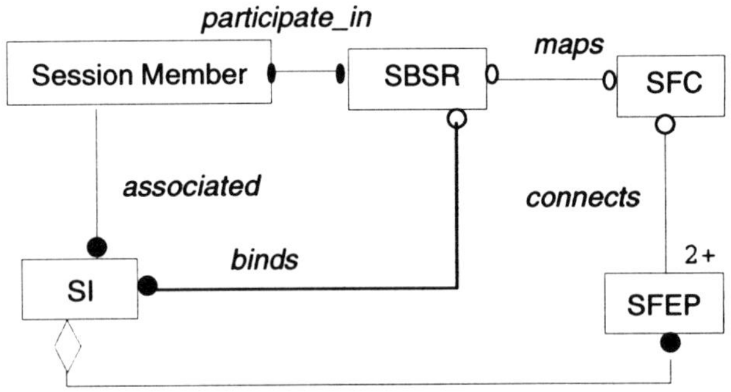

Figure 2 SI,SFEP,SFC and their relations

It may be noted that SI, SFEP and SBSR are *logical* representations. The SFEPs are to be later mapped to physical devices which are (possibly dynamically) bound (by terminal flows) to network termination points. Thus TINA describes the stream binding independent of the underlying network technology. Type, QoS etc. are used on this level to describe the binding requirements as seen from the service level.

Depending on the type of service, there may be different needs for explicit manipulation of individual SFEPs and SFCs. We will describe the participant oriented and the flow oriented stream bindings in this paper.

2.2.1 Participant oriented SBSR

This is the high level approach; stream bindings are described in terms of participating session members, type and QoS. Thus this may be seen as the multi-party-to-multi-party concept, with focus on the parties (or participants) and less focus on the individual end points (SFEPs).

The focus is on the participants and on the 'participate_in' relationship, and the 'binds' and 'maps' relationships are established from 'participate_in' (and from description of the SIs -not shown).

This type of stream binding specification allows a high level request of multi-party-multi-media stream bindings. The type and QoS parameters may be used to implicitly specify multimedia requirement.

1. Stream binding and stream binding session relationship (SBSR) are synonyms.

2. The third and simplest model is a restriction of the second, and will not be described here.

2.2.2 Flow oriented SBSR

This approach to stream bindings uses stream flow connections (SFCs) to specify the stream binding. In this case, a stream binding may be considered an aggregation of SFCs and the session members participate indirectly by exporting their SI information before any request to establish a stream binding is made.

This approach to stream bindings is more oriented towards the stream flow connections(SFC). Thus the focus is more on the 'maps' relation, and on the 'connects' relation. From these two relations (and the 'associated') we may then derive the 'participate_in' relation.

3 Computational view: service usage part and streams in focus

The differences between the two approaches to streams are quite visible on the computational level. In the participant oriented approach the operations sent are operations like 'add_participant', while in the flow oriented approach the operations manipulates more directly the SFEPs, SFCs etc. Hence in the latter approach the SIs of the other members must be known in advance. The StreamInterfaceFS feature set offer the necessary extra operations in the flow oriented case (see Section 3.2).

Only the service usage part is handled here. First the access part needs to be performed (as described in detail e.g. in [5] and [4], and in another paper to this conference). The 'hooks' to the network (through CSM and ConS) are indicated, and further described in [3] (and in the case of internet as TN: by another paper to this conference).

3.1 Participant oriented SBSR feature set (PaSBFS)

The binding may be initiated by the SSM, or by an end-user. In the scenario shown here the binding is initiated by an end-user, while in the example given in Section 4 the SSM is the initiator. Each participant is then requested for suitable stream interfaces which are to be bound by the stream binding. This high level description can be mapped by the service logic to a number of stream flow connections. These stream flows may be of different types and QoS. Special resources, such as bridges, may also be introduced to support the binding, in which case the bridge may be introduced by a party, or by the SSM itself. The stream binding may be manipulated by modifying the type, QoS and participants.

A simple setup of a stream binding may look like the following (when coming from and end-user). (The numbers below refers to Figure 3).

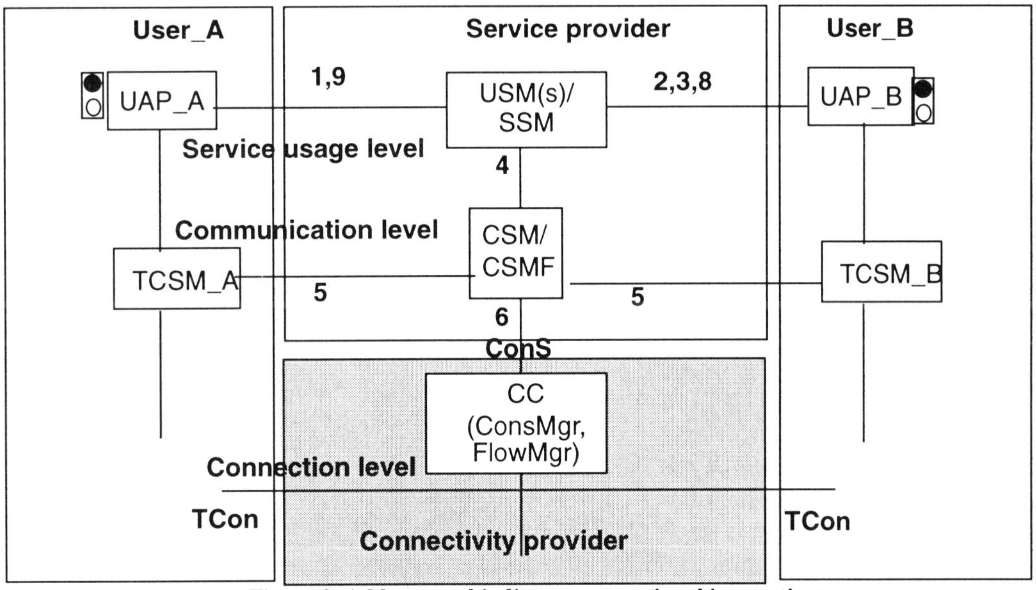

Figure 3 Add stream binding: computational interactions
The figure shows only service usage level and connectivity level, for access and connectivity levels the readers are referred to other papers at TINA'97

1. User A makes an addPaSBFSreq()
2. optionally make addPaSBFSInd() operations
3. joinPaSBFSExe()

   Once each participant has returned their SFEPs, then the provider determines if the stream binding can proceed (based on success criteria) and runs a stream binding algorithm to determine which SFEPs need to be connected.
4. SSM requests the SFC(s) from the CSM
5. CSM makes a resolveSFEP() towards the TCSM(S)
6. Interactions over ConS to set up the network communications.
7. return of 1
8. confirm() or notify() towards other entities (which are involved in the SBSR or owning it).

The main part of the signatures of the main operations from Figure 2 are given here for reference:

```
void addPaSBFSreq(
    in t_ParticipantsSecretId myId,
    in t_SBType reqType,
    in t_MediaDescList media,
    in t_ParticipantDescList reqMembers,
    in t_SFEPDescList requesterSIs;
    in t_SBSuccessCriteria criteria,
    out t_SBBindState status)
```

```
void joinPaSBFSexe(
    in t_SessionId sessionId,
    in t_SBSRId sbId,
    in t_SBType reqType,
    in t_MediaDescList media,
    in t_ParticipantIdList others,
    in t_ReqParticipantDesc
                 reqParticipation,
    in t_RequestId reqId,
    out t_SFEPDescList participantSIs)
```

3.2 Flow oriented SBSR feature set (FlowSBFS)

Since the 'maps' relation is now explicit, service level special resources must be specified by the stream binding requester. The stream binding may be manipulated by adding, deleting and modifying the SFCs. Each SFC is specified in terms of type, branches, state, and QoS. The SFCs may be manipulated by adding, removing and modifying branches. Note that before this type of stream binding can take place the requester must know the SFEPs of the other members. This can be done through operations (from UAP to USM/SSM) like registerSIs and RegisterSFEPs, or operations (from SSM/USM to UAP) like getSIsExe

The signature of one of the operations in StreamInterfaceFS is given here for reference:

```
void RegisterSIs(
    in t_ParticipantsSecretId myId,
    in sequence t_SFEPServDescList my SFEPs
    out t_RetElementIdList identifiers)
    raises...
```

Note that in this case the end-user have much more detailed view of the branches in the SFC, and hence operations for detailed manipulations of the SFC will be given as well in FlowSBFS. This feature set is dependant on the StreamInterfaceFS as already described: First the operations in StreamInterfaceFS are invoked so that the different parties can see each others SFEPs, and then operations for creation and modification of SFC between these SFEPs are invoked (e.g., the operation 'addSFlowSBFSReq' given above).

The main part of the signature of one of the FlowSBFS operations are given here for reference:

```
void addSFlowSBFSReq(
    in t_ParticipantsSecretId myId,
    in t_SBType reqType,
    in t_SBTopology reqTop,
    in t_AdministrativeState reqState;
    in t_SFlowDescList flows,
    in t_SBSuccessCriteria criteria,
    in t_SBRecover recActions,
    in t_ParticpantIdList owners,
    out t_SBBindState status)
```

No detailed computational scenario is given for the flow oriented SBFS, but details can be found in [4]. It should, however, be obvious that more operations are needed by this flow oriented approach, then by PaSBFS.

3.3 Comparison of the approaches

When we examine the operations and parameters we see that the flow oriented approach will need more operations in order to establish the stream binding (i.e., first registration of SIs from all parties, and indications back to the other users about this information; then establishment of the flows). It will also require the requester to know the SFEPs of the other party (which is good when detailed end-user control is wanted), but there may be cases where the (other) end-user(s) do not want to show these details to all the other end-users.

In the latter case the PaSBFS is a good alternative. The PaSBFS gives less information to the UAP of the end-user, so even a 'simple' UAP should be capable of handling participant oriented stream binding (and the terminal's TCSM will handle the SFCs related to this terminal, through resolve operations etc.).

The 'simplicity' of the UAPs may then be 'compensated' by a strong and powerful SSM that does all the mapping from SBSR to actual SFCs. We may also note that some of the users may support PaSBFS, while other users (e.g., a 'chairman' in a conference) may support the more detailed FlowSBFS. This will be possible as long as the SSM supports both PaSBFS and FlowSBFS.

4 Example: tailored tele-education

This section describes an example of a tele-education service tailored to handle also a deaf student. Typically a number of ordinary students will participate in the service session as well. For simplicity only one 'normal' student is depicted, and represented by party-N in SSG and by UAP-N (and possibly USM-N) in the computational view. The figures can be easily extended by adding Party-M, UAP-M where M=1,2,3 etc. The teacher is not shown, we show only a resource 'AV_source', but in real examples we should have a party-Teacher which is controlling the audio and video resources (or possibly controlling together with the students).

We assume in this example that the deaf student is known to this service (i.e. the service is tailored for this student), so that when the special deaf student (Party-S) joins the SSM automatically configures a special resource, representing the audio to sign language (video) interpreter. Of course other ways of handling the deaf student exists as well. E.g., by using service composition we may allow for cases where the interpreter is in 3rd party service domain but we wanted to keep this example simple.

For the sake of illustration a few assumptions are made in this example:

- 'Video plus audio' is modeled by one SI with two SFEPs
- 'Video and video (sign language)' is modeled by one SI with one SFEP

In fact it might turn out that video&audio has stronger synchronization requirement that the video&video for the sign language, and that the opposite modeling is as good, but that is irrelevant for the explanations given in this example.

As already noted the SBSR and SI concepts are logical (network technology independent) representations of underlying physical entities. Thus this example may be realized by several technologies. One option is to use ATM as transport technology for all the streams. However it may also be realized by using a number of ISDN B-channels as transport network (TN) for the streams (and possibly to realize the operation invocations over the kTN by using GIOP specialized to SS#7 as an alternative to IIOP based on IP). Since TINA is now studying the use of TINA streams that use Internet transport (as described in an other paper to this conference), the example may also be realized by using Internet (with video QoS) as the transport network for the streams.

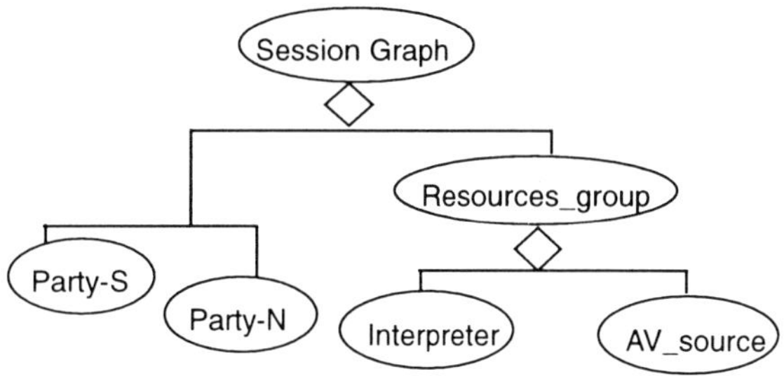

Figure 4 SSG as viewed by SSM

Example before participant oriented stream binding setup
after invitation and joining of the parties
(and after setup of the resources)

We do not show any student-to-student interactions in this example (to keep the complexity low), but in the case of (audio) communication to the deaf student the interpreter resource should be involved as well.

Is is also likely that the deaf student may have service specific operations towards the interpreter (like stop, rewind and request for specific word explanations), but such details will not be shown.

Before any type of stream binding can take place, the parties (and resources needed) must be set up and made known to the SSM (and possibly to the other users as well). After this the SSM may have the view of the SSG shown in Figure 4.

At this stage either the SSM or any party can initiate session control operations, in this example we will assume that the SSM initiates the stream binding. We may also assume that the students at least support the multiparty feature set.

4.1 Participant oriented stream binding

The deaf student will have a view different from the SSM. The view of the UAP-S before the stream binding is established is shown in Figure 5. Here we assume that UAP-S has visibility of the interpreter in order to invoke service specific operations.

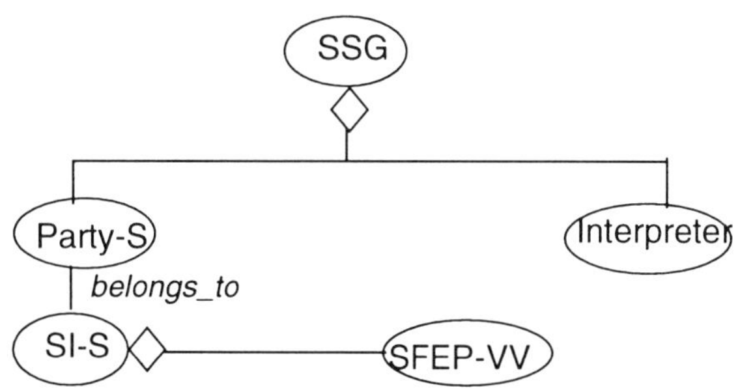

Figure 5 SSG as see from UAP-S (before)

Each normal student will have a similar view, but with two SFEPs each, (and no visibility of the interpreter). This is shown in Figure 6.

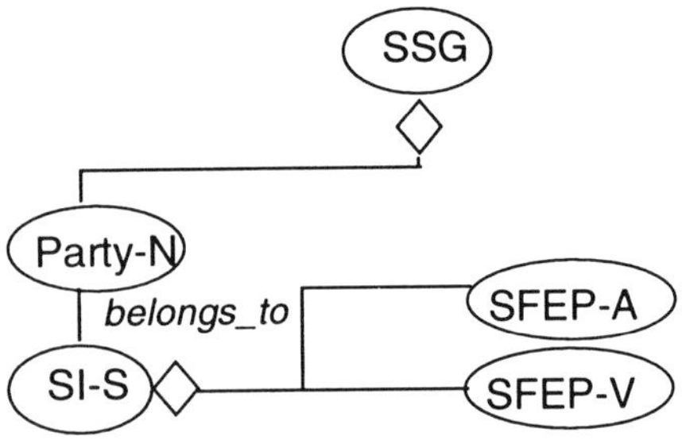

Figure 6 SSG as see from UAP-N (before stream setup)

UAPs sees their own SI etc. before the SSM does, e.g. after requesting them from TCSM as response to the join-PaSBFSexe operation. (Refer to scenario in Section 3.1).

The SSM now have a view of all the parties and their stream interfaces and corresponding SFEPs. The SSM may now do the mapping to SFCs. In this example the task of the SSM may be simply to match SFEP sinks and SFEP sources, resulting in the SFCs shown in Figure 7.

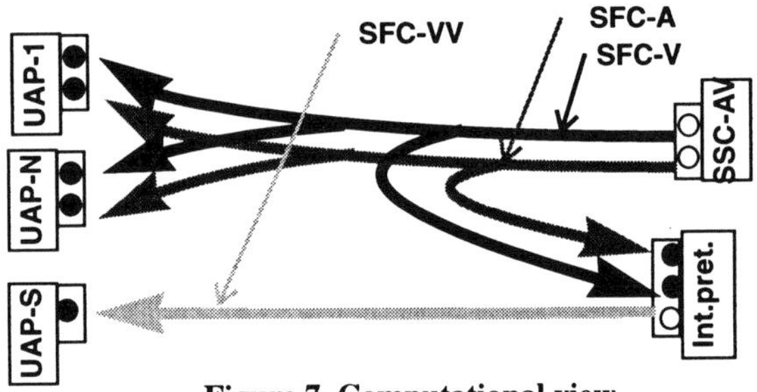

Figure 7 Computational view after stream setup ('global' view)

Figure 7 corresponds to the complex view of the SSG as shown in Figure 8.

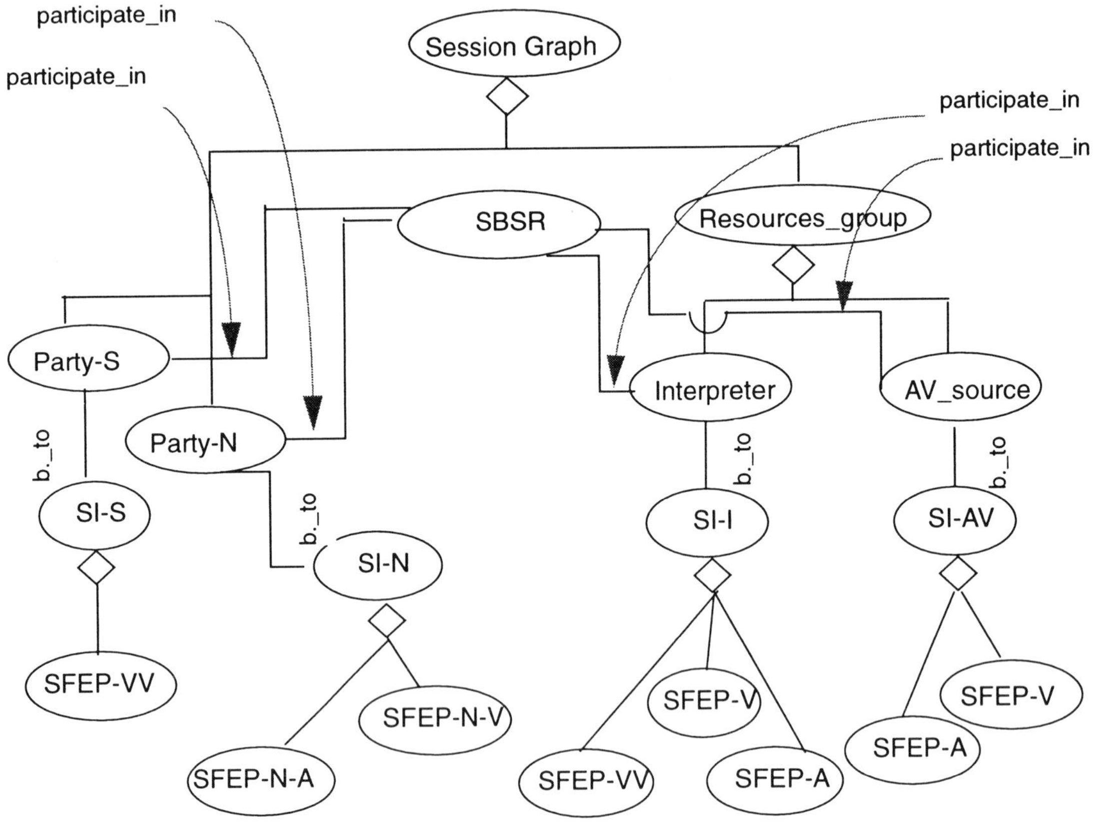

Figure 8 SSG as viewed by SSM (after stream setup)

Here in this participant oriented setting only the 'participate_in' relations are shown in the figure, though the SSM will have detailed views also of the 'bind', 'maps' and 'connects' relations. The relation between 'participate_in', 'binds', 'maps' and 'connects' are shown in Figure 2, but these relations have been omitted from this figure, since this figure is already quite complex. Also for the sake of reducing complexity no control session relationships are shown (while it is very likely that some will exist)

Figure 8 illustrates the case where SSM have global view. We may also note that in the case of service federation the different SSMs involved will only have partial views of the SSG, so TINA allows for decentralized service control as well, though that is not what is illustrated in this example.

The end-user, however, does not see all the details. We will illustrate what the deaf student will see in Figure 9 (with the simplification that the SSG and the corresponding aggregation is not shown).

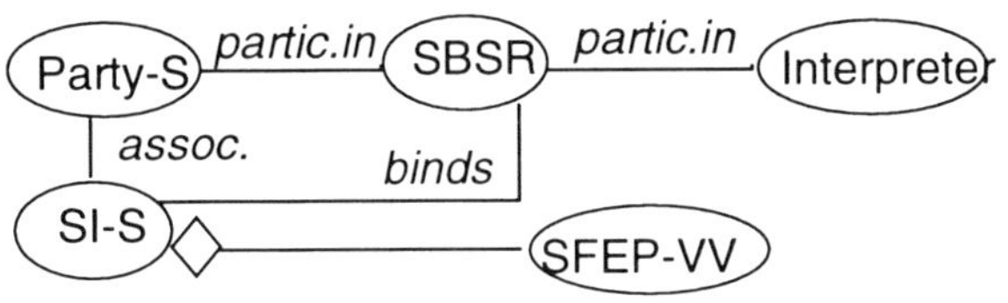

Figure 9 SSG as see from UAP-S (after stream setup)

Though not shown in this example, it is likely that there might be tele education examples where all the students and the teacher all have both audio and video SFEPs for both sink and source, so that all students can talk to the teacher, and the teacher to all the students. In such cases the QoS of the SFEPs will match in many different ways, also in ways where the stream flows from the students are connected to other students, and not to the teacher.

In such cases the SSM may contain service specific logic, so that the mapping from stream binding to stream flows really connects all the students to the teacher, and the teacher to all the students, and not just connects some students to some other students, (possibly leaving the teacher without connections!). This service specific logic may be obtained by specializing the role of 'party' into 'student' and teacher', and write the corresponding 'mapping' algorithm internal to the SSM to reflect this.

4.2 Flow oriented stream binding

For comparison we will show what the UAP-N will see in case of flow oriented stream bindings. This is shown in Figure 10 (again with the simplification that the SSG and the corresponding aggregation is not shown). We may also note that the illustration of what the SSM sees will be (almost the) same in the two cases. UAP-S will also have this detailed view in this case, seeing the different SFCs.

It might be noted that the interpreter is not visible to the ordinary student (since the interpreter is of no relevance to him). This non-visibility is the choice of this particular service logic, or it might be explicitly requested by the Party-S (by setting up a control session relationship).

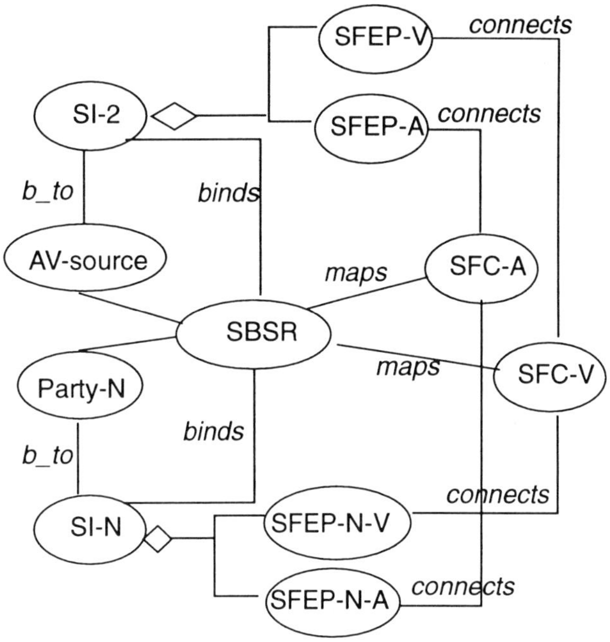

Figure 10 SSG as viewed by UAP-N (after stream setup)

5 Conclusion

The two different approaches (participant oriented and flow oriented feature sets) presented here, allows for flexible stream binding control, depending of the needs in different services. Quite sophisticated configurations may be set up without to much computational operations. The participant oriented approach allow for lightweight stream setup, but without to much detailed control to the end-user. FlowSBFS used together with StreamInterfaceFS allows for more detailed control to the enduser.

Even within one service session some of the parties may choose PaSBFS, while some other (e.g. the 'chairman' of a conference may use FlowSBFS). This is possible as long as the service provider (SSM) is providing both options.

The lightness of the computations have been obtained by decoupling the computational operations from the direct manipulation of 'small' entities in the SSG information model.

The feature set approach also allows 'simple' parties not to offer advanced features like control session relationship functionality, as well as it allows 'simple' services not to offer any stream binding functionality at all.

References

[1] C. Abarca (ed.) "Service Architecture delta95," TINA-C internal document, April 1996.
Available through http://www.tinac.com

[2] R. Minetti (ed.) "Service Architecture," TINA-C baseline document, Ver.4.0 Oct. 1997.
Publicly available through http://www.tinac.com

[3] F. Steegmans (ed.) "TINA Network Resource Architecture," TINA-C baseline document, Ver. 3.0, Febr. 1997.
Publicly available through http://www.tinac.com

[4] P. Farley (ed.) "The Ret Reference Point," TINA-C internal document, 1997. Available through http://tinac.com:4070/97/integration/rfrs/RFR-96-01/interim/draft0.4/status.html

[5] L. Kristiansen (ed.) "Service Architecture," TINA-C baseline document, Ver. 5.0, June 1997.
Publicly available through http://www.tinac.com

Acronyms

CSM	communication session manager
CSMF	CSM factory
FlowSBFS	flow oriented SBFS
GSC	global session control (generic part of SSM used in [1])
kTN	kernel transport network
PaSBFS	participant oriented SBFS
SBFS	stream binding feature set
SBSR	stream binding session relationship
SFC	stream flow connection
SFEP	stream slow endpoint
SI	stream interface
SSM	(provider) service session manager
TCSM	terminal communication session manager
TN	transport network
UAP	user application
USC	user session control (generic part of USM used in [1])
USM	user service session manager

Note: The term 'user application' or UAP may be a bit misleading. UAP and USM may be called 'user (or member) usage service session manager' and 'provider usage service session manager' respectively. Likevise SSM may be called 'provider service session manager', but SSM may as well be considered as an 'application'.

Acknowledgments

This paper is based on previous and ongoing work in the service stream in TINA core team. The author wishes to thank all authors of all versions of the TINA Service Architecture and corresponding specification documents. The work is also based on input from Vital auxiliary project to the Ret document. The author is particularly grateful to Stephanie Hogg, Telstra/TINA core-team and Marcel Mampay, Alcatel/Vital.

A Networking Architecture for Mobility Services Using Mobile Agent Approach

Choong Seon Hong
Korea Telecom Telecommunication Networks Research Lab.
62-1 Whaam-dong, Yusung-gu, Daejon, 305-348, Korea
cshong@tnrl.kotel.co.kr

Yuzo Koga, Yutaka Matsushita
Matsushita Lab., Faculty of Science and Technology, Keio Univ.
3-14-1, Hiyoshi, Kouhoku-ku, Yokohama, 223 Japan
yuzo@myo.inst.keio.ac.jp, on@myo.inst.keio.ac.jp

Abstract

In this paper, we propose a scalable service networking architecture as a TINA-like environment for providing flexibly various mobility services. The proposed architecture provides an environment that enables the advent of service providers and rapidly introduces multimedia applications, considering networks scalability. For supporting customized mobility services, this architecture adopts a new service component, which we call Omnipresent Personal Environment Manager (OpeMgr). In order to support mobile users who move between heterogeneous networks, for instance, between the TINA-like environment and the Internet environment, we propose a structure of a gateway. In addition, the proposed architecture uses the fixed and mobile agent approaches for supporting the user's mobility, and we evaluated their performances with comparing those approaches.

1. Introduction

The recent advances in the distributed computing and telecommunication technologies, and the users' needs for new services are driving the new direction in developing telecommunications networks which have open networking features to openly introduce new service applications into public networks. Therefore, the necessity of the integration and cooperation of the telecommunications and the distributed computing technology is gradually increasing. An open networking environment in order to realize this requirement would be a "Networks Plug-In Environment" as shown in Figure 1.

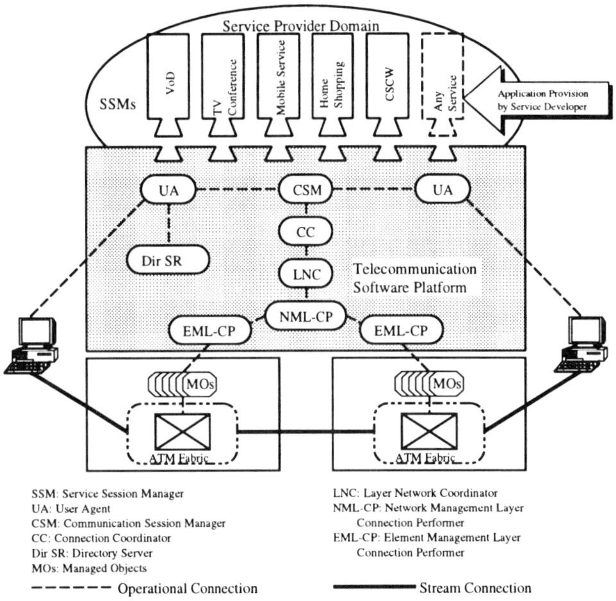

Figure 1. Networks plug-in environment

This new network architecture framework can provide various multimedia services. It also can provide an open networking environment for service providers to develop new services easily, flexibly and quickly. These reasons lead to the fact that TINA is becoming the main stream of the future public network architecture.

The representative architecture for this environment is TINA (Telecommunications Information Networking Architecture)[14]. TINA-Consortium has been studying the service and connection managements architectures and issued deliverable specifications of them. However, the networks mobile computing on such as personal mobility, session mobility, terminal mobility, and their services are in a rudimentary phase[12][13], and their studies are in the rudimentary step. Therefore, we propose a new service networking architecture to realize mobility services that are defined in TINA[11]. Especially, among the various mobility services, we concentrate on personal mobility and session mobility services that have an influence mostly on the service architecture. The proposed architecture is implemented by using MINDS [2] which we realized as a TINA-like model.

2. Networks Scalabilities

Service providers that have open networks environments hold various scalabilities. In this paper, we define networking scalabilities as follows:

The viewpoint of networking flexibility which means the networking capability of service providers possessing software platform that provides new services to users through flexible accommodation of multimedia applications supplied by service developer. To support this capability, service providers need to use resources of network providers. In other words, this viewpoint is networks openness.

The viewpoint of networks services capability which is the networks service capability of service providers in order to provide services to users. Therefore, users may select a contractual service provider by considering its service provision capability.

The viewpoint of extensibility which means the networking extensibility for providing modified services and interworking services. For instance, this scalability includes capabilities that interwork between heterogeneous environments, and federate between service providers in service levels.

Our architecture proposed in this paper guarantees the properties of scalabilities that are described above.

3. Proposed Service Architecture

TINA service architecture introduces a session concept[10]. A session is a collection whose goal is to satisfy the purpose of a service. The goal is achieved by performing activities during a specific period of time. The session is associated with the allocation of resources that are necessary to execute the service with the participating parties. The session is classified into four types, access session, service session, communication session, and user session.

We defined several computational objects on our new networking architecture, TINA-like model for these sessions. Our model realizes the flexible service architecture for a future public networks. These components on this model are defined more simply and integrated than TINA's one, to concentrate more on the service architecture, and as mentioned in the following sections.

3.1. User/Access Session Components

Computational objects on our TINA-like model for the user and access sessions can be defined as follows:

User Application (UAP): An object which is differentiated according to the kind of service and, located in the user system. In addition, it models service application, for instance, provides a user interface for an application.

Terminal Session Manager (TSM): An object which is independent on specific services. It is defined to model the minimal set of capabilities as an end-point of service session control and as an access session. A UAP interacts with a user agent (UA) via a TSM for making access to services. The TSM also coordinates with a MMCM to transfer user requirements for reserving and using terminal resources. The TSM in our architecture is corresponding to the GSEP in TINA.

User Agent (UA): A computational object which represents an end user in a service provider. It acts as a single contact point to control and manage, for example, creating, suspending, resuming the life-cycle of the user sessions and service sessions. This object takes into account to user restrictions by referring a UCxt and a PPrf.

Personal Profile Object (PPrf): A computational object which manages users' preferences in order to execute the service access and service session. In our proposed architecture, it also manages personal schedules and user's mobile states to support various mobile services.

Usage Context Object (UCxt): A computational object which manages the pool of knowledge for the association of an end user and his/her related resources. It keeps track of resource configuration information of end user systems by interacting with a TE-A. Accordingly, it always holds the information of end systems and access points.

Terminal Equipment Agent (TE-A): A computational object which represents user terminals in a service provider domain. It also monitors the user's login state.

Subscription Agent (SubAgt): A computational object which is the contact point for accessing user's subscription information. In our architecture, we use X.500[4] directory system in SubRgs (Subscription Registrar) to manage users' profiles.

3.2. Service/Communication Session Components

Computational objects on our TINA-like model for the service and communication sessions can be defined as follows:

Service Session Manager (SSM): A computational object which provides a specific service to users such as video on demand, home shopping and CSCW (Computer Supported Cooperative Work). It is managed by a service provider, and the functionalities are such as adding/deleting/joining participants and the controlling session of specific services. It also performs operations on users using logical connection graphs of a service session. An SSM supports negotiation capabilities among users and requests users' connection requests to a CSM.

Communication Session Manager (CSM): A computational object which performs the establishment, modification and release of connections that are requested by a UA or an SSM. It converts logical address to physical address in order to control networks resources for end-to-end connections. For efficient connection control, a CSM manages the information of network topology. It also coordinates with a connection coordinator (CC), layer network coordinator (LNC) and connection performer (CP) to establish end-to-end connections.

Multimedia Control Manager (MMCM): An object which manages terminal resources and coordinates with a CSM to provide networks connections. Its functionality is a superset of a TCSM in TINA, which reserves local terminal resources before passing a user requirement to the network side[1].

4. TINA Mobility Services

TINA defines mobility, and classifies it into personal mobility, discrete terminal mobility, continuous terminal mobility, session mobility and application mobility[11]. We particularly concentrate on personal mobility and session mobility.

- **Personal mobility** provides communication services to a user that are personalized with their preferences and identity ubiquitously, independently of both physical location and specific equipment, where the level of service obtained is dependent only on the capabilities of the access equipment and/or method. This mobility corresponds to classes of services such as Universal Personal Telecommunication (UPT), or card calling. TINA describes the capability set 1 of mobility for access session and service session. To receive mobility services, a CPE (Customer Premise Equipment) must be TINA-compliant and a terminal may need a specific application related to the service session.
- **Session mobility** enables a session to be maintained independently of the equipment used to support it. Session mobility of capability set 1 in TINA must terminate an access session from the same CPE (Customer Premise Equipment) and NAP (Network Access Point) that he/she initiated it from. Therefore, the user cannot resume a session that has been suspended globally, but only a session he/she had previously suspended individually can. In addition, the QoS of the new CPE/NAP must meet a set of minimal requirements of the existing session.

TINA service architecture[11][10] specifies the roles of DPE services[9] related to mobility services. When a user makes access to a UA interface, DPE supports the user in one of the following three ways: 1) create a new UA, 2) create a replica of the original UA by using DPE migration and replication services, 3) access the original UA by using DPE trading service. In a multi-carrier environment, networks provide a DPE federation service to support users over multi-domains. In case of 3) that is described above, we adopt an Intelligent Mobile Object (IMO) to support the users to use an environments such as personal profile, etc.

5. Omnipresent Personal Environment Service

The OPE (Omnipresent Personal Environment) service provides a networking functionality that supports an environment for mobility from the network point of view.

Users think that they can receive services without depending on their terminals and locations. Some services among users' contract services are not necessary for mobile users. For instance, the home shopping service may be little used by users on business trips. Such service can be temporary suspended for cost effectiveness. However, the information such as the user's schedule should be moved to his/her destination according to the user's movement because it is important regardless of locations. Therefore, we must process tasks for mobile users as follows:

- Previous service registration and automatic installation of applications at the destination.
- Temporary suspension of services to be not used.
- Construction of the user's login environment on their terminal at the destination.
- Transfer necessary user's data to the destination (e.g. user's schedule, etc.).

The services described above are performed by the TINA-like platform with an OPE Manager (OpeMgr) which we propose as an SSM that provides OPE service. Thus, a novel mobile computing and mobility support environment will be a scalable platform coping with diverse users' behaviors.

In order to realize the OPE services, we can consider two approaches. One approach is the fixed agent approach.

The fixed agent does not move, and interacts with other agent via networks. In this case an OpeMgr gathers and processes necessary information by interacting with other computational objects. The second approach is the mobile agent approach. The mobile agent moves to where the other agent exists, and interacts directly with the agent. An OpeMgr issues a mobile agent, and the agent moves several places and executes tasks related to the mobile user at each place.

The OPE service has to consider the move between terminals in heterogeneous backbone environments. For instance, the move between a terminal on the TINA in user's home and a terminal on the Internet in his/her office would be a representative case for applying mobile service. Therefore, in this paper we concentrate on the service networking architecture for supporting the personal and session mobility between the TINA-like environment and the Internet environment. Accordingly, the OPE services must be realized not only in the TINA-like environment but in the Internet environment. For the purpose of effective coping with these environments, we consider a mobile agent using a script language that we named Intelligent Mobile Object (IMO). Platforms (e.g. JAVA, Telescripts, etc) for an IMO would have compatibility and interoperability to execute mobile tasks.

6. Interworking between the TINA-like environment and the Internet environment

6.1. The Future Way to Utilize the Internet

On the next generation networking infrastructure, we considered three ways to utilize the Internet.

1. The case in which a user is subscribed to an ISP (Internet Service Provider). In this case, when a user utilizes the Internet service, he/she connects to the gateway that is provided by ISP, and the IP address is temporarily allocated to the user terminal.

2. The case in which a user is subscribed to the network that can provide the Internet service, and he/she can utilize the Internet directly. In this case, IP address is allocated semipermanently.

   With this case, two kinds of networks which he/she is subscribed can be considered.

   (a) The network which can provide only the Internet service.

   (b) The network which can provide the Internet service as the one of the several services, such as B-ISDN and TINA.

   When FTTH (Fiber to the Home) is realized, both types of networks will use ATM (Asyncronous Transfer Mode) as a data transfer protocol. In the latter case (b), because ATM network will not be used only for the Internet Service, a certain bandwidth will have to be allocated between the terminal and a network provider for the Internet Service. Whether bandwidth will be allocated permanently or temporarily depends on a contract between user and network provider. If a user utilize the Internet Service so long time, it had better to allocate bandwidth permanently. Otherwise bandwidth should be allocated only for the period time that a user utilize the Internet Service.

   To have a certain bandwidth permanently allocated is, in other words, to have a PVC (Permanent Virtual Connection) established between user terminal and a network provider as a dedicated line. With this connection, a user will send/receive IP packets. On the other hand, if a certain bandwidth is allocated temporarily, a SVC (Switched Virtual Connection) will be used for this connection.

3. The case in which no IP address is allocated to the user terminal, but the IWU, which is the entrance of the Internet environment, guarantees to transfer the data to the user terminal as a proxy server.

Except for utilizing at a user's home, the Internet will be utilized via dedicated line at his/her company or school.

6.2. Technology Aspects for Realizing Interworking

6.2.1 Data Tranfer Level

Considering the interworking, we must consider the data transfer level as follows:

- The ATM network, which is managed/controled by TINA, must support IP. However this is realized by some ways, such as MPOA (Multiprotocol Over ATM)[6], IPOA (Classical IP and ARP over ATM)[7].
- It is also studied to apply IP to TINA-kTN (TINA kernel Transport Protocol) as a basic protocol.
- It is better that a user can ubiquitouly transfer the data using IP.

To realize the interworking between next generation networks which are based on ATM, there are some proposed/standardized technologies as described above, and they are now still being studied.

6.2.2 Service Level

Therefore, we consider the interworking of the service level. But, what is the interworking of the service level?

We defined it as "each network can make use of each other's services", that is, supposing that it is possible to interwork at the data tranfer level, user on the TINA-like environment can make use of services on the Internet, and user on the Internet environment can also make use of services on the TINA-like environment. By realizing such the sevice level interworking, it will become possible to utilize these two networks seamlessly.

6.2.3 Case to Utilize the Services on the Internet Environment from the TINA-like environment

If it will be realized to interwork at data transfer level, it will be possible to make use of the services on the Internet environment from TINA-like environment. In this case, because there is a function on the TINA-like environment to get the IP address temporarily if needed, or, because the proxy server guarantees to transfer the data to/from the user terminal, or because the ATM network itself will be able to support the IP, it will be possible to transfer the data from the Internet environment to the TINA-like environment.

Moreover, there is a significant difference between the TINA-like environment and the Internet environment, such as the charging system. But if the user on the TINA-like environment is the subject to make use of the services, because payment is charged to the calling party, that is, to the user on the TINA-like environment, TINA's charging system can be used. Furthermore, it is one of the reasons that make use of the services on the Internet easier that the ability of the TINA-like environment is higher than the Internet environment, in a sense that the TINA-like environment can be applied to the ability of the Internet environment.

6.2.4 Case to Utilize the Services on the TINA-like Environment from the Internet environment

There are some problems to make use of services on the TINA-like environment from the Internet environment. First of all, because there is no charging system on the Internet environment in the present circumstances, it is impossible to charge the user on it, who is the subject to make use of the services on the TINA-like environment. Secondly, in relation to the first problem, because there is no way on the TINA-like environment to authenticate the user on the Internet environment, it is impossible to charge or to limit the user on the Internet environment when he/she makes access to the TINA-like environment.

Moreover it is also the problem that there is no concrete way on the Internet environment to designate the services on the TINA-like environment. How can the user on it make use of the services on the TINA-like environment? This may be the most difficult issue because the ability of the Internet environment is lower than the TINA-like environment.

6.3. Protocol Stacks

Considering to transfer the signaling messages between the node (or the terminal) on the TINA-like environment and the terminal on the Internet environment, protocol stack can be represented like Figure 2.

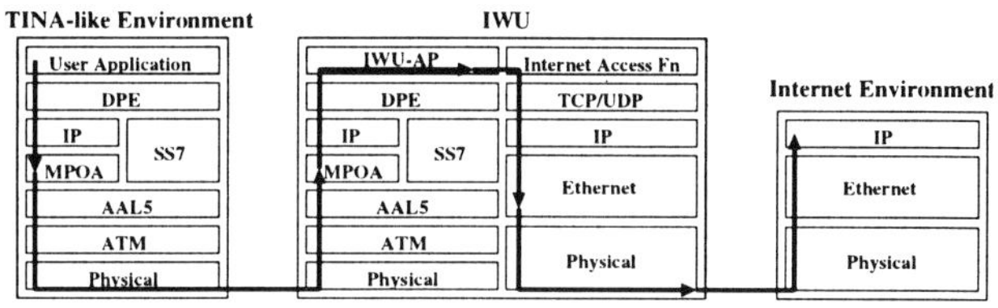

Figure 2. Signaling protocol stack

Again considering to transfer the user data between the terminal on the TINA-like environment and the terminal on the Internet environment, protocol stack can be represented in two cases as follows:

1. The case in which the terminal on the TINA-like environment behaves as an Internet terminal, and sends user data using IP directly.

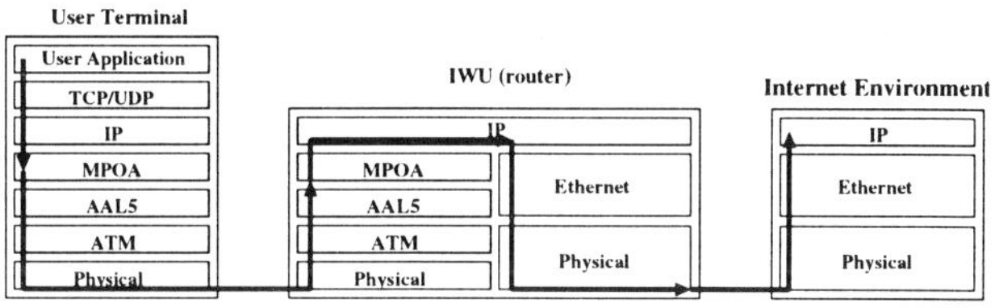

Figure 3. User data protocol stack (1)

2. The case in which the connection is established between the terminal on the TINA-like environment and IWU, and then protocol is conversed to IP in IWU, and the IWU sends the user data to the Internet environment.

6.4. Functionality of IWU

Considering the interworking between the TINA-like environment and the Internet environment, we introduced the IWU (Interworking Unit) in order to be cooperated these two networks seamlessly

The functions of the IWU that plays the role of service providers are defined as follows:

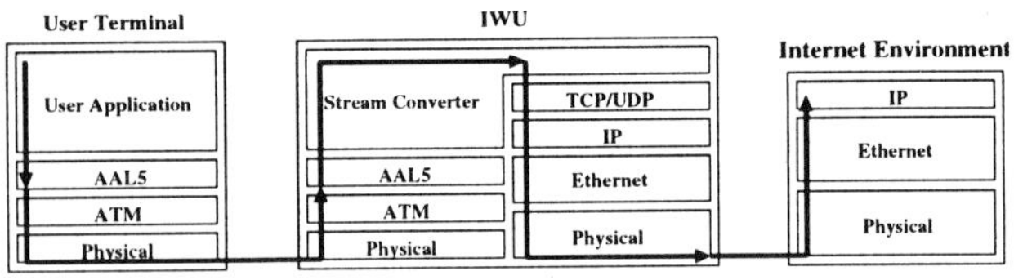

Figure 4. User data protocol stack (2)

IW-UA: An object that belongs to the TINA-like environment in the IWU. To interact with services in the Internet from the TINA-like networks, firstly, an object in the TINA-like environment has to make access to the IW-UA. For example, as for a mobile user moving toward the Internet environment, the IW-UA is the entrance point. This object also interacts with an object in the Internet terminal via IW-AP (Interworking Application). Inversely, this object is the entrance point to make access to the TINA-like environment from the Internet environment.

IW-TEa: An object that is created when a terminal on the Internet is connected to the TINA-like network. It includes the information for access points such as terminal access points and network access points.

IW-AP: An object that converts the protocol based on requirements between calls on the TINA-like environment and calls on the Internet environment. It also transfers requirements to IW-UA from the Internet.

IW-MMCM: An object that controls and manages resources of an interworking unit and establishes connections by the request of the CSM on the TINA-like environment.

TINA to Internet Streams Converter: An object that performs conversion between ATM cells as streams on the TINA-like environment and IP packets as streams on the Internet environment.

Internet Access Fn: An object that performs functions such as IP routing, etc., according to the IW-AP request. Requirements from the Internet are transfered to an IW-AP.

Figure 5 shows the proposed signal flows between each components when interworking the TINA-like model with the Internet.

7 Information Model for Mobility Services

We show the relation of information objects. Figure 6 shows this information model based on OMT diagram[5]. In this figure, Access Session is managed by UA, Service Session and Session Profile are managed by SSM, Terminal is managed by TE-A, $PPrf_{IO}$ is managed by $PPrf_{CO}$, and $UCxt_{IO}$ is managed by $UCxt_{CO}$.

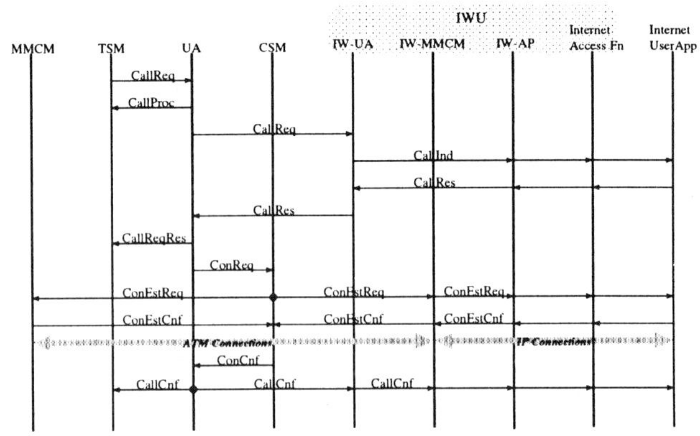

Figure 5. Signal flows when interworking between the TINA-like environment and the Internet environment

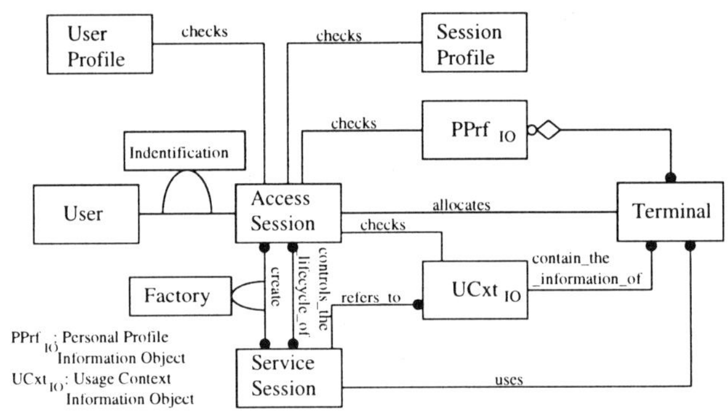

Figure 6. Information model for Mobility Services

8. Scenario for Providing Mobility Services

In order to model a service networking architecture for supporting personal mobility and session mobility as we described in the previous section, we consider a session mobility between the TINA-like environments and a personal mobility between the TINA-like environment and the Internet environment, respectively. Therefore, the architecture needs an IWU for interworking with the Internet.

8.1. Realization of Personal Mobility

Personal mobility is achieved by changing the information of the PPrf and UCxt. When a user reaches his/her destination, the PPrf and UCxt information are changed, based on the terminal at the his/her destination. When a call for the user reaches UA, it makes access to PPrf to get the information of which terminal to deliver it to, and then makes access to UCxt to get the terminal's resource configuration information, and finally delivers the call to the terminal which the user is on.

8.2. Realization of Session Mobility

Session mobility (suspend/resume of a session) is performed by an SSM. There are two possible cases. The first is the case when the whole session is suspended, and the second is when only a particular user is suspended but the whole session stays in progress. The state of the suspended session/user is kept by SSM as a profile. When resuming, UA gets the user's location by interacting with the PPrf and UCxt, and, if needed, another connection will be established by a CSM.

8.3. Movement from TINA-like Environment to the Internet Environment

The following proposed scenario assumes that a user has an account on the Internet environment (refer Figure 7).

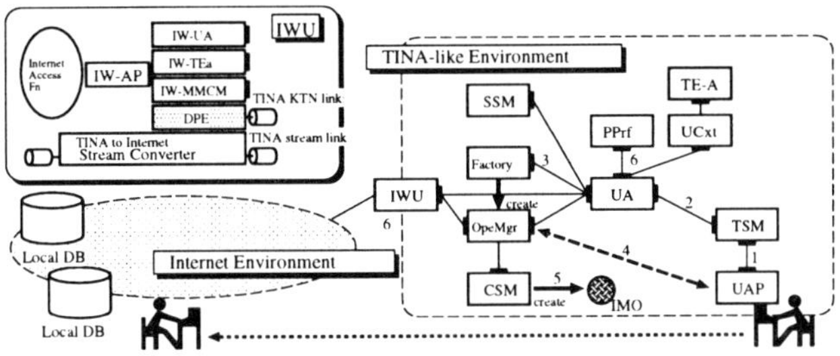

Figure 7. Movement between the TINA-like environment and the Internet environment

1. An end-user (UAP) sends a request for personal mobility service to a TSM.
2. The TSM transfers the OPE service request to a UA (User Agent).
3. The UA requests the creation of an OpeMgr to a service factory.
4. After an OpeMgr is created, a connection for interactive service between an OpeMgr and the mobile user is established.

   Through the interactive connection, the user sends requirements for personal mobility services such as destination address, preferring service environment and so on, to the OpeMgr.
5. The OpeMgr creates an IMO. The IMO that holds interface references of a TSM, PPrf, IW-UA and SSM moves according to sequent order to those objects to execute tasks and get necessary information for the mobile user, then moves to an IWU for interworking with the Internet. The reason why IMO moves to an SSM is to get service specific information for the user.
6. The IW-UA executes the IMO's task. In case that the service by the IMO is an OPE service in the Internet, the IW-UA sends a protocol conversion request to IW-AP.
7. The IW-AP having the IP address sends a routing request to the Internet Access Fn.
8. The IMO that moved to a local domain of the Internet performs the setting of mobile user's terminal and his/her service environments. Using the mobile user's requirements and preferences obtained from steps 4 and 5, the terminal environment for a mobile user is established, that is, he/she can use the terminal and services with his/her preferable environment. All the mobile user have to do is to make access (login) to the terminal.

If a user does not have an account in the Internet domain, an IMO has no destination (i.e. user's terminal). Therefore, the IMO waits for the mobile user's request in an OpeMgr. The OpeMgr can be made access via WWW by the mobile user on the Internet environment with, for example, a guest account. The OpeMgr should be created so that it can be made access from both the TINA-like environment and the Internet environment. If a mobile user's request occurs, the OpeMgr issues an IMO toward an IWU, and then the tasks from step 6 are carried out.

8.4. Movement between the TINA-like Environments

Mobile users normally use personal mobility and session mobility in the TINA-like environment. Therefore, the federation service between different operational domains is needed. The following shows the proposed scenario to realize these services (refer Figure 8).

1. A user requests OPE services to a UA as a personal and a session mobility service for TV conference or VoD service session.
2. The UA sends a request to create an OpeMgr to a service factory, and the OpeMgr is created.
3. The OpeMgr issues a request to establish an OPE service connection to a CSM.

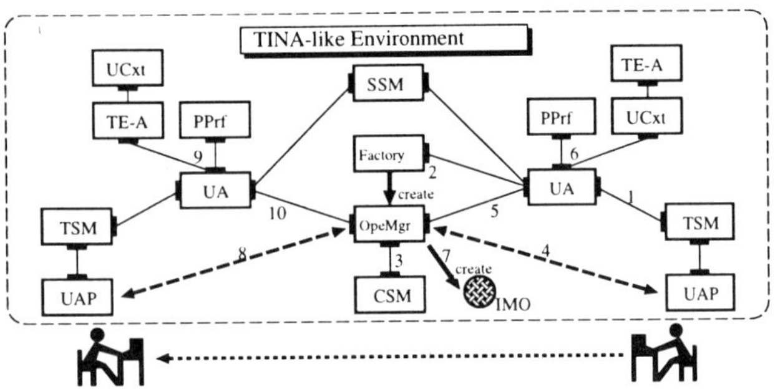

Figure 8. Movement between the TINA-like environments

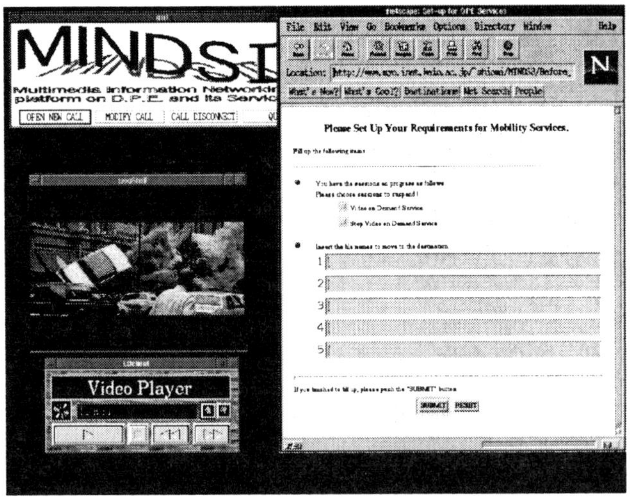

Figure 9. TINA-like model implementation

4. Using the OPE service connection between an OPE application and an OpeMgr, the negotiation for the OPE service is performed.

5. After the negotiation, the OpeMgr sends a suspension request for the session on progress to a UA.

6. The UA sends suspension the request to the appropriate SSM, modifies the session profile, and sends the change request of the TE-A profile.

7. The OpeMgr creates an IMO, and the IMO moves to the user terminal to gather the necessary information of the user's terminal environment and then returns to the OpeMgr.

8. After the user makes access (login) at the destination, a new UA issues an OPE service request to the OpeMgr. Service negotiation between the user's OPE application and the OpeMgr is performed.

9. The new UA sends a request to modify the terminal profile information to a UCxt.

10. Finally, an OpeMgr sends a UA a resume request for the suspended session in a SSM. The suspended session is resumed by the re-establishment of the connections. The IMO in step 7 constructs the user's terminal environment in case that a mobile user knows the destination address in advance.

9. Implementation of the Proposed Architecture and Evaluation

9.1. Implementation

We implemented our proposed TINA-like model for mobility services using the Orbix platform[3] that is CORBA[8] compatible, an ASX-200 switch of Fore System for ATM connections, and Parallax XVideo cards as a video tool. With this implementation, we realized the session and personal mobility services. As for personal mobility, we realized the OPE services that perform several tasks in place of the mobile user, and adopted a mobile agent approach for its realization.

To realize the mobile agent, we constructed the IMO and its engine on an existing platform for mobile agents. The realization of mobile agent is now studied using ,for example, Java or Telescript at many places by many groups. However, considering to apply it to mobile services between heterogeneous backbone, the most important thing is not what language is used for realizing mobile agent, but how the platform that can understand and execute mobile agent is realized. The things we paid attention to is the impact of mobile agent for mobility services between the heterogeneous backbone. In this sense, we concluded that whatever language (or script) is used to realize mobile agent, there is no relation to the impact of it. The engine for the IMO in our implementation was constructed using JAVA language, coping with the TINA-like environment and the Internet environment. The reasons for why we used a de facto standard such as JAVA is as follows:

- In case that a network environment has no mobile engine, it is easy to construct its environment by down loading a mobile engine composed of JAVA language.
- Mobile agent operation environment is interoperable on heterogeneous networks platforms.

A mobile agent operation environment is represented in Figure 10. As for the session mobility, the sessions' suspending/resuming for a TV conferencing service and a VoD service are realized.

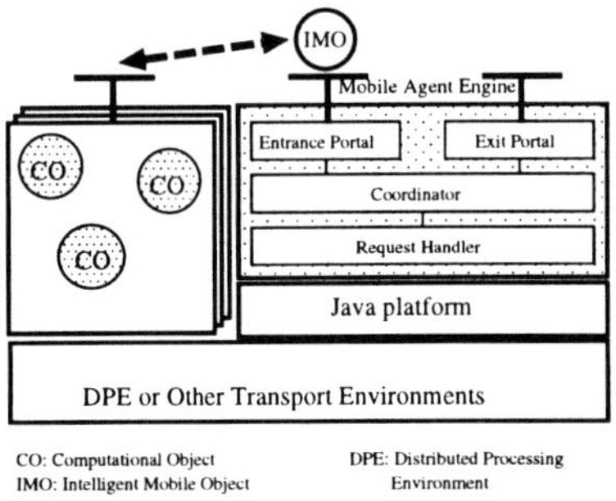

Figure 10. An example of mobile agent operation environment

Figure 9 shows the implementation of our proposed TINA-like model. The web window is for submitting the mobile user's requirements.

9.2. Evaluation of our Architecture

In order to realize the OPE services, the fixed and mobile agent approaches could be used. We adopted the mobile agent approach in the load distribution and scalability point of view. The mobile agent approach using script languages has a high interoperability and could solve load distribution problems. Table 1 compares the two approaches in realizing the OPE services.

In future public networks, users will make contract with various service providers, and service specific information will be distributively placed on each SSM. Therefore, if a user wants to make use of services after moving to another place, he/she needs the service specific information for each service he/she wants to use.

To evaluate our approach, we carried out a computer simulation using M/M/1 queuing system to compare the utilization of the OpeMgr with service request arrival rate, according to the number of SSM's. The simulation parameters and their implementation measured values are shown in Table 2. In this simulation, we concentrated only on the utilization of the OpeMgr, so these parameters are the time that the OpeMgr itself took to perform the events.

In the fixed agent approach, an OpeMgr first interacts with the user terminal to get the user environment information. Then, it interacts with SSM's to get service specific information. Finally, it interacts with the terminal which the user makes access (login) after his/her movement. In these parameters, n is the number of SSM's that the OpeMgr must interact with.

While in the mobile agent approach, an OpeMgr first creates an IMO and lets it go. Then, it receives the IMO which has gotten various information, that is all registered in an OpeMgr. Finally, the OpeMgr recreates an IMO to take the information to the terminal, which the user makes access (login) after his/her movement.

Table 2. Simulation parameters

Fixed agent approach	
Interact with the user terminal	2177 [ms]
Retrieve service specific information	2188 [ms] x n
Transfer information to the destination	2243 [ms]
Mobile agent approach	
Create an IMO (before user move)	2257 [ms]
Receive an IMO	2441 [ms]
Recreate an IMO (after user move)	2259 [ms]

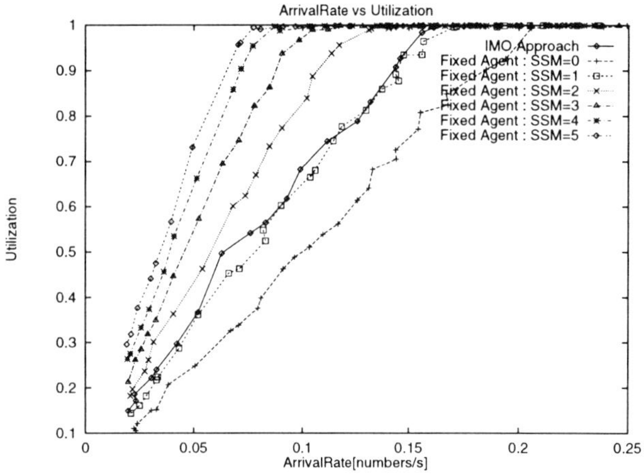

Figure 11. Utilization vs arrival rate in an OpeMgr node

Figure 11 shows the simulation results of the fixed and mobile agent approaches.

With the fixed agent approach, an OpeMgr, which is regarded as a fixed agent, must interact with each SSM that the user makes contract with, to get service specific information. As the number of SSM's increases, the utilization of the OpeMgr reaches 1.0 (the state that the CPU of the OpeMgr is in full use) with a smaller arrival rate. Accordingly, the queuing delay time is diverged.

On the other hand, with the mobile agent approach, because mobile agent itself moves to the place where the SSM exists and directly interacts with it, the utilization of the OpeMgr does not depend on the number of SSM's. Though the utilization of the whole system may increase, the utilization of the OpeMgr could definitely be kept at a low level.

Table 1. Comparison of fixed agent and mobile agent approaches

	Fixed agent approach	Mobile agent approach
Move between Heterogeneous Environments	Protocol conversion between heterogeneous environments in IWU	Realized by a moving script, using Java for interoperability, with mobile agent engine constructed on each environment
OPE Service Upgrade	Network module dependent on each backbone	Modifying OpeMgr script, creating rules
Traffic Volume	Same as the data amount	Data+Script
Load Distribution	Concentrating on the OpeMgr's node	Distributed to the OpeMgr node and other service nodes

The performance of the mobile agent approach is almost the same as the fixed agent approach when the number of SSM is 1. If the number is more than 2, it is evident that the mobile agent approach has a better performance. As a result of the simulation, concentrating on one particular agent, the mobile agent approach is more effective as the number of interaction, which this particular agent must do, increases.

10. Future Work

10.1. Requirements for Realizing Interworking

Considering the things we described in the section 6, some requirements for realizing the interworking can be listed.

1. **How to treate the signaling messages of the TINA-like environment in the Internet environment.**

   To realize the multiparty call in which users on the TINA-like environment and users on the Internet environment coexist, the signaling messages must be sent from the TINA-like environment to the Internet environment. If the TINA-kTN (TINA kernel Transport Network) is based on IP, there will be no problem if only transferring the messages. However signaling messages on the TINA-like environment cannot be sent to the terminal on the Internet environment without modification, because there is specific information for the Internet environment. Therefore we will define the platform and its architecture that enables users on the Internet environment to make use of the services on the TINA-like environment. We will also define the way of how to handle the signaling messages on the Internet environment.

2. **How to authenticate users who make access from the Internet environment to the TINA-like environment, and how to manage the information for this authentication.**

   When opening the call, TINA refers user's profile and makes session profile for this call, but it is impossible to do so without authenticating the joining user. Therefore we will introduce the authentication mechanism using the user pesonal ID.

10.2. Security for Mobile Agent

An IMO traverses several domains, the firewall for security will be a barrier to realize OPE service. In this case IMO needs to be authenticated with some way. One simple considerable way is that an OpeMgr that creates an IMO, previously holds security keys that had been received from contract service providers. The IMO obtains a ticket from the service provider of the destination domain using this security keys. Therefore, the IMO can pass the firewall using this ticket. But this is not enough fot the system to authenticate an IMO, so more secure system is needed for authentication of IMO.

11. Conclusions

In this paper, we proposed a scalable service networking architecture that provides various multimedia mobility services. Especially, we proposed a gateway that has interworking functionalities between the TINA-like environment and the Internet environment and discussed of mobility service scenarios. To realize the OPE services that we defined, we introduced an OpeMgr as an SSM and adopted an IMO (Intelligent Mobile Object) that moves user's data and constructs a new service environment for a mobile user. Finally, we evaluated proposed architecture using the fixed and mobile agent approaches. Our evaluation lead to the result that enhance the platform of mobile agents is necessary.

Our evaluation lead to the result that performance enhancement of the platform for the mobile agent is necessaly.

References

[1] C. S. Hong, D. Kashiwa, Y. Koga and Y. Matsushita. A Novel Service Networking Architecture Using TINA Model and Its Realization to VoD Manager. *Proceedings of SIG DPS*, DPS-76-29:169–174, 1996.

[2] C. S. Hong, S. Honda, K. Kawauchi, and Y. Matsushita. Distributed Computing Architecture for Effective Management of Multimedia Streams on DPE. *International Journal of Multimedia Tools and Applications*, 2(3):233–152, 1996.

[3] IONA Technologies Ltd. The Orbix Architecture. http://www.iona.ie/Orbix/Arch/index.html, 1996.

[4] ITU-T. Data Communication Networks, Directory. ITU-T Rec. X.500-X.521, ITU-T, Dec. 1992.

[5] James Rumbaugh, Machael Blaha, William Premerlani, Frederick Eddy, and William Lorenson. Object-Oriented Modeling and Design. Prentice Hall, Inc. A Simon & Schuster Company Englewood Cliffs, 1991.

[6] Juha Heinanen. Multiprotocol Encapsulation over ATM Adaptation Layer 5. RFC 1483, Telecom Finland, Jul. 1993.

[7] M. Laubach. Classical IP and ARP over ATM. RFC 1577, Hewlett-Packard Laboratories, Jun. 1994.

[8] Object Management Group. The Common Object Request Broker: Architecture and Specification. CORBA Version 2.0, OMG, 1995.

[9] TINA-C. Engineering Modeling Concepts (DPE Architecture). TINA Draft Doc., TINA-C, Dec. 1994.

[10] TINA-C. Service Component Specification. TINA Draft Doc., TINA-C, Nov. 1994.

[11] TINA-C. Service Architecture. TINA-C Report, TINA-C, Apr. 1996.

[12] TINA-C. Support of User Mobility in the TINA Service Architecture Version: 1.0 (Draft). TINA-C Engineering Note, TINA-C, Jun. 1996.

[13] TINA-C. Terminal Mobility Version: 2.0. TINA-C Report, TINA-C, Mar. 1997.

[14] W. J. Barr, T. Boyd, and Y. Inoue. TINA Initiative. *IEEE Communicaions Magazine*, pages 70–81, Mar. 1993.

TINA-Based Network Resource Information Model for Next-Generation Mobile Systems

Roberto Minetti
CSELT S.p.A.
Via G. Reiss Romoli 274
I-10148 Torino, Italy
Roberto.Minetti@cselt.it

Flavio Piolini
CSELT S.p.A.
Via G. Reiss Romoli 274
I-10148 Torino, Italy
Flavio.Piolini@cselt.it

Abstract

The paper describes how TINA concepts can be applied to the Universal Mobile Telecommunications System (UMTS) to achieve flexible connectivity control in the Radio Access Network. It presents a general Network Resource Information Model (NRIM) for the UMTS, defined as a strict extension of the TINA model. The required enhancements to the TINA NRIM are introduced, and the application of the resulting model to the UMTS is described in detail. In particular, the multi-bearer and macro-diversity features and the handover procedure are analyzed. Examples are provided to illustrate the concepts presented.

1. Introduction

The definition of "third-generation" mobile systems could take advantage of emerging technologies and architectures, in particular the Telecommunications Information Networking Architecture (TINA) [1]. TINA enables the fast introduction of future services by separating service aspects from the underlying technology [8], thus meeting important requirements of advanced mobile networks. This paper considers the Universal Mobile Telecommunications System (UMTS), the planned European third-generation mobile system.

So far, studies on UMTS systems have aimed at the identification of its requirements and at the definition of its procedures. Procedures are specified in terms of functional models, identifying atomic entities performing simple operations, and information flow diagrams between functional entities [6].

Applying TINA to UMTS implies to map UMTS functional entities onto TINA computational objects. However, a direct mapping prevents from taking advantage of the object-oriented approach: in fact, functional entities (or their decomposition) would be mapped directly onto computational objects, thus keeping a "masked" functional approach. Furthermore, possible overlapping between native TINA functionality and UMTS functional entities makes it difficult to identify where it is useful to extend the TINA architecture to support terminal mobility. Hence, the first necessary step to apply the TINA approach to UMTS is to define an object oriented **Network Resource Information Model (NRIM)** of the UMTS Radio Access Network. An object-oriented model of network elements facilitates the mapping of UMTS functionality onto TINA objects, by reusing and extending the TINA model. The second step is to derive a **Computational Model**, relying on the NRIM, that implements UMTS functionality in terms of computational objects; this is achieved by following the TINA Connection Management Architecture, which is part of the TINA Network Resource Architecture.

This paper focuses on the first step. First, it briefly describes the UMTS system and the TINA model, then it introduces the extended TINA NRIM for mobile systems, illustrating it with examples. Finally, it describes the setup of a mobile connection and the handover procedure.

2. The universal mobile telecommunications system (UMTS)

This section recalls the key features of the Universal Mobile Telecommunications System (UMTS) as defined in the European RACE-II project *MoNet* [7].

In a cellular system two different networks are identified: an Access Network that permits the mobile users to access the system, called **Radio Access System (RAS)**, and the Core Network, providing fixed transport capabilities to supply the requested services.

In the UMTS Radio Access System some new features are introduced in order to increase the system quality and capacity with respect to present mobile systems; they are described in Section 2.2. These features significantly impact on the establishment of a connection with the terminal, suggesting the application of the TINA Network Resource Architecture [2], or a solution derived from it. In particular, management of different kinds of channels on the radio interface (both logical and physical), as well as the handover procedure, can be handled by the connection

0-8186-8335-X/98 $10.00

management architecture. The following subsections describe the UMTS RAS architecture and the main aspects that are benefited by the introduction of TINA concepts.

2.1. UMTS radio access network architecture

Since the specification of the UMTS system is not stable yet, in this paper we consider the architecture defined in the European ACTS project *Rainbow* [12], that is derived from the one defined in the *MoNet* project [7]. The UMTS RAS architecture varies according to the different types of environment. The environment is the classification of the situation in which the mobile service is provided. It is determined on the basis of the user traffic and mobility levels, network functionality and set of services. In the most complete configuration the UMTS network architecture is shown in Figure 1.

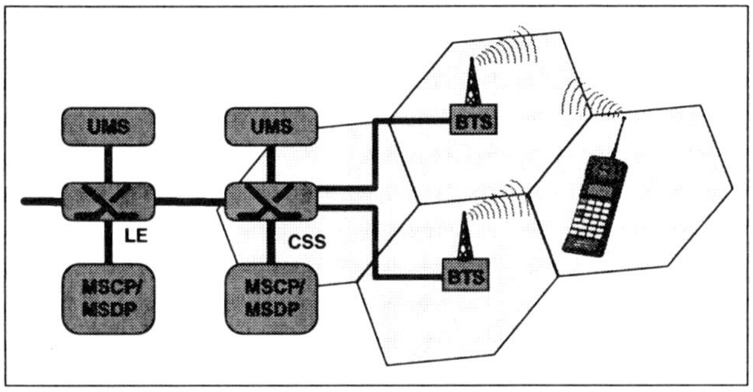

Figure 1. UMTS network architecture.

UMTS network elements are described below.

MT (Mobile Terminal): it comprises the functions allocated to the cordless device constituting one side of the radio interface.

BTS (Base Transceiver Station): it represents the functionality necessary to set up, maintain and release the radio link, and constitutes the other side of the radio interface.

CSS (Cell Site Switch): it includes the basic switching functionality in the Access Network. It provides the links between two or more BTS and can be directly connected to another CSS.

UMS (UMTS Mobility Server): it supports the mobility related functionality, as the combining and multicasting function related to macro-diversity and multi-bearer features explained in Section 2.2.

MSCP (Mobile Service Control Point): it contains the functionality needed to control mobility procedures and service operation in a given area; it can act on the information contained in the MSDP. In the access network, it contains functionality that allow to manage locally some procedures (like handover), avoiding to load the upper hierarchical layers in the network. In the core network, it allows an efficient use of existing backbone infrastructures with minimum impact or modifications on them. This helps to keep a generic, non mobile-specific core network, which reduces costs and allows maximum flexibility in the provision of mobility services on fixed networks.

MSDP (Mobile Service Data Point): it can be seen as a node of the UMTS Distributed Database (DDB). It stores information concerning terminal, subscriber, network, location data, and so forth. It also contains the functionality related to the communication control in the DDB.

LE (Local Exchange): it represents the existing switching infrastructure upon which the UMTS services and procedures are supported.

2.2. UMTS features

A brief description of the peculiar features of the UMTS is given in the following subsections.

2.2.1. Radio interface. A radio interface is characterized by a frequency, a modulation scheme (e.g. FM, PSK) and an access technique scheme (e.g. TDMA, CDMA). The access technique scheme defines two kinds of channels: logical channels, characterized by the logical meaning of the carried information, and physical channels, constituting the physical radio support on which the information is transported. Both logical and physical channels are classified into dedicated channels, common control channels and system control channels. A description of each is given below. Information related to a logical channel is mapped, possibly after an encoding process, on the corresponding physical channels. Logical channels may be multiplexed in a flexible way, and can be mapped either on a single physical channel or on multiple physical channels.

Dedicated channels. Dedicated channels are uniquely assigned to a specific link between MT and BTS (up-link towards the BTS or down-link towards the MT) when a connection is established. Logical dedicated channels are classified into traffic channels and dedicated control channels. Traffic channels carry the user data to be transmitted on the radio interface (i.e. encoded speech, video or data). Dedicated control channels carry control information to be exchanged between MT and network (connection control, mobility control, radio link control). Traffic channels and dedicated control channels can be multiplexed on a single physical dedicated channel; this paper makes this assumption.

Common control channels. The broadcast channel, the paging channel and the access grant channel are logical common control channels used on down-link only and available to all MTs. They broadcast information specific to the radio cell and network, as well as paging and access grant messages. The logical common control channels can be multiplexed on a single physical common control channel. The only logical common control channel on the up-link is the random access channel used by a MT for initial access to the system.

System control channels. System control channels are provided on the down-link to facilitate BTS monitoring and identification, synchronization and channel estimation in the MT.

Common control channel and system control channel are not administered by the connection management functions, since they are not assigned to a specific link between a MT and a BTS.

2.2.2. Handover. Handover is a procedure involving a change of physical radio channels and/or terrestrial channels involved in an association between a MT and the fixed network, while maintaining this association. The change in physical channels may be required due to movement of the terminal or capacity reasons, or ordered by the network management functions.

2.2.3. Multi-bearer. The multi-bearer technique is introduced in the UMTS system in order to support new generation services with an increasing bit rate demand. A flow of $n*y$ KBit/s can be split, for example, in n flows related to n bearers supporting each y KBit/s. The n flows are then recombined in order to obtain the original information stream in the MT and in the fixed network. The point in the network where the multi-bearer is resolved can be either in the BTS itself or in a special network element (UMS) in the fixed network. This paper assumes that multi-bearer is always resolved in UMS entities.

2.2.4. Macro-diversity. The macro-diversity technique is introduced in order to improve the quality of the radio connection between the MT and the UMTS system. It allows the MT to be connected through more than one BTS at the same time for the same connection in progress. In this way, the same information (traffic and signaling data) is carried in parallel along a number of different paths between two common points, one located in the MT and the other in the fixed network. In these points, the received information is combined according to suitable algorithms and synchronization requirements, while the information that has to be sent is replicated onto the different paths. Macro-diversity is resolved in the UMS.

Multi-bearer and macro-diversity techniques, as well as the handover procedure, are only applied to dedicated channels.

3. The TINA approach

The Telecommunications Information Networking Architecture (TINA) integrates information technology techniques (distributed processing, object-oriented modeling and design) with the typical telecommunications approach, giving a framework for flexible design of advanced telecom systems. In particular, UMTS can profit from the TINA approach of separating service and delivery aspects, maintaining a uniform computational (software) model.

3.1. TINA methodology

The TINA design methodology adopts the Open Distributed Processing (ODP) viewpoint separation [10]. First, an **Information Model** is specified, defining how the system information is represented and organized. Then, a **Computational Model** is derived, capturing the processing aspects on the information itself; this forms the basis for implementation. Both models are object-oriented, that is are expressed in terms of **Information Objects**, the atomic units of information, and **Computational Objects**, the atomic units of processing, respectively. Computational Objects interacts by means of **Computational Interfaces**, accessed through a Distributed Processing Environment (DPE). Information modeling is described in detail in [4], computational modeling in [5]. The information model for the TINA network aspects is the TINA Network Resource Information Model (NRIM), described in [3]. It defines the information object classes (types) that model network resources. Correspondingly, the computational model for the network functions constitutes the TINA Network Resource Architecture [2], whose main part is the Connection Management Architecture. The TINA Connection Management Architecture (CMA) is based on computational objects associated to the information objects specified in the NRIM. Each computational object type in the CMA manages (i.e. controls and administers) a set of information objects classes of the NRIM. Each computational object instance has the same kind of relationship to information object instances. In this way a mapping is established between the ODP Information and Computational viewpoints.

3.2. Network resource information model

As currently defined, the TINA Network Resource Information Model does not support macro-diversity and multi-bearer features. The NRIM is the basis for the model proposed in Section 4.

The TINA NRIM provides a view of the transmission and switching equipment independently of the specific realization technology [3]. Network elements are modeled as information objects [4] and aggregations of information objects. The NRIM is a pure resource model, independent of the functional architecture of the resource management functions.

A network modeled with the TINA NRIM (TINA

network), viewed by a specific stakeholder offering connectivity services (Connectivity Provider), is called a Connectivity Layer Network (CLN). Furthermore, the NRIM relies on two concepts, first defined in ITU-T [9], that enable a structured decomposition of a CLN: **layering** and **partitioning**.

Layering defines different **Layer Networks** (LNs), representing all resources concurring in the delivery of a given *characteristic information*, i.e. of information in a given format; examples of characteristic information are ATM VP, ATM VC, TCP/IP. Layer networks can have two types of relationship with each other: **client-server** and **peer-to-peer**. In the client-server relationship, the server LN provides the underlying transport for the client LN; for example, the situation of a TCP/IP network on top of ATM VCs is modeled by the ATM VC LN, taking the server role, and the TCP/IP layer network, taking the client role. In the peer-to-peer relationship, information flows from one LN to the other by undergoing adaptation; for example, the situation of an information flow between an ATM VC and a Frame Relay LNs is modeled by two peer-to-peer LNs.

Partitioning is applied to each layer network independently. It divides each LNs into **subnetworks** interconnected by **links**. In this way the network topology is modeled. Partitioning into subnetworks is recursive, i.e. a subnetwork can be divided into lower-level subnetworks, and so on. At the end of this decomposition process, there are "atomic" subnetworks, which correspond to network elements.

In order to master complexity, the model is divided into **fragments**, each defining object classes related to a specific problem domain. The fragments considered in this paper are the Network Fragment, the Connectivity Fragment and the Termination Point Fragment; a short description of each is given below.

3.2.1. Network fragment. The Network Fragment models the overall structure of a TINA network using the layering and partitioning concepts. The following information object classes are defined in this fragment: Customer Premises Equipment (CPE), Connectivity Layer Network (CLN), Layer Network (LN), Subnetwork (SNW), Link, and Topological Link (TL). The CPE, CLN, LN and SNW classes reflect the concepts explained above; the Link and TL classes are used to model interconnection between subnetworks, from a virtual and physical point of view, respectively.

3.2.2. Connectivity fragment. The Connectivity Fragment models connectivity across a TINA network. The information object classes relevant to the paper defined in this fragment are: Network Flow Connection (NFC), modeling an end-to-end connection between CPEs across a CLN; Trail, modeling and end-to-end connection within a layer network; Subnetwork Connection (SNC), modeling a switched connection across a subnetwork (point-to-point or involving multipoint topologies); Link Connection (LC), modeling a connection between two subnetworks; Tandem Connection (TC), modeling a chain of subnetwork connections and link connections; Edge, modeling a leg of a subnetwork connection. Furthermore, in order to model multipoint topologies, the Root and Branch relationships (modeled as object classing in accordance with [11]) are defined between a SNC and its edges.

3.2.3. Termination point fragment. The Termination Point Fragment models termination points in a connectivity layer network. This fragment models the termination points at the network edge, as well as inside the network (such as terminations of switching systems). The relevant object classes defined in this fragment are Network Flow End Point (NFEP), Network Trail Termination Point (NWTTP), Network Connection Termination Point (NWCTP) Topological Link termination Point (TLTP) and Link Termination Point (LTP), which terminate network flow connections, trails, link connections/subnetwork connections, topological links and links, respectively. The Stream Flow End Point (SFEP) is also defined, to model sources and sinks of information flows.

4. NRIM for the UMTS radio access system

This section describes the application of the TINA NRIM to the UMTS. The paper presents a general model, which describes virtually any UMTS system. Subsets of this model can be used for systems where only some of the UMTS features are supported (e.g. only multi-bearer or only macro-diversity).

First, the necessary enhancements to the TINA NRIM are introduced in order to make it applicable to mobile systems; subsequently, the resulting modified model is presented.

4.1. Extensions to the TINA NRIM for mobile networks

In order to use the TINA NRIM to model UMTS systems, some extensions to it are required. They allow to describe the multi-bearer and macro-diversity features, explained in Section 2.2. These extensions impact both Network and Connectivity Fragment. Otherwise, the TINA NRIM is applied as defined in [3].

4.1.1. Network fragment. In this fragment, the **Multi-Bearer Subnetwork (MB-SNW)** and **Macro-Diversity Subnetwork (MD-SNW)** object classes are introduced, by inheritance from the Subnetwork class, to model network elements specific for the UMTS. These classes model the systems, located in the MT and in the UMS, that resolve multi-bearer and macro-diversity, respectively. Note that the mobile terminal (MT) is modeled by information objects of the Subnetwork class. Subnetworks modeling the mobile terminal can be created and deleted dynamically in a layer network domain, when a mobile terminal accesses the system; this does not happen to subnetworks as defined in [3].

4.1.2. Connectivity fragment. In order to model macro-diversity and multi-bearer connections, it is necessary to allow connectivity in a layer network to be provided by *parallel connections*. This means that a trail in a layer network can be composed by multiple correlated chains of SNCs and LCs. In case of macro-diversity, the different chains composing a trail carry a copy of the same information; in case of multi-bearer, the total information is the sum of the information carried by each chain.

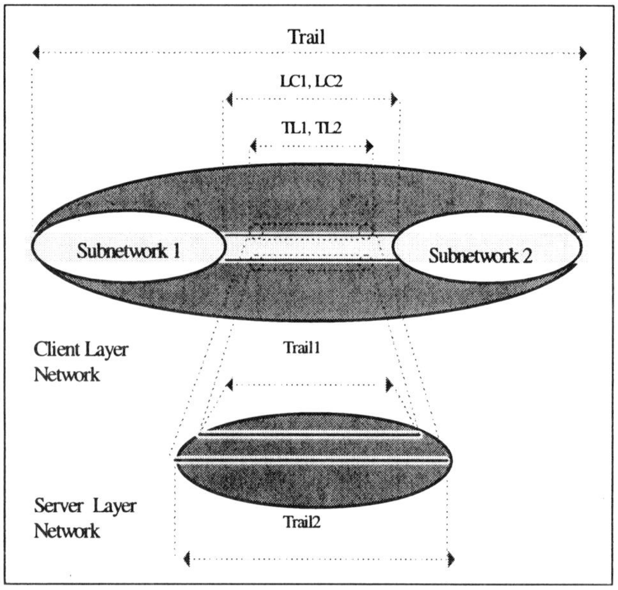

Figure 2. Parallel connections in a trail.

For simplicity, the following constraint is also introduced: a link connection (LC) in a client LN maps to exactly one trail in a server LN. This means that a LC in the client LN corresponds to exactly one topological link which in turn corresponds to a trail in the server layer network[1] (see Figure 2).

In order to model assembling and disassembling of multi-bearer and macro-diversity flows, new objects classes have to be introduced, as shown in Figure 3. They specialize the subnetwork connection and edge classes. Note also that the exact connection topology is not static, but can change during the connection lifetime because of handover.

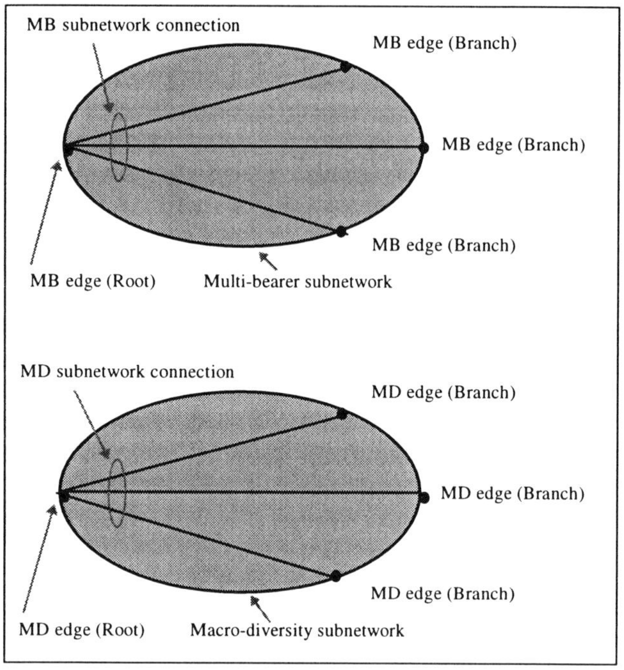

Figure 3. Multi-bearer subnetwork connection and macro-diversity subnetwork connection.

The top diagram represents a subnetwork connection in a multi-bearer subnetwork. The bottom diagram represents a subnetwork connection in a macro-diversity subnetwork. In both cases, the Root edge is bound to the NWCTP where the "compound" flow is handled, while the Branch edge is bound to the NWCTP where the "component" flows are handled. The topology attribute in these specialized subnetwork connection classes has to take the value `point-to-multipoint bidirectional` (not supported by [3]), corresponding to a bidirectional flow in multi-bearer or macro-diversity.

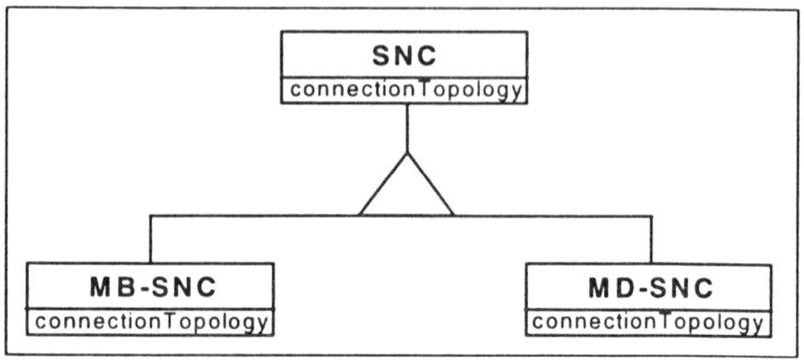

Figure 4. OMT inheritance diagram for the Subnetwork Connection (SNC) information object class.

[1] There is an exception to this constraint, discussed in Section 4.2.5: multiple LCs in the RAS logical LN can be multiplexed over a single TL in that layer network corresponding to one trail in the Physical Radio LN.

Figure 4 shows the inheritance diagram (in OMT notation, see [11]) for the SNC; similar diagrams could be drawn for the Subnetwork and Edge object classes, also specialized for multi-bearer and macro-diversity.

Modeling the handover procedure does not require any additional object class (handover will be described in detail in Section 7).

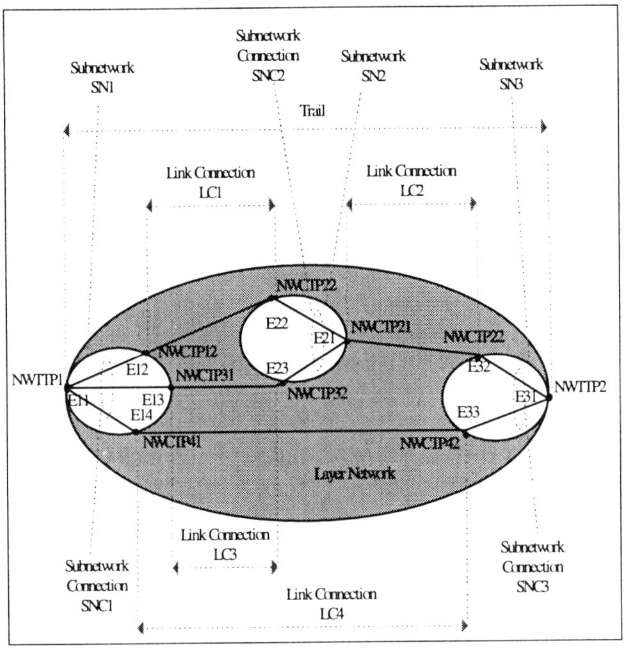

Figure 5. Usage of MB or MD subnetworks.

Figure 5 shows an example of a trail composed of multiple parallel component paths; the example applies to both the multi-bearer and the macro-diversity situation[2]. The trail in figure connects the network trail termination points NWTTP1 and NWTTP2. The root edge is associated to one, the branch edge to the other termination point; information flows in both directions point-to-point. Let us assume that the subnetwork SN1 represents the mobile terminal, while SN2 and SN3 represent two UMS systems where multiple macro-diversity or multi-bearer flows are resolved. The subnetwork connection SNC1 models the splitting of the information flow into three parallel paths at the network termination point NWCTP11 (colocated with NWTTP1). The first path goes through NWCTP12, NWCTP22 and NWCTP21; the second goes through NWCTP31, NWCTP32 and NWCTP21, where it merges with the first one; the third path goes through NWCTP41, NWCTP42 and merges with the other two at NWCTP2. The subnetwork connection SNC2 models the merge of the two upper paths; similarly, SNC3 models the merge of the "compound" upper path with the lower path.

[2] But the two features are not visible in the same layer network, as it will be explained in Section 4.2.

Since we assume that the trail is bidirectional (as it is the case for mobile systems) flow splitting in one direction corresponds to merging in the opposite direction. The multiple subnetwork connection legs are associated to corresponding branch edges: for example, in SNC1 the root edge E11 is bound to NWCTP1, and branch edges E12, E13 and E14 are bound to NWCTP12, NWCTP31 and NWCTP41, respectively. Figure 6 shows all information object instances and their relationships in OMT-like notation [11].

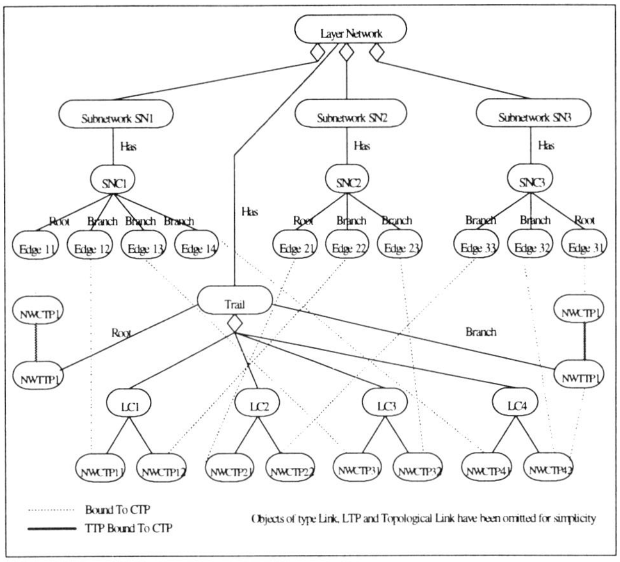

Figure 6. Instance diagram for Figure 5.

4.2. Resulting model

The peculiar structure of a RAS and of its connections suggests to rely heavily on **layering**, rather than partitioning, to model the simultaneous presence of multi-bearer and macro-diversity elements, of fixed and mobile connections. In fact, layering enables to take into account all information adaptation (in terms of rate, coding and format) taking place to resolve macro-diversity and multi-bearer and to go from radio transport to transport in the fixed part of the access network.

The complete NRIM for the UMTS Radio Access System is shown in Figure 7 and Figure 8. For drawing simplicity, Figure 7 does not show objects of the connectivity fragment (i.e., dependent on the established connections); conversely, Figure 8 omits termination point fragment and link objects, showing connectivity fragment objects.

Figure 7 shows that the mobile terminal equipment is modeled as a CPE part (with stream flow end points modeling the source and sink of the information flows) *plus* a MT part, which is a virtual extension of the network domain. In accordance with [3], the CPE part contains stream flow end points (SFEPs), which are mapped dynamically onto network flow end points (NFEPs) in the connectivity layer network. The connectivity layer network view (unchanged with respect to [3]) corresponds to the top-level view. Within the connectivity layer network, the various layer networks defined are associated to the different characteristic information types involved, that is to the different information formats.

A Radio Access System consists in five different layer networks, related by interworking and client-server relationships:

- **RAS Multi-Bearer Layer Network (RAS-MB LN)**: this LN deals with the characteristic information associated to the multi-bearer feature;
- **RAS Macro-Diversity Layer Network (RAS-MD LN)**: this LN deals with the characteristic information associated to the macro-diversity feature;
- **RAS Logical Layer Network**: this LN deals with the characteristic information associated to the logical radio channels;
- **Physical Radio Layer Network**: this LN deals with the characteristic information associated to the physical radio channels over the radio interface;
- **Fixed Layer Network**: this LN deals with the characteristic information associated to connections over the fixed network in the RAS.

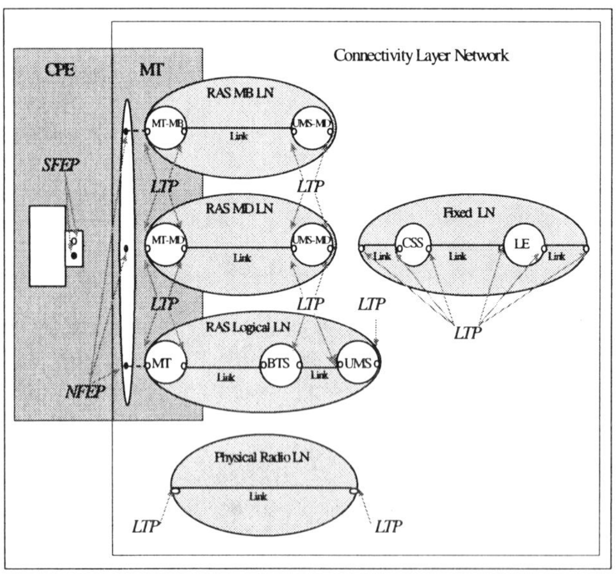

Figure 7. Connectivity layer network.

The various client-server relationships between these layer networks model the adaptation functions that are necessary to convey information flows over the radio interface and to handle the macro-diversity and multi-bearer features. In a client-server relationship, the server layer network takes as "payload" the information flow from the client layer network, and adapts it to the type of transport specific of the server. This adaptation takes the form of addition of identifiers or of quality information, splitting/merging of data units, and so forth.

In each layer network (except for the Physical Radio LN), UMTS network elements are modeled by subnetworks. Since the same physical network element (MT, UMS...) performs a variety of functions, related to multiple characteristic information types, it corresponds to different subnetworks in different layer networks.

Links represent the "possibility" to establish link connections, that is the available bandwidth between network elements, both in the mobile and in the fixed network. In accordance with [3], link termination points (LTPs) contain network connection termination points (NWCTPs), not shown for simplicity in Figure 7. End-to-end access points to a layer network are modeled by network trail termination points (NWTTPs), bound to NWCTPs, also not shown in Figure 7.

Figure 8 presents the model of connectivity aspects, focusing on client-server relationships between layer networks. To fix ideas, we assume that both multi-bearer and macro-diversity are present, and macro-diversity is resolved earlier (i.e. closer to the MT) than multi-bearer. In this case, the RAS-MB LN is client of the RAS-MD LN, which is client of the RAS Logical LN. The RAS Logical LN, in turn, is client of the Physical Radio LN for connections over the radio interface, and of the Fixed LN for terrestrial connections. In case the multi-bearer and/or macro-diversity features are not supported by the particular system, the corresponding layer network is not present; in case multi-bearer is resolved earlier than macro-diversity, the client/server relationship between the two layer networks is reversed.

Each layer network defined for the UMTS Radio Access System is described below.

4.2.1. RAS multi-bearer LN (RAS-MB LN). The RAS Multi-bearer LN hides the number of bearers that constitute the end-to-end connection (trail) in the RAS-MB LN. In this layer network, the functionality related to the multi-bearer feature is modeled introducing the multi-bearer subnetwork class (see Section 4.1.2). This object class, together with the termination point classes, models the network element that performs the splitting of the input streams into one or more output streams and that combines more input streams into one output stream. Splitting and merging are performed by segmenting/reassembling information from all bearers and converting the overall format. The multiple bearers are associated to a single trail, which therefore consists of multiple parallel connection segments. These segments originate and terminate in a multi-bearer subnetwork, and are modeled by chains of SNCs and LCs each associated to one bearer.

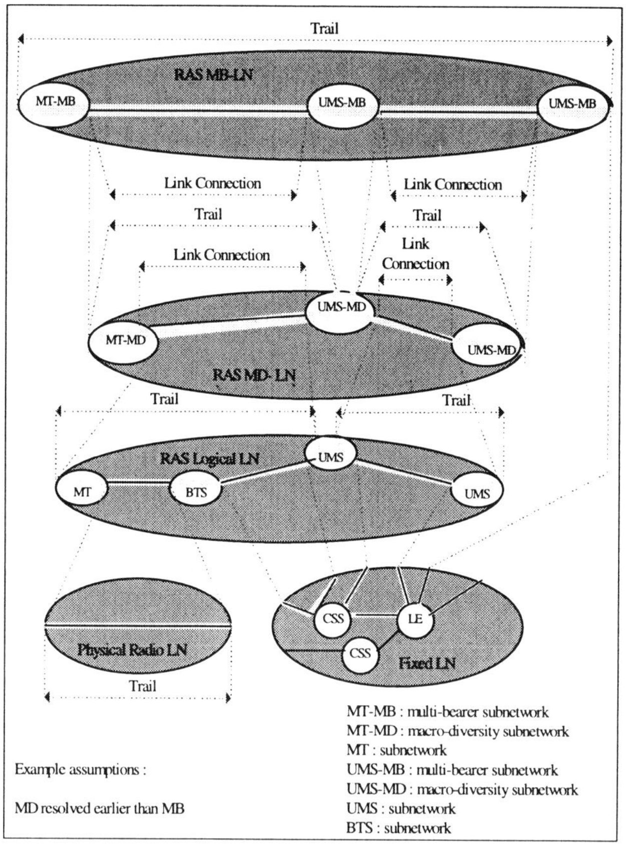

Figure 8. RAS network resource information model.

4.2.2. RAS macro-diversity LN (RAS-MD LN). The RAS Macro-diversity LN hides the fact that each bearer of a multi-bearer connection can be in macro-diversity. In this layer network, the macro-diversity subnetwork class is introduced, to model functionality related to the establishment of a macro-diversity connection (see section 4.1.2). This object class, together with the associated termination point classes, models the network element performing the multicasting of input streams into identical output streams, and recreating a stream in output by selecting the best quality among all available inputs. The multiple copies of the information stream are associated to a single trail, consisting of multiple parallel connection segments. The difference with respect to the RAS-MB LN is that macro-diversity segments contain copies of the same information, whereas multi-bearer segments contain a different portion of the total information.

4.2.3. RAS Logical LN. The RAS Logical LN provides transport service to the RAS Macro-diversity LN or to the RAS Multi-bearer LN. It does not depend on how logical channels are mapped onto physical channels. The end points of this LN allow to identify the logical channels described in section 2.2.1.

In this layer network no modifications to the SNW object class as defined in [3] are necessary. Connections in these subnetworks (modeled by plain SNCs as defined in [3]) allow to map a logical channel, represented by a trail in this layer network, onto a physical channel, represented by a trail in the Physical Radio LN, and so by a Topological Link in the RAS Logical LN. This association can be changed during an intra-cell handover, when the same logical channel is mapped onto a new physical channel (see Section 7).

4.2.4. Fixed LN. The Fixed LN represents the transport resources of the fixed part of the RAS and of the core network. It is modeled independently of the technology used over the fixed network. Its server layer networks (not considered in the paper because they are not specific to mobile systems) model different transport technologies (ATM VC, PSTN, etc.).

4.2.5. Physical Radio LN. This LN provides transport for the RAS Logical LN, modeling how logical channels are mapped onto physical channels. Its characteristic information depends on the information encoding on the physical channels. The end points of this LN allow to identify the physical channels described in section 2.2.1. More than one link connection in the RAS Logical LN can be provided on a single topological link, corresponding to one trail in the Physical Radio LN (see Figure 9).

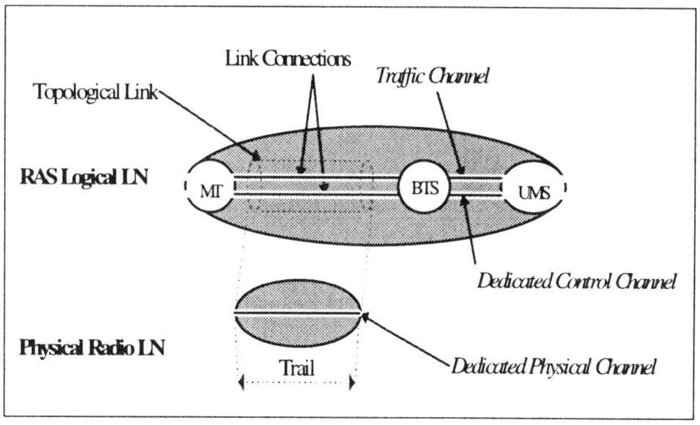

Figure 9. RAS logical/physical radio LNs client-server relationship.

This means that, in the RAS Logical LN, the constraint that one link connection in the client LN corresponds to exactly one trail in the server layer network is not valid if the latter is the Physical Radio LN. In fact, a trail in the Physical Radio LN corresponds to a physical channel, while a trail in the RAS Logical LN corresponds to a logical channel. Therefore, all logical channels that are carried over one physical channel on the radio interface are represented in the RAS Logical LN by link connections bundled together in a topological link; the topological link corresponds to a trail (physical channel) in the Physical Radio LN.

These topological links in the RAS Logical LN are created and deleted dynamically.

5. Example of RAS resource information modeling

In order to clarify the NRIM for the UMTS Radio Access System, an example is presented in Figure 10. The MT has a connection in multi-bearer; the two bearers are indicated with 1 and 2. Bearer 2 is in macro-diversity (2A and 2B). The "single" bearer (1) is through BTS-1; the macro-diversity bearer (2A and 2B) is through BTS-2 and BTS-3. Macro-diversity is resolved in UMS-2; multi-bearer is resolved in UMS-3.

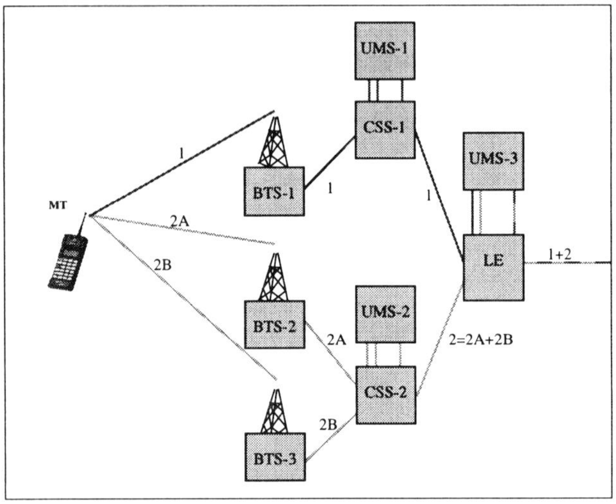

Figure 10. Example of connection with the network.

The resulting model for the three upper layers is shown in Figure 11. Since macro-diversity is resolved earlier (i.e. closer to the mobile terminal) than multi-bearer, the RAS-MB LN is client of the RAS-MD LN; the latter is client of the RAS Logical LN; the Physical Radio LN is not shown for simplicity. In the RAS-MB LN there is one trail, composed by the parallel paths 1 and 2. The MB-SNCs in the MT-MB and UMS3-MB subnetworks model the splitting/merging of the two bearers. The remaining portion of each parallel path is composed by LCs and "normal" SNCs. Each LC in the RAS-MB LN corresponds to a trail in the RAS-MD LN[3]. Let us examine the SNC between MT-MB and UMS2-MB (RAS-MB LN, path 2): this corresponds, in the RAS-MD LN, to the trail between MT-MD and UMS2-MD. This trail models a macro-diversity connection, composed by the paths labeled 2A and 2B; it is made of two MD-SNCs (in the MT-MD and UMS2-MD, respectively) and by two LCs corresponding to 2A and 2B, respectively. Finally, each LCs in the RAS-MD LN corresponds to a trail in the RAS Logical LN[3].

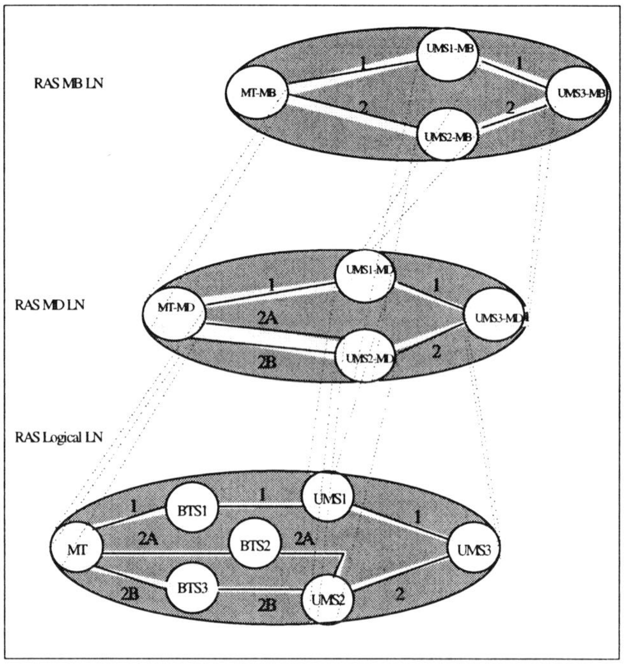

Figure 11. Network resource information model for the RAS example.

6. Connection setup

This section describes the steps that are necessary to set up a connection in a UMTS system, between a mobile terminal and another terminal, either mobile or fixed. The steps are described for the information modeling point of view, that is in terms of creation/deletion of information object instances.

First (Section 6.1) we describe the basic connection setup; then, the macro-diversity and multi-bearer features (Section 6.2 and 6.3, respectively).

6.1. Set-up of a connection during system access

The following steps are executed during the set-up of a connection, supposing that the system supports both macro-diversity and multi-bearer and that macro-diversity is resolved earlier (i.e., nearer to the mobile terminal) than multi-bearer. During the initial setup only a single connection is established. Other macro-diversity paths or other bearers are added in a later phase, to increase the capacity and the quality of the connection.

[3] Actually, it corresponds to exactly one link, which is configured by exactly one topological link, which corresponds to exactly one trail in the server layer network.

Preconditions. The following information objects are already present:

- the connectivity layer network and all LN objects;
- a MB subnetwork in the RAS-MB LN for each UMS;
- a MD subnetwork in the RAS-MD LN for each UMS;
- a subnetwork in the RAS Logical LN for each BTS and UMS;
- a subnetwork in the Fixed LN for each CSS and LE.
- NFEPs for each fixed terminal involved.

Assumptions. The Fixed LN and all its server LNs have already been configured (this paper does not describe the steps necessary to configure them, which are exposed in [2]). As such, we assume that the following information objects are also already present:

- topological links in the Fixed LN between the BTSs and the UMSs and between the UMSs at CSS level and the UMS at LE level;
- NWCTPs for all topological links in the Fixed LN.

It is also necessary that the LTP, NWCTP and NWTTP objects modeling the BTS edge of the radio interface are already present in the Physical Radio LN. The initial state is shown in Figure 12.

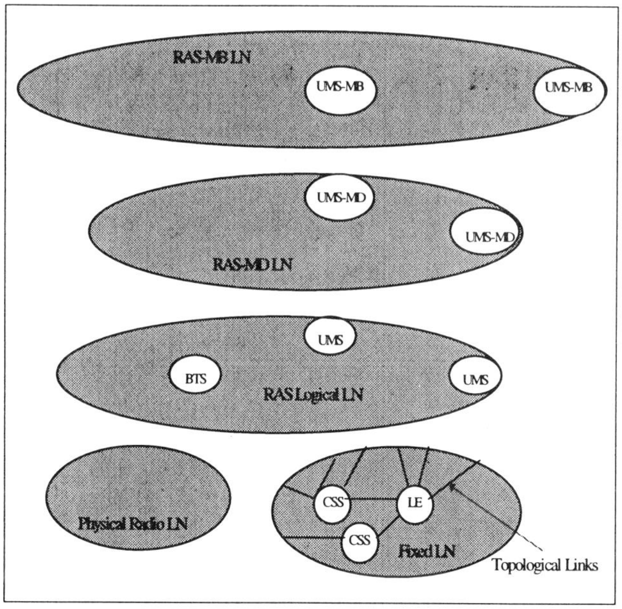

Figure 12. System access connection setup: initial state.

Step 1: Configuration of the layer networks and channel assignment for the radio link. (Figure 13).

- A bidirectional NFEP corresponding to the MT is created in the Connectivity Layer Network.
- A MB subnetwork corresponding to the MT is created in the RAS-MB LN.
- A MD subnetwork corresponding to the MT is created in the RAS-MD LN.
- A subnetwork corresponding to the MT is created in the RAS Logical LN.
- In the Physical Radio LN, a LTP and its NWCTPs and associated NWTTPs are created at the MT side; the NWTTPs are associated to the NFEP in the MT. A Link is created between the LTP on the MT side and that on the BTS side.

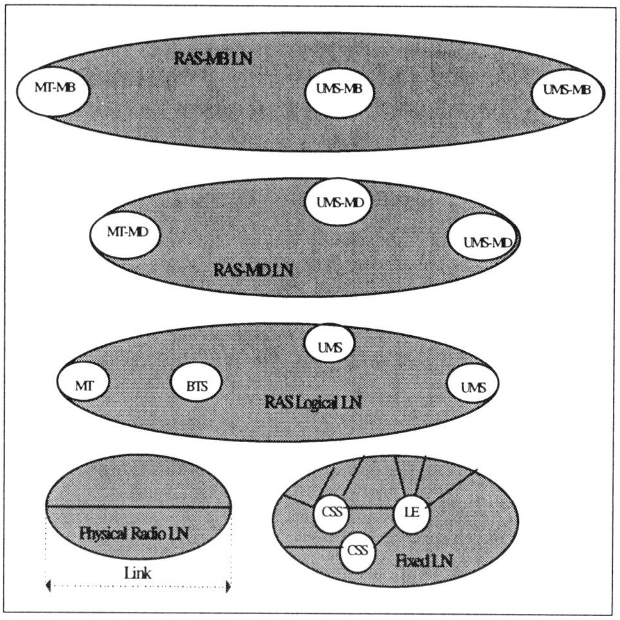

Figure 13. System access connection setup: step 1.

Step 2: Connection establishment in the Physical Radio and Fixed Layer Networks. (Figure 14).

- A Link Connection (LC), corresponding to a physical radio channel, is created between the MT and the BTS in the Physical Radio LN (we assume that only one physical radio channel is set up) relying on the link set up in Step 1.
- A Trail connecting the NWTTPs in the MT and in the BTS is established in the Physical Radio LN, composed only by the LC set up before. As a result, a topological link (TL) with its termination points (TLTPs) is created in the RAS Logical LN between the MT and the BTS subnetwork.
- Other trails connecting the BTS to the nearest UMS and this to the other UMSs are established in the Fixed LN, relying on non-mobile specific layer networks (e.g. ATM VC layer network). As a result, the corresponding TLs and TLTPs are created in the RAS Logical LN.
- The newly created TLs and TLTPs in the RAS Logical LN are used to configure (create) a Link object and two Link Termination Points (LTPs) in the same LN. The NWCTPs in the LTP on each side and the associated NWTTPs are also created.

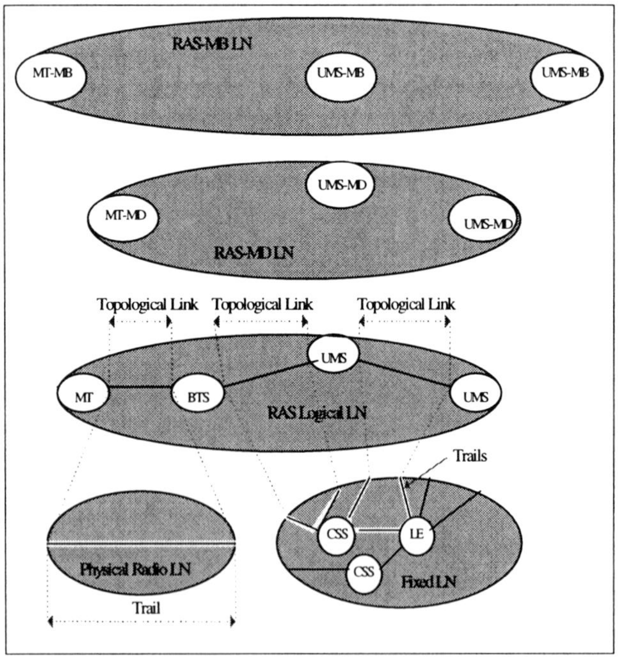

Figure 14. System access connection setup: step 2.

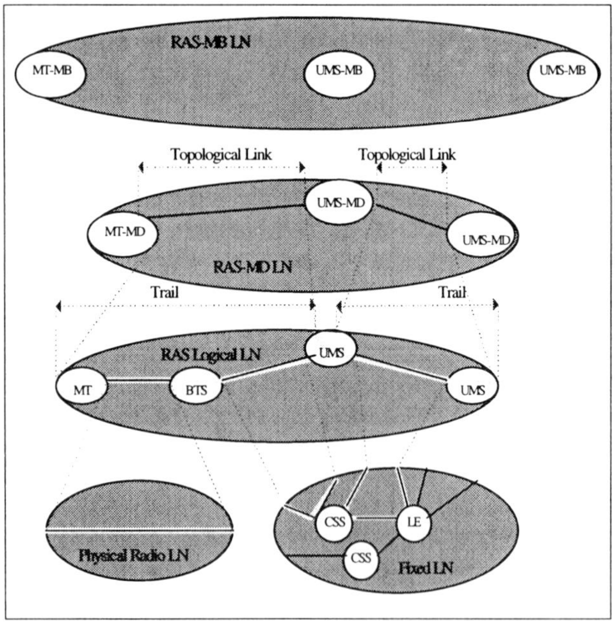

Figure 15. System access connection setup: step 3.

Step 3: Connection establishment in the RAS Logical Layer Network. (Figure 15).

- A Trail in the RAS Logical LN between the NWTTPs in the MT subnetwork and in the UMS subnetwork at CSS level is established; this requires the creation of subnetwork connections (SNCs) in the MT, BTS and UMS subnetworks, and LCs between them. The LCs are set up between the NWCTPs configured in step 2, and are contained in the links created in step 2, both for the radio part (server layer: Physical Radio LN) and for the fixed part (server layer: Fixed LN). As a result of the trail setup, a TL and its TLTPs are created in the RAS-MD LN.
- A Trail in the RAS Logical LN between the NWTTPs in the UMS subnetwork at CSS level and in the UMS subnetwork at LE level is established, by setting up SNCs and LCs as above (now involving only the fixed network). Also here, the corresponding TLs, and TLTPs are created in the RAS-MD LN.
- The newly created TLs and are used to configure (create) Links (one for each TL), and LTPs (one for each side of the link) in the RAS-MD LN. One NWCTP for each LTP is also created. NWTTPs are created and associated to the NWCTPs at the layer network edge.

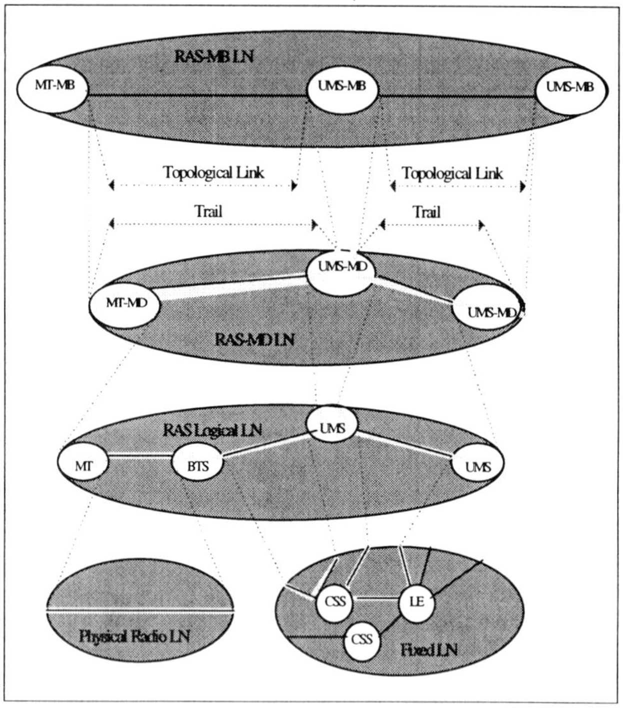

Figure 16. System access connection setup: step 4.

Step 4: Connection establishment in the RAS-MD layer network. (Figure 16).

- A Trail in the RAS-MD LN between the NWTTPs of the MD subnetwork in the MT and of the MD subnetwork in the UMS at CSS level is established. Initially, only one macro-diversity path is present; hence, the trail is composed by MD-SNCs with only one branch in each subnetwork (they can potentially evolve to multiple branches) and LCs contained in the links created in step 3. As a result of the trail setup, the corresponding TLs and TLTPs are created in the RAS-MB LN.

- A Trail in the RAS-MD LN between the MD subnetwork in the UMS at CSS level and the MD subnetwork in the UMS at LE level is established, similarly as above. As a result, the corresponding TLs and TLTPs are created in the RAS-MB LN.
- The newly created TLs and TLTPs are used to configure (create) Links (one for each TL), and LTPs (one for each side of the link) in the RAS-MB LN. One NWCTP for each is also created. NWTTPs are created and associated to the NWCTPs at the layer network edge.

Step 5: Connection establishment in the RAS-MB layer network. (Figure 17).

- a Trail in the RAS-MB LN between the MB subnetwork in the MT and the MB subnetwork in the UMS at LE level is established.
- a Trail in the Fixed LN between the MB subnetwork in the UMS at LE level and the end user (if fixed) is established.

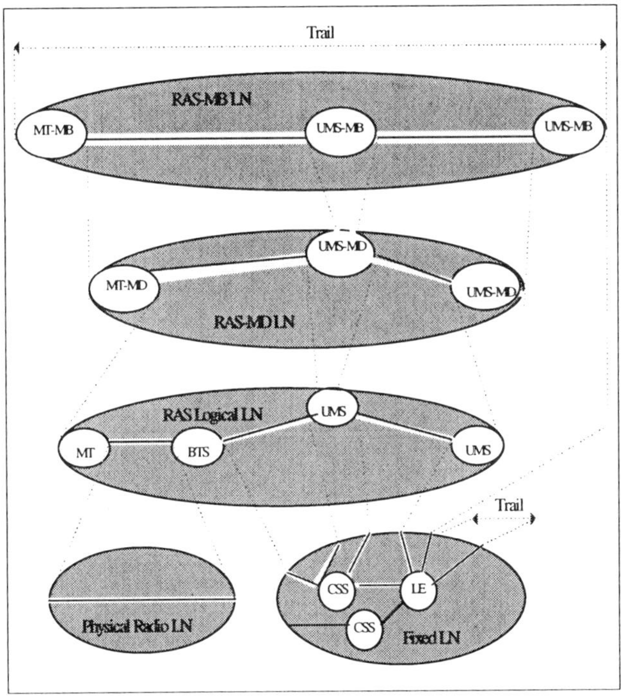

Figure 17. System access connection setup: step 5.

Step 6: Establishment of the end-to-end connection in the Connectivity Layer Network.

- A Network Flow Connection (NFC) is created between the NFEP in the mobile terminal and the NFEP in the other terminal. The NFC contains the trails in the RAS-MB and Fixed LNs set up in step 5. The NFEPs are associated to the corresponding NWTTPs.

6.2. Adding a macro-diversity path

This phase does not have any impact on the MB LN, since we assume that macro-diversity is resolved earlier than multi-bearer.

Step 1: New channel assignment for the radio link.

- A new Link between the MT and the new BTS is created in the Physical Radio LN, corresponding to a new physical radio channel.

Step 2: New connection establishment in the Physical Radio and Fixed layer networks.

- A new Trail connecting the MT and the new BTS is established in the Physical Radio LN. As a result, a TL with its TLTPs is created in the RAS Logical LN between the MT and the new BTS subnetwork.
- Other trails connecting the new BTS and the UMSs, if required, are established in the Fixed LN. The corresponding TLs and TLTPs are created in the RAS Logical LN.
- The newly created TLs and TLTPs are used to configure (create) a Link and its LTPs in the RAS Logical LN. The NWCTPs in the LTP on each side and the associated NWTTPs are also created.

Step 3: New connection establishment in the RAS Logical layer network.

- A new Trail in the RAS Logical LN between the MT subnetwork and the UMS subnetwork at CSS level is established. The corresponding TLs and TLTPs are created in the RAS-MD LN.
- If the new BTS is not connected to the same CSS as the other BTS, a Trail in the RAS Logical LN between the UMS subnetwork at CSS level and the UMS subnetwork at LE level is established. As a result, the corresponding TLs, and TLTPs are created in the RAS-MD LN.
- The newly created TLs and TLTPs are used to configure (create) Links and LTPs in the RAS-MD LN. NWCTPs and NWTTPs are also created.

Step 4: New connection establishment in the RAS-MD layer network.

- In the MD subnetwork in the MT and the MD subnetwork in the UMS (at CSS or LE level, depending on where macro-diversity is resolved) a branch is added to the MD-SNCs. At this point, two parallel connections are present in the same trail in the RAS-MD LN.

6.3. Adding a bearer

The following steps are executed to add a bearer, creating a multi-bearer connection.

Step 1 to 3: Same as Section 6.2.

Step 4: New connection establishment in the RAS-MD layer network.

- A new Trail in the RAS-MD LN between the MD subnetwork in the MT and the MD subnetwork in the UMS at CSS level is established. As a result, the corresponding TLs and TLTPs are created in the RAS-MB LN.
- If the new BTS is not connected to the same CSS as the other, a Trail in the RAS-MD LN between the MD subnetwork in the UMS at CSS level and the MD subnetwork in the UMS at LE level is established. The corresponding TLs and TLTPs are created in the RAS-MB LN.

Step 5: New connection establishment in the RAS-MB layer network.

- In the MB subnetwork in the MT and the MB subnetwork in the UMS (at CSS or LE level, depending on where macro-diversity is resolved) a branch is added to the MB-SNCs. At this point, two parallel connections are present in the same trail in the RAS-MB LN.

7. Handover procedure

Handover procedures can be classified into several types depending upon the network elements involved in their execution. The various handover types are described in Section 7.1, while the impact of the different handover types on each LNs defined in the NRIM is analyzed in Section 7.2.

7.1. Handover types

Handover types are classified according to the network elements involved in the execution of the procedure.

Intra BTS (intra-cell): it consists in a change of radio physical channel within one cell.

Inter BTS - Intra CSS: it involves a switch between radio physical channels belonging to two different BTSs connected to the same CSS.

Inter CSS - Intra LE: it involves a switch between radio physical channels belonging to two BTSs of two UMTS access networks, controlled by the same LE.

Inter LE - Intra TX: it involves a switch between radio physical channels belonging to two BTSs controlled by two different LE, connected to the same transit exchange (TX) in the core network. This type of handover requires handover control functionality in the core network.

Inter TX: it involves a switch from radio physical channel in one cell to radio physical channel in another cell belonging to two UMTS access networks, controlled from two different LE, connected to different transit exchanges. Also this type of handover requires handover control functionality in the core network.

7.2. Examples of handover

In this section some simple examples of handover are shown.

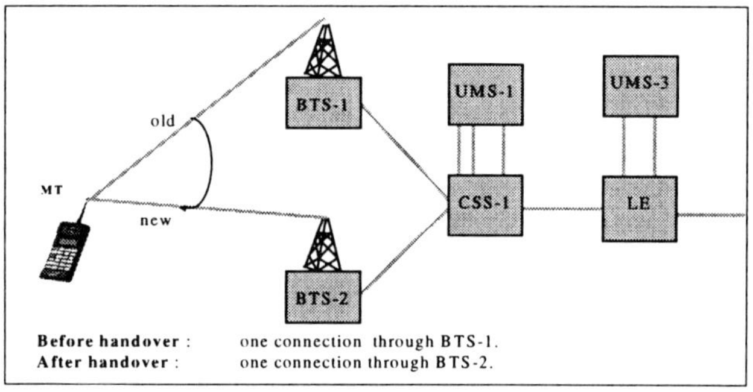

Figure 18. Intra CSS handover for a MT with a single connection.

In Figure 18, a MT has a single connection with the network. Both BTS-1 and BTS-2 are connected to the same CSS, so that the handover is handled by UMS-1 (UMS access), while coding and decoding are always performed in UMS-3 (UMS core).

During handover the following actions are performed:

- New LCs in the RAS Logical LN are established for the connection through the new BTS; they are supported by new trails in the Physical Radio and Fixed LNs.
- New SNCs are established in the UMS-1 subnetwork and in the MT subnetwork, in order to switch on the new path.
- Old SNCs are released.
- Old TLs in the RAS Logical LN are released; the corresponding trails in the Physical Radio LN and in the Fixed LN are released.

The information objects for this example are shown in Figure 19.

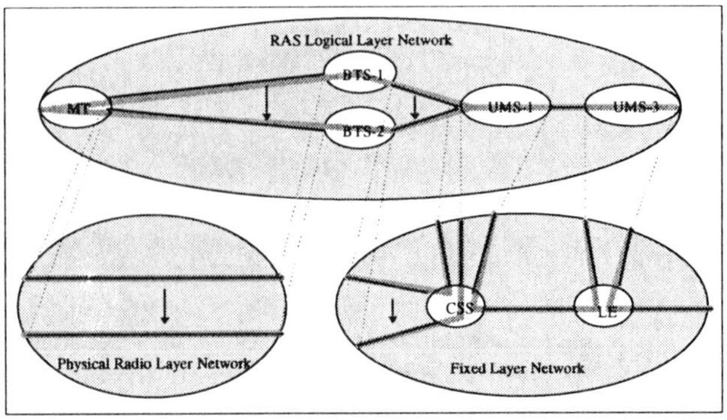

Figure 19. Network resource information model for the example in Figure 18.

In Figure 20, a MT has a single connection with the network. BTS-1 and BTS-2 are connected to different

CSSs, so that the handover is handled by UMS-3 (UMS core).

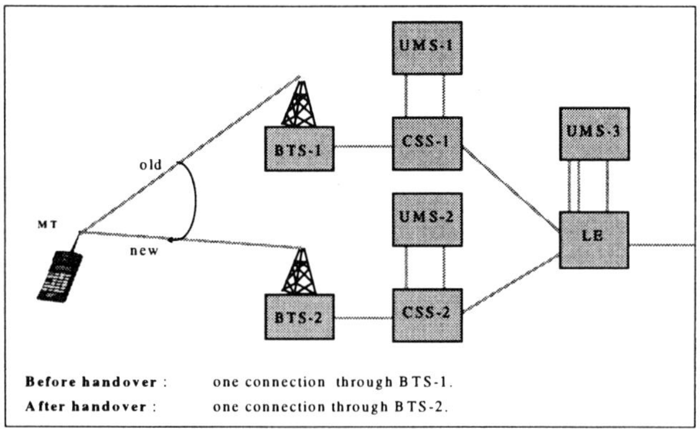

Figure 20. Inter CSS handover for a MT with a single connection.

During handover, similar actions as the previous example are performed. The involved information objects are shown in Figure 21.

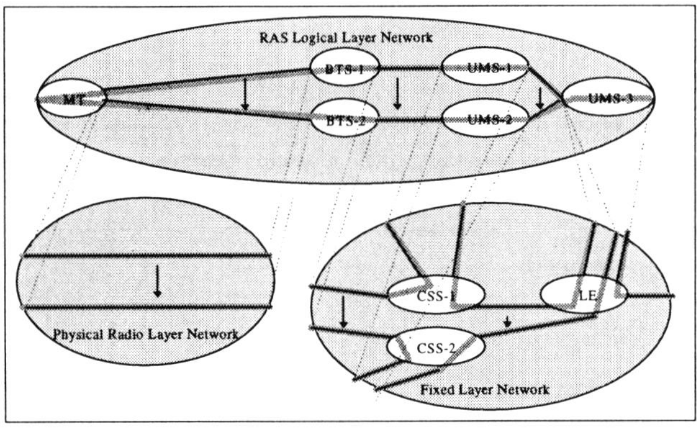

Figure 21. Network resource information model for the example of Figure 20.

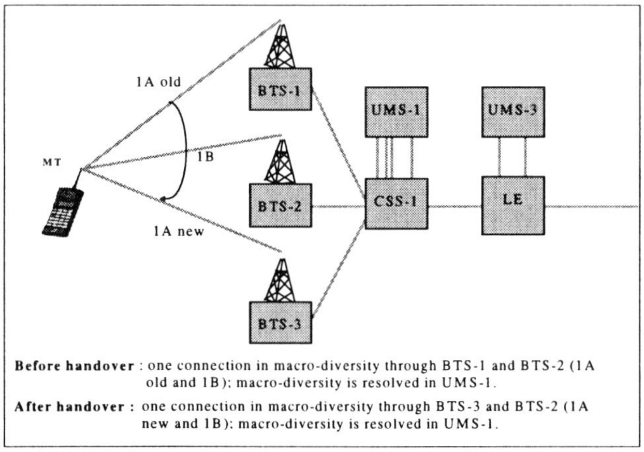

Figure 22. Intra CSS handover for a MT in macro-diversity.

The example in Figure 22 concerns a MT having a macro-diversity connection with the network. All BTSs are connected to the same CSS, so that macro-diversity is handled by UMS-1 (UMS access), while coding and decoding are always performed in UMS-3 (UMS core).

During handover, the following actions are performed:

- A new macro-diversity path is established (see Section 6.2).
- the old path is released (see Section 6.2).

The information objects are shown in Figure 23.

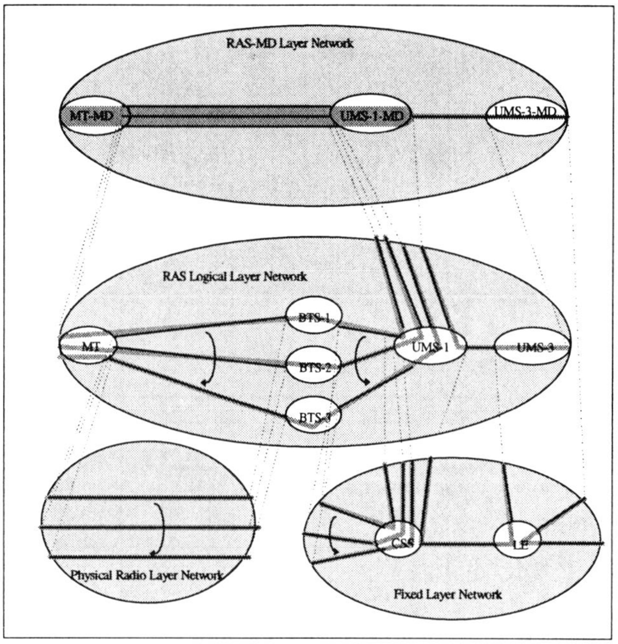

Figure 23. Network resource information model for the example of Figure 22.

7.3. Impact of handover on the different layer networks

The number of LNs involved in a handover depends both on the handover type and on the kind of connection the MT has with the network. Thus, a handover could involve only the Physical Radio LN, or involve all levels. The possible cases are summarized in the following tables.

Type of handover	New Trails	New LCs/SNCs
Intra BTS	Physical Radio LN	-
Inter BTS - Intra CSS	Physical Radio LN / Fixed LN	RAS Logical LN
Inter CSS - Intra LE	Physical Radio LN / Fixed LN	RAS Logical LN
Inter LE - Intra TX	Physical Radio LN / Fixed LN	RAS Logical LN
Inter TX	Physical Radio LN / Fixed LN	RAS Logical LN

Table 1. Impact of handover on LNs for a mobile terminal with a single connection with the network.

Type of handover	New Trails	New LCs/SNCs
Intra BTS	Physical Radio LN	-
Inter BTS - Intra CSS	Physical Radio LN / Fixed LN / RAS Logical LN	RAS-MD LN
Inter CSS - Intra LE	Physical Radio LN / Fixed LN / RAS Logical LN	RAS-MD LN
Inter LE - Intra TX	Physical Radio LN / Fixed LN / RAS Logical LN	RAS-MD LN
Inter TX	Physical Radio LN / Fixed LN / RAS Logical LN	RAS-MD LN

Table 2. Impact of handover on LNs for a mobile terminal in macro-diversity.

Type of handover	New Trails	New LCs/SNCs
Intra BTS	Physical Radio LN	-
Inter BTS - Intra CSS	Physical Radio LN / Fixed LN / RAS Logical LN	RAS-MB LN
Inter CSS - Intra LE	Physical Radio LN / Fixed LN / RAS Logical LN	RAS-MB LN
Inter LE - Intra TX	Physical Radio LN / Fixed LN / RAS Logical LN	RAS-MB LN
Inter TX	Physical Radio LN / Fixed LN / RAS Logical LN	RAS-MB LN

Table 3. Impact of handover on LNs for a mobile terminal in multi-bearer.

Type of handover	New Trails	New LCs/SNCs
Intra BTS	Physical Radio LN	-
Inter BTS - Intra CSS	Physical Radio LN / Fixed LN / RAS Logical LN / RAS-MD LN	RAS-MB LN
Inter CSS - Intra LE	Physical Radio LN / Fixed LN / RAS Logical LN / RAS-MD LN	RAS-MB LN
Inter LE - Intra TX	Physical Radio LN / Fixed LN / RAS Logical LN / RAS-MD LN	RAS-MB LN
Inter TX	Physical Radio LN / Fixed LN / RAS Logical LN / RAS-MD LN	RAS-MB LN

Table 4. Impact of handover on LNs for a mobile in macro-diversity and multi-bearer (macro-diversity under multi-bearer).

Type of handover	New Trails	New LCs/SNCs
Intra BTS	Physical Radio LN	-
Inter BTS - Intra CSS	Physical Radio LN / Fixed LN / RAS Logical LN / RAS-MB LN	RAS-MD LN
Inter CSS - Intra LE	Physical Radio LN / Fixed LN / RAS Logical LN / RAS-MB LN	RAS-MD LN
Inter LE - Intra TX	Physical Radio LN / Fixed LN / RAS Logical LN / RAS-MB LN	RAS-MD LN
Inter TX	Physical Radio LN / Fixed LN / RAS Logical LN / RAS-MB LN	RAS-MD LN

Table 5. Impact of handover on LNs for a mobile in macro-diversity and multi-bearer (macro-diversity over multi-bearer).

8. Conclusions and further work

This paper considers the application of the TINA Network Resource Architecture to next-generation mobile systems, analyzing the European initiative UMTS.

The TINA approach can bring significant added value to the management of connections in the UMTS Radio Access System (RAS). In order to apply TINA, it is not sufficient to map functional entities on computational objects, but it is necessary to define a Network Resource Information Model (NRIM). The paper defines an enhancement to the TINA Network Resource Information Model that can be used as a general model for UMTS systems. Enhancements (in terms of strict superset of the NRIM defined by TINA-C) are necessary to take into account specific features of mobile systems: multi-bearer, macro-diversity, handover.

The next step is to define a general TINA computational model for UMTS systems, that relies on the NRIM presented in this paper.

9. Acknowledgments

The authors would like to thank Marco Lenti (CSELT) for helpful discussion in the initial phase of the work, as well as the members of the TINA-C Core Team for their interesting feedback to the ideas at the basis of this paper.

References

[1] *Overall Concepts and Principles.* TINA C, February 1996.

[2] *Network Resource Architecture*, Version 3.0. TINA C, February 1997.

[3] *Network Resource Information Model Specification.* Draft. TINA C, June 1997.

[4] *Information Modeling Concepts.* TINA C, April 1995.

[5] *Computational Modeling Concepts*, Version 3.2. TINA C, May 1996.

[6] *Baseline Document.* MoNet Project Deliverable R2066/PTT NL/MF/DS/P/099/b2.

[7] *UMTS Functional and Network Architecture.* MoNet Project Deliverable R2066/SEL/GA3/DS/P/073/b1.

[8] *Opportunities to utilise TINA Concept within UMTS Standards / UMTS Mobility Requirements on TINA.* EURESCOM P608 Project Deliverable, February 19 1997.

[9] *Recommendation G.803.* ITU-T.

[10] ISO/IEC JTC1/SC21 N7524, ISO Committee. *Draft on Basic Reference Model of Open Distributed Processing - Part 2: Descriptive Model.* International Organization for Standardization and International Electrotechnical Committee, December 1992.

[11] J. Rumbaugh, M. Blaha, W. Premerlani, F. Eddy, W. Lorensen. *Object Oriented Modeling and Design.* Prentice Hall, USA 1991.

[12] *Rainbow demonstrator architecture and services.* Rainbow project deliverable AC015/BELL/CIT/DS/P/002/b1.

Session 13

Connection Management

Pier Giorgio Bosco, CSELT

Enhancements on TINA Connection Management

Drissa Houatra and Laurent Leboucher
France Télécom CNET
38-40 rue du Général Leclerc
92794 Issy Moulineaux Cédex 9, France
{drissa.houatra, laurent.leboucher}@cnet.francetelecom.fr

Abstract

This paper describes a contribution to connection management. The contribution is motivated mainly by shortcomings in the specifications of the TINA connection management architecture : in particular, it is considered to be too difficult to use. The TINA specifications are based on the ODP binding model. In this model, connection management is viewed as an explicit binding service, that is, a binding service that can be controlled from the service components. Starting from this binding model, we propose a way to structure TINA connection management services. As a result, we achieve important simplifications on the current specification.

1 Introduction

The ODP architectural principles [1] are being applied to information networks, in the TINA specifications. The information networking principles of TINA can be summarized as follows :

- The introduction of object oriented distributed platforms (Object Request Brokers) in telecommunication networks.
- The development of object oriented software frameworks that will be supported by the distributed platform (e.g. the service architecture [2], in order to ease the development of large applications involved in networks (control and signaling, administration, end-user services). One of the effects of these frameworks is that they reflect functions and components of the underlying transport networks.

Connection management is a possible sub-framework. A connection management service must be seen as a part of the reflection of functions and components of the transport network. In fact the sub-framework "connection management" constitutes a high level abstraction of some transport services.

This paper proposes a new approach for TINA connection management sub-framework. The proposal brings simplifications and clear statements on its relationship with the explicit binding concept of the ODP computational model.

2 Overview of related work, our contribution

The strength of applying ODP principles to the design of information networks have been discussed in [4]. In particular, this paper stresses some basic requirements on (generic) connection management : it should enable identification, arbitrary communication semantics (point-to-point, point-to-multipoint, etc.), dynamic reconfigurability and extensibility. Since then, propositions and/or experiments have been done. Some of the most relevant to our work are presented in the following paragraphs.

TINA connection management Roughly, the requirements above correspond to what the TINA connection management is trying to meet. The information description provides a complete modeling of network connections using connection graphs and the composite elements : flow Connections, flow Endpoints and flow fonnection franches. In the later version [5] (or something near) there have been important modifications, compared to the version of 1995. For example, the interfaces to the service architecture have changed, as well as the connection management component objects. The connection management architecture is considered to have the following disadvantages (in particular):

Usablility : Service components are faced with multiple interfaces, making the connection management service difficult to use.

Description : The relationship of connection management with the conceptual model (ODP) is not

0-8186-8335-X/98 $10.00 © 1998 IEEE

completely established. In particular, the interactions between (service architecture) computational objects and the connection management service to setup and use stream interfaces are missing, or delayed for further study.

Our work is concerned with the design of CSM and TCSM services, their interaction with service components and the relationship the basic ODP explicit binding principle. The work is not concerned with network connectivity services (e.g. the Connection Coordinator (CC), the Connection Coordinator Factory (CCF) and the Flow Connection Controller (FCC). These services are internal and network oriented. Viewed from the service architecture, they do not need to comply to any principle, provided that they implement the adequate connectivity service.

Flexible ORBs A work on flexible ORBs is presented in [3]. It is achieved within the ReTINA project. The ReTINA project defined and implemented a prototype of flexible ORB that allows a direct realization of the ODP computational model. In particular, it allows the realization of arbitrary binding objects with advanced communication semantics. Among the binding functions developed by ReTINA, there are a library of multimedia streams. This library provides an elementary support for multimedia streams based on RTP and multicast IP. It can be used to build a complete framework for the control and management of multimedia flows such as those envisaged by the Object Management Group (OMG). Our work can be viewed as a part of such a framework.

Multimedia System Services (IMA) Another proposition in this area is the Interactive Multimedia Association - Multimedia System Services (IMA-MSS)[6, 7]. This specification is very complete and it is consistent with the ODP explicit binding principles. Its features are summarized in [3]. Some less successful aspects in the IMA-MSS are the following :

- The limitation of the type of connections. Only two types of connections are possible : virtual connection (point-to-point) and virtual connection multicast (point-to-multipoint).
- It is multimedia terminal - oriented : the clients of the connection system are supposed to be end user personal computers (PC). A virtual device interface (a producer/consumer of stream flows) is provided by terminal objects, within the User Application (UAP). The description of interfaces to the connection system covers diverse aspects (stream flow formats, adapters, QoS, ports, etc.) but it does not address problems related to information networking.

CORBA streams The OMG issued a "request for proposal"(RFP) on the theme "Control and Management of Audio/Video streams". A joint specification is under the process of being adopted as a CORBA service(see [8]). This specification is similar to that of the IMA-MSS, at the highest interfaces. There is also interesting contributions on device modeling, stream flow modeling (flow connections and stream connections) and QoS attributes. In fact, this contribution addresses most of the issues identified in the RFP but concerning the topology of streams, the overall design (architectural choices) addresses only point-to-point and point-to-multipoint connections.

Our solution The connection management system we propose extends the propositions above in diverse aspects. The most important are the following :

- We organized our system in a way that conforms to the ODP explicit binding principle, and extend some choices that are left open in ODP. This aspect is not considered enough in the TINA connection management architecture. We also implemented a prototype of our system, in order to consolidate the architectural principles it is based on.
- We did not make limitations on topologies as it is done in most of the propositions (except that of TINA). A simple mechanism will be used to compose complex topologies starting from basic topologies.
- Even with (potentially) complex topologies, the external interfaces are simple to use and described clearly. We therefore limit the problems with manipulating complex topologies.

3 The connection management framework

3.1 Information model

The core components of the information elements are those of the two global connection graphs considered in TINA : the Physical Connection Graphs (PCG) and the Logical Connection Graphs(LCG). In our opinion, the distinction between the LCG and the PCG is particularly important for flexibility. It is a conceptual principle that will facilitate the separation

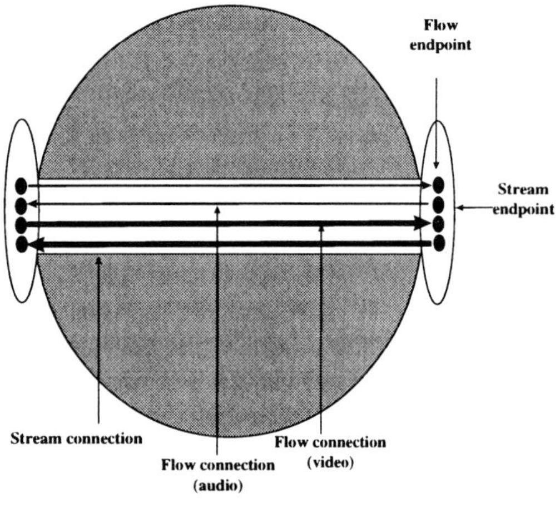

Figure 1: Flows, streams and endpoints

of call and connection control. The PCG describes network connections, while the LCG describes end-to-end connections between service objects. This distinction allows to manipulate separately physical connections and logical (end-to-end) connections. The graphs LCG and PCG are built using four information elements : stream connection, flow connection, stream endpoint and flow endpoint. Information elements are represented in the figure 1.
The nodes of a PCG are stream endpoints, and the links between the stream endpoints are stream connections. Similarly, the nodes of a LCG are a combination of media device identifiers (audio/video, input/output) and the links are end-to-end connections between these media devices.
We do not consider information elements related to connectivity session (connection establishment), network partitioning or layering, since we rather focus on the higher level aspects of connection management. For our prototype, the connectivity service is provided by service components of the connection management system. Layering services are not taken into account.

3.2 Computational description

This section describes the object model, the binding principle, the service components and the way arbritrary topologies can be established. Interface signatures and their operations will be described in the following section.

Object model and the binding principle We make use of the ODP general object model and its explicit binding mechanism, we adopt the following conventions :

- Objects can support multiple interfaces. Examples of such objects are (service or user domain) computational objects that can support :
  1. An interface for resource management.
  2. An interface for stream control.
- Concurrency is not allowed for the use of interfaces that involve distributed execution (binding for example). This assumption has important implications on the underlying engineering mechanisms : it simplifies the needs to ensure consistency in the concurrent use of such an interface.
- Interfaces can be created dynamically, at any moment during the lifetime of the supporting objects. Examples of such interfaces are stream control interfaces that are created after a binding is established.

Basically, the explicit binding system involves the following entities :

- A set of computational interfaces to be bound. These interfaces belong to the service or user domain. They are used by distributed multimedia applications to enable multiple bindings within a terminal and the manipulation of stream flows.
- A binding factory, which creates and maintains instances of binding objects.
- The binding objects themselves, which support the binding of computational interfaces. A binding is said to be explicit if the supporting binding object can be controlled by the service components. A binding object is typically distributed. In this case, it involves objects on the different terminals, that will ensure the distributed implementation (especially local binding and network connections).

Service components This paragraph describes our connection management service with respect to the design choices and its relationship with some of the service design presented in paragraph 2, mainly the TINA connection management.
The binding system is presented in figure 2. The service components are named using the TINA terminology. Even if the semantics are a little different from those of TINA, the general features are very close. A box represents a group of objects, typically implemented by a (CORBA-like) server. The interfaces of the system (binding_control, controlled_stream, ...)

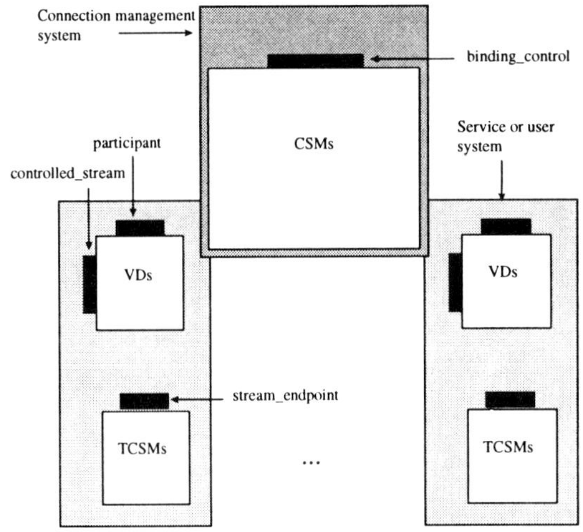

Figure 2: Service components of a binding

are represented on the service components that contain the supporting object.
Normally, all these service components provides an interface "factory", supported by a single factory object, but only the interfaces supported by the embedded objects are represented. An interface "factory" provides operations for the instanciation (or more generally the lifecycle) of the objects embedded in a service. We shall describe only the factory interface for service components CSMs. The factories for service components VDs and TCSMs follows the same principle.
Objects and interfaces contained in these services are described as follows :

- Virtual Devices(VD) : Related to the binding model, a VD is an object supporting the interfaces to be bound. A VD is an abstraction of some capabilities of an audio/video terminal, namely stream endpoints and audio/video devices. It allows the representation of a terminal (producer/consumer of a stream flow) within the service logic, or an UAP. A VD also provides operations for managing the local bindings (to audio/video devices) and for sending or receiving flows. Resource indexes in a VD will be used to create stream endpoints. A stream endpoint can manipulate up to four stream flows, depending on the values of the resource indexes:

    1. video input flow,
    2. video output flow,
    3. audio input flow,
    4. audio output flow.

  VDs are inspired from the IMA-MSS virtual devices but they have not exactly the same semantics : we do not suppose that VDs belong to the UAP. The location of the VDs are left to the service logic. The recent CORBA streams specification [8] also have a similar construct, with semantics that very close to those of IMA-MSS.
  An interface *participant (part for short)* provides operations to manipulate resources to be used by VDs. A VD can be used for multiple and simultaneous bindings (case of composition of a complex topology). Thus, the resources accessible at a given time will be used for the next binding operation.
  Interfaces *controlled_stream* are used to control stream endpoints and the local binding. A VD may have multiple such interfaces, one for each binding it is involved in. They are created dynamically (resulting in a bind operation). If a binding is established, references to interfaces *controlled_stream* are obtained through an operation on an interface *part.*
  The separation between interfaces *part* and *controlled_stream* is not absolutely necessary but it enables to limit the availability of an interface *controlled_stream* to the lifetime of the binding.

- CSM factory(CSMF) : Related to the binding model, CSMF is a binding factory. A CSMF object creates and maintains binding objects, on request of service objects. A CSMF is created for a service session. It will be used to manage multiple communication sessions. An interface to a CSMF provides principally an operation to create CSMs, and some optional facilities (e.g. list of current binding objects, etc.).

- Communication Session Managers (CSMs) : The CSMs are the ODP binding objects. The number of VDs to be bound and their topology are not limited in the model. Limitations may only come from the capabilities of the infrastructure. Unlike the design choice in TINA, a CSM ensures only one binding (one connection graph at a given time). In the TINA connection management, a CSM manages multiple connection graphs. The connection graphs are created incrementally, with the same generic interfaces. These design choices make CSMs heavy, difficult to implement and use. The interface *binding_control* results in the creation of a CSM. It allows the binding/unbinding of VDs, and some additional operations (e.g. adding/removing VDs, changing QoS parameters).

- Terminal CSMs (TCSMs) : A TCSM ensures the terminal part of the binding mechanisms on each involved terminal. It ensures the participation of a terminal, for setting up connections, modification of QoS, local bindings to audio/video devices. A TCSM object supports an interface called *stream_endpoint*. This interface will be used by a CSM (the creator) for a binding establishment. Setting up connections is performed by Connection Coordinators in TINA. We have no objection on this design, but our choice is motivated by simplification of the connectivity service, in order do concentrate on high level aspects of connection management. The TCSM functions we propose can easily be replaced by those of TINA, without perturbation on the overall design choices. The interface *stream_endpoint* is therefore used for both (entire) connection establishment and local binding.

Composition of complex topologies We stated that the binding system should allow any topology. A question arises : how to build and manage CSMs with complex topologies, relatively easily ? TINA also adopts such an approach, the ability to manage connection graphs of any topology, but the way connection graphs are managed is very complex. We adopt the following principle :

- First, we consider elementary CSMs, which consist of (bi-directional) point-to-point and point-to-multipoint CSMs. Unidirectional point-to-point and point-to-multipoint CSMs can be obtained from these elementary CSMs by setting up the adequate flow resources to *Null*.

- Any topology can be obtained by "placing" an adequate CSM between sub-graphs of the entire connection graph.

The evolution of the connection graph (add/delete branches) is expressed in term of adding or deleting CSMs between the vertices. Normally the two elementary CSMs suffice to build any complex graph, but we can also consider other currently used CSMs (cliques, stars, rings, ...) as well. An implementation can typically build a library of CSMs that reflect a class of application.

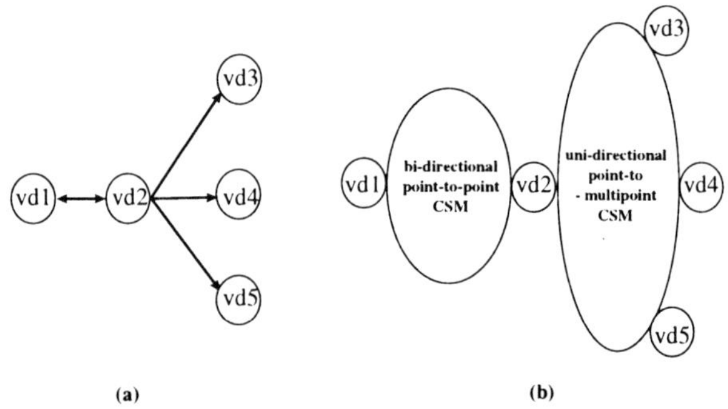

Figure 3: Composition mechanism

For example, consider the graph in the figure 3 (b). This connection graph is realized according to the figure 3(a). A typical sequence of operations for building this topology is the following :

1. Create a CSM for point to point connections

2. Setup the resources (ports numbers, device indexes and QoS information) of *vd1* and *vd2* so that the appropriate connections will be performed automatically. In this example, we need to setup audio/video and input/output resources for both *vd1* and *vd2*, for a bi-directional audio/video connection.

3. Bind *vd1* and *vd2* using the previously created CSM

4. Create a CSM for point to multipoint connections

5. Setup the resources of *vd2*, *vd3*, *vd4* and *vd5*. In this example, we need to setup audio/video output resources for *vd2* and audio/video input resources for *vd3*, *vd4* and *vd5*. The other resources should be set to *Null*, to indicate to the CSM service that the corresponding flow connections will not be used.

6. Bind *vd2* (as the source) with *vd3*, *vd4* and *vd5* (as sink) using the previously created CSM

3.3 Interface signatures and operations

This section describes the interface signatures of the connection management system using the CORBA Interface Definition Language (IDL). It also describes the behavior of operations.

QoS data structure We use the QoS defined in the ReTINA specifications [9] as such. It consists of a structure with attributes related to throughput, transmission delays, reliability, setup and release of network connections. A QoS structure is passed as a parameter for each flow endpoint, indicating the quality of service to be used to setup, or modify this connection.

```
struct FlowQoS {
  frame_sizeType frame_size;
  frame_rateType frame_rate;
  peak_rateType peak_rate;
  burst_sizeType burst_size;
  delayType delay;
  jitterType jitter;
  loss_rateType loss_rate;
  loss_intervalType loss_interval;
  setup_delayType setup_delay;
  release_delayType release_delay;
  setup_failure_probType setup_failure_prob;
  release_failure_probType release_failure_prob;
};
```

The QoS data structure is very general and oriented to the capabilities of an ATM network. The reason is that we have in mind the use of such a system in (principally) ATM networks. For some traditional network (e.g. ISDN), most of the parameters may not have a signification, but there is no implication on the overall design of the system. The QoS structure can be tailored to meet the needs of a specific network and taken into account easily.

Virtual Devices interfaces A VD supports two kind of interfaces :

1. An interface *part*: this interface provides control over the network and audio/video resources used for the binding,
2. One or many interface *controlled_stream*. An interface *controlled_stream* provides an abstraction for manipulating stream interfaces : it allows the control of stream flows and their link to the audio/video devices (local binding).

An interface *part* is described as follows :

```
interface part{
  NetworkAddr GetHost() ;
  PortIndex GetPort(in Flow fl) ;
  opStatus SetPort(in Flow fl, in PortIndex port) ;
  FlowQoS GetQoS(in Flow fl) ;
  opStatus SetQoS(in Flow fl, in FlowQoS qos) ;
  DeviceIndex GetDeviceIndex(in FlowType ft) ;
  opStatus SetDeviceIndex(in FlowType ft, in DeviceIndex idx) ;
  controlled_stream GetCtlStream(in binding binding_obj) ;
  void destroy();
};
```

A flow is an element of the enum type *enum Flow {audio_input,audio_output,video_input,video_output}*. The Flow type is defined by : *enum FlowType {audio, video}*.

The operation *GetHost* returns the network address of the audio/video terminal that is reflected by the supporting VD object. This information will be used by CSMs (binding objects) for network connections, when establishing a binding.

The operations *GetPort* (respectively *SetPort*) are used to read (respectively write) a port index associated to a flow. The clients of the operation *GetPort* are CSMs. Port indexes will also be used for network connections. The operation *SetPort* is called by service architecture object to setup port indexes before a binding operation.

The operations *GetQoS* (respectively *SetQoS*) are used to read (respectively write) QoS attributes associate to a flow. QoS attributes are used in the same way as the network address and the port indexes to establish a binding. They are also stored by service architecture objects.

The operations *GetDeviceIndex* and *SetDeviceIndex* concern device indexes. Their role is similar to that of the operations on port indexes and QoS attributes.

The operation *GetCtlStream* returns the stream control interface associated to a binding. When a binding is established, the CSM that supports the binding creates and stores a stream control interface reference for each involved VD object. The operations on these interfaces are related to that CSM. Therefore, a stream control operation will act on the stream flows established by the supporting CSM. Due to complex topologies, a VD can provide multiple such interfaces.

The operation *destroy* eliminates the supporting VD object from the system. References to interfaces *part* and *controlled_stream* supported by this object are no longer valid.

An interface *controlled_stream* can be described as follows :

```
interface controlled_stream {
  opStatus receive( in part sender) ;
  opStatus start( in FlowSet fls);
  opStatus suspend(in FlowSet fls);
  opStatus resume(in FlowSet fls);
  opStatus stop(in FlowSet fls) ;
};
```

The data structure *FlowSet* allows the composition of stream flows. A client will use this construct to pass a combination of stream flows as a parameter. For example, such a parameter can be a single stream flow (e.g. {audio_input}) or, at the other hand, the complete set of stream flows encapsulated by a VD : {audio_input, audio_output, video_input, video_output}. In case of the use of stream flows of the same type (e.g. {audio_input, audio_output}), the operation acts on these stream flows in a synchronous manner. This is the classical manipulation of stream connections.

The operation *receive* starts the reception of stream flows from the VD supporting the interface *sender*. This operation is necessary in case of the possibility for the calling VD to receive stream flow from multiple senders. Such an example is a VD that is the "source" of a bi-directional point-to-multipoint connection.
The operations *start* and *stop* (respectively *suspend* and *resume*) starts/stops (respectively suspends/resumes) the (end-to-end) streaming mechanism of a flow set.

Interface to a CSM factory An object CSMF supports an interface called *factory*, mainly for the life-cycle of CSMs and some facilities for the management of CSM objects. At this time, only a simplified operation is provided, that returns the list of CSMs in the system.
The interface *factory* has the following structure :

```
interface CSMfactory{
  opStatus create(in csm_type type_id, out binding binding_obj);
  opStatus list(out csm_list the_list);
};
```

When creating a CSM, a type identifier is passed as an input parameter. The resulting CSM is stored in an output parameter of type *binding*. The type *binding* is a generic type that can contain binding objects of different topologies.

Interfaces to a CSM A CSM supports a binding control interface. Interface *binding_control* is responsible of the binding mechanism itself, that is (mainly):

- Establishment and the release of network connections,
- Establishment and the release of local bindings,
- Modification of QoS attributes,
- Modification of the binding when it is possible : add and remove participants.

The operations available at an interface depends on the binding topology. In the following section, we give interface signature for point to point and point to multipoint binding objects.

An interface for a point to point binding object is described as follows :

```
interface Pt2PtBinding {
  opStatus bind(in part user1, in part user2);
  opStatus unbind();
  opStatus ChangeQoS(in part user, in Flow fl, in FlowQoS qos);
  opStatus destroy();
};
```

Interfaces for a point to multipoint binding object have a slightly different signature. They provide supplementary operations to take into account multiple participants :

```
interface Pt2MptBinding{
  opStatus bind(in part source, in sequence <part> sink);
  opStatus ChangeQoS(in part user, in Flow fl, in FlowQoS qos);
  opStatus AddPart(in sequence <part> consumers);
  opStatus RemovePart(in sequence <part> consumers);
  opStatus unbind();
  opStatus destroy();
};
```

Since services provided by the interface *Pt2PtBinding* are a subset of those provided by the interface *Pt2MptBinding*, we shall only describe the later.

The operation *bind* establishes the distributed binding itself : it establishes the point to multipoint network connection between the "source" stream endpoint and the "sink" stream endpoints, and the local connections

(within the terminal) between each of the stream endpoints and the audio/video devices.
The operation *unbind* performs the reverse mechanism : it releases network connections and the local bindings.
The operation *destroy* removes the binding object from the service.
The operation *ChangeQoS* changes the quality of service associate to a flow connection, within a VD. This operation could imply the re-establishment of the connection associated, depending on the underlying network. There is also an important issue related to QoS attributes : they involve an agreement between senders and receivers of stream flows, in order to ensure that the QoS expressed in both sides are compatible. A compatibility criteria should therefore be adopted in the future.
The operations *AddPart* (respectively *RemovePart*) add new participants in the binding (respectively remove participants). These operations are not available for a point to point binding.

Interfaces to a TCSM A TCSM supports an interface for the control of stream endpoints. This interface *stream_endpoint* is an abstraction of a stream endpoint. This vision of stream endpoint includes both network connections and links to devices.

An interface *stream_endpoint* for point-to-point connections has the following structure :

```
interface stream_endpoint {
  opStatus connect(in FlowSet fls, in part remote_user);
  opStatus disconnect(in FlowSet fls);
  opStatus chQoS(in Flow fl, in FlowQoS req, out FlowQoS rsp);
  opStatus lbind( in FlowSet fls);
  opStatus lunbind(in FlowSet fls);
  opStatus start(in FlowSet fls);
  opStatus stop(in FlowSet fls);
  opStatus suspend(in FlowSet fls);
  opStatus resume(in FlowSet fls);
  opStatus destroy();
};
```

The operations on a point-to-point stream endpoint reflects the overall involvement of a terminal in a binding. When an operation is invoked on a CSM interface, this operation is in part traduced into operations on interfaces *stream_endpoint* those supporting objects are running on the involved terminals.

3.4 Engineering description

This paragraph discusses how streams can be implemented on top of a network infrastructure and how the abstract protocol presented in [3] can be used to establish a binding.

Requirements Implementing such a system requires at-least two end-user nodes with multimedia capabilities (typically PCs) and a stream host system. The multimedia nodes should be connected through a network that can transfer audio/video flows at an acceptable QoS (typically ISDN or ATM). All the involved nodes needs to be connected through a data network to provide the kernel transport network (KTN) function. Notice that KTN and the (audio/video) transport network can be supported by the same network. Communications between the DPE-nodes are achieved through the KTN, and the stream management service is implemented on top of the DPE.

The mapping of service components TCSMs are replicated on any node that can manipulate audio/video flows. They are realized by a service that runs on each node (on the DPE). The TCSM services provides a capsule for stream endpoints.
CSMs and VDs can be mapped in different ways. CSMs are typically placed on a special purpose node (server), or a group of nodes. Each node runs a service (CSM service), as a capsule for binding objects. VDs follow the same logic as CSMs. However, the VDs seams (from a first analysis) to be better placed on the reflected nodes (end-user terminals) for performance considerations.

Binding establishment The protocol presented in [3] can be used to establish a binding in the following way : let us consider a binding of *N* VDs *VD1*, ..., *VDN*. The binding object CSM is created by the factory CSMF. Each involved DPE node *DPE1*, ..., *DPEN* runs a TCSM service. From the CSM logic, the binding is established in the following manner :

1. The resources in *VD1*, ..., *VDN* are analyzed to detect the connection topology and QoS requirements, and check if these are possible.
2. The CSM asks TCSMs on *DPE1*, ..., *DPEN* to create stream endpoints.
3. The channels are established.

Local binding through controlled stream interface mechanisms :

3. Operations *start* causes the local binding of the adequate flow endpoint. A connection is established between the flow endpoint and the device index associated, by the TCSM service associated to the calling VD. It also launches the associated device driver if necessary.

2. Operation *stop* causes the local unbinding of the adequate stream endpoint and stops the associated device driver if necessary.

3. Operations *suspend* and *resume* cause the local unbinding (respectively the local binding) of the adequate flow endpoint.

4 Implementation

We prototyped a first version of our connection management system with point to point and point to multipoint connections. The network infrastructure we used is composed of :

- An ATM local area network (LAN) built with ATM switches from Fore Systems. Links between the switches have a transfer rate of 2Mb/s. The ATM links are used by LAN emulation. The terminals are PCs equipped with ATM cards (25Mb/s) from First Virtual Corporation. We also installed an ATM video-conferencing software tool (from Zydacron Inc.) on PCs. A Multipoint Control Unit (MCU) will be installed for point-to-multipoint audio/video streams over ATM. At this moment, ATM network is used as a control network (IP over ATM) : the KTN.

- A 128Kb/s links ISDN wide area network (WAN). Terminals are PCs with Ethernet cards and an ISDN video-conferencing software. An ISDN MCU is available for the point-to-multipoint audio/video streams. The ISDN network is used as the transport network for audio/video stream.

The figure 4 shows this infrastructure and the service components of our prototype. The DPE we used is Orbix 2.0 from Iona Technology. Currently TCSM services and point-to-point CSMs are implemented. The design of point to Multipoint CSMs software is ready and an implementation is being done using the ISDN MCU. The connection management system will be extended to the ATM network and used for group communication services.

5 Concluding remarks

We discussed various design of connection management systems, followed by the design and implementation of our connection management system. The design is inspired from the ODP explicit binding model, and extends the description in term of architectural choices. A prototype is built using an ISDN transport network with multimedia PCs. Our system contributes mainly to the following features :

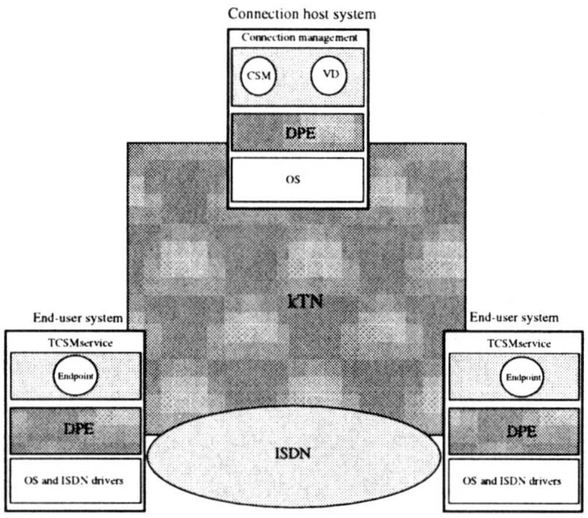

Figure 4: Prototype connection management system

- Overall design choices and description,

- Construction of complex topologies in a simple way,

- Engineering mechanisms and prototyping.

The design choices presented in this paper can be taken into account in TINA connection management. We are currently working on the prototype. The system will be refined incrementally, by taking advantages of results from the prototypes. Notice that the goals of this work is not to give a complete specification and implementation of an industrial system. Therefore we did not consider features such as multiple protocols, adaptations (multiple coding systems), security, etc. The primary goal is to validate our architectural choices. Then, other issues can be integrated easily, in order to implement a full connection management system.

Acronyms

- ATM : Asynchronous Transfer Mode

- CC : Connection Coordinator

- CCF : Connection Coordinator Factory

- CORBA : Common Object Request Broker Architecture
- CSM : Communication Session Manager
- CSMF : Communication Session Manager Factory
- DPE : Distributed Processing Environment
- FCC : Flow Connection Controller
- IDL : Interface Definition Language
- IMA : Interactive Multimedia Association
- IP : Internet Protocol
- ISDN : Integrated Service Digital Network
- KTN : Kernel Transport Network
- LAN : Local Area Network
- LCG : Logical Connection Graph
- MCU : Multipoint Control Unit
- MSS : Multimedia System Services
- ODP : Open Distributed Processing
- OMG : Object Management Group
- ORB : Object Request Broker
- PC : Personal Computer
- PCG : Physical Connection Graph
- QoS : Quality of Service
- RTP : RealTime Protocol
- ReTINA : Realtime Telecommunication Information Networking Architecture
- TCSM : Terminal Communication Session Manager
- TINA : Telecommunication Information Networking Architecture
- UAP : User Application
- VD : Virtual Device
- WAN : Wide Area Network

References

[1] ITU-T — ISO/IEC Recommendation X.903, ODP Reference Model : Architecture, *International Standard Organization*, 10746-3, 1995.

[2] C. Abarca, P. Farley, J. Forslow, T. Hamada, P.F. Hansen, H. Hegeman, S. Hogg, H. Kamala, L. Kristiansen, M. Mampaey, R. Minetti, H. Mulder, *Service Architecture Version 4.0*, TINA-C Baseline Document Label TB_RM.001_4.0_96, 28 Oct. 1996.

[3] F. Dang Tran, V. Perebaskine, J-B. Stefani, B. Crawford, A. Kramer, D. Otway, "Binding and Streams : the ReTINA Approach," *Proc. TINA'96*, pp. 101-113.

[4] J-B. Stefani, "Open Distributed Processing : An Architectural Basis For Information Networks", *Computer Communications*, Vol.18, No11, Nov. 1995.

[5] C. Abarca, J. Forslow, T. Hamada, S. Hogg, H. Beon Jeon, D. Seok Kim, H. Young Lee, N. Natarajan, F. Steegmans, *Network Resource Architecture Version 3.0*, TINA-C Baseline Document Label NRA_v3.0_97_02_10, Feb. 10, 1997.

[6] IMA Recommended Practice, *Multimedia System Services - Part 1 : Functional Specifications*, First edition - Second draft, 23 Sept. 1994.

[7] IMA Recommended Practice, *Multimedia System Services - Part 2 : Multimedia Devices and Formats*, First edition - Second draft, 23 Sept. 1994.

[8] IONA Technologies, Lucent Technologies, Siemens Nixdorf, *Control and Management of Audio/Video Streams*, OMG RFP Submission - Revised Submission telecom/97-05-07, Version 1.0, 30 May 1997.

[9] ReTINA, *DPE Kernel Architecture Specifications V1*, ReTINA deliverable D1.03, Sept. 1996.

Network Provider Domain Federation in TINA

Juan Pavón
Alcatel Corporate Research Center
Ramírez de Prado, 5, 28045 Madrid (Spain)
pavon@alcatel.es

Edgar Montes
Alcatel Corporate Research Center
Route de Nozay, 91460 Marcoussis (France)
montes@aar.alcatel-alsthom.fr

Peter Komisarczuk
Ericsson Ltd.
WC3, Burgess Hill, West Sussex, RH15 9UB (UK)
etlpkzk@etlxdmx.ericsson.se

Abstract

Federation in TINA CMA (Connection Management Architecture) provides the mechanisms for cooperation between different interworking network domains possibly owned by different administrators. In order to be able to offer services to their users, these administrators must cooperate. In this paper we present an implemented and validated architecture including the federation techniques necessary. We illustrate the problem based on experience from a User Trial, in which different operators, and suppliers with different equipment participate.

1. Introduction

The CMA (Connection Management Architecture) defined by TINA, specifies the functions necessary to manage network transport resources. The CMA gives a uniform view of these resources and provides a model for their control and management that is independent of the underlying network technology. These functions are implemented as distributed software components that are made available to applications that need to provide communication and network management services.

The Architecture is divided into three levels:

Communication Session level providing service independent interfaces so that service components can establish end-to-end communication in an abstract way.

Connectivity Session level providing technology independent interfaces for the above level so that it can interconnect network termination points. It also handles the interworking of different network technologies, our basic concern here.

Layer Network providing an abstract view of the specific network technology.

Federation in the TINA CMA provides the mechanisms for cooperation between different interworking network domains. Different administrators own and manage sets of resources or domains. In order to be able to offer services to their users, these administrators must cooperate so that connections can be set up, information exchanged, and management assured across these domains.

In TINA, the definition of inter-domain reference points guarantees that inter-operation between ADs (Administration Domains) will be possible. Specifically, in this paper we study and validate the LNfed interface (the Layer Network federation inter-domain reference point) that allows different Connectivity Providers to implement their own version of the Connectivity Platform and be able to interact to establish connections across their respective domains.

0-8186-8335-X/98 $10.00 © 1998 IEEE

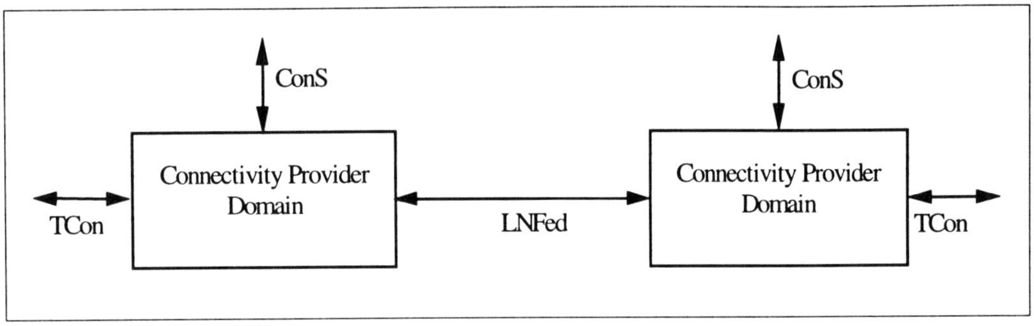

Figure 1. LNFed reference point between Connectivity Provider domains

Federation between Connectivity Provider domains provides more flexibility for routing, it de-couples different implementation strategies for each domain, and it provides a means to support interworking with legacy systems and protocols. As we will show later in this paper, the architecture implemented here not only allows interoperability between TINA compliant systems but is also a basis for the integration of legacy systems and protocols such as the existing broadband signaling systems.

Federation must not only establish connections between ADs, but it must also provide solutions to the control of management functions or systems across these boundaries. In this first implementation of the LNFed interface we only establish connections between multi-provider domains, but in the next phase we will investigate the problem of management services and present a solution which integrates these aspects.

In this paper we present an implemented and validated architecture *vis-à-vis* federation (another paper presents the overall project [2]). To achieve this we illustrate the problem based on experience from a User Trial, in which different operators, and suppliers with different equipment participate. Figure 2 shows the system developed by two manufacturers, Ericsson (on the left), and Alcatel (on the right), interacting through the LNfed interface.

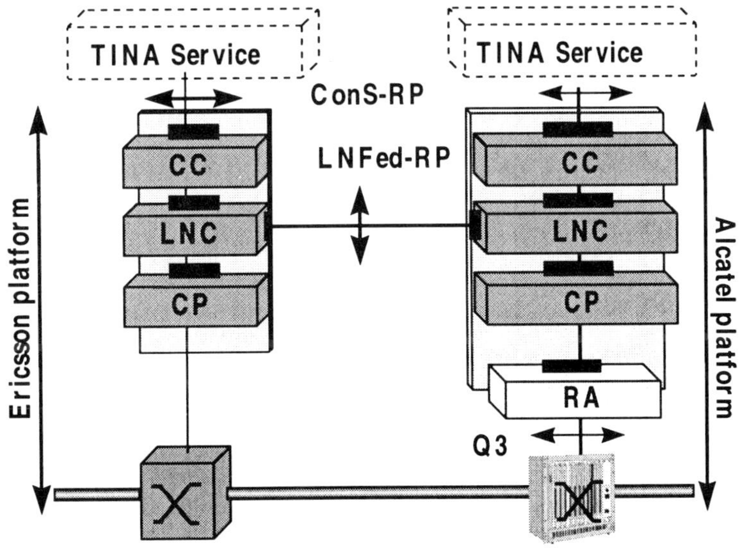

Figure 2. TINA Connection Management Architecture in SPOT

In section 2 we present the implemented components and illustrate how they work with an example scenario for the setup of a trail over two federated ADs. We then briefly describe, in section 3, the procedure used to test federation. In section 4 we present some original solutions to interworking with legacy systems. Finally, in the conclusion we review the results obtained through this work, the possible extensions and the issues yet to be resolved.

2. Description of components

The TINA architecture has adopted an object-oriented approach that permits the division of the system into different computational objects (COs) that interact through well defined interfaces. Their interactions are supported by a distributed processing environment (DPE), which is based on CORBA. Thus, the Connection Management Architecture defined by TINA [1] consists of several COs of two different types, as shown previously in Figure 2:

- those that offer an interface to the TINA services, rendering them independent of the underlying network structure: basically these COs are the Communication Session Manager (CSM), and the Connection Coordinator (CC).
- those that allow connection management within a layer network; basically these COs are the Layer Network Coordinator (LNC), and the Connection Performer (CP).

Through the use of this architecture the services can request connections specified in the form of a connection graph (CG). The CC is basically responsible of giving an abstract view of the network connection, whereas the LNC serves trail and tandem connection requests on a layer network. The tandem connection requests allow LNCs of different domains in the same layer network to federate and establish a trail. The LNC acts as a single point of access to a layer network within an AD, through which is provided a simple means of access to the entire federated layer network, spanning many ADs.

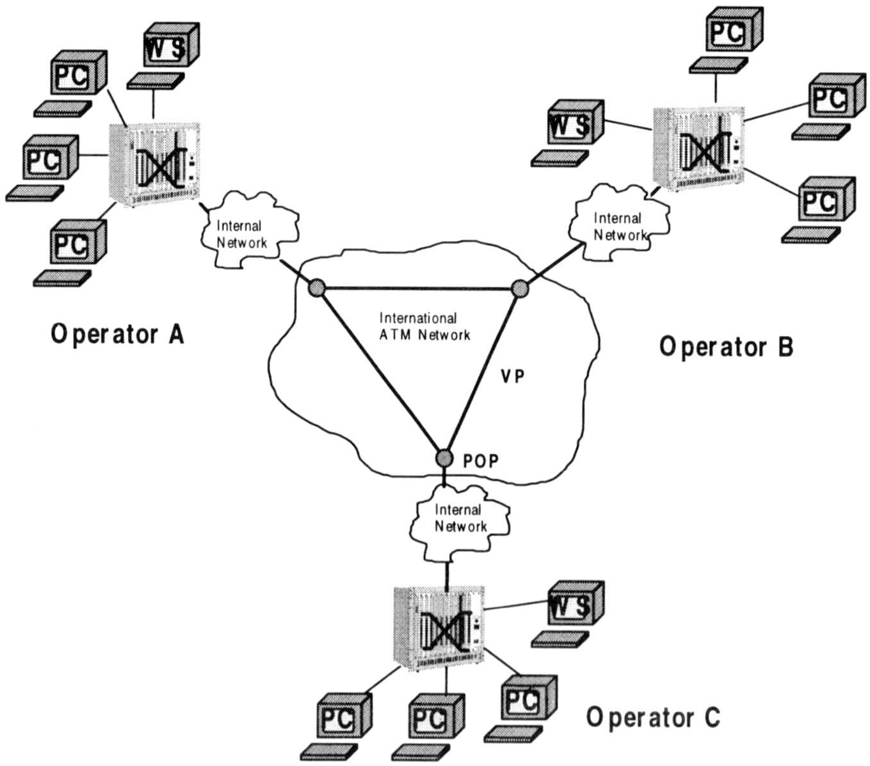

Figure 3. Scenario for LNC federation in SPOT

2.1 LNC and federation interfaces

When establishing a trail, portions of it may belong to different administrative domains; these, in turn, may belong to different connectivity providers. Network federation involves cooperation between different administrative domains so that the trail can be set up. To be able to implement the federation of layer network domains, i.e., between LNC computational objects, we must first define the operations that are needed at the federation interface, which has been named LNFed. To formally define this interface we use the OMG Interface Description Language (IDL) [4] but could also use the extended version of this language defined by TINA-C (ODL) [5].

Figure 3 shows the network infrastructure that has been used to explore the LNfed issues. There are three ATM VC layer network domains (called Operator A, B, and C). These are interconnected through an international ATM network. This is the configuration considered in the SPOT project with Unisource, where the three operators correspond to Telia, Swiss Telecom, and KPN, all interconnected through the JAMES European ATM Network. In the first two domains, Alcatel 1100 HSS series 1000 ATM switches are used, and a FORE ASX 200WG switch is used in the third domain. Note that this configuration involves a heterogeneity of physical equipment, and the interconnection of a diversity of networks.

2.2 Configuration of LNC

For each domain there is one LNC computational object which serves requests for trails from clients (e.g., the CC), and serves requests for tandem connections from neighbor LNCs. Each LNC has to be configured with the information required to determine with which domains it is federated and to which domains it must delegate in order to set up tandem connections. Each LNC is a client of a NML-CP which will setup and establish subnetwork connections among points visible at the top level of the partitioning in the respective domain. In the figure some of these points are shown.

Initially, each LNC is configured in order to know which is the top NML-CP in its layer network domain, and which end points are visible at its borders (these include the access and transit end points). Information is also provided on the neighboring LNCs. More explicitly, at each of these LNCs the following information is provided:

- The federation interface reference of the neighbor LNC .
- The topological links [3] between the domains. Topological links between domains can be defined by pairs of link termination points (e.g., a121-c211, or a133-b132) including the bandwidth associated to the topological link.
- Routing information (e.g., based on prefix or explicit enumeration of ranges of end point identifiers, which is the solution that has been implemented).

For the neighbor LNCs that are to be served the LNC creates a federation interface on which to receive tandem connection requests. Note that neighbor LNC objects should be configured in a consistent way. Another possibility was to define an LNC-Federation server object, which is asked for by LNC objects when this kind of information is required. The problem with this second approach is how to determine the owner of such an object, and the risks of having to centralize this information.

Several open issues have been identified at this time, these include: how federation at the Resource Configuration Management level takes place, and what form of dynamic routing information may be exchanged with regard to network reachability. In the present trial scenario the planning of the configuration of federation interfaces, topological links, and routing information is considered off-line as part of the federation negotiation mechanisms.

2.3 Scenario for establishment of a trail between end points in domains A and C

The following scenario is based on the network configuration shown in Figure 3, and explains how a trail between different ADs is established. Operations that illustrate the scenario have been simplified to consider only the parameters which are relevant to the purpose of discussing federation.

A request to establish a trail can be accepted by any of the LNCs in the three domains. A way of improving communication resources usage is to make that an LNC analyzes the root, and delegates the construction of the trail to the LNC of the root's domain. In this example, let's consider that the LNC in domain A (LNC-A) receives a request from a CC to create a trail between two network access points a1 and c2. The CC requests the LNC to:

create_trail (a1, c2, type, conn_desc)

where *a1* is the root (origin) of the trail, which is a network access point associated to a terminal in the domain A, and *c2* the leaf (destination), which is a network access point associated to a terminal in domain C. The *create_trail* primitive indicates the type of trail required (point-to-point uni- and bi- directional, or point-to-multipoint unidirectional) and the connection description structure which contains bandwidth description parameters.

LNC-A will determine first whether *a1* and *c2* are in its domain. In this example, *a1* is in its domain, but not *c2*. There should be some mechanism (a simple static routing table was used in the trial) which determines through which other domain the connection is to continue. In the example, LNC-A can route directly to domain C, or it can choose a route transiting domain B (for instance, when all topological links with domain C are busy).

Once LNC-A has determined to use domain C, it should decide which topological link to use. This can be determined by LNC-A and dictated to LNC-C, or it can be negotiated with LNC-C. A tandem connection primitive containing a negotiation parameter is used to create the tandem connection. If LNC-A, for example, chooses topological link *a121-c211* (*a121* being a link termination point in domain A and *c211* in domain C) but is prepared to negotiate, LNC-A can request LNC-C for a tandem connection using the following primitive:

create_tc (c211, negotiable, c2, type, conn_desc)

LNC-C will provide LNC-A with a confirmation of the use of *c211* (note that the link termination point is given as parameter) or propose the use of another topological link, e.g., *c212* (assuming there is a topological link *c212-a113*, for instance). In the case where LNC-A wants to use *a121-c211* and no other topological link, it would set the flag to non-negotiable status. As a result of the execution of this operation, LNC-C will return the following information:

- whether it has succeeded or not in creating (setting-up and establishing) the tandem connection
- which topological link has been used
- which link connection has been allocated in the topological link
- which network trail termination point (NWTTP) has been allocated to *c2*.

If the chosen topological link is different, LNC-A will have to confirm to LNC-C its use or release the tandem connection.

With this information, LNC-A can request its top NML-CP to establish the corresponding subnetwork connection in domain A.

Note that several strategies can be envisaged in order to improve the connection setup time. For instance, if LNC-C returns the link connection for the topological link between the domains without waiting to complete the tandem connection establishment, the subnetwork connection in domain A can be setup and established while the subnetwork connection is being setup in the other domains. However, the release would take more time in the case of a failure to create the tandem connection.

LNC-A will be able to return the network trail termination point (NWTTP) associated to *a1* and *c2* to its client. The NWTTP for *a1* is managed by its own domain whilst the NWTTP for *c2* was obtained as a return parameter of the operation invoked on LNC-C. In case that there is interactions with TLA objects (see section 4 below), LNC-A provides this information to TLA in domain A, and LNC-C to TLA in domain C. Note that this requires that references to TLA objects should be passed also in the interface together with the network addresses.

It is important to consider the similarity of these procedures with current signaling protocols.

3. Procedures used for testing LNFED

Several steps were followed to assure, as much as possible, that the federation would work when the final field trial was set up. First an agreement was reached concerning the LNFed IDL. Next local tests were performed by configuring locally defined domains. These test also permitted the elaboration of scenarios to be used for the final test between the different systems (Ericsson's and Alcatel's). But before establishing a direct connection through the international network, executable LNC-Fed servers were exchanged including dummy CPs (see Figure 2) so that the test scenarios could be performed and the results in form of traces could be verified. By dummy we mean a component that has de same behavior than the final component but only prints traces of the generated requests to the underlying element. The following figure (figure 4) illustrates this test configuration. Here we desire to create a connection from termination point A to B where A is in one domain and B is in the domain of the other network provider. The resulting behavior tested was as follows:

CC requests a trail from the LNC. The LNC determines that the end point is in another network provider domain or that it can be reached through this other domain. The LNC creates the trail up to the network interworking point B'. The LNC requests a tandem interface from the other LNC. The LNC requests a trail from the interworking point terminating at the remote termination point B.

In this way we can ensure that most of the problems that may arise have been eliminated before establishing a physical connection and also verify that the system is properly configured.

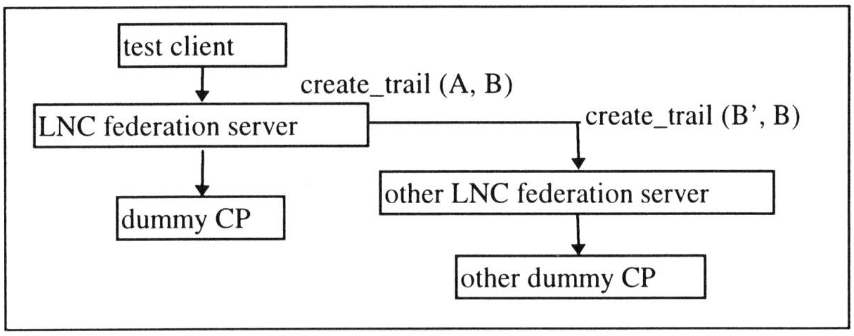

Figure 4. Configuration of initial test

4. Interworking with legacy systems

Interworking of LNC is considered with the customer premises equipment when connected to UNI based networks, and, secondly, at the interworking points between legacy and TINA networks. Two interworking scenarios between legacy and TINA networks are considered, these are interworking between broadband networks and the TINA CMA, and between TMN systems and the TINA CMA.

4.1 LNC at the Customer Premises Network (CPN-LNC)

An interesting consequence of this study is the possibility to consider that the customer premises equipment (e.g., the terminal) may have the capability to offer an LNC federation interface located on the TLA (Terminal Layer Adapter) [3]. Previously the TLA was called CPE-LNC (Customer Premises Equipment-Layer Network Coordinator) and one exists for each type of layer network in the consumer domain so that network flow endpoints can be created and managed. This is one potential way to provide interworking with current access signaling systems. Note that in the case UNI signaling is used, an adapter should be used in the same sense as discussed in the next section.

Current TINA-C specifications consider that this interface consists of three operations, from LNC to TLA, in order to propose the use of a trail termination point and some bandwidth and quality of service parameters. On reply from TLA, LNC should check whether TLA accepts or propose a new trail termination point and connectivity description parameters, and confirm the agreement of the TLA parameters.

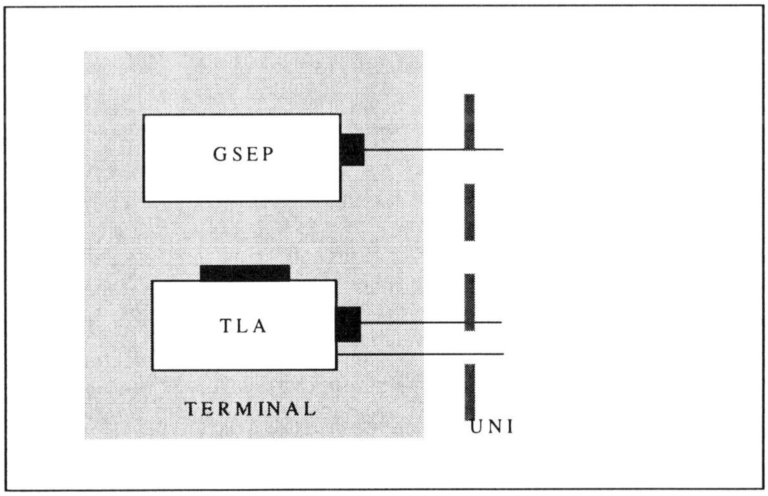

Figure 5. LNC in the terminal

4.2 Broadband Signaling and LNC interfaces

Current broadband signaling protocols are described in ITU-T Recommendations:

- Q.2764 (point-to-point)
- Q.2722.1 (point-to-multipoint)

The main difference with respect to the TINA approach is that both service and connection management information are described together (although there are some claims to separate call and connection models). Anyway, some primitives of the above protocols can be considered similar to those primitives we have identified for layer network federation.

The scenario for interworking TINA CMA (and also TINA service architecture) with current signaling can be based on some signaling adapter object as depicted in Figure 5. Note that current signaling protocols transport both call control and connection control information. Therefore, the Signaling Adapter computational object could offer interfaces to both LNC and Service Session Management objects, although the possibility of using TINA call control on an end-to-end basis can also be employed. In this proposal the LNC interface with the Signaling Adapter is the same as for federation with other LNCs.

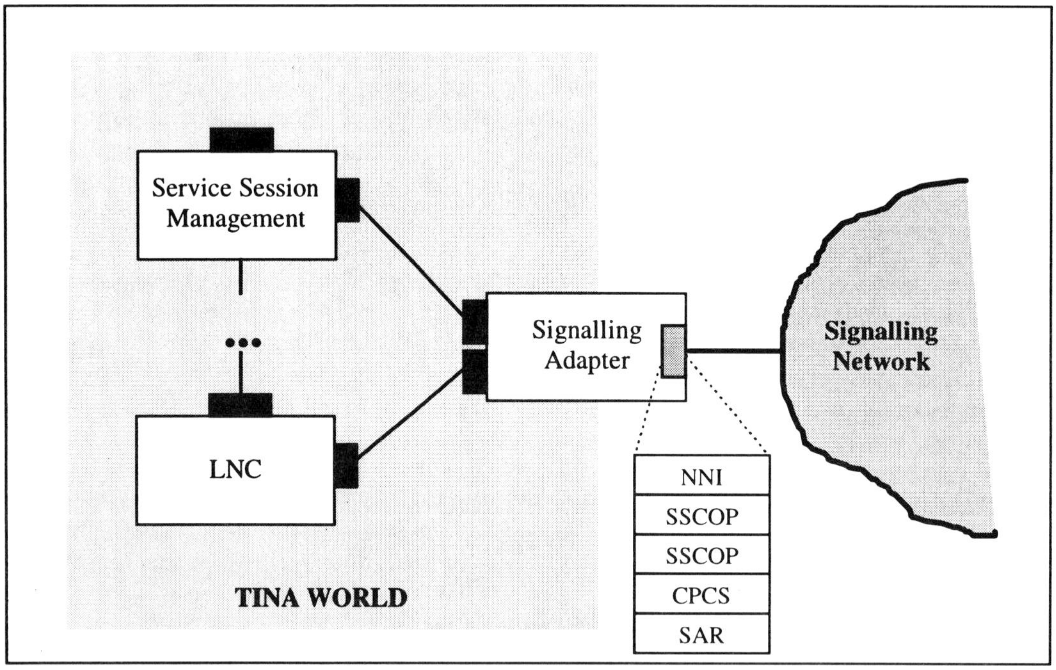

Figure 6. Interworking with current signaling protocols

4.3 TMN and LNC interfaces

An important issue to consider is interworking with TMN systems. The TINA Connection Management Architecture has been defined with a hierarchical nature similar to the one applied for TMN layering [1]. In the case of federation between TINA and TMN domains, we are interested in two types of interfaces for federation between domains:

- Federation for the establishment of a trail.
- Federation for configuration.

It is the federation requirements and interfaces for configuration which require a detailed analysis. This is needed also in the case that we plan to use federation using signaling, as each domain needs to be configured in order to determine through which neighbor domain it has to create a tandem connection.

A possible solution is to consider X-Coop. The next figure shows an interworking and federation scenario using an X-Coop adapter.

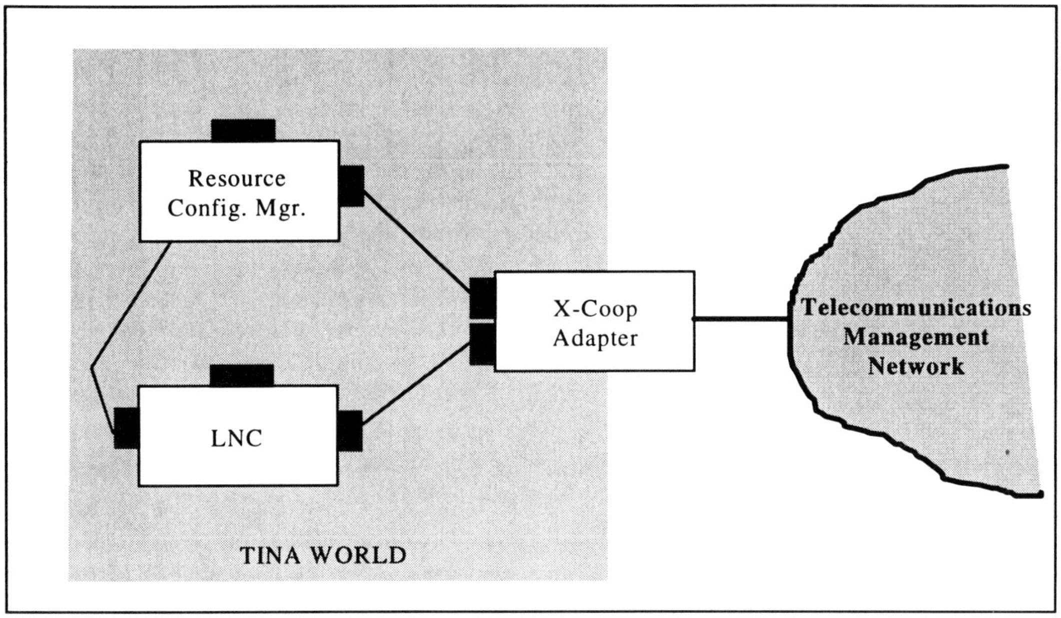

Figure 7: Interworking with management systems

5. Extensions to the LNFed interface

The basic LNFed interface described in the preceding sections has provided a mechanism for connection creation across Layer Network domains, and some potential legacy system solutions have been proposed. However connection creation is not the only function that the LNFed interface must deal with. These issues include routing information exchange, federation management, and topological link management (i.e. the application of FCAPS). From experience with work on the ConS reference point (TINA reference point between the components of the Service Architecture and the Connection Coordinator introduced in section 2) the creation of interfaces including FCAPS is non trivial and it has not been considered in this paper. Likewise federation management interfaces are not considered, however the exchange of routing information is briefly discussed in this section.

From a connection management perspective the routing infrastructure within the Layer Network will require information about network reachability, or better still it should be in a position to obtain, from neighboring domains, or from some third party, the information required to choose a route. The routing information obtained plus the local fault and performance data, and any local routing policies will affect the routing choices within the Layer Network. The fault and performance information will also have some bearing on routing information propagated by the Layer Network for use by other Layer Networks.

Within this prototype, simple static routing has been used in order to test the LNFed connection management interface, however this is clearly not the most practical mechanism for real networks. Two routing architecture choices are briefly proposed in this section, firstly a hierarchical model and secondly a peer-to-peer model. Both models require the existence of a "Router" within the LNC that acts as the source and sink of the inter Layer Network routing information.

The Hierarchical Model

In the hierarchical model a routing information hierarchy is created with knowledge of all the layer networks and their interconnections. The routing hierarchy can be based on the same principles as the hierarchical TINA Connection Management Architecture where each Layer Network forms a managed element in the routing hierarchy. Routing information updates are propagated up through the hierarchy, originating at routing elements, and periodically processed routing tables are disseminated back to the routing elements. The routing information updates can be based on fault conditions and/or CAC traffic levels exceeding predefined levels, as well as on the internal Layer Network congestion levels which could be passed as a "cost" parameter for types of through traffic. The routing hierarchy could also be provided with dynamic pricing information with the Layer Network routing information, so that shifting price considerations can be taken into account when routing connections. The scope of the routing information could be worldwide or localized to the immediate neighboring networks. The availability of worldwide information implies that the preferred route of a connection could be specified in the LNFed tandem connection request.

Obviously the routing hierarchy would need to be part of a neutral telecommunication infrastructure, similar to the Broker architecture to ensure fair play. Alternatively the architecture could be employed for a set of federated Network Providers. The scope of the information provided by the hierarchy could range widely, depending on the information that Network Providers are willing or able to contribute to the routing hierarchy.

The Peer-to-Peer Model

In the peer-to-peer model routing information between adjacent networks is propagated between the Routers. This model reflects the routing paradigm found in the Internet between Autonomous System (AS) which use Inter Domain Routing Protocols (IDRP) to exchange routing information about network reachability. The IDRP routing protocols used are classified as either distance vector or link state routing protocol. The BGP, border gateway protocol, defined in RFCs 1105, 1163, and 1267 is the most versatile IDRP protocol in general use on the Internet. It is a distance vector protocol with a path string to eliminate routing loops and policy based routing. It provides a full routing information exchange between AS's so that optimal routes can be chosen and load sharing enabled. It is feasible to port the BGP to an IDL interface for routing information exchange between Layer Networks. The Routing interface would require the addition of a federation parameter to limit the scope of the routing information propagation so that Federation boundaries could be honored.

6. Conclusions

The purpose of the work presented in this article was to define and implement a solution on how to manage federation between different domains and determine the extensions that are necessary. Its intention is to contribute to the TINA requirements to define what is needed concerning federation. The initial aim of this work, which was to achieve connectivity across Network Provider domains, has been achieved.

The first version of the system developed and used in the trials contained a number of simplifications. We concentrated mainly on allowing different network providers to federate in order to provide a complete connection through different domains. There was no topological link negotiation, no accounting, fault management and recovery. These issues will be treated in future work since they form part of our desire to show that the TINA architecture can be a framework for providing industrial quality services and connection platforms that adapt well to existing networks.

It was found that although the model can solve the trail creation problem at the software level, there are always hardware problems to consider in order to interface with legacy networks. Thus, it is also our intention to further study the implications, with respect to interworking with legacy systems, that might exist on the federation interface.

7. References

[1] J. Bloem et al., *TINA-C Connection Management Architecture*, in TINA'95, Melbourne, Australia, 13-16 Feb. 1995, conference proceedings vol.1, pp. 485-494

[2] J. Bengtsson, P. Hellemans, L. Lehman and N. Mercouroff, *Implementation of Services for Computer Supported Cooperative Work on TINA: The SPOT Project,* to be published in ISS'97, Toronto, Canada

[3] *Network Resource Architecture*, version 3.0, TINA-C Baseline NRA_v3.0_97_02_10, February 1997.

[4] *OMG, The Common Object Request Broker: Architecture and Specification*, rev. 2.0, July 95

[5] *TINA-C, ODL Manual*, ver. 2.3, June 96

Transaction Concepts in Connection Management Applications

Ennio Grasso
CSELT
Via Reiss Romoli, 274
10148 Torino, Italy
ennio.grasso@cselt.it

Nathalie Perdiguès-Charton
Alcatel Corporate Research Center
Alcatel Alsthom Recherche
Route de Nozay
F-91460 Marcoussis, France
perdigue@aar.alcatel-alsthom.fr

Abstract

Connection management is the activity of coordinating the allocation and relinquishment of network resources on behalf of a manager. This activity is inherently distributed as the resources involved are spread in the network. But, was not the transaction concept conceived by the database community really to mean coordination? So why not to exploit the results and experience achieved in the database domain and apply those ideas in connection management?

Transactions in the database world are a means of allowing each concurrent client with the illusion that it has the sole access to any data it reads or writes and to guarantee that, in case of failure, the data will be restored to a consistent state. In connection management there is a shift of concern on where coordination is to be applied: from data to processing. Indeed, in connection management transactions must be thought of as an instrument to guarantee global consistency of the connection set up and tear down activities.

The work presented in this paper has been partially funded by the European Union in the context of the ACTS ReTINA project (AC048).

1. Introduction

Transaction management has been traditionally investigated in the context of conventional database applications. More recently, the transaction concept has been extended in the broader context of distributed object computing. In this respect, the Common Object Request Broker Architecture (CORBA) defined by OMG [OMG91] is gaining wide consensus as de-jure standard. CORBA specifications are now being populated with general-purpose object services, called CORBAservices, which support basic functionalities useful in a broad range of application domains [OMG95]. Among these services, the Object Transaction Service (OTS) provides support for controlling the scope and completion of distributed transactions.

Yet, there are relatively few studies that address transaction issues in the context of connection management. Although transactions are a useful instrument to guarantee consistency of applications, other compelling requirements are relevant in telecom applications. In a highly demanding environment the ORB must exhibit high performance and predictability and, more importantly, provide the necessary "hooks" that allow the application programmer to fine-tune the behaviour of the system in terms of memory allocation, thread priority, communication management, and concurrency level. In particular, concurrency is a paramount issue in connection management applications, where a trade-off exists between availability on one hand and consistency on the other.

In this paper we argue that traditional transactions are too restrictive for telecom applications and a more general model is needed for allowing more parallelism of applications. We propose an eXtended Transaction Service (XTS) that supports open nested transactions that relax Atomicity and Isolation but increase concurrency and allow the programmer to tailor abort recovery according to application semantics. The IDL specification of the XTS is derived from OMG OTS thus exploiting inheritance of interfaces.

2. Transaction Modelling Concepts

A transaction is an atomic unit of consistent and reliable computing composed of a sequence of operations involving one or more objects. Traditional transactions have the following properties collectively referred to as ACID properties [GRAY93]:

- **Atomicity**: either the transaction completes successfully or it is as if it never began;

0-8186-8335-X/98 $10.00 © 1998 IEEE

- **Consistency**: upon the completion of the transaction all objects accessed by the transaction are in a consistent state;
- **Isolation**: the effects of (i.e. object state changes made by) a transaction are invisible to other concurrent transactions until the transaction completes;
- **Durability**: the effects of a transaction persist after the successful completion of the transaction.

A transaction model can be characterized by the *transaction structure* and the *correctness criterion* of the allowed transactions [BUCK92], [GRAY93].

- the transaction structure specifies the structural dependencies between related transactions; a complex transaction consists of a set of constituent transactions and a set of constraints on the execution structure of those constituent transactions.
- the correctness criterion determines the histories that result from the concurrent execution of a set of transactions which preserve the consistency of the objects accessed.

The basic approach for specifying correctness criteria is based on the definition of conflicting operations: two operations A and B on a shared object O conflict if the effects of executing A after B differ from executing B after A. The most common correctness criterion for concurrent transactions is *serializability*. An interleaved execution of transactions is serializable if there is a possible serial execution of the same set of transactions such that the order of conflicting operations is the same in both executions.

2.1. Nested Transactions

Traditional database applications rely on flat transactions operating on disk pages by simple read and write operations. The need for extended transaction models to support advanced applications has been recognized for some time and several models have been proposed in the literature.

An important step in the evolution of the traditional transaction model is to extend the flat transaction structure by defining *nested transactions* [MOSS85]. In this model a transaction may contain any number of subtransactions which may recursively contain other subtransactions giving rise to a tree of nested transactions. The tree structure of nested transactions induces the following structural dependencies between related transactions:

begin-dependency:
A subtransaction must start after its parent transaction and must complete before it;

commit-dependency:
The commit of a subtransaction is conditionally subject to the commit of its parent and all its ancestors up to the top-level. A nested transaction can finally commit only if the top-level transaction commits.

abort-dependency:
If a transaction (at any nesting level) aborts, the tree of subtransactions whose root is the failed transaction is also aborted whether or not some of the descendent transactions have performed their local commit. Conversely, if a subtransaction fails its parent is not required to abort but can perform its own recovery by:

1. ignoring the failure;
2. retrying the execution of the subtransaction;
3. attempting to execute an alternative subtransaction;
4. aborting itself if the failed subtransaction is vital.

The abort-dependency provides a powerful mechanism for fine-tuning the scope of rollback. In contrast to flat transactions where a failure necessarily undoes the transaction to its start point, nested transactions allow the rollback at the granularity of a single subtransaction. This confines the failure to a smaller part making it less costly to recover.

Since a committed subtransaction may later be rolled back, the nested transaction model achieves global serializability by requiring subtransactions to transfer all the locks acquired during their execution to their parents at commit. Unfortunately, this **transfer rule** may be detrimental for real-time applications because locks are held until the top-level transaction completes.

Note that there exists a close relationship between the modularization concept of distributed object architectures and the nested transaction model. The effect of tallying the invocation hierarchy of object operations with nested transactions (i.e. a new subtransaction per object invocation) produces a grace modularization design that confines failures on a per object basis. In the TINA Computational Model object operations can be marked with a "transactional" attribute. During the execution of a transactional operation an object may invoke other transactional operations giving rise to a tree of subtransactions, one for each invoked transactional operation [BOSC95].

2.2. Open Nested Transactions

Although the requirements of full isolation and global serializability may be mandatory for traditional database applications, in real-time and telecom domains the need to selectively relax ACID properties according to application semantics and improve concurrency is a paramount issue.

Open nested transactions are a generalization of *sagas* and multilevel transactions. *Multilevel transactions* [WEIK91,92], [BEER88] impose a strict layered hierarchy of subtransactions capable of preserving serializability. Sagas [GARC87] introduce a two-level hierarchy between a saga and its child transactions.

Open nested transactions bear a strong structural resemblance with closed[1] nested transactions by retaining the same tree structure and the *begin*, *commit*, and *abort* dependencies between related transactions. Yet, open nested transactions relax *Isolation* by allowing the effects of committed subtransactions to be visible to other concurrent transactions thus waiving the transfer rule of closed nested transactions. An open nested transaction spawns a new "sphere of control" [GRAY93], meaning that all its effects become visible to other transactions when it completes. On the other hand, a closed nested transaction simply extends the "sphere of control" of its parent so that its effects are still isolated against concurrent transactions.

Since the locks acquired during the execution of an open subtransaction are relinquished at commit, the abort-dependency induced by a parent transaction on its open subtransactions cannot rely on traditional rollback: restoring the old state of data modified by an open subtransaction may disrupt the effects of other possibly interleaved transactions. Hence, abort recovery of a committed open subtransaction must be performed by a *compensating transaction* that semantically reverses the effects of the open subtransaction. This relax *Atomicity* since the compensating transaction need not necessarily undo all effects of the transaction to be compensated, but can result in a semantic specific repair action. A formal approach to recovery by compensating transactions can be found in [KORT90].

As an example, consider a subtransaction that inserts a record into a B-Tree causing the tree to split. In the closed model, the subtransaction holds the locks on the B-Tree for the duration of the entire transaction family thus severely limiting access by other concurrent transactions. In addition, if any of the ancestor transactions aborts, the abort recovery scheme will undo every single write on every page changed, ultimately undoing the B-Tree split. While the recovery must remove the inserted record, undoing the B-Tree split is unnecessary burdensome. Conversely, in the open model the locks on the B-Tree will be held only for the duration of the subtransaction, not for the duration of the entire transaction family. Besides, if the parent transaction aborts, the compensating transaction need only remove the new record without undoing the B-Tree split, which is an efficient use of the B-Tree. In general, there might be operations that cannot be compensated. However, since compensation is a generalization of traditional rollback, this problem is not specific to open nested transactions but affects also traditional transactions.

[1] By "closed" nested transactions we mean nested transactions with traditional semantics.

3. OMG Transaction Service

OMG OTS defines the scope of a transaction by means of a transaction context shared between the objects that participate in the transaction. The OTS supports two modes of operation: implicit and explicit. When operating in the implicit mode, the transaction context is associated with the current thread and is implicitly transmitted as part of an operation invocation. When executing in the explicit mode, the client directly controls the propagation of the transaction context as an explicit parameter of a request. The OTS distinguishes between these entities:

- *transactional client*: invokes operations on transactional objects;
- *transactional object*: is an object whose behaviour is effected by being invoked within the scope of a transaction;
- *recoveral object*: is a transactional object whose data is affected by committing or rolling back a transaction. A recoveral object must participate in the transaction protocol by registering a Resource object with the OTS.

The IDL interfaces of the OTS are defined in the *CosTransactionService* module and are:

- **Current**: defines operations that allow a client of the OTS to manage the association between threads and transactions and is defined as a pseudo object[2]. The most important operations of the Current interface are `begin`, `commit` and `rollback` with straightforward semantics.
- **TransactionFactory**: provides a `create` operation that begins a new top-level transaction and returns a Control object.
- **Control**: allows a client to explicitly manage and propagate the transaction context. A Control object is associated with exactly one transaction and provides two operations, `get_terminator` and `get_coordinator`, which respectively return a Terminator object and a Coordinator object for that transaction.
- **Terminator**: supports the operations `commit` and `rollback` to complete the transaction.
- **Coordinator**: supports operations used by the participants in the transaction to query the transaction

[2] Pseudo objects reside in the same address space of their client and their references cannot be propagated in remote invocations (refer to [OMG91]).

about its status and relationship with other transactions, e.g. `get_status`, `is_same_transaction`, `is_related_transaction`, `is_top_level_transaction`, etc. The `register_resource` operation allows a recoveral object to register a Resource as a participant in the transaction. If the transaction is a nested transaction, the resource is implicitly registered with the top-level transaction. When the top-level transaction commits, the Resource objects will participate in the two-phase-commit protocol to commit or rollback the updates performed as part of the transaction family. The `register_subtran_aware` operation registers a resource with a subtransaction such that it will be notified when the subtransaction commits or rolls back. The `create_subtransaction` creates a child transaction of the current transaction and returns a Control object of the new subtransaction.

- **RecoveryCoordinator**: provides the `reply_completion` operation invoked by a recoverable object to determine the state of the transaction after the associated resource has been prepared. A reference to the RecoveryCoordinator object is returned by the `register_resource` operation.
- **Resource**: defines operations invoked by the OTS to drive the two-phase-commit protocol. The operation supported by the Resource interface are very similar to the ones defined by the X/Open XA interface, namely `prepare`, `rollback`, `commit`, `commit_one_phase` and `forget`.
- **SubtransactionAwareResource**: provides two operations, `commit_subtransaction` and `rollback_subtransaction`, invoked by the OTS to notify a resource about the completion of a subtransaction. In general, a resource may have rules that are specific to ancestry and must perform some work as all or some of the ancestors complete. For example, a recoverable object can use this interface to establish a "committed with respect to" relationship useful for implementing the transferal of locks by a subtransaction to its parent (transfer rule).
- **TransactionalObject**: by supporting this interface an object implicitly requires that the transactional context associated with the current thread be propagated on remote invocations.

4. The eXtended Transaction Service

The XTS extends the OTS by allowing any combination of open and closed nested transactions. Note that open subtransactions behave just like closed subtransactions while they are active: the different behaviour of the two models lays only on the commit semantics. In view of the above, the XTS allows the programmer to procrastinate the decision of the nesting semantics up to transaction commit so as to gain the maximum amount of flexibility.

The IDL definition for the *ExtendTransactions* module is given below. In the remainder we will elaborate in same details the extensions and the differences with respect to the OTS.

```
#include <CosTransactions.idl>
module ExtendedTransactions {

 interface Compensator {
   void compensate();
 };

 interface Current : CosTransactions::Current {
   void definite_commit(in boolean
          report_heuristic, in Compensator c)
   raises(
   CosTransactions::SubransactionsUnavailable,
   CosTransactions::HeuristicMixed,
   CosTransactions::HeuristicHazard);
   };

 interface Terminator :
      CosTransactions::Terminator {
   void definite_commit(in boolean
          report_heuristic, in Compensator c)
   raises(
   CosTransactions::HeuristicMixed,
   CosTransactions::HeuristicHazard);
   };
};
```

Since the XTS is designed to be fully compatible with the OTS most interfaces extend by inheritance the interfaces defined in the CosTransactions module. As will be shown, the implementation of the XTS can leverage on the capabilities of an implementation of the OTS already available. The ExtendedTransaction module defines these interfaces:

- **Current**. This interface inherits from the CosTransactions::Current interface and adds the `definite_commit` operation used for definitely committing a subtransaction thereby implying the open semantics for that subtransaction. In contrast to the commit operation, the `definite_commit` operation forces a definitive commit by triggering the two-phase-commit protocol for the subtransaction so that all its effects are made permanent and visible to other transactions. All locks acquired by the subtransaction and by its closed descendent transactions are relinquished in this phase. The `c` parameter specifies the Compensator object that will be responsible for compensating the committed subtransaction in the event that the parent transaction aborts.
- **Terminator**. This interface extends the CosTransactions::Terminator interface with the

operation `definite_commit` with exactly the same semantics as in the Current interface.

- **Compensator**. This interface defines the `compensate` operation that will be invoked by a superior Coordinator to ask the compensation of a committed open subtransaction. The XTS does not pose any constraints on the compensation policy for an open subtransaction. In a typical usage the Compensator object will create a new top-level transaction within the scope of which the compensation is to be performed.

In papers [GRAS96a], [GRAS96b] we advocate the importance of a Programmable Concurrency Control Service (PCCS) in which the developer can specify, for a particular lock set, the number of lock modes and their conflict relations. By combining the features of the XTS (definitive commit) and the PCCS (semantic conflicts) the concurrency degree of the application increases considerably.

4.1. Implementation Issues

In this section we outline a possible implementation of the XTS assuming that an implementation of the OTS is available. The XTS can delegate much of the work to the OTS while retaining the responsibility for managing the behaviour that realizes the open semantics of subtransactions. Indeed the XTS has been implemented in the context of the ReTINA project whose aim is the definition of a real-time ORB for the telecom domain.

As for the commit phase, the only difference between the XTS and the OTS is the implementation of the `definite_commit` operation. In the OTS the root Coordinator is responsible for driving the two-phase-commit protocol when the top-level transaction completes. The participants in the protocol are the Resource objects registered with the transaction and the subordinate Coordinators as a result of *interposition*[3]. When a subtransaction is completed using the commit operation, the Coordinator that represents that subtransaction notifies all registered SubtransactionAwareResources and registers any resources registered using `register_resource` with its parent Coordinator, or registers a subordinate Coordinator to relay any future requests to the resources.

In the XTS, when a subtransaction is completed using the `definite_commit` operation, the Coordinator that represents that subtransaction actually begins the commitment protocol involving all the Resources and the subordinate Coordinators registered with it as if it were a top-level transaction. If the commitment protocol completes successfully, the Coordinator registers with its parent a Compensator object that will be responsible for compensating the committed subtransaction in the event that its parent transaction aborts.

As for the abort recovery scheme of the XTS, when a transaction aborts all its child subtransactions are rolled back or compensated; specifically:

- every active subtransaction is aborted and rolled back;
- every committed closed subtransaction is rolled back;
- every committed open subtransaction is compensated by invoking the `compensate` operation of the Compensator object. The inverse operation must contain enough semantics to undo the effects of the whole subtree of transactions: no nested compensation will be invoked. In a typical design the compensate operation creates a new top-level transaction and invokes other operations provided by different objects.

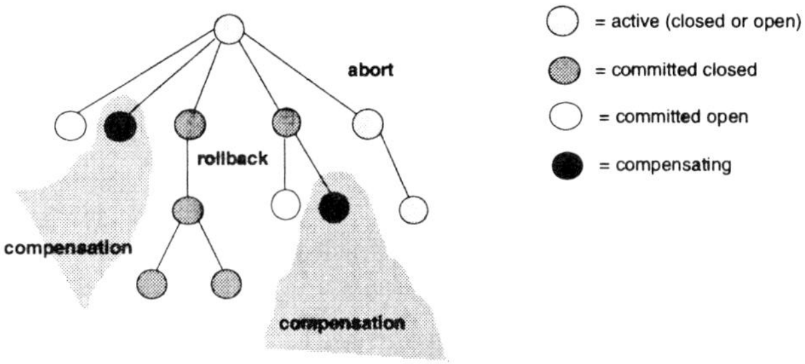

Figure 1. The abort recovery scheme

5. XTS in Connection Management

The mission of Connection Management is to support telecommunication services in their need for connectivity. So it provides basic functions to set-up, control and release connections within a Network.

By essence Connection Management (CM) would take benefits from a transactional behavior. Indeed *Atomicity*, *Consistency*, *Isolation* and *Durability* properties (see 2. Transaction Modelling Concepts) would improve the CM efficiency.

- **Atomicity**: From the client point of view either the whole connection has been set up or nothing has happened. This property is fundamental in particular for the establishment of point-to-multipoint connections.
- **Consistency**: The CM state remains consistent. For instance changes of the amount of available bandwidth on resources are made in a consistent way during the establishment process.
- **Isolation**: Multiple clients are able to request connections simultaneously, without interfering with

[3] Interposition implies that when the transaction context is propagated beyond a transaction domain, the importing domain creates a new Coordinator object that acts as subordinate of the calling domain.

each other. This property seems crucial since telecommunication environments are often characterized by a high number of potential simultaneous clients.

- **Durability**: once created, connections survive until explicit deletion.

The introduction of transactional properties into an existent CM service is currently being experimented in the context of the ReTINA ACTS project [RETINA]. The main goal of the ReTINA project is the development of an industrial strength TINA compliant Distributed Processing Environment (DPE) that meets requirements related to the support of telecommunication and multimedia interactive applications. A TINA compliant Connection Management service has been specified and implemented as a DPE (telecom) service. The objective is now to improve the efficiency and reliability of this Connection Management service while introducing transactions together with other engineering mechanisms such as parallelism and concurrency.

5.1. Overview of Connection Management

In the ReTINA (or TINA) context, a network is modelled as a number of Layer Networks (for instance, an ATM network modelled with a VC layer and a VP layer networks). A layer network is made up of (partitioned in) Subnetworks interconnected via links. Each Subnetwork, in turn, can be partitioned into smaller Subnetworks. At the lowest partitioning level Subnetworks model individual network elements (e.g., ATM switches).

The ReTINA Connection Management is defined by a set of computational objects. Each object supports a number of operational interfaces which specify groups of operations that can be performed by that object. Furthermore each object of the architecture is responsible for the management of a network infrastructure element. In particular, a Connection Performer (CP) object is responsible for a Subnetwork, either an intermediate partitioned Subnetwork or one of the lowest Subnetworks (modeling the network elements). In the former case, the CP is called a NML-CP (Network Management Layer - CP) while in the latter it is called an EML-CP (Element Management Layer -CP).

Any CP object is responsible for the establishment, control and release of Subnetwork Connections (Snc) within its own Subnetwork (Snw). The overall connection set-up process is therefore performed following a hierarchical approach. In particular hierarchical routing is performed by the NML-CP objects architecture, i.e. routing is performed at each partitioning level, by each involved NML-CP. Figure 2 presents the Connection Performers architecture with a three-partitioning-level network. The lowest 5 EML-CP objects are attached to the 5 switches.

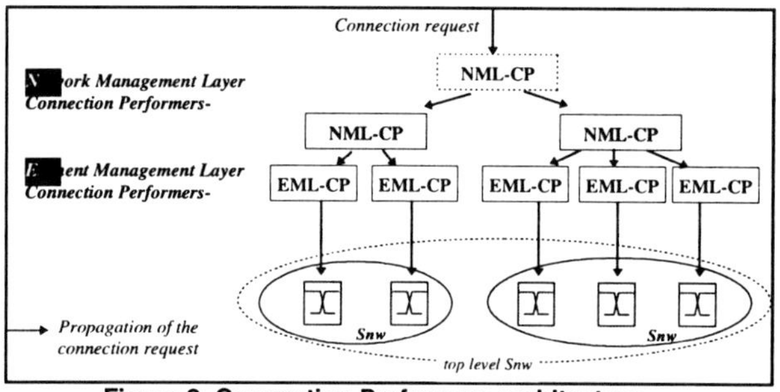

Figure 2. Connection Performers architecture

5.2. Transactions in Connection Management

As shown in Figure 2, a connection request sent to the top-level NML-CP is propagated recursively down until the leaf EML-CPs are reached. Each EML-CP then converts the operations into the ones defined on the controlled switch element.

Due to the hierarchical approach, the connection set-up can really benefit from transactional support that guarantees *Atomicity* and *Isolation*. Indeed the establishment of some of the lowest Subnetwork Connections along the end-to-end path - even for a point-to-point connection - may fail. Atomicity thus guarantees that either the whole connection set-up procedure succeeds or any partial allocation is torn down. Moreover for performance reasons (mainly shorter set-up delays) concurrent connection requests are allowed and requests are propagated in parallel along the CPs hierarchy. Therefore the connection requests must be supervised in order to avoid allocating the same resources. Isolation becomes a fundamental property.

In the following, code fragments illustrate how a transactional Connection Management service may be specified and implemented.

Below is the code fragment describing interfaces involved in the connection set-up process[4]. CM_i_CPControl is a CP interface that supports set-up (`setup_connection`) and release (`release_connection`) operations. This interface is specified as transactional (inherits from TransationalObject).

In addition the CM_i_SNCContol interface provides access to a given Subnetwork Connection (Snc) and CM_i_TPPool to a given TPPool. TPPools (Termination Point Pools) are end-points characterized by an amount of

[4] This specification fragment is extracted from the ReTINA specifications currently being developed.

available bandwidth, in each direction (either going in the Snw or out of it)[5]. Resources are reserved allocating (removing) bandwidth and are released freeing (adding) bandwidth from TPPools.

```
...
interface CM_i_CPControl :
     CosTransactions::TransactionalObject {

  // setup a new point-to-point or
  // point-to-multipoint connection.
  void setup_connection(inout CM_TinaName root,
                        inout CM_TinaNames leaves,
                        out CM_i_SNCControl conn)
  raises (...);

  // release a connection.
  void release_connection(
                         in CM_i_SNCControl conn)
  raises(...);
};
...
interface CM_i_SNCControl {...};

interface CM_i_TPPool {
  readonly attribute BandwidthType availBWin;
  readonly attribute BandwidthType availBWout;

  void removeBandwidth(in BandwidthType bwin,
                       in BandwidthType bwout);

  void addBandwidth(in BandwidthType bwin,
                       in BandwidthType bwout);
};
```

The `setup_connection` operation may be implemented as shown below (pseudo C++ code). The set-up process is realized within the context of an "eXtended" subtransaction. First a Route has to be selected among the lower Subnetworks of the current Subnetwork. Once the Route is selected the resources along that Route are reserved; this implies the allocation of bandwidth in the involved TPPools. Then the selected inferior CPs (corresponding to the inferior Snws along the Route) are requested to establish a Snc within their own Snw. These requests are made in parallel.

Finally the subtransaction is definitely committed (`definit_commit`), with the suitable compensator (`snc_compens`) as a parameter. If the parent transaction decides it, the current subtransaction may be compensated which would simply correspond to the deletion of the established Subnetwork Connection.

```
CM_i_SNCControl CM_i_CPControl::setup_connection(
             TINAName root, TINANames leaves) {

  txn_ctx.begin;
  snc=new CM_i_SNCControl;
  snc_compens=new Snc_Compensator(snc);

  // selection of the route and bandwidth
  // allocation
  txn_ctx.begin;
  .... (see details below)
  txn_ctx.definite_commit(...);
  // route selected and bandwidth allocated

  // ask for inferior Snc, in parallel
  for (int i=0; i<route.length; i++)
       sncs[i]=cps[i]->setup_connection(...);
  ...
  definite_commit(0,snc_compens);

  return snc;
}

// Snc_Compensator inherits from XTS Compensator
// interface: Snc_Compensator
// CosExtendedTransactions::Compensator{...};

Snc_Compensator::compensate() {
   my_snc.cp->delete_snc(my_snc);
}
```

The selection of the route and the bandwidth allocation are themselves executed within the context of one subtransaction. Thus the parent subtransaction (Snc subtransaction) keeps the control on its "sub-subtransaction" and is able to request its compensation in case of a problem. In particular the resource reservation can be compensated, adding a given amount of bandwidth of the corresponding TPPools. Locks are obtained on the set of topological resources (set of TPPools) before the selection of the route. Indeed admission testing functions have to decide whether the requested connection can be accepted or not on the basis of the TPPools available bandwidth. Therefore concurrent requests should not be able to modify the bandwidth information. Once selected the TPPools are allocated and the obtained locks are automatically released when the subtransaction definitely commits. A possible implementation is presented below.

```
  ...
// selection of the route and bandwidth
// allocation
  txn_ctx.begin;
  my_lockset=lock(mode);

//route selection
  route=find_route(root, leaves);
  bw_compensator=new Bw_Compensator(route);

// bandwidth allocation
  for (int i=0; i<...; i++)
       TPPools[i]->remove_bandwidth(...);
  ...
  txn_ctx.definite_commit(0,bw_compens);

// route selected and bandwidth allocated
  ...
```

Other locking policies could be applied to control the access to shared resources. Indeed to ensure correctness of the program but also to achieve acceptable performances there is basically a subtle trade-off between the overhead imposed by fine-grain locks and expected benefit from concurrency [CHAR97].

[5] The way parallelism, i.e. multithreading, is dealt with is irrelevant here.

The use of `commit` rather than `definite_commit` to complete the allocation functions would give closed semantics to the subtransaction. The execution of concurrent connect operations would then be serialized because the lock that protects the set of end-points would be transferred by the allocation subtransaction to its parent at commit. Indeed when the allocation is executed as an open subtransaction, as in the code above, the locks are held only for the duration of the subtransaction itself, not for the duration of the entire connection set-up. In this way we gain the maximum amount of concurrency and still guarantee a level of consistency tuned to the connection management application.

A different alternative for improving concurrency is to execute the bandwidth allocation outside the scope of any transactions. However, the Atomicity property of the whole connection set-up procedure would not be extended to the allocation operation so that allocated termination points will not be automatically released as part of the abort recovery scheme.

6. Conclusion

This paper aimed at addressing transaction issues in the context of connection management. Connection Management services support telecommunication application in their need for connectivity, providing basic functions to set-up, control and release connections within a Network. The introduction of transactions would improve the efficiency and reliability of Connection Management.

Support of specific requirements such as performance requirements (e.g. short set-up delays) and scaleability requirements (e.g. support of high number of clients) also implies the support of parallelism and concurrency within CM services. Therefore the use of traditional transactions would be too restrictive.

The proposed eXtended Transaction Service (XTS) that supports open nested transactions relaxes the pure Atomicity and Isolation properties. However it increases concurrency and allows the programmer to tailor abort recovery according to application semantics. Integrating XTS into Connection Management services we gain the maximum amount of concurrency and still guarantee a level of consistency tuned to these services. Such an integration process is currently being experimented within the ReTINA project.

7. References

[BEER88] C. Beeri, H.J. Schek, G. Weikum; "Multilevel Transaction Management, Theoretical Art or Practical Need?"; in Advances in Database Technology - EDBT'88, Springer-Verlag, 134-154, 1988.

[BOSC95] P.G. Bosco, E. Grasso, C. Moiso, "An Extended Transaction Model for an Object Distributed Architecture", 1st Int. Workshop on High Speed Networks and Open Distributed Platforms, St. Petersburg, 1995.

[BUCK92] A. Buckmann, M. Tamer Ozsu, M. Hornick, D. Georgakopoulos, F.A. Manola; "A Transaction Model for Active Distributed Object Systems"; in Database Transaction Models for Advanced Applications; Morgan Kaufmann Publishers, 1992.

[CHAR97] N. Charton, Y. Hervé, N. Mercouroff, "Engineering of a Broadband Connectivity Service: The TINA approach"; submitted to 2IN'97 conference, Paris, September 97.

[GARC87] H. Garcia Molina, K. Salem; "Sagas"; in Proc. ACM SIGMOD Int. Conf. on Management of Data, May 1987.

[GRAS95] E. Grasso, G. Spinelli; "Exploiting an Extended Transaction Model for Connection Management Applications"; in Proc. TINA Conf., Melbourne, 1995.

[GRAS96a] E. Grasso; "A Programmable Concurrency Control Service for Real-Time Object Request Broker"; in Proc. 8th Euromicro Workshop on Real Time Systems, L'Aquila, 1996.

[GRAS96b] E. Grasso; "A Programmable Concurrency Control Service CORBA"; in Proc. 3rd International Workshop on Services in Distributed Networked Environments, Macau, 1996.

[GRAY93] J. Gray, A. Reuter; "Transaction Processing: Concepts and Techniques"; Morgan Kaufmann Publishers, 1993.

[KORT90] H. Korth, E. Levy, A. Silberschatz; "A Formal Approach to Recovery by Compensationg Transactions"; Proc. od 16th Conf. VLDB, 1990.

[OMG91] "The Common Object Request Broker Architecture and Specification", OMG doc. n. 91-12-1, December 1991.

[OMG95] "CORBAservices: Common Object Services Specification", OMG doc. n. 95-3-31, March 1995.

[MOSS85] E. Moss; "Nested Transactions"; MIT Press, 1985.

[RETINA] "DPE Services Specifications v2 (Telecom Services)"; ReTINA ACTS Project AC048 Deliverable, to be delivered in June 97.

[WEIH88] W. Weihl; "Commutativity-Based Concurrency Control for Abstract Data Types"; IEEE Trans. Comp. 37(12):1488-1505, December 1988.

[WEIK91] G. Weikum; "Principles and Realization Strategies of Multilevel Transaction Management"; ACM Transaction Database Systems 16(2):132-180, June 1991.

[WEIK92] G. Weikum; "Concepts and Applications of Multilevel Transactions an Open Nested Transactions"; in Database Transaction Models for Advanced Applications; Morgan Kaufmann Publishers, 1992.

Session 14

Application Development Tools

Nicolas Mercouroff, Alcatel Alsthom Recherche

Towards a Behavioral Description of ODL

Nils Fischbeck, Joachim Fischer
Humboldt University Berlin, Dept. of Computer Science
Axel-Springer-Str. 54a, 10099 Berlin, Germany
Tel. +49 30 20181 231, Fax. +49 30 20181 234
{fischbeck | fischer}@informatik.hu-berlin.de

Marc Born, Andreas Hoffmann, Mario Winkler
GMD FOKUS Hardenbergplatz 2, D-10623 Berlin
Tel. +49 30 254 99 385, Fax. +49 30 254 99 202
email: {born | a.hoffmann | winkler}@fokus.gmd.de

Abstract

This paper presents a methodology to provide behavioral description starting with TINA ODL. It defines a mapping from TINA ODL to the ITU-T specification and description language SDL. SDL guarantees a formal and executable semantics for behavior and a smooth development process from computational viewpoint to engineering viewpoint. Experience, using this technology for implementing parts of the TINA service architecture is provided.

1. Introduction

TINA ODL [18] is a language designed to develop specifications of TINA systems from the ODP computational viewpoint [20], [17]. However the audience for ODL as a description technique goes far beyond the TINA community. Advanced concepts of TINA ODL like multiple interfaces, groups and stream interfaces influence the development of other definition languages. The Object Management Group is on its way to allow multiple interfaces in OMG IDL [11] and has recognized the importance of stream flows by requesting a stream control interface language binding [12]. The international standardization body ITU-T has established a new question to develop an ITU-ODL based on TINA ODL [15].

Despite many advantages of TINA ODL open issues remain. Compared to the initial proposal for a description of quality of service in ODP standardization [5] TINA ODL has only a weak concept for quality of service. Already identified by TINA but not solved is the ambiguous description of behavior in TINA ODL.

2. Why behavioral descriptions?

A TINA ODL description serves at least two purposes:

- to describe the usage of TINA system components and
- to give a starting point for an implementation of TINA system components.

The first objective addresses a system designer who wants to model a TINA system from computational viewpoint. The second objective addresses a system designer (or programmer) who wants to implement a TINA conform component. Given that a TINA ODL specification is the only source of information it is likely that the implemented component does not satisfy the needs required by the user. This is caused by the unsatisfying means to describe behavior in TINA ODL. It is not possible to describe formally

- which component supplies a supported interface of a group,
- which operations yield the availability of new interfaces,
- under which constraints an operation is successfully executed, like
  - what operation shall be invoked before
  - what conditions shall be satisfied (attribute values, timing conditions).

Behavior description could be supplied by two means:

1. to extend ODL with behavioral description concepts and
2. to supply additional descriptions in another language.

The first approach, an extension of ODL, could include:

- distinguished object states: operations are only executable in certain states,
- description of static and dynamic object and interface creation: what objects/interfaces exist at creation time of the TINA system, who creates additional objects/interfaces,
- interface identification: what group interfaces identify which object interfaces and

0-8186-8335-X/98 $10.00 © 1998 IEEE

• object decomposition: in contrary to groups independent to distribution.

This paper addresses the second approach: we describe behavior of TINA ODL objects in the object-oriented language SDL [23]. The ITU-T specification and description language SDL is developed for design and implementation of telecommunication applications. Communicating extended finite state machines are the main foundation of SDL.

Using SDL guarantees a behavioral description with formal and executable semantics in the computational viewpoint as described in [21]. It can be used to simulate prototypes of a TINA system. In addition it guarantees a smooth design process from computational viewpoint to engineering viewpoint since SDL can be used as an implementation language. Such a smooth viewpoint transformation reduces the total development costs.

To represent ODL components in SDL a mapping has to be defined. The mapping and an evaluation of the mapping is contained in the following sections. Using ODL in connection with SDL is successfully deployed in several projects like TANGRAM [14], TOSCA [19] and Camouflage [4]. An example of the TANGRAM context is presented in this paper.

Note, that an extension of the methodology described above can also lead to an extended ODL. In that case, the behavior clause of current ODL syntax can be used to place SDL like behavior description elements (states, transitions ...) there. This would be a new language and a semantic has to be defined for it. The main drawback of that method is, that existing tools cannot be easily used and a consensus about the language must be found - a process which may be time consuming.

3. Methodology

In the following section our tool-based methodology for the design of a computational model is shown. Figure 1 shows the overall development process by using SDL. The most important steps of the methodology are:

Step (1): Using the information and the enterprise specification, partially developed in OMT, a computational ODL specification has to be defined and realized by a graphical OMT/ODL editor like Y.SCE [22].

Step (2): The ODL specification is mapped into a structural equivalent SDL specification. This can be done automatically and is founded by a language mapping which is discussed in the next section. The resultant SDL specification is an abstract one in such a sense that it can neither be simulated nor executed. It includes interface and object templates without any behavior descriptions or instance definitions.

Step (3): The SDL inheritance feature is applied to enrich the SDL specification generated by step 2 with behavior descriptions for both the interface and object templates. That means, that for all generated block and process types sub types are defined. The behavior is than added in the sub type specification. Defining the behavior there insures an easier reengineering, because only the super types have to be generated in case of changes of the ODL the behavior descriptions are preserved. Possible incompatibilities can be detected then by the checker tools. Furthermore, inheritance guarantees the possibility of repetition outlined in step 5. The behavior description is based on the specification of states and transitions in dependence of a required support of sequences of operations at the object interfaces and defined exceptions.

Step 3 can be repeated to achieve different levels of abstraction in the design of object composition and behavior. The result is an executable computational model which is not the engineering solution. It defines only a simulation model to observe and evaluate special aspects of object behavior which are visible from outside. Especially complex object configurations can be studied.

Step (4): The SDL specification from the last step can be developed using the SDL Integrated Tool Environment SITE [9], [13] or other tool packages which provide tools to edit SDL specifications, to check the correctness of syntax and static semantics, and to generate C++. The generated C++ code can be linked with a simulation library, which produces a simulator for the SDL system which represents an ODL specification added by computational behavior aspects.

Step (5): A simulation of the computational model is used to detect design errors prior to implementation. The components of the SDL simulator can be distributed using CORBA. An SDL debugger [10] is a component which supports that kind of validation. If design errors are detected, a repetition of the steps 1 to 4 could be necessary.

Step (6): The accepted computational model is used for deriving an engineering one. Decisions concerning the engineering and technology viewpoints like the physical distribution of the objects or of the communication mechanism must be made at this point. The engineering specification cannot be derived from the computational model by refinement in each case, so that the structure and behavior parts cannot derived by SDL inheritance in each case too. However the SDL specifications which describes the engineering model can also be analysed by the SITE tool (step 4 etc.). Furthermore the target code can be generated and linked with corresponding runtime systems.

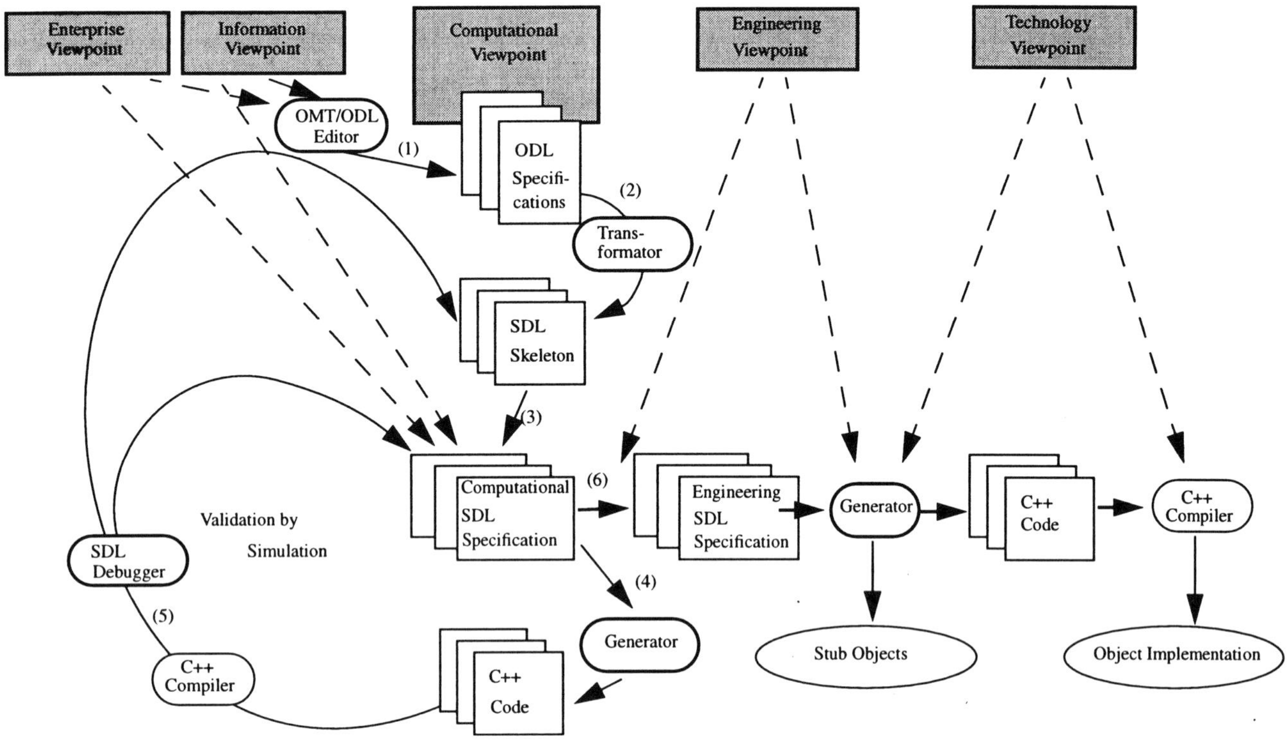

Figure 1: Overall development process using SDL

4. Mapping

As addressed in step 2 of the methodology (Section 3) mapping rules are used to convert an ODL specification to SDL. A first approach was published in 1995 [1]. It defines a language mapping from OMG IDL to SDL and was the foundation for further developments which lead to [16] as a contribution to the ITU-T standardisation. In parallel it was contributed and included as an issue in [21]. Based on [1] and [16] the mapping rules were also extended to handle TINA ODL [3]. A similar mapping from OMG IDL to SDL was developed by Telelogic [2]. They both map operations onto remote procedure calls which leads to following problems:

- **Deadlock situation**
  Since the client is blocked after calling a remote procedure there is no possibility for the client to (re)obtain control if the server is not responding within a certain period of time.

- **Call rejection not possible**
  Due to the restrictions of remote procedures in SDL it is not possible for the server to reject a remote procedure call in a certain state.

- **No adequate exception mapping**
  Another issue is that the mapping [1] offers no possibility for raising an exception in the case of an unsuccessful termination of a remote procedure by the server. The mapping for exceptions mentioned in [16] is too sophisticated and not efficient enough.

As a result of these restrictions it is not possible to describe the overall behavior of an interface on a high level of abstraction because there is no facility to specify allowed sequences of operation invocations and the exception handling. The exception handling for itself is absolutely necessary when describing distributed telecommunication systems of today in an easy way. The authors are convinced, that a new software development methodology has a chance on the market only if it can reduce time and costs of the development cycle.

It should be noted that all these restrictions and problems are not caused by the basic SDL concepts itself but by the remote procedure concept. The solution for all these problems is to substitute a remote procedure by a set of signals which contains the invocation and the different terminations (success and exceptions). This more flexible approach allows to:

- throw and catch exceptions by sending resp. receiving the appropriate signals,

- specify timeouts at the client side and so avoid deadlocks,
- reject procedure calls without invoking the procedure implementation if the server is not in an appropriate state,
- describe high level interface behavior independently from procedure implementation,
- use the full flexibility of the communicating extended finite state machines.

Table 1 shows the improved mapping rules. However it should be noted that the ITU-T is currently working on an enhancement of the SDL remote procedure concept. This new concept allows exception handling and raising as well as call rejection and specifying time constraints on client side. It's model in SDL is similar to the signal exchange concept described above. So, it is explicitly noted here that the outlined signal exchange concept is used as an currently available workaround only. If the new remote procedure concept is integrated in SDL, the language mapping will be changed to use it again.

ODL Construct	SDL Construct
basic datatypes	ASN.1 datatypes (e. g. **INTEGER**)
structured datatypes	structured ASN.1 datatypes (e.g. **SEQUENCE**)
constants	SDL synonyms
interface references	SDL **PId**'*s*
interfaces	SDL process types
modules	no construct - only represented in the naming rules
operations	SDL procedures, signals and signal-lists
exceptions	ASN.1 datatypes and signals
interface attributes	SDL variables and remote procedures for access
flows	parameterisized signals
trading attributes	SDL variables and remote procedures for access
objects	SDL block types
groups	SDL block types
behavior and usage texts	SDL comments
quality of service attributes	SDL variables and remote procedures for access

Table 1: ODL constructs and their SDL representations

Table 1: ODL constructs and their SDL representations

The practical experiences made within the context of the TANGRAM project and the ACTS project TOSCA have shown that this language mapping is suited to describe formal and executable behavior which is needed to improve robustness of components as well as to enable interoperability between components developed by different partners. Furthermore, the foreseen transparent replacement of these components which is necessary in the project's scenarios requires behavioral compatibility.

The introduced methodology including the language mapping allows to specify external visible behavior at the interfaces of an object which is a major requirement resulting from the practice. This is provided by three kinds of behavioral definitions:

1. definition of valid sequences of operation calls which covers the raising of exceptions in the case of a breach,
2. definition of valid operation parameters (exceptions are raised if a failure occurs),
3. specification of detailed operation behavior inclusive the interaction with other interfaces at other objects.

The first kind of definition is not limited to one interface. Several interfaces can be involved in a valid sequence of operation calls. The last kind of definition represents a complex behavior and it is difficult to verify it without the means of simulation provided by SDL tools.

The following example explains the first kind of definition in more detail. The interface specification is part of the TINA Retailer Reference Point specification [6] and shows the RetailerAccess interface which is shortened for simplicity reason. The allowed sequence of operation calls is originated from the project's requirement. It is not prescribed in the TINA architecture.

```
interface i_RetailerAccess {
        void setConsumerCtxt (// for the purpose of initiali-
zation
                in t_AccessSessionSecretId asSecretId,
                in t_ConsumerCtxt consumerCtxt
        ) raises (
                e_AccessError,
                e_ConsumerCtxtError);
        void endSession (// to terminate a service
                in t_AccessSessionSecretId asSecretId,
```

```
            in t_SessionId sessionId
        ) raises (
            e_AccessError,
            e_SessionError);
        void startService (// to establish a service
            in t_AccessSessionSecretId asSecretId,
            in t_ServiceId serviceId,
            in t_ApplicationInfo app,
            in t_SessionModelReq sessionModelReq,
            in t_StartServiceUAProperties uaProperties,
            in t_StartServiceSSProperties ssProperties,
            out t_SessionInfo sessionInfo
        ) raises (
            e_AccessError,
            e_ServiceError)
};
```

Figure 2 shows the desired sequential order of operation calls at this interface as a state overview diagram which can be transformed easily into an SDL specification.

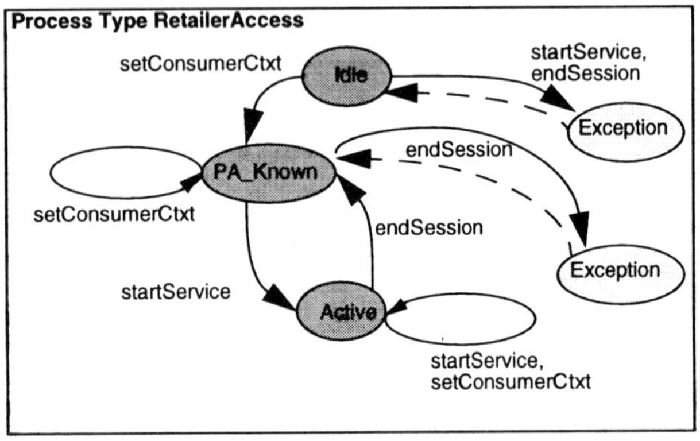

Figure 2: SDL state overview diagram

SDL gives even more support to define behavior due to its concept of extended states: an SDL process contains variables which can be used to refine the transition between states. Applying this concept to the example would allow to associate every creation of a service to its termination. In a simple specification this could be done by a session counter which is incremented at creation time and decremented at termination.

The defined states have to be taken into account during the later implementation of the objects and their interfaces.

As one result of the interface behavior specification in SDL the missing of exceptions in the ODL specification of TINA has been identified.

5. Conclusion

The behavioral description in TINA ODL is insufficient to describe enough detail for the use or implementation of TINA system components. An approach to solve this problem is to extend TINA ODL. Such extensions could include distinguished object states, object composition, object creation constraints and group interface identification.

Another approach enriches TINA ODL with additional behavioral description in another language. In this paper a mapping from TINA ODL to the ITU-T specification and description language SDL is presented. This approach guarantees a smooth development process from the computational viewpoint to the engineering viewpoint. It provides constructs to describe the behavior of TINA objects on a high level as: the order of operation invocation, the rejection of operations in selected states, the specification of timeouts and the handling of certain exceptions. More detailed descriptions of object behavior can be added in later development steps. The problems identified in this paper regarding SDL remote procedure calls are currently addresses by ITU-T study group X.

6. References

[1] Born, M.; Fischer, J.; Winkler, M.: Formal Language Mapping from CORBA IDL to SDL'92 in Combination with ASN.1, Informatik-Bericht Nr. 44, Humboldt-University Berlin, 1995

[2] M. Björkander: Mapping IDL to SDL, Telelogic AB, 1996

[3] Born, M.; Winkler, M.: SDL-92, CORBA-IDL und TINA-ODL im ODP-Entwurfsprozeß, Master Thesis, Humboldt-Univ. Berlin, 1996

[4] Project Camouflage: Humboldt-University Berlin, http://camouflage.informatik.hu-berlin.de/camouflage

[5] SO/IEC JTC 1/SC 21/WG/ N1192: Working Document on QoS in ODP, Canberra, Jan. 1997

[6] Farley, P.; Hogg, S.; Kristiansen, L. et al.: Ret Reference Point Specifications, Snapshot 1, Version 0.4, TINA-Consortium, May 14, 1997

[7] Fischbeck, Nils: Experiences with ISDN Validation Models in SDL and Proposal for new SDL Features, SDL Forum, Evry, 1997

[8] Fischbeck, N., Fischer, J. et al.: Improving the Development and Validation of Viewpoint Specifications, FMOODS Canterbury, 1997.

[9] Behnke, U.; Geipl, M., Kurzbach G.; Mundstock, R.; Schröder, R.; Fischbeck, N.: Development of Broadband ISDN Telecommunication Services using SDL'92, ASN.1 and Automatic Code Generation, FORTE'95, Quebec, 1995

[10] v.Löwis, M.: Using CORBA in an SDL Simulation Environment. DOCT Frankfurt, 1997.

[11] OMG: Control and Management of A/V Streams RFP (Telecom RFP1), telecom/96-08-01, OMG 1996

[12] DSTC, IONA: Revised Submission on Multiple Inter-

faces and Composition, orbos/97-01-02, OMG, 1997

[13] Project SITE: http://www.informatik.hu-berlin.de/Themen/SITE

[14] Project TANGRAM (Distributed, Integrated Platforms for Multimedia Telecommunications, Management, Mobility and Security) is commissioned by DeTe-Berkom, a subsidiary of Deutsche Telekom Group, http://www.fokus.gmd.de/ovma/general/projects/tangram/

[15] ITU-T: Report on the Question 2/10 Meeting, TD57 , 28. April - 6.May 1997, Geneva, 1997

[16] ITU-T SG10: Formal Language Mapping from CORBA-IDL to SDL'97 in Combination with ASN.1, Temporary Document 64, 1996

[17] TINA-C: Computational Modelling Concepts, Version 3.3.0, 1997

[18] TINA-C: TINA Object Definition Language MANUAL, Version 2.3, 1997

[19] Project TOSCA (TINA Open Service Creation Architecture): ACTS (Advanced Communications Technologies and Services) Project

[20] ITU-T Rec. X.903 | ISO/IEC 10746-3: 1995, Open Distributed Processing - Reference Model Part 3

[21] ITU-T Rec. X.904 | ISO/IEC 10746-4: 1995, Open Distributed Processing - Reference Model Part 4

[22] GMD Fokus Berlin: Y.SCE Tool, 1997

[23] CCITT: SDL - Specification Description Language, International Standard Recommendation Z.100, Genf,1992

[24] CCITT: SDL in combination with ASN.1, International Standard Recommendation Z.105, Genf, 1996.

[25] CCITT: Message Sequence Charts, International Standard Recommendation Z.120, Genf, 1996

Specifying and Developing TINA Applications with ACE

Piergiorgio Bosco, Gianni Canal, Giovanni Martini, Corrado Moiso
CSELT
Via Reiss Romoli, 274
10148 Torino, Italy
{piergiorgio.bosco, gianni.canal, giovanni.martini, corrado.moiso}@cselt.it

Abstract

*The objective of the paper is to present an innovative environment (**A**pplication **C**onstruction **E**nvironment - ACE) for the specification, the development and the generation of Telecommunication applications according to the TINA and OMG standards. The functionalities offered by the environment are outlined in the following sections, in addition to experiences of usage of ACE for specification activities.*

The work presented in this paper has been partially funded by the European Union in the context of the ACTS ReTINA project (AC048).

1. Introduction

The motivations for developing ACE (Application Construction Environment) were to allow the specification, development and the generation of applications according to the computational model of TINA, with a particular emphasis on the aspects which nowadays are more difficult to be found in commercial tools.

In the IT world the available object-oriented CASE tools for distributed applications currently are more oriented to code generation than to support the specification process. However there is some attempt at improvement due to *defacto* standard architectures such as CORBA [6] and MS OLE (distributed OLE) [13] which allow to consider architectural principles in the early stages of specification. In fact some CASE tools are able to generate applications based on these architectures. These typically generate description languages such as IDL for CORBA and OLE types for MS OLE, from the analysis models (GUI, architecture early phases, application logic and Data logic are forced to be kept separate); finally they also generate the stubs that deal with the distribution mechanics. This approach offers a good starting point for later design/development phases, however it represents only a small step further, since it still lacks the expression of constructs or "services" such as transactions, persistence, concurrency, typical of distributed object environments. Not to say, other requirements traditionally well considered in the telecom industry, such as more formal behavioural description, validation of specifications, performance evaluation etc. are, usually, not addressed in a integrated way. Notably, only SDL-based CASE environments, approximate this level of requirement coverage.

However standard languages familiar to the telecommunication community such as SDL or LOTOS and their latest evolution (that at least addresses some of the behavioural description requirements) do not have sufficiently matched the object-oriented models and behavioural concepts inherent of the recently emerged architectures mentioned above.

In general the lack of a complete formalism for object-oriented distributed processing models is one of the causes of the absence of CASE tools suitable to support the design phases of a system according to such models. This problem can be addressed to some degree by following a pragmatic approach to apply and extend existing notations such as OMT, SDL, CORBA IDL and TINA ODL etc. to distributed computational architectures.

ACE therefore aims at providing functionalities to support main application lifecycle phases, and to represent, in terms of development, a nut-shell for a number of tools, specification editors, development/compiler tools, performance evaluation tools and code generators, etc., from within the analysts/designers/developers all together orchestrate.

2. ACE Functionalities

In this section the main functionalities supported by ACE are outlined. For the sake of clarity, in Figure 1 an overview of the main phases of the development lifecycle supported in ACE is illustrated. From the set of the ACE editors it is possible to produce code both for the simulation environment and for the real platform. The specifications (aligned with the TINA Computational Model [1]) can be executed in a simulated environment, for a first functional validation of the service logics, that

0-8186-8335-X/98 $10.00 © 1998 IEEE

can be traced directly on the specification editors in order to achieve a better comprehension of their execution.

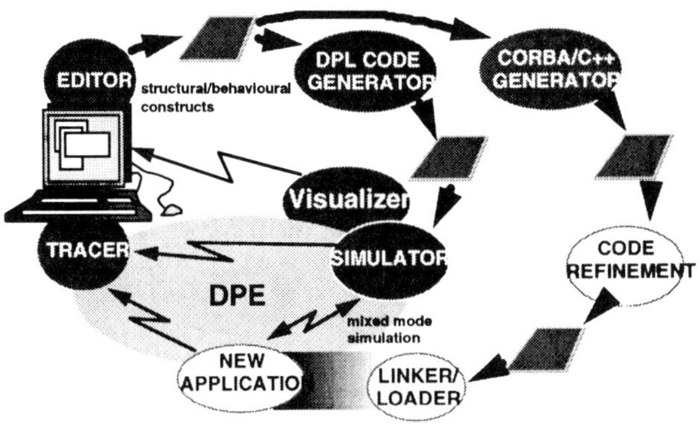

Figure 1: overview of the development cycle

The code generated for the real platform, after a refinement process of the implementation skeletons, is ready to be installed and deployed on the DPE. In the next sections the main components and features of the environment are described.

2.1 ACE Computational Model

As far as conformance objectives towards TINA Computational Model [1][8], the following computational model is realised in ACE:

- objects have multiple interfaces;
- creation/deletion of objects and object-groups;
- interactions among objects, through interfaces:
  - invocations of operations: invocations can be either blocking or non-blocking;
  - accesses (read/write) to attributes;
- concurrence control in multi-threaded objects: guards and semaphores.

2.2 Editing Component

The editing component consists of a suite of graphical editors. They support the detailed specification of templates for object-groups and objects (including their behavioural parts) and interface templates.

Editors for the analysis and design offer support for the standard OMT object model notation, in addition to the possibility of devising object interaction diagrams. Previously defined components in TINA ODL [7] and CORBA IDL [6] files can be imported in ACE, and hereby used and extended.

Each graphical editor is associated to a (graphical) language, consisting of a set of dedicated and customized icons and rules on their composition and interconnection through (labelled) links. The icons can have attributes that are specified through dialogue-boxes: the attributes are introduced by a sequence of dialogue-boxes, where the structure of a dialogue-box depends on the attributes introduced in the previous ones.

Objects' methods are specified by means of a behavioural editor in a flow chart language, where the icons, derived from SDL symbols, are specialised to represent behavioural constructs of the computational model (e.g. invocation of an operation), control constructs (e.g. if-then-else constructs) and synchronisation constructs.

2.3 Syntactic and Semantic Checking

ACE performs global syntax and type checks of the specified templates (e.g. all the constructs and declarations are type-checked, the syntax of the attributes of the icons is controlled). The global checks can detect errors even if no errors were signalled during the editing of a template: in fact, for example, after the change of a variable declaration all the expressions where it occurs may be badly typed.

If the global analysis successfully terminates, it generates an intermediate code representation of the template, that holds all the information concerning the semantics and the typing of the templates, and is used as input to all the final code generators.

2.4 Simulation and Visualization of Specifications

The user specification can be automatically translated into a multiparadigm language that can be used to execute them in a (sequential) simulated environment.

In this way, it is possible to model behavioural evolutions of objects dynamics, and, with the provision of an internal scheduler, to support the execution of concurrent entities within the simulation environment, by providing in this way interleaving semantics.

Mixed mode simulation

Though the provision of a simulated environments has proved to be sufficient for the validation of user specified service logics, it shows a certain degree of inadequacy when developing service components making intense use of already deployed services.

This issue has been tackled devising a mixed mode simulation extension, that allows simulated components to interact with applications and services executing on the real CORBA platform. The interaction of simulated and real world allows the possibility of validating more complex service logics specified by the user (involving interactions with already deployed services), by executing them in a controlled and secure environment.

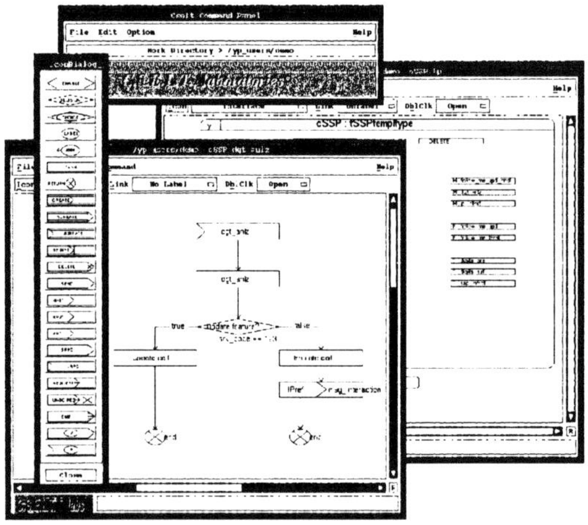

Figure 2: behavioural and object template editors

Visualisation and tracing of executions

The execution of specifications during a simulation can be visualised in two different ways:

- model oriented: the executions are traced in terms of concepts of the computational model (e.g. interactions through interfaces);
- application oriented: the specifications are coupled with scenarios that represent the agents and the artefacts involved in the application (e.g., telephones, switches,...).

The model oriented visualisation is based on the user selectable trace of behaviour diagrams and of object states. These functionalities are tightly integrated with the editing component, in order to trace the executions on the same graphical representation of specifications. The users can interact with the execution by invoking announcements of the objects, define break-points or inspect objects' state.

The application-oriented visualisation (explicitly programmed by the specifier) is based on an interface through which the users can introduce inputs to the application (e.g. by means of dialogue-boxes) and the application can provide outputs in a graphical way (e.g. through animations).

An additional output resulting from a simulation session consists of the automatic production of object interaction diagrams. They ease the task of comparing required output with the effective implemented results.

2.5 Validation and Performance Evaluation of Specifications

At present, functional validation of specification is assumed to be carried on in a user-based interactive way (see 2.4), although a more automated approach has also been devised. It uses technology for exhaustive state space exploration based on SPIN+[2], an extended model checker based on a timed version of PROMELA.

The technical approach chosen for the performance evaluation of an application is the simulation of its specification, augmented with a non-functional characterisation of the significant activities, i.e.:

- duration of activities: single actions or groups of actions can be given a duration (possibly stochastic);
- mapping of objects and messages on computing and communication resources: each object-group is "deployed" on a physical processor;
- weighting of non-deterministic choices.

The same compiler used for the validation is currently used for performance evaluation. A number of executions of the PROMELA model are run and, out of them, estimates are calculated for the desired system parameters. The definition of the performance indexes is embedded within the system specification, as well as validation properties.

2.6 Code Generation for Real Distributed Execution

ACE provides a code generator to produce CORBA/IDL and C++ for deploying and running the applications on real CORBA platforms (Iona Orbix [4] and Chorus COOL Orb [3]).

The actual mapping tackles and solves the differences between the object model of the TINA computational model and the one supported by the CORBA [6] binding with C++ language. One of this differences concerns the support for objects having multiple dynamic interfaces, as prescribed by the TINA computational model and not equally present in the CORBA C++ mapping. It is in the process of the code generation that this semantic gap is fulfilled by proper object composition and dedicated "glue code" that reduce the mismatch between the two different models.

The code generation covers the production of both the declaration part of templates and the behaviour parts which are specified through the graphical diagrams.

As far as interfaces are concerned, CORBA IDL code is generated automatically starting from the graphical specification of an interface, as well as the contrary (i.e. it is possible to import externally defined interfaces specified in CORBA IDL, and generate the corresponding graphical specification, in order to be later re-used in the development of a component). Likewise, starting from the object specifications (structural and behavioural sections), skeletons C++ implementations are generated automatically from the SDL-like behavioural notation.

3. Experiences in using ACE

ACE is the reference environment within CSELT to specify object-oriented distributed systems for the control and the management of services and network resources. In particular, it has been exploited, in traditional IN and

TINA contexts, just as a specification environment or also as a development environment.

In the context of IN-like service specification, ACE is used as a tool to specify the components of services allocated on the different IN systems (e.g., SSP, SCP, SMS) and their interactions. In addition, in order to visualise the behaviour of a service from the point of view of the users, it was exploited the animation tools provided by the environment.

The TINA components that, at the moment, have been specified through ACE are:

- the session control, in particular the GSC component;
- the access session (i.e., the objects that perform user registration, service browsing, service invocation and user invitation;
- some modules of the specifications of the Connection Management, developed in the context of the TINA World Wide Demo [11].

The specifications of these components refined the definition provided by TINA Core Team and extended it by providing the behaviour in a formal language. During their development several feature of the environment were successfully exploited; in particular:

- the import in the environment (textual) IDL specification;
- the SDL-like formalism to express the behaviour of the system;
- the simulator to verify the correctness of the specifications and the tracer to perform their "debugging";
- the code generator to produce a first version of the implementation on the Iona/Orbix CORBA platform, to be refined to obtain the final code.

The possibility to formally define the behavioural aspects of the specification pointed out several aspects that were not properly covered by a specification of structural aspects and a description of the behaviour in natural language. In fact, the use of SDL-like formalism, aligned with the computational concepts, and forces the production of easily implementable specifications on a distributed platform. A possible example concerns the configuration of a set of objects: ACE forces to specify the actions and the invocations necessary to pass the necessary interface references to all the instances.

The production of computationally complete specifications is stressed also by the possibility to run the specifications in the simulated environment which covers the main aspects of a distributed platform. It was experimented the possibility to describe the behaviour in ACE at different level of detail: for instance, the specification of TINA GSC component includes objects whose behaviour was defined in a very detailed way (similar to the detail level of traditional implementation language), while the specification of TINA access session was produced through a process of incremental refinements to reach the satisfactory level of detail.

The analysis phase was performed on the design editor that provides a superset of the functionalities of analysis one. In this phase was detected the lacking of a support to the definition of object interaction diagrams.

The previously described specifications were mainly written by persons which were not involved in the development of ACE. The training period was about one week, provide that they have some knowledge of the computational model concepts and constructs underlying ACE. This training period does not consider the tools supporting the definition of animation scenarios: at the moment the training period of these tools is tightly related to that required by Metacard. In addition to available documentation/manuals the training required some interactions with the designers of the environment. The development of TINA specifications of session control and access session considered in this section took about two months each.

These experiences detected some weak points of the current version of ACE. The principal ones concern the simulation and the code generation. In particular, it was detected some limits in the simulation of complex systems with very detailed object behaviour; for instance it was not possible to simulate some objects of TINA GSC, whose methods were defined at a C-like level of details. While the code generation is satisfactory for the aspects concerning the structure of the applications (e.g., the production of CORBA servers from BB specifications and the handling of objects with multiple interfaces), at the moment it presents some limits in the generation of C++ methods, that require a general revision of the produced code. Some improvements are also required to the generation of documentation, mainly related to the aspects concerning the documentation of object behaviour.

As mentioned before, strong points that were confirmed are the possibility to specify both structural and behavioural aspects and to simulate the specifications. The simulation was exploited both to "debug" the specifications and to demonstrate the evolution of the specified systems; at this aim, the experiences showed the benefits of having both the tracing of specifications and the animation of application scenarios. Another strong point is the possibility to generate code for a real distributed platform, also with the current limitations, that contributes to speed up the implementation of specifications.

ACE is also used within the TINA Core Team for the validation [10] of some of the Reference Points, providing support for rapid prototyping and specifications in terms of IDL, ODL files and OMT diagrams.

Finally, in the lifetime of the ACTS ReTINA project[12], ACE is the core environment providing support functionalities for the definition, validation, and code generation of service components.

4. Conclusions

ACE is still evolving, and in the short term extensions are expected, in order to increase the offered functionalities. Among them, the support for the Java platform, at language and infrastructural level. The usage of alternative platforms (e.g. DCOM) for the distribution aspects will also be object of extensional activities.

At the present stage ACE is however an advanced integrated environment for distributed object oriented applications compliant to TINA computational model, and its support has proved of value in the development of new applications.

A public domain version of ACE can be obtained from the authors: `http://andromeda.cselt.it/ace/ACE.html`.

5. References

[1] TINA-C deliverable: *Computational modelling concepts, vers. 3.2*, TINA Consortium, 1996.

[2] E.Chiocchetti; R.Manione, P.Renditore: *Specification based performance evaluation of distributed systems for telecommunications*, Tools'94, 1994.

[3] Chorus Systeme: *COOL-ORB Programming Model*, Technical Report, 1994.

[4] Iona: *Orbix 2 Programmer Guide*, 1995.

[5] J.Rumbaugh et al.: *Object-Oriented Modeling and Design*, Prentice Hall, 1991.

[6] OMG: *The common object request broker architecture and specification revision 2.0*, OMG Document Number 95.07.01, 1995.

[7] TINA-C deliverable, *TINA Object Definition Language*, TINA Consortium, 1995.

[8] P.G.Bosco; G.Giandonato; C.Moiso: *A distributed processing model for telecommunication services and operations software*, Proc. TINA'93, 1993.

[9] N.Bersia, P.G.Bosco, G.Canal, R.Manione, C.Moiso, M.Spinolo: *A Case Environment for TINA-oriented applications*, Proc. ISS'95, 1995.

[10] R.Westerga, M.Hedberg, E.Darmois: *Using ACE for the validation of TINA Reference Points*, Service Creation Workshop, Como, 1997.

[11] G.Spinelli, W.Takita, G.Martini, S.Chikara: *TINA Connection Management Implementation over Distributed Processing Platforms*, Proc. NOMS '96, 1996.

[12] ReTINA: *http://www.chorus.com/Documentation/retina.html*.

[13] K.Brockschmidt, *Inside OLE 2*, MS Press, 1995

Integrated Environment for Service-type Repository Management (IE-STREM)

Hiroki Tanaka
NTT Optical Network Systems Laboratories
Distributed Network Systems Laboratory
9-11, Midori-cho 3-Chome, Musashino-shi, Tokyo 180, Japan
h-tanaka@exa.onlab.ntt.co.jp

Abstract

This paper proposes an environment for operating and managing the service-type repositories advocated by TINA-C, ISO-ODP, and OMG. The environment enables users of telecommunication services and network management services to easily find the servers that provide services they want to use. It also provides a mechanism that ensures the response time of operations performed by distributed service-type repositories. Moreover, service-type data replication and repository failure recovery are processed automatically. A prototype implementing IE-STREM has already been developed on a CORBA-compliant product and its performance is satisfactory.

1. Introduction

An increasing number of telecommunications and network management services are being developed and are functioning on advanced software platforms. The Object Management Group's (OMG) Common Object Request Broker Architecture (CORBA)[1] is currently one of the most promising software platform technologies for effectively executing distributed application software components of those services. The Telecommunications Information Networking Architecture (TINA)[2] consortium has been developing specifications for the Distributed Processing Environment (DPE)[3], which is closely related to CORBA and to the Reference Model of Open Distributed Processing (RM-ODP)[4][5] specifications. Type repositories are key DPE components. They support the execution and management of service applications by providing service applications and other DPE components with information such as the data-type, object-type, interface-type, service-type, and binding-type. The OMG and ODP trading service specifications[6][7] prescribe a service-type repository that stores the service-type data used for the trading service. Because service-type repositories are necessary for the trading service, effective operation and management of the repositories are essential.

When many new kinds of telecommunication and network management services become available, users of those services often need to quickly and easily check if services that satisfy their needs are available. They may also want to choose and access servers that offer the desired services with appropriate conditions pertaining to fee, Quality of Service (QoS), business hours, and so on. On the other hand, service administrators may need support tools for DPE operation and management; for example, tools for the handling of DPE configuration information and for the failure recovery of DPE components. They may want to provide service users with reliable and high-performance services using such tools.

Three useful functions should be included to enhance the value and usefulness of service-type repositories. These are: service-type navigation, which enables service users to easily retrieve their preferred service-type information; service-type repository interworking, which ensures the operation response time; and service-type repository management, which supports the life cycle management and fault management of service-type repositories. These functions are expected to promote service utilization by service users and service provision by service administrators.

This paper proposes an integrated environment for service-type repository management (IE-STREM) that provides components for achieving these three functions. The results of a performance evaluation of a prototype IE-STREM system implemented on a CORBA-compliant product are also reported.

0-8186-8335-X/98 $10.00 © 1998 IEEE

2. Requirements for a service-type repository management environment

When operating and managing service-type repositories, the following three requirements should be considered.

2.1. Assisting service users in retrieving service-type information

Before client applications used by service users issue query operations to traders, the service-type name and the user's constraints and preferences for each service property must be specified as arguments of those operation requests. Service users find it difficult to keep up with the names and related properties of the latest services because new services are introduced continuously and the number of services available to service users will continue to increase. Thus, service users may often need to obtain certain desired service-type data, which includes a service-type name and names and data type kinds of service properties, from service-type repositories.

However, it is difficult for service users to efficiently obtain such information by using only the operations provided by the OMG- and ODP-compliant service-type repositories. For example, the service-type repositories provide no convenient operations for checking the subtypes of a given service-type. Hence, a mechanism is needed to enable service users to easily obtain their preferred service-type information.

2.2. Service-type repository interworking and assurance of the operation response time

When implementing service-type repositories, the following two items should be taken into account.

(1) *Need for service-type repository interworking*
It is preferable to deploy service-type data over several service-type repositories to achieve load balancing and fault tolerance. However, when this is done, a service-type repository may receive an operation request for data that it does not have. In this case, the repository must forward the request to another service-type repository. We call this behavior "service-type repository interworking".

(2) *Ensuring the response time for operation requests*
It is desirable to keep the response time for operation requests to service-type repositories below a certain value. There are three kinds of software components that are operation requesters: client applications used by service users, traders, and service-type repository management tools used by service-type repository administrators. We assume the following, regarding the use of operations provided by service-type repositories.
(i) Operation requests are much more frequent from traders and client applications than from service-type repository administrators. The largest number of requests comes from traders.
(ii) Ensuring the response time of the operation requests from traders and client applications has priority over requests from service-type repository administrators in order to satisfy service users' requirements.
Because both service users and traders use the describe_type() and fully_describe_type() operations, as shown in Table 1, the response time of the requests of these two operations should be ensured.

Accordingly, a mechanism is needed for service-type repository interworking that ensures a reasonable response time for describe_type() and fully_describe_type() operation

Table 1 Service-type repository operations used by service-type repository users.

Operations \ Service-type repository users	Service user (through client applications)	Trader	Service-type repository administrator (through management tools)
add_type()			✔
remove_type()			✔
list_types()	✔		✔
describe_type()	✔	✔	✔
fully_describe_type()	✔	✔	✔
mask_type()			✔
unmask_type()			✔

requests. With a mechanism such as "linked traders", as described in Refs. 6 and 7, it is difficult to limit the number of interworking service-type repositories below a predetermined small value, so ensuring the response time is not possible.

2.3. Facilitating management of service-type repositories and service-type data

Typically, service-type repository administrators are service providers. It is desirable to provide them with a graphical user interface (GUI) environment to help them create and delete service-type repositories and manipulate service-type data. Also needed is a mechanism for achieving fault management for service-type repositories.

3. Integrated environment for service-type repository management (IE-STREM)

As shown in Fig. 1, IE-STREM provides solutions that fulfill the three requirements listed in Sec. 2. IE-STREM is characterized by the three proposed parts: service navigation

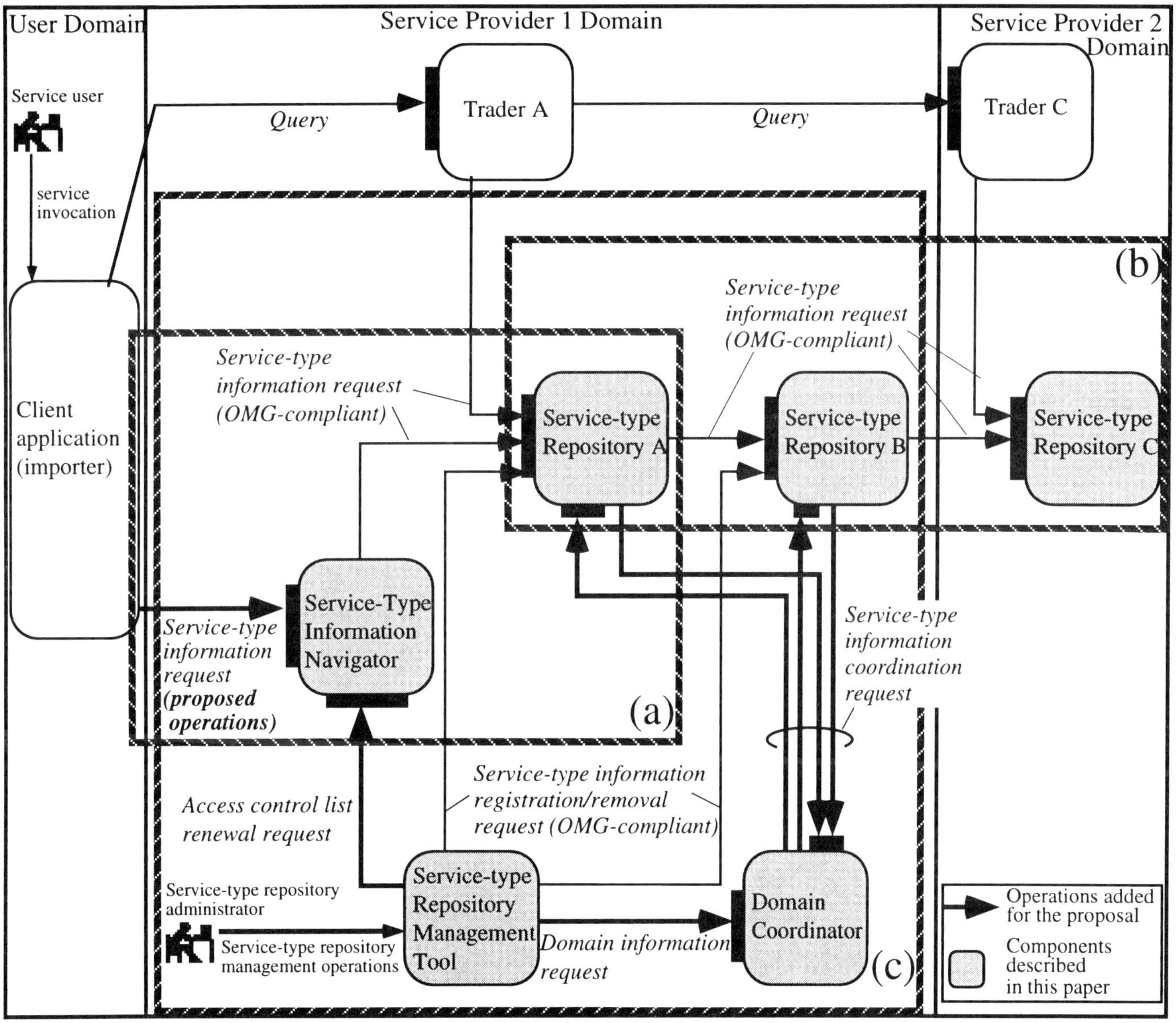

(a) Service navigation (b) Service-type repository interworking (c) Service-type domain management

Fig. 1 Proposed integrated environment for service-type repository management.

(fulfills the Sec. 2.1 requirement), service-type repository interworking (fulfills the Sec. 2.2 requirement), and type domain management (fulfills the Sec. 2.3 requirement).

3.1. Service navigation (Fig. 1(a))

The service execution procedure is shown in Fig. 2. Service users obtain service-type data, execute trading, and then invoke the desired operations. The Service-type Information Navigator (STIN) helps the users obtain the desired service-type information by using service-type trees that maintain the supertype/subtype relationships between service-types, as shown in Fig. 3. STIN provides the operations shown in Fig. 3 to enable users to do the following:
- check the services provided at the time of retrieval,

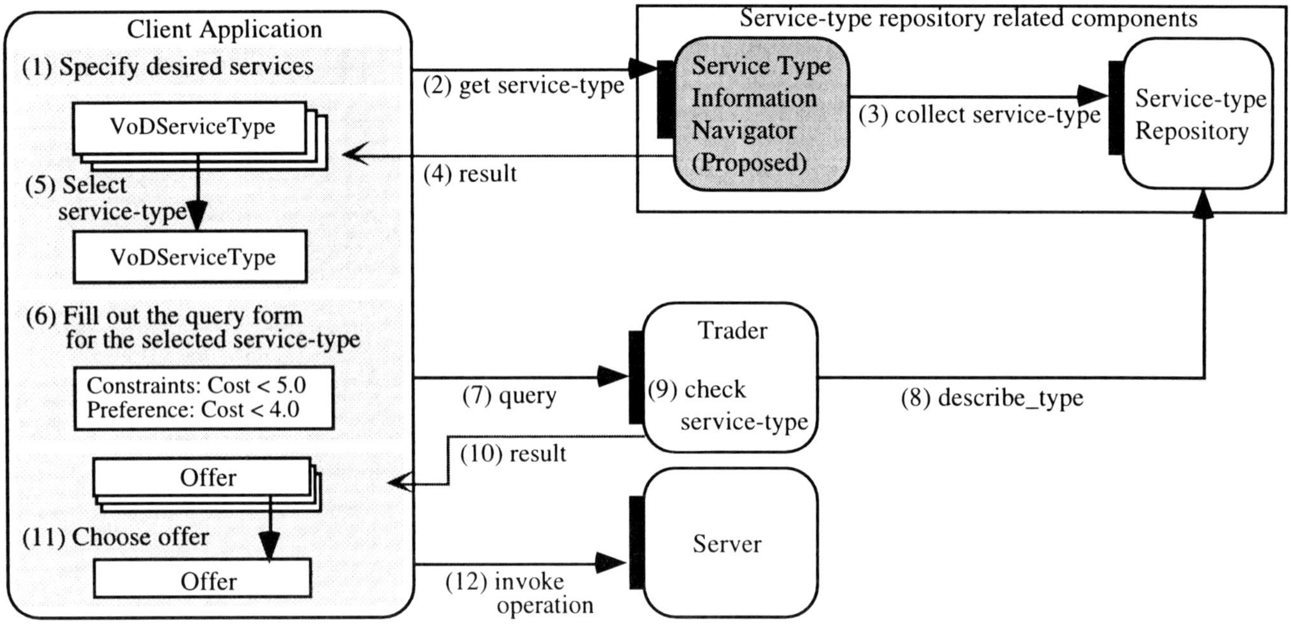

Fig. 2 Procedure for executing services using service navigator, service-type repository, and trader.

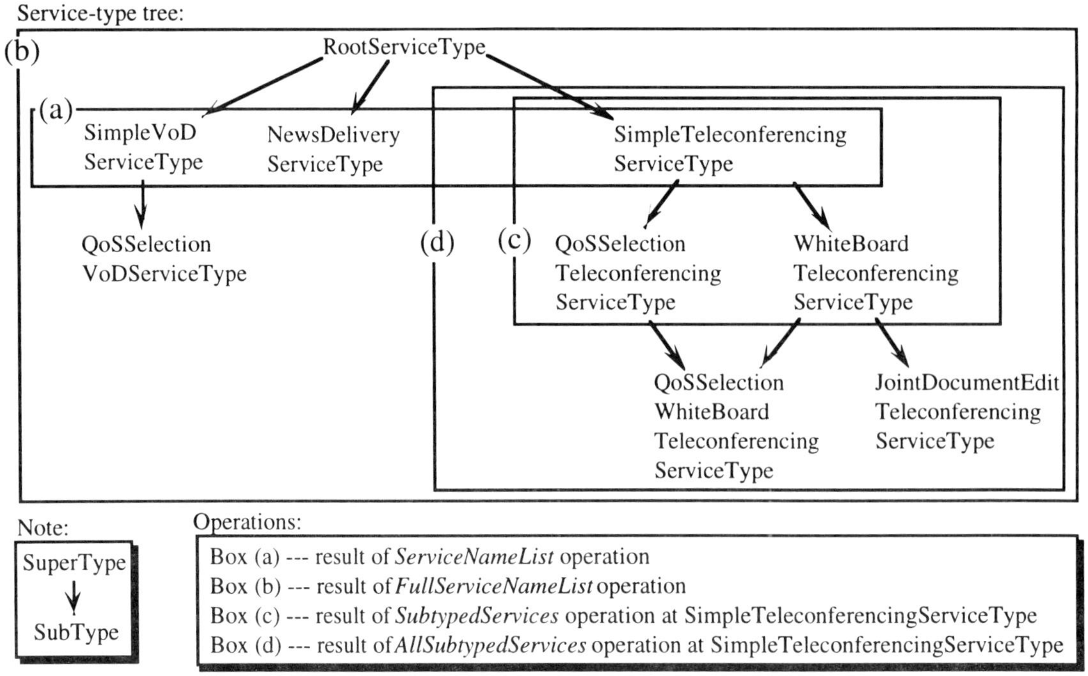

Fig. 3 Service-type tree and operations supported by the service-type information navigator.

- refine the demand to choose a desired service-type, and
- acquire information on the service-type that matches the desired services.

Our prototype also provides GUI applications that work on client terminals to access STINs. Examples of GUIs displaying service-type and offer data are shown in Figs. 4(a) and (b), respectively. The desired operation is invoked by selecting an offer and clicking the operation box (Fig. 4(b)).

In the current version of our prototype, we assume that users begin their search at the most supertyped service-types (e.g., SimpleTeleconferencingServiceType in Fig. 3) and then descend through the subtyped service-types (e.g., QoSSelectionWhiteBoardTeleconferencingServiceType via QoSSelectionTeleconferencingServiceType in Fig. 3). We are now developing a retrieval method that uses key-word matching. Combining these two methods offers the users efficient and convenient retrieval of their preferred service-type information.

3.2. Service-type repository interworking (Fig. 1(b))

A diagram illustrating service-type repository interworking is shown in Fig. 5. The domain coordinator (hereinafter referred to as the coordinator) supports type information location management. The ODP type repository

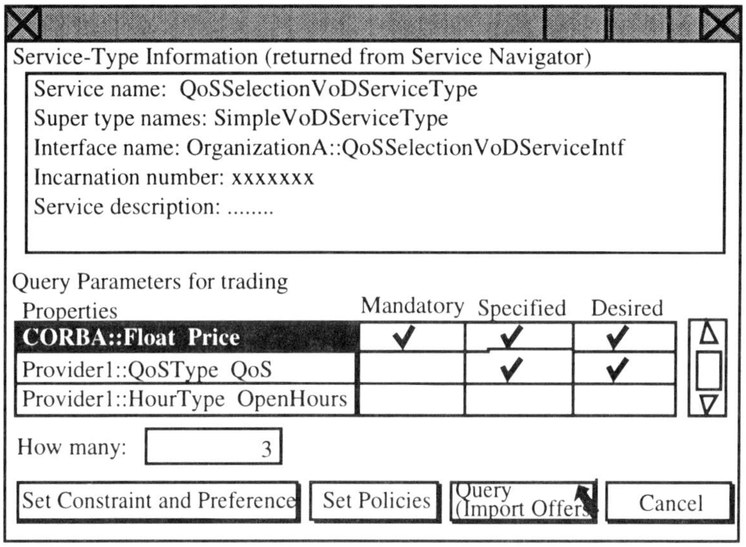

Fig. 4(a) Client-application GUI displaying service-type information.

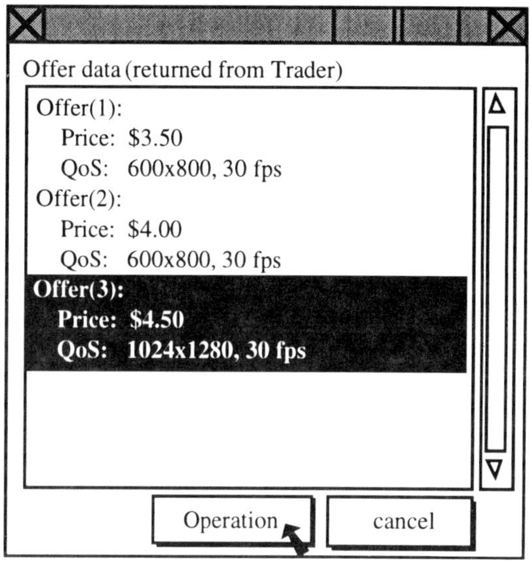

Fig. 4(b) Client-application GUI displaying service offer information.

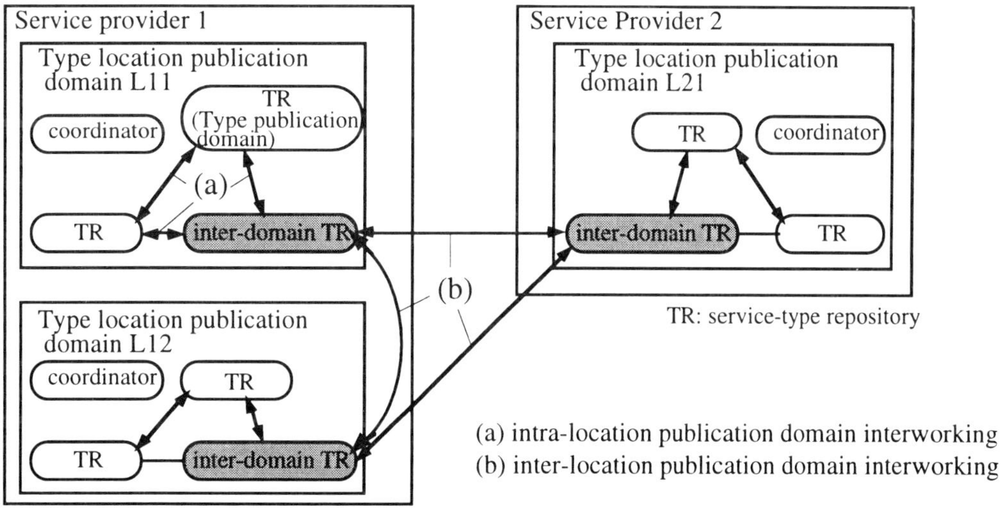

Fig. 5 Service-type repository interworking.

Table 2 Type domain in ISO ODP type repository specifications.

Type domain (ISO ODP)	Definition	Corresponding type domain specified for our implementation
Type publication domain	a set of type descriptions published by a type repository	Type publication domain
Type specification domain	a set of type descriptions given by a type specifier	Type specification domain (to be considered)
Type language domain	a set of type descriptions described using a type language	Type language domain (Only OMG type expression is considered.)

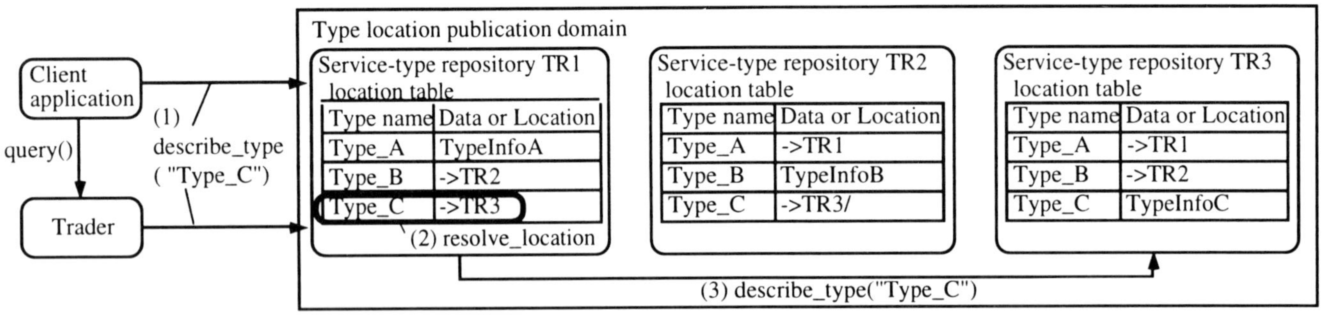

Fig. 6 Service-type repository interworking within a type location publication domain.

specification[8] proposes three kinds of type domains: type publication domain, type specification domain, and type language domain. These represent mainly the differences in type information characteristics, such as type specifiers or type expression. An explanation of these domains and the corresponding domains in our prototype are shown in Table 2. In addition to these, we propose the type location publication domain to represent the differences in run-time service-type repository management policies. A set of service-type repositories is called a type location publication domain if each service-type repository of the set knows what service-type information every other service-type repository in the set stores. For example, in Fig. 6, TR1 knows that information of Type_A, Type_B, and Type_C is stored in TR1, TR2, and TR3, respectively. Because it is not practical to publish type location information to other service providers, we assume that one type location publication domain is possessed by only one service provider.

Two kinds of service-type repository interworking are considered here: interworking within one type location publication domain and interworking between two type location publication domains. To ensure the response time of service-type information manipulation requests from client applications and traders, the implementation method should be chosen according to the kind of interworking used.

3.2.1. Interworking within a type location publication domain

Figure 6 shows the proposed mechanism for service-type repository interworking within a type location publication domain. All type location information open to the public within a type location publication domain is recorded on location tables kept by each service-type repository. In the figure, TR1 receives a describe_type() or fully_describe_type() operation request. Then, using its location table, it finds out which service-type repository stores Type_C information and forwards the request to that repository (TR3). With this mechanism, the number of interworking service-type repositories is sure to be either one or two if the required service-type data exists in the location publication domain in which the repositories are located. This is because the service-type repository accessed by client applications (e.g., TR1 in Fig. 6) is sure of which service-type repository has the required service-type data (e.g., TR3 in Fig. 6). Thus, this mechanism satisfies the requirements listed in Sec. 2.2. In contrast, when using a method such as linked traders, it is difficult to limit the number of interworking service-type repositories and thereby ensure the response time of the describe_type() or fully_describe_type() operation requests.

3.2.2. Interworking between type location publication domains

It seems natural to deploy two service-type repositories in separate location publication domains if there is no need to open service-type information kept by one of the repositories to the other. For example, the service-type repositories may be owned by different service providers or accept accessing from different kinds of users (e.g., service users and service-type repository administrators). In this case, infrequent interworking is expected. Dividing the type location publication domains prevents location tables from becoming large, thereby helping to retain the scalability of service-type repositories.

Combining these two interworking methods helps fulfill the various requirements for implementing service-type repositories.

3.3. Service-type repository management (Fig. 1(c))

In this section we present two key components for managing service-type repositories and service-type information.

3.3.1. Service-type repository management tool

This component provides service providers with a GUI environment to:
(1) create or delete coordinators and set or get domain information possessed by coordinators, and
(2) add or remove service-type data to or from service-type repositories.

3.3.2. Coordinator

This component exists in each location publication domain and provides the following three functions.
(1) Maintains the domain information table needed for managing the type location publication domain (Fig. 7)
(2) Coordinates data stored in all location tables in a location publication domain
If service-types are added or removed, or their location is changed, the type location data in all location tables must be renewed. The coordinator supports this renewal by providing a function for multicasting renewal requests to all service-type repositories in the domain to which the coordinator belongs. Figure 7 shows the procedure for processing an add_type() operation request. When a coordinator receives a coordinate_add_type() operation request, it checks the value of the item "Number of the replication data" in the domain information table and determines the service-type repositories at which replicated service-type data is registered. This mechanism balances loads between service-type repositories. The coordinator then asks all service-type repositories in the domain to add an entry, including location data or replicated service-type data (Fig. 7, (5)-(8)).
(3) Detects and recovers from failure of service-type repositories
Our IE-STREM prototype can also deal with service-type repository failures. Each coordinator periodically exchanges messages with all service-type repositories in the type location publication domain to which the coordinator belongs. This is done to check if there is anything wrong with any service-type repository in the domain. If the coordinator detects a time-out on any reply message and the message exchange is not successful after a predetermined number of tries, the coordinator notifies the service-type repository management tool of the problem. If any problem with a service-type repository is found by its administrator, the administrator may deal with the problem (e.g. terminate

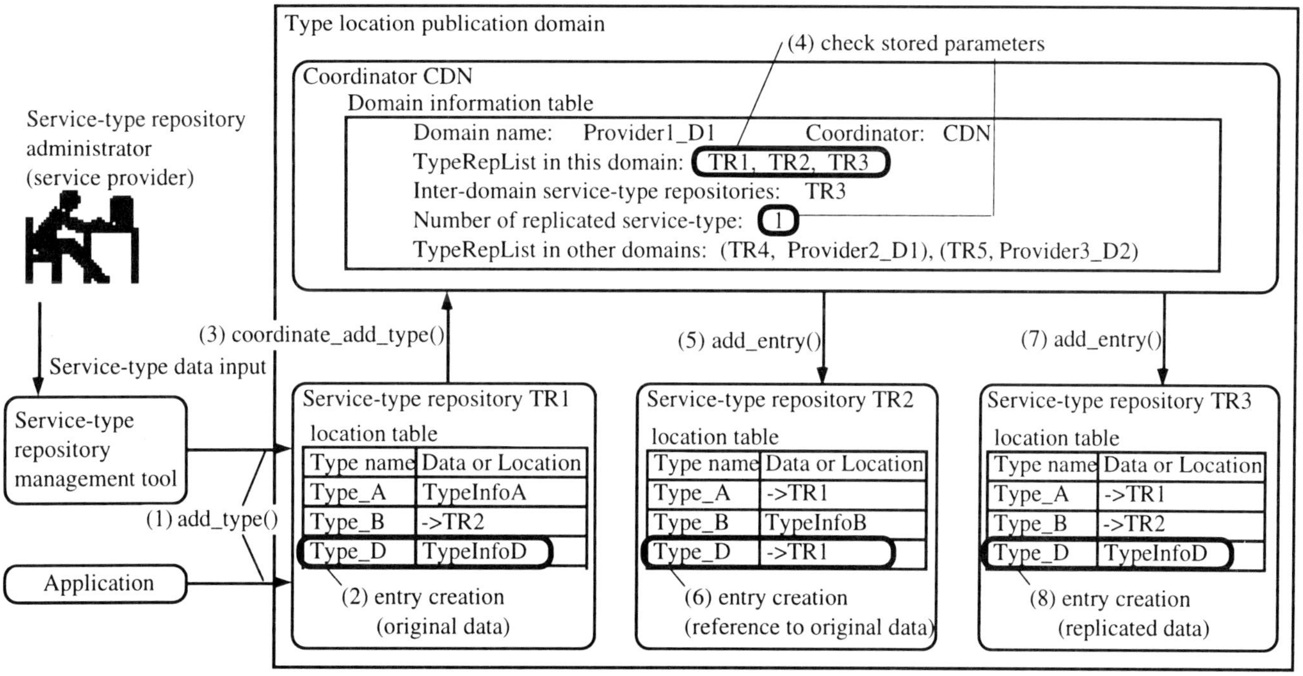

Fig. 7 Procedure for adding an entry to each location table.

the unreachable repository) and request the service-type repository management tool to create a substitute for the unreachable repository. After the coordinator detects the problem, it requests all the service-type repositories (excluding the unreachable repository) within the domain to change the data in the location tables so that each repository does not issue operation requests to the unreachable repository when processing service-type repository interworking.

In addition, we are currently developing a method for detecting and recovering from coordinator failure.

4. Performance evaluation

The performance of our prototype system is currently under evaluation. The actual measurement environment is shown in Fig. 8. Fig. 8(a) shows the relationship between the number of interworking service-type repositories and the response time of the operation requests from client applications or traders. If the interworking is done within a single type location publication domain, the method proposed in Sec. 3.2.1 is able to keep the response time under 25 ms because the number of interworking service-type repositories is two or less. In contrast, if a method like linked traders is used for service-type repository interworking, the response time varies widely and is difficult to ensure. Fig. 8(b) shows the relationship between the number of service-type data items searched in a service-type repository and the time required to finish processing a service-type retrieval operation. This time is under 3 ms if the number of service-type data items stored in one repository is less than 1000, which seems to be a reasonable assumption. Hence, this time is not a main factor in determining the maximum value of the response time for our prototype (25 ms). The time required for communication between service-type repositories seems to be the main factor determining the response time. Therefore, although the maximum value of the response time is satisfactory, we would expect a substantial improvement in the maximum response time if our prototype is implemented on a real-time high-performance platform such the Distributed Object-oriented Network Architecture (DONA)[9] platform.

5. Summary

In this paper we proposed an integrated environment for service-type repository management (IE-STREM) that provides functions for service-type navigation, service-type repository interworking, and service-type repository management. IE-STREM is useful for both service users

Environment and conditions (for both Figs. 8(a) and 8(b))

Environment:
- Software platform:
  Orbix 2.0.1 (multi-thread) on TCP/IP, Solaris 2.5
- Hardware
  Ultra Sparc1, Sparc Station 10, Sparc Station 20

Conditions:
- All client applications and service-type repositories are running on separate hosts.

Conditions (Fig.8(a) only):
- Each service-type includes three properties of C++ struct type.
- Each service-type repository possesses 1000 service-type data items.

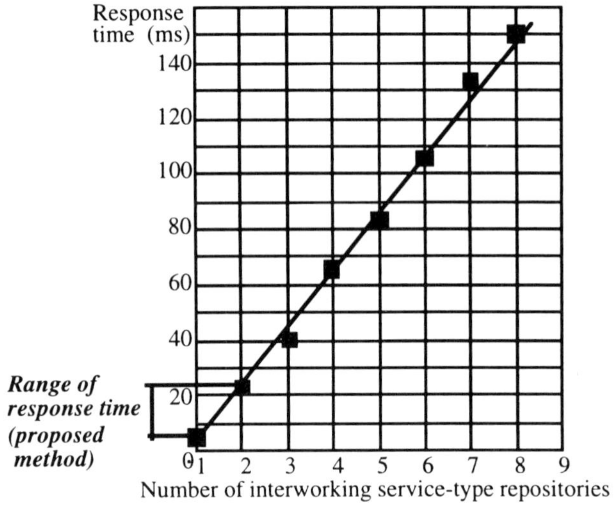

Fig. 8(a) Response time of operations invoked by client applications.

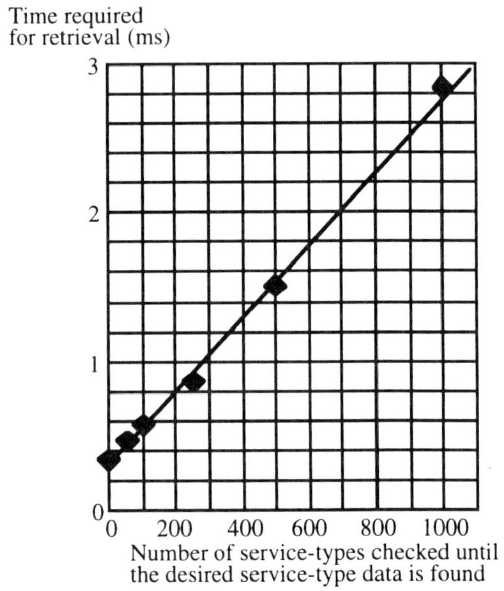

Fig. 8(b) Time required for retrieving the desired service-type data from a service-type repository.

and service-type repository administrators and consequently enhances the usefulness of service-type repositories that comply with OMG and ODP specifications. A prototype implementing IE-STREM has already been developed on a CORBA-compliant product. Although the maximum value of the response time for the operation requests from client applications or traders is satisfactory, it can be improved if our prototype is implemented on a real-time high-performance platform such as the Distributed Object-oriented Network Architecture (DONA)[9] platform when required.

Although our prototype currently handles only service-type data, other kinds of types, such as object types, will also be supported. Object-type management is useful for application component management, such as SOMSE[10]. We plan to implement IE-STREM on a DONA platform, which complies with CORBA and provides TINA DPE components.

References

[1] The Common Object Request Broker: Architecture and Specification, revision 2.0, July 1996.

[2] G. Nilsson et al., "An Overview of the Telecommunication Information Networking Architecture", TINA 95, pp. 1-12.

[3] TINA-C Engineering Modelling Concepts (DPE Architecture), version 2.0, December 1994.

[4] ITU-T X.901 | ISO/IEC 10746-1 ODP Reference Model Part 1: Overview, June 1995.

[5] ITU-T X.901 | ISO/IEC 10746-1 ODP Reference Model Part 3: Architecture, June 1995.

[6] OMG Trading Object Service, version 1.0.1, July 1996.

[7] ISO/IEC SC21 N10770 ODP Trading Function: Specification., January 1997

[8] ISO/IEC JTC1/SC21 N10389 ODP Type Repository Function, May 1996.

[9] S. Suzuki et al., "DONA: Distributed Object-Oriented Network Architecture for Revolutionary Network Reconstruction", ISS 97 (to be printed).

[10] H. Tanaka and H. Ishii, "Service Operation and Management Architecture using Surveillance of Application Software Elements (SOMSE)", GLOBECOM '95, pp. 1977-1981 (Nov. 1995).

Author Index

Notes